International Institute of Philosophy
Institut International de Philosophie

La philosophie contemporaine

Chroniques nouvelles

par les soins de
GUTTORM FLØISTAD
Université d'Oslo

Tome 7

Philosophie asiatique

SPRINGER SCIENCE+BUSINESS MEDIA. B.V.

Contemporary philosophy

A new survey

edited by
GUTTORM FLØISTAD
University of Oslo

Volume 7
Asian philosophy

SPRINGER SCIENCE+BUSINESS MEDIA, B.V.

Library of Congress Cataloging-in-Publication Data

Asian philosophy/edited by Guttorm Fløistad.
p. cm. — (Contemporary philosophy, a new survey; v. 7)
English, French, and German.
Includes bibliographical references and index.
ISBN 978-94-010-5107-1 ISBN 978-94-011-2510-9 (eBook)
DOI 10.1007/978-94-011-2510-9
1. Philosophy, Oriental. I. Fløistad, Guttorm. II. Series.
B804.C573 vol. 7
[B121]
190'.9'045 s—dc20
[181]

ISBN 978-94-010-5107-1

Published under the auspices of the International Council of Philosophy and Humanistic Studies and of the International Federation of Philosophical Societies, with the support of UNESCO.

Publié sous les auspices du Conseil International de la Philosophie et des Sciences Humaines et de la Fédération Internationale des Sociétés de Philosophie, avec le concours de l'UNESCO.

Printed on acid-free paper

Contents/Table des matières/Inhalt

TAIWAN

KOREA

JAPAN

Preface

The present publication is a continuation of two earlier series of chronicles, *Philosophy in the Mid-Century* (Firenze 1958/59) and *Contemporary Philosophy* (Firenze 1968), edited by Raymond Klibansky. As with the earlier series the present surveys purport to give a survey of significant trends in contemporary philosophical discussion.

The need for such surveys has, I believe, increased rather than decreased over the last years. The philosophical scene appears, for various reasons, rather more complex than ever before. The continuing process of specialization in most branches, the emergence of new schools of thought, particularly in philosophical logic in the philosophy of language, and in social and political philosophy, the increasing attention being paid to the history of philosophy in discussions of contemporary problems as well as the increasing interest in cross-cultural philosophical discussion, are the most important contributory factors. Surveys of the present kind are a valuable source of knowledge about this complexity and may as such be of assistance in renewing the understanding of one's own philosophical problems. The surveys, it is to be hoped, may help to strengthen the Socratic element of modern philosophy, the world wide dialogue or *Kommunikationsgemeinschaft*.

So far, six volumes have been prepared for the new series. The present surveys in Asian Philosophy (Vol. 7) follow the surveys in the *Philosophy of Language and Philosophical Logic* (Vol. 1), *Philosophy of Science* (Vol. 2), *Philosophy of Action* (Vol. 3), *Philosophy of Mind* (Vol. 4), *African Philosophy* (Vol. 5), and *Medieval Philosophy Part 1–2* (Vol. 6).

Each volume contains, as a rule, fifteen surveys, each 25 pages long. However, this rule has been broken in all volumes. In some cases it turned out to be impossible to receive surveys in time for the date of publication. In others the authors, if they thought it necessary in view of their topic, were allowed to exceed the allotted number of pages. This is certainly the case with the present volume in which, for instance, the opening article covers over 80 pages. It also turned out to be exceedingly difficult to get a Buddhist scholar from Asia to write an article on present research in Buddhist philosophical traditions. Three different agreements were made. Finally I had to ask a Norwegian scholar to write the contribution. It likewise turned out to be difficult to get someone to cover philosophical research in China. A number of topics in Chinese philosophy are, for obvious reasons, taken care of by scholars from other Asian countries. A general account of Chinese philosophy is, however, given by a Swiss scholar.

Asian philosophy is not always Asian philosophy, just as European philosophy is not always specifically European. A number of articles in the present volume clearly show the influence of European tradition in philosophy. The distinctiveness of Asian philosophical traditions is, however, made quite clear by most authors. That some of these traditions have a strong religious and theological flavour is also well known from European philosophy.

Most of the chronicles, as is to be expected, are written in English, some in French and German. The rendering of Asian philosophy in European languages may present some difficulties. The basic notions in Asian philosophy and the corresponding terminology are deeply rooted in cultures that in many ways differ from the European counterpart.

The bibliographies are selected by the authors themselves and as a rule contain only those works they take to be of special interest for the topics under discussion.

I am most grateful to a number of persons who have assisted in various ways in the preparation of the new series. My thanks are first of all due to the Secretariat of the Institut International de Philosophie, especially to Cathrine Champnier. The Secretariat has also been responsible for the contact with Unesco and Kluwer Academic Publishers in Dordrecht. My thanks are also due to Unesco for their financial support and to the publishers for their willingness to Publish these surveys. The staff in charge of the series at the publishers have provided invaluable help in the final editing of the surveys.

Finally, I want to thank all the scholars, from various countries, whose contributions have made it possible to complete the new series of surveys.

University of Oslo, March 1992 Guttorm Fløistad

Introduction

From time to time we all need to revise and reconstruct our problems. The present volume on Asian philosophy is an opportunity to acquire new perspectives.

The volume covers a variety of topics. Most of them are specifically Asian in that they are related to the great religious and philosophical traditions. Some of them bear witness to the influence of Western philosophy, first and foremost the philosophical traditions of Germany and the United States. This should be no cause for wonder since a considerable number of philosophers in the Asian countries have received their university training in Europe and the United States. In some countries (South Korea, for instance) it even happens that the majority of dissertations are discussions of topics in Western philosophy, which may cause some wonder in view of the richness of Eastern traditions. However, this is of no concern here, except in one respect: the study of Western philosophy is, at least in fact, the study of philosophies that are influenced by Eastern traditions.

We know that Alexander the Great and his army marched right to India and that, ever since, there has been an extensive trade and exchange of commodities. The Chinese Silk Road is famous, even today. The exchange of ideas is more difficult to trace. In the case of Plotinus there are two schools of interpretation. In view of the similarity, or near-symmetry, between the world structure found in Plotinus and in Buddhism, one of the schools maintains that a mutual influence must have occurred; the other that the similarity, or symmetry, due to lack of reference, must be explained in terms of a co-occurrence of world views in the two cultures. We also know that Buddhist scholars travelled in Europe in the 9th century. In the 16th century the first Jesuit missionaries settled in China and Japan and other countries, and they wrote extensive letters to their authorities in France, for instance on Japanese culture and belief systems and the result of their missionary work (1587). In the following century, especially from 1630 onwards in the Netherlands, dozens of books appeared from various European expeditions to Asian countries, reporting extensively on various aspects of Asian cultures, including religion, moral convictions and customs of daily life. New embassies were established and the East-Asian trading companies flourished as never before.

No wonder, then, that philosophers of the 17th century became engaged in the religious and moral systems of Asia. Spinoza and Leibniz are the exponents of this influence. The presence of Taoist and Buddhist principles in Spinoza's philosophy of man and nature is obvious, as is shown in several interpretations (J. Wetlesen's,

Contemporary philosophy. A new survey. Vol. 7, 1–8.

for instance). Spinoza is often regarded as the Aristotle of the Renaissance and was, incidentally, temporarily in charge of his family's East-Asian trading company. And Leibniz wrote a book on China and is the great Confucian scholar of European philosophy. The reason is not difficult to understand. The decline of religion in the era of the European enlightenment made it imperative to search for a new foundation for moral principles.

In the 19th and 20th century the exchange of ideas increased rapidly. After the Meiji restoration in 1868, for instance, when Japan was reopened to the West, Nishi Amane started introducing Western philosophy to Japan. He invited professors from several European universities and from the United States to lecture on positivism, utilitarianism and various other positions.

Soon, however, German idealism became the dominant Western philosophy in Japan, whereas Buddhism appeared to continue its invasion of Western thought. Schopenhauer obviously regarded himself as a Buddhist scholar, although he seems to have lacked any proper understanding of the notion of *nirvana*. And Hegel, the main exponent of idealism, demonstrates his knowledge of Eastern philosophers and religions in his philosophy of history and other writings. The way, or ways, in which they have excerted an influence on him may be difficult to trace. But his dialectical principles no doubt reflect the yang and yin principle of Taoism. It is the unity of a diversity.

The upshot of these sketchy remarks is simply to indicate that the metaphysical systems of the West and in the East certainly have a genuine identity of their own. They are, however, not entirely different. They exhibit some common features which make them accessible to each other. The linguistic barriers are, of course, considerable. One's mode of thinking is, however, equally decisive for their mutual understanding.

In his long opening survey, R. Panikkar discusses the meaning, or various meanings, of the term 'philosophy'. He observes that philosophy, also in India, is heavily influenced by various traditions in Western philosophy and that philosophy as pursued in India sometimes has been regarded as being synonymous with philosophy in the West. However, both the fact that Indian ways of thinking and experiencing the world have, to some extent, been known in Europe since ancient times, and especially over the last three hundred years, there is every reason to reconsider the meaning of the term 'philosophy'. The author insists that these are genuine Indian traditions of thinking that have their own radical starting point from "a differentiated but unbroken experience of Reality." The key word is *karman*, the fundamental sense of which is to work "one's own salvation, thus contributing to the liberation of the entire universe." *Karman* thus has an existential meaning. This notion of philosophy is, *mutatis mutandis*, not unknown in the West either. Most classical systems of philosophy, from Plato onwards, may be characterised as moral systems, in which knowledge leads to the liberation and unification of the knower and the known.

D. Krishna continues the discussion of new approaches in the study of classical Indian philosophy. His starting point is the same: recent philosophical thinking in India has to a large extent been occupied by Western models. In opposition to this philosophical imperialism, Indian scholars have in recent years gathered to strengthen philosophy in the traditional Indian manner. Securing the old texts and promoting the dialogue between different traditions in India are important tasks. Some of the results are promising, also in regard to a discussion with Western traditions.

In her survey M. Chatterjee focuses on three prominent Indian thinkers from the 20th century: Jehangir N. Chubb, Kalidas Bhattacharyya and Nikunja Vihari Banerjee. The first two are both trained in Indian as well as in Western philosophy and combine topics from both traditions in their philosophies. The metaphysical quest, which is central to both traditions, is a quest for unity. It is a quest not only for a certain type of knowledge, but for a personal experience, sometimes called a mystical experience. Bhattacharyya stresses metaphysical philosophy and experience as 'a way', a metaphor that is also well known in the religious philosophy of the West (that of Bonaventura, for instance).

The experiential character of Buddhism is also stressed by Jens Braarvig in his survey. The variety of positions in Buddhist philosophy also point to a variety of experience. The author focuses on two main topics, the problem of selflessness and of the negative ontology which has strong tendencies towards nihilism. Thi notion of a negative ontology is, perhaps-although with a different meaning-well known from Greek philosophy (Plato) and from the Christian mystics. Braarvig shows how the experiential character of Mahayana Buddhism also expresses itself in a positive morality; positive in the sense of being directed towards all living beings. Buddhism is a philosophy of liberation of all beings.

Indian philosophy is often regarded by Western philosophers as synonymous with metaphysical and religious philosophy. Purusottama Bilimoria shows in his paper on Pramāna epistemology that this view is untenable. Even in early Indian philosophy one can find extensive studies in logic, epistemology and philosophy of language. That is the reason why S. Sankaranarayanan opens his survey by stating that "Indian Logic is vast," and that the available material "looks like an ocean." Logic borders, as usual, on epistemology. An interesting, though controversial, issue concerns verbal authority: does verbal authority (e.g. the Upanisadic 'Great Sentences') produce direct or only indirect cognition? Some interpreters maintain that verbal authority leads only to indirect cognition and needs to be supplemented by meditation to lead to direct knowledge.

Communication has attracted the general attention of Indian thinkers. What are the 'powers and limitation' of language in communication? Above all, is language capable of transmitting personal, even mystical experience, and of dealing with ultimate reality? Is language a barrier or a bridge?

Theories of meaning are naturally an essential part of the discussion. K. Kunjunni Raja surveys the variety of theories, comprising theories of word meaning, of sentence meaning, contextual meaning and intentional, emotive and metaphorical meanings. One even finds in ancient Indian philosophy theories of meaning from the hearer's point of view which, in modern Western philosophy, is sometimes called the hermeneutic of the receiver.

Vom Mythos zum Logos is a famous book on the emergence of Greek philosophy from mythical thinking (Nestle). A similar process may be observed in ancient India. Modern Western philosophy is rational, conceptual and abstract only. Classical philosophy both in the East and West is conceptual *and* experiential. Myth and mythical thinking appear to be experiential only. A myth is something that is lived in the practice of rituals. It is revelatory knowledge. A myth is an expression of an ultimate experience of a transcendent reality. It is accessible in ritual only. Myth and ritual are therefore inseparable.

Baidyanath Saraswati gives a fine description and analysis of the content of various myths. The themes are similar to those found in other cultures, the creation of the cosmos, cosmology in general, man's embodiment in nature and his close relationship with animals, plants, trees and stones, the role of priests in maintaining order among men and in nature in general, and the battle between good and evil. The first principles of the universe according to the tribal myth is One and the Many — the central topic in classical metaphysics. The myth, however, is, through rituals, made more intelligible and accessible than any other mode of knowing reality. Hence, it places itself above philosophy.

The philosophy of ancient China is no less rich than ancient Indian philosophy. One catalogue lists all in all 189 philosophers in the earliest centuries, who together seem to have produced 4500 texts of various length. The main figures are known world wide as initiators of religious, moral and philosophical systems: Laotse and Confucius (Kung-Fu Tse). Jean-Paul Reding, who gives an outline of some of the main lines of thought, concentrates, however, in his presentation, on logic, on geometry and to some extent also on moral and political thinking. The basic text on logic is *le Canon Moïste*. Mozi is, however, also a moral and political philosopher, giving rise to the legislation schools of thought so strongly opposed by the Taoists. Each individual, according to the Taoist, is fully capable of mastering his own life without being administered by a legislative system. The geometries found in Mozi are above all a theory of measurement, based on axioms and theorems. There are both differences and similarities with the Greek counterparts in philosophy. Parts of their geometry, for instance, are similar to some of the expositions in Aristotle's *Second Analytic*.

Taoist philosophy is further presented and discussed by Peter Kun-yu Woo. He points to the political and social degeneration in China at that time and Taoist philosophy as a moral and educational system. Like Confucius, Lao-tzu concerned himself with "the passion of the people" and tried to build up a new social order. This objective was also part of Plato's concern. Whereas Confucius' method was to strengthen interpersonal relationships, the Taoists preferred to focus on the individual and his self-perfection.

Taoism is, above all, a practical philosophy designating "the Way to self-improvement". The ethic of Taoism is, however, clearly related to a world view and ways of knowing, to ontology and epistemology. Although these are three aspects of the same issue, for purposes of comparison with Western philosophy they may be separated into three disciplines.

In the classical metaphysics of the West, ontology, epistemology and ethics are certainly interrelated. But when it comes to modern philosophy the unity of the disciplines is more questionable. Kun-yu Woo focuses on the relation of Taoist philosophy to the phenomenology of Husserl, Heidegger and Merleau-Ponty. Despite some similarities, a comparison encounters great difficulties and needs futher elaboration. A major difficulty lies perhaps in the basically experiential character of Taoist philosophy, whereas phenomenology — with the possible exception of parts of Heidegger's philosophy of Being — is primarily conceptual in its approach.

Taoism is also the topic of Dr Sang-Rak Nam. He surveys research into the origin and development of Taoism in China and its transition, together with Buddhism, to Korea in the 4th century AD. The philosophy of Lao-tzu and his main follower Chuang-tse was, in the 7th century, transformed into a religion and found wide acceptance. A remarkable feature of the Korean intellectual and spiritual life in these centuries and onwards was the close coexistence of the three main philosophical and religious systems, Taoism, Buddhism and Confucianism. They partially fused into one another, both on the level of belief and of research.

Confucianism has many faces and is interpreted in a variety of ways. A main reason for this is its adaptation to the national cultures in the various countries. This is most abvious in the case of Korea. Youn Sa-Soon surveys the various interpretations of Neo-Confucianism (from the 12th to the 19th century), and shows how it overcame its religious and dogmatic characteristics and developed into a fully fledged philosophy. Confucianism is a practical philosophy. The author argues that as a moral theory it has had a major impact on education in that it stresses the principle of lifelong learning. In Korea Confucianism also came to strengthen the loyalty to the family and to the country and its traditions, as well as the need for harmonious relations between man and nature.

Philosophy of religion is a central topic in Asian philosophy. It should also include philosophical aspects of Asian folk religion, Shamanism. Bou-Young Rhi surveys some of the literature dealing with Shamanism in general and, more specifically, with Korean Shamanism. M. Eliade's, C. S. Jung's and Freud's interpretation of Shamanism (including Korean Shamanism) is well known in the West. The author surveys several aspects of Shamanism, such as the initiation disease, the concept of illness and the healing cult and ethical value, a main purpose of which is to secure a community's cohesion. The original philosophical base, however, is the idea of harmony and balance in life with the cosmic order.

Tomonubu Imamichi introduces Japanese philosophy by pointing to its dynamic structure and its close relationship with language. It is also closely related to the Japanese culture, especially the religious culture which is dominated by Buddhism and Shintoism. These features account for some major differences between the philosophy of the East and the West.

Traditional Japanese philosophy is primarily concerned with salvation in which the notion of negation and nothingness plays an important role, whereas Western philosophy is mainly concerned with positive thinking and objectivity. This difference yields, moreover, differences in the concept of truth. In the objective thinking of the West truth is located in propositions and their relationship to a given reality. In Japanese philosophy one speaks of truth in terms of a perfect situation, often involving both a theoretical and a practical aspect. Action is a way of completing a situation. Imamichi also draws attention to natural phenomena of organic life and thinks that they may contribute to an enrichment of future philosophy.

"Homo est animal loquens: Man is a speaking animal." This is the opening words of the contribution by *Noriko Hashimoto*. In her discussion of language, that is, the concept of *Ma* in the Japanese language, she relates the origine of language to the act of creation. Language is thus primarily to be regarded as a transformation of energy into reality. The primary form of this creation process is the imperative—just as in the creation story in the Bible. The imperative form of language has two aspects, an ethical aspect relating to the authority in the creation process, and an esthetic one referring to the creation of a beautiful island.

Noriko Hashimoto applies this original imperative form of language to the dialog and also to the 'reduction' of this imperative to the desires and wishes of our daily life in modern society.

It belongs to a religious philosophy to discuss the notion of creation and the relation between the Creator and the created, between God and the world. Ryosoke Inagaki reviews the concept of Creation in the Philosophy of Kitaro Nishida. It turns out that Kitaro Nishida does not accept the traditional view of God as a transcendent Creator. He somehow identifies God with the world. God is the immanent Creator of the phenomenal world. Creation is nothing that occurred sometime in the past. Creation is rather an ongoing process. The world is, consequently, to be regarded as the continuous manifestation of God's creativity. This position is obviously similar to European pantheism (for instance in Spinoza).

The most topical notion in Japanese religious philosophy is no doubt the *Tao* or 'the Way'. The most outstanding philosophy of *Tao* is to be found in *Dogen* from the 13th century. Kakichi Kadowaki first presents the varieties of meaning of the term *Tao*. It means not only the Way but also the right coherence of one's thinking and doing, and the source of the myriads of things in the Universe and also the method and theory. *Tao* is above all *practice*. Through practice one's (existential) truth comes into being in the unity of mind and body. The truth is 'the freedom of the Way'. This Way, which is the Way of Buddhism, is, according to Dogen, to be entered without fail through practice, that is, through discipline and enlightenment.

The history of the notion of freedom is closely related to general social and political development in Japan. Kanji Tanemura begins his survey by pointing to the original meaning of 'freedom' in Buddhism, that is, the autonomy of the individual self. In the beginning of the 17th century the Christian missionaries brought to Japan European notions of freedom, for instance 'behaving freely' and 'freedom of the will'. In the following century and with a growing trade, a new social class of citizen emerged. And freedon became the freedom to trade and to earn money.

However, in the new civil state, the freedom of the individual may only be realized in duty towards the state. Rights and duties became part of the content of the notion of freedom. Freedom, according Hajime Tanabe, one of the most important philosophers of modern Japan besides Nishida, is self-sacrificing, and at the same time, self-realizing. In the 20th century, after World War II, the philosophical discussion of freedom in Japan was strongly influenced by European discussions, particularly existentialism, Marxism and pragmatism.

The European influence on Japanese culture, needless to say, creates tensions. And it creates tensions in various fields: in philosophy, in art and imagination (or creativity), and in cultural identity in general. Megumi Sakabe reviews the position of Tetsuro Watsuji on modern Japanese philosophy. Watsuji is well versed in European literature and philosophy. He has written essays on Blake, Strindberg, Ibsen and Schopenhauer, Nietzsche, Kierkegaard and Kant. He has written major studies in the philosophy of culture and cultural conflicts and several volumes on ethics. In his nationally-oriented philosophy of culture he takes as his starting point Herder, who maintained that each culture should be judged according to its own values. And from this point of view he analyses what happens when culture encounters culture.

Watsuji sets out to clarify both the building of a culture and cultural identity, as well as how a mutual understanding (intelligibility) between different cultures is constituted. He points to the significance of 'cultural space' and finds that this topic is greatly neglected, for instance in the philosophy of Heidegger. In his description of cultures and cultural encounters, however, Watsuji makes extensive use of Heidegger's phenomenological hermeneutic.

Modernization in Japan after the Meji restoration is the topic of Yoshida Masatoshi's survey of political philosophy in modern Japan. Japan, a feudal society right up to this century, was a late starter. Modernization was Westernization. The process, however, was not so much directed towards 'democratization' and the 'autonomy of the individual' as towards industrialization. Under an authoritarian regime the main objective was 'to increase production and promote industrial development'.

The intellectuals found themselves, for obvious reasons, in a dilemma: between human rights and the authority of the state, between tradition and modernization. What course should Japan take 'to best serve the interest of the people?'.

Yoshida focuses on three themes of discussion among thinkers and opinion leaders in modern Japan. He surveys the works of three political philosophers and writers, Fukuzawa Yukichi, the nationalist philosopher Watsuji Tetsuro (mentioned above) and Maruyama Masao.

Yoshida observes that, economically and socially speaking, Japan is on a par with the West. He also observes that the tension between democratization and an authoritarian industrial culture has not yet quite been resolved, and that the modernization process has created many contradictions, both in Japan and in the West. He advocates the need to develop a political philosophy relevant to both Japan and the West 'that sublimates modernity itself'.

The influence of Western philosophy in Japan is mainly the influence of German philosophy. In the middle of the 19th century, the intellectuals took a certain interest in English and French philosophy, mainly in Mill, Comte and Rousseau. At the end of the century, however, German philosophy became dominant. Chukei Kumamoto gives both a survey and a catalogue of Japanese philosophers and their works. Kant is without doubt the dominant figure, particularly his moral philosophy. Kumamoto puts it in the following way: the development of German philosophy in Japan is the development of Kant studies.

Kant is followed by Fichte, Hegel, Nietzsche, Husserl, Heidegger and Jaspers. It is remarkable that their works, almost *in toto*, have been translated into Japanese. In 1955, for instance, the works of Karl Jaspers appeared in 26 volumes. Some interest has also been taken in Leibniz, Schelling, Schopenhauer and Hartman and, after World War II, also in Marxism and pragmatism.

To study Western philosophy is one thing, to communicate it to traditional Japanese philosophy with a view to helping structure the modernization process is quite another. The challenges to philosophy from the modern industrial societies are far from being mastered, either in Japan or in the West. A main challenge is how to preserve a personal and cultural identity in a world dominated by economy and technology.

Asian philosophy

Śatapathaprajñâ: Should we speak of philosophy in classical India? A case of homeomorphic equivalents

RAIMON PANIKKAR

Contemporary philosophy. A new survey. Vol. 7, 11–67.

d) Worldview
   22. *Darśana*
   23. *Dṛṣṭi*

e) Systematisation
   24. *Mokṣaśâstra*
   25. *Dharmaśâstra*
   26. *Hetuśâstra*
   27. *Tarka*
   28. *Nyâya*
   29. *Vâda*
   30. *Siddhânta*

III. *Śatapathaveda*
   1) *The wisdom of the hundred paths*
   2) *The meta-semantic problem*
   3) *The philosophical query*

Following a wish of the author some of the key words in the text are capitalized to indicate a special meaning. Thus Man is written as Man (and not man) when denoting the human being (and not the male). Furthermore, the capitalization of God demands to capitalize Gods. For similar reasons, sometimes Nature, Earth and Reality are equally capitalized. For the particular topic of this article the word Philosophy, when carrying the cross-cultural meaning which is being discussed, is also capitalized. The abbreviations used in the article are as follows:

AV	Atharva Veda
B	Brâhmaṇa
BG	Bhagavad Gîtâ
BGBh	Bhagavad Gîtâ Bhâṣya
BS	Brahma Sûtra
BSBh	Brahma Sûtra Bhâṣya
BU	Bṛhdâraṇyaka Upaniṣad
CU	Chândogya Upaniṣad
Dham	Dhammapada
KthU	Kaṭha Upaniṣad
KenU	Kena Upaniṣad
MB	Mahânârâyaṇa Upaniṣad
Manu	Mânava Dharmaśâstra (Manu Smṛti)
MundU	Muṇdaka Upaniṣad
NBh	Nyâya Bhâṣya
NS	Nyâya Sûtra
NSBh	Nyâya Sûtra Bhâṣya
RV	Ṛg Veda
S	Samhitâ
SB	Śatapatha Brâhmaṇa
TB	Taittirîya Brâhmaṇa
U	Upaniṣad
V	Veda
YS	Yoga Sûtra
YSBh	Yoga Sûtra Bhâṣya
YV	Yajur Veda (Vâjasaneyi Samhitâ)

I. Introduction

All the terms used by the Modern Natural Sciences can be translated into any given language. If need be, a scientific discipline introduces new names by defining them with precision. This does not mean that they are universal, as some would claim. It means that they belong to a particular field of accepted or postulated references[1]. Modern scientific names are *terms*, i.e. *signs* which designate empirically verifiable or mathematically definable entities. This is not the case with *words*, i.e. those *symbols* in which complex experiences of peoples and cultures have crystallized[2]. *Terms* (such as 'table', 'triangle', 'sulphuric acid') aim at univocity and can be determined by empirical or purely conceptual experiments. *Words*, on the other hand (such as 'justice', 'beauty', 'God'), are polysemic and can only be understood against particular dialogic experiences[3]. The fact that words at home in one tradition have been carried over to other cultures — with the intention of understanding them, or perhaps ruling over them — has been the source of grave misunderstandings[4]. The problem begins with words[5].

There are words which, in certain times, possess a great force and radiate an extraordinary energy. They are symbols of living myths. Different religions and ideologies share in those myths and tend to use their symbols.

The word 'science', in the sense of the Modern Natural Sciences, is one such example. We see, as a result, that most modern methods of cognition lay claim to being 'scientific'[6]. 'Democracy' is another example, exploited by ideologies at opposite ends of the political spectrum. Yet no one has the right to monopolize words[7]. It is solely within the dialogue taking place in a given situation that the sense of words can be understood[8].

Of course, one can also restrict the meaning of certain words to a particular field. This enables a particular group to express certain matters unambiguously. All forms of scholasticism create such a language for internal use, "ad usum nostrorum", precisely because they move in a self-contained world. 'Grace', for example, was such a word in the Catholic post-Tridentine universe. As a result, catholic theology would not find it permissible to translate the Śaiva Siddhânta doctrine, using the word 'grace'[9].

It is imperative, however, to warn against the danger of superficiality under the pretence of generality or universal validity. Were grace merely to be understood as a certain inspiration from 'above' this would be doing injustice to both christian and śaivist intuitions. Grace is not a term with a univocal meaning, but a word, a symbol of polysemic value[10].

We could multiply examples of political domination or cultural influence by means of words. Why do my editors, for instance, resist my writing 'Gods' while they write 'God'? Why do they discriminate between 'animists' and 'Hindus'? Or, who dares, today, in places where television dominates, to speak or write about gay colors and gay knowledge? Words have a life of their own, but they are also subject to manipulation.

1. *The question of an indic philosophy*

Something similar occurs when we seek to use the name 'philosophy' in a transcultural sense. Is the transcultural use of words possible? Everything that is spoken is, *ipso facto*, incorporated into the culture of that language, regardless of whether or not this culture transcends its own boundaries in order to incorporate the new element. Inter-cultural values may exist, but trans-cultural ones do not. All values are inextricably bound up in a culture[11]. No value exists *in vacuo*. There are, in actual fact, no 'global perspectives'[12]. We are continually directed back to our own (particular) points of view[13]. What we can do is to become aware of this and live with our prejudices, while simultaneously engaging our perspective with as much breadth and depth as possible, preserving our openness in order to be able to learn from other attitudes[14]. There is no *lingua universalis*. We can only speak in our own language(s), while at the same time being aware of the fact that our language does not exhaust the universal range of the human experience.

What, in our case, is Philosophy? Is it an achievement of the western world or a more universal symbol? Is it a technical name belonging to a single culture or a word whose meaning can also be partaken of by other linguistic universes? Is Philosophy a term, like 'sulphuric acid' (we produce and sell it), or a word, like *dharma* (it encompasses and overwhelms us)? Were the latter to be the case, could the meaning of Philosophy then be extended to and enriched by other cultures? But how far can or may one stretch the meaning of words?

Philosophy may have many meanings. And in fact, there have been multiple views on the nature of Philosophy. We use, of course, the word Philosophy in its most embracing sense, which includes what has also been called metaphysics, theology, theosophy, and the like. But, until recently, those diverse opinions were within western culture, as broad as this culture is[15]. If we want to apply this notion to other cultures we should justify the extrapolation or show that it is a cultural universal[16].

Underlying this question there is a grammatical problem (which indic grammarians tackled, probably for the first time). Is Philosophy a *proper name*, although, of course of a second order (for it does not refer to individuals), or is it a *common name*, a universal?[17]. Does it belong to a single culture (or group of subcultures) as a proper name belongs to an individual, or is it a general notion denoting different entities sharing in a common nature? Is it formally descriptive, like Man, *anthrôpos*, or does it also claim some content, like justice, *dharma*?

Let us insist on the insidiousness of just hurrying to answer questions uncritically, accepting the assumptions of one particular culture as apodictically valid. To ask almost rhetorically whether there is *also* Philosophy in India is on a par with a child asking her companion if her parents have a colour television set. We assume that Philosophy is something as concrete as a TV set, and that obviously one cannot live without television, at least in black and white (in case Indians do not have the sophistication of a western colour Philosophy). What, in fact, do we ask when we put the question of an 'Indian Philosophy'? Whether India has a Locke, a Spinoza or a Hegel? What idea of Philosophy do we presuppose?[18]. And further, why should the concept of 'philosophy', in whatever sense, be the paradigm for a question which, as question, claims to transcend its own boundaries?

Diatopical hermeneutics does not ask whether there is also Philosophy in India, but starts with a double entry: Indians learning from western culture and westerners from indic culture, see what happens in the encounter, and then forge their respective questions.

We could recall here some insights from Sociology of Knowledge and quote abundant texts of modern indian scholars telling us that Philosophy could not be missing in the glorious past of India[19]. Scores of books written in English bear that title[20]. We also have a television set, and even better than yours. Ours has three dimensions. You have forgotten holiness, liberation, happiness in your Philosophy. We have it all included. Ours is a more complete Philosophy[21].

It has been fateful, due precisely to the renown of 'philosophy' and the prestige of modern western culture, that a majority of indic thinkers use this word without analyzing it critically and interculturally[22]. Yet we also hear certain misgivings[23]. Or nostalgic disclaimers[24].

It is not the intention of this essay to analyze the meaning of the word 'philosophy' and all its multifarious homeomorphic equivalents in indic culture. We shall cite only a few expressions which are used in Sanskrit literature, and concentrate on the cross-cultural problem[25].

It is a well-known fact that a significant portion of contemporary indic scholars have enjoyed an anglo-saxon education. As a result of this, many of them will desire to prove that indic Philosophy can also show positivist, analytical and rationalist traits[26]. Indeed, there are some who take pains to argue that in India there are and have been not only logicians, skeptics, materialists, and even marxists, but also positivists, pragmatists and language analysts[27]. On the other hand, there are others who, as a reaction to this, will emphasize that, although indic Philosophy is indeed Philosophy, it has little in common with 'western philosophy'[28]. The problem would have taken on quite a different appearance if modern indians were more directly exposed to the philosophical and theological traditions of ancient Europe[29].

Contemporary literature displays a confusing vagueness regarding the translations of the 'philosophical' expressions within the indic tradition[30]. This problematic state of affairs is compounded by the fact that corresponding western words are often only understood in an extremely narrow sense[31]. In addition to this, one has to consider that the usual cognitive strategies employed in the West to understand other cultures must be re-examined[32]. Distinctions and concepts, such as ontology, metaphysics, ethics, epistemology, aesthetics, cosmology, theology, anthropology, as well as such modern turns of phrase as educational philosophy, religious philosophy, etc., are often taken over and employed all too uncritically[33].

If we apply such strategies to a cultural world like the indic, then the studied material is not only ravaged, but it also remains incomprehensible, or still worse, it is misunderstood[34].

Western scholarship has shown as increasing interest in indic Philosophy, but our specific problem is seldom mentioned[35]. Indian scholars have by and large followed the same line, if with reservation and warnings[36]. Many people are beginning to think differently today[37]. To be sure, as long as we speak and write in a given language it is inevitable that we should apply that language's range of intelligibility as a yardstick[38]. But one ought always to bear such limitation well in mind[39].

The mutual influence between India and the West is not of yesteryear. It goes as far back as the greeks[40]. In spite of their differences, both cultures still belong to the indo-european world[41]. We should not overstress the diversity. Philosophy may well be a legitimate name for a common endeavour[42]. The differences are striking when we compare with african culture, for instance[43]. African intellectuals have tackled the problem with an uncommon depth[44]. This may be due to the fact that african thinkers are more distant from the greek tradition than the cultivators of indic 'philosophy'[45]. The problem is, of course, not limited to Africa[46].

2. *A typology of positions*

We shall summarize what has been understood by Philosophy in view of our cross-cultural problematic. We stress the present-day perspective, because we should not forget the other meanings that the word Philosophy has had also in the West[47].

The answers run from the most restricted to the most general conceptions.

1) A clearly unequivocal opinion will state that: "The frequently heard talk of 'western Philosophy' is, in fact, tautological. Why? Because 'philosophy' is, in essence, greek"[48]. One would find no 'indian philosophy' here[49]. This opinion does not say that the greeks were the only thinking people. It says that the peculiar way of thinking, namely the thinking about the essence, the *quidditas*, which is assumed to be the essential question of Philosophy, is greek and not universal[50]. Few others, on the contrary, will affirm that practically only the western mind can argue and thus properly philosophize[51].

2) With the well-intentioned desire to open up the 'greek' notion, a second interpretation will recognize a kind of 'universal philosophy' tailored along the greek model, but without reducing it to western moods. There is no doubt, this opinion will hold, that the origins of Philosophy have been greek, yet there is no difficulty in recognizing the blossoming of other branches of that european trunk all over the world. To be sure, Philosophy is a kind of universal human reflection about the nature of things, but this has been mainly cultivated by the european mind[52]. The model here is the nature of Modern Science. There is a specifically modern scientific approach to Reality which, in spite of later accretions and even refinements, remains basically the creation of the european genius, although now it claims to be universal. Similarly, there would be a universal Philosophy in the same sense that it is claimed that a modern universal Science exists, in spite of its particular birth-place. A certain evolutionary mould is clearly detectable here.

3) The third opinion will blow up the greek paradigm. It will accept, as it were, the several candidates to that prestigious name of Philosophy and proceed quite democratically to chop out divergences in order to find, through consensus, the nature of Philosophy. It will look for a non-commital common denominator and abolish any bias. Philosophy would then be an umbrella notion, neutral and univer-

sal because it has eliminated any particular opinion[53]. We may perchance find common structures and a merely formal character. People will then speak of a "Global Phenomenology of Philosophy" which will bring out a general and formal concept of Philosophy: a kind of logical understructure.

4) There is still another 'philosophical' reading of Philosophy which will draw its inspiration from a certain ('philosophical') History of Philosophy, and look not for a common denominator and formal concept, but the possible harmony of the different ratios (to follow the same mathematical metaphor) among the corresponding different numerators and denominators. There may emerge a kind of genus of several species of Philosophy. The genus will not be, properly speaking, Philosophy, but only a kind of abstract notion re-presented and re-enacted each time in its respective species. Each Philosophy will claim to be ultimate, and at the same time 'merit' the name of Philosophy precisely because it incarnates in its own manner that otherwise indefinable genus. This 'generic Philosophy' will be a kind of 'open universal' in transcendental relation with the each time corresponding 'philosophy'. We will then never be able to define what Philosophy is, except in the language of one particular Philosophy making efforts to open up to other understandings. For this reason we said 'open universal', because there are no definite notes to describe it except the notes which may appear when we consider a particular 'philosophy'.

A century ago hardly anybody would have dreamed of including an 'african philosophy' in the philosophical club. African 'wisdom' was not deemed to fit into the parameters of what 'is' or 'should be' Philosophy. The situation has changed today, and the change is mainly political. Yet this fact has 'philosophical' repercussions. It makes us take seriously the impact of the african contribution to the cultures of the world. It changes the very parameters of human (philosophical?) self-understanding. The influence is mutual. The political situation changes the conception of what Philosophy is or could be, and Philosophy in its turn changes the notion of culture.

5) The other end of the spectrum would be furnished by the opinion of a merely nominalistic label. 'Philosophy' would then be irreducible to any kind of formalization and intelligibility. We would be dealing here with a radical multiversity which would contradict itself the moment it is formulated in any univocal or analogical way. Significantly enough this opinion practically coalesces with the first one inasmuch as the word 'philosophy' could be applied to one single culture and only abusively or equivocally to others. In fact there would be no 'philosophies' because the plural would be an empty word, devoid of meaning.

The word 'philosophy' would then be both a particular name for a concrete activity of the human spirit and at the same time an umbrella notion holding whatever candidates to this purely formal notion. Only by convention could we speak of Philosophy. The african example again comes to mind.

* * *

Comparing this scheme with the example of languages, we would have the following instances:

1) There is only one language.
2) There is a single language with diverse dialects.
3) There are many languages, each of them independent and connected only by a kind of *kata-language* or pre-language (of a merely formal structure) without any linguistic meaning in itself; a kind of algebra. The kata-language is not a language.
4) There are diverse independent and linguistically irreducible languages, which, however, in their own particular fashion, articulate the otherwise unspeakable of a *meta-language*. The meta-language is language only in the respective languages.
5) There are diverse phenomena rendered by us with the empty name 'language' which are mutually unbridgeable. These represent genera in themselves, so that they can in no way be reduced to a higher unity.

We may further bring the analogy to the sample scheme:

1) There is one single individual set (language or Philosophy).
2) There are many subindividual sets (languages or Philosophies) of one single, theoretically definable individual set (the greek paradigm).
3) There are many individual sets of a mutually irreducible nature, except on the level of an abstract sub-structure — each Philosophy being a monad.
4) There are many species (sets) of a theoretical genus, which confers a certain intelligibility to the indefinite sets of species.
5) There are independent and irreducible entities which can in no way be brought together, except by abusively applying the name of one entity to the others.

* * *

We have here again a political issue echoing the past, reflecting the present and influencing the future. Each of the five options depends on a political panorama which permeates the culture sponsoring a particular option. Eurocentrism will favour the first, neo-colonialism the second, sophisticated western scientism the third, pluralism the fourth, and skepticism the fifth. To call Philosophy only the greek model, either rigidly or flexibly, entails a certain predominance of the european or the western culture. To call Philosophy an aseptic classification of the philosophical opinions, reducing ourselves mostly to figuring out the meaning of their respective utterances, implies another political outlook: pragmatic and antimetaphysical. Or we may struggle to defend the peculiarity of every single human effort and recognise a certain pluralism. Or, finally, we may renounce making any sense of the intellectual enterprise outside the particular limits where it defines itself.

In point of fact all the opinions are possible and plausible within their respective political views. Yet, in the present day intellectual and political constellation, we may be forced to criticise four of them.

The *first* position is coherent and avoids the ambiguity of the others. Philosophy is the very question about the essence of Philosophy itself: *Was ist das—die Philosophie*?[54]. But it has two weaknessess. It precludes a cross-cultural enlarging and an eventual deepening of the meaning of the word. In a sense it freezes the meaning of Philosophy as if it were a scientific term. The widening of the notion of Philosophy seems imperative after almost two centuries of a perhaps improper but effective use of the word by other cultures.

Another weakness of this first opinion consists in assuming that the strict philosophical question about Being present in beings, or the ontic power of language (*logos*) is exclusively greek[55]. First of all, we cannot know it *a priori* and, secondly, we are inclined to find in India a very similar enterprise[56].

The *second* interpretation, wanting to correct the restrictions of the first, makes it worse inasmuch as it erects one particular way of Philosophy, namely the greek, as the paradigm of all philosophising.

The *third* opinion empties Philosophy of its ultimate meaning, and, converting it into a merely common denominator, it saps from Philosophy any philosophical value. It falls under the spell of 'scientific' abstraction, which, while being legitimate in modern sciences, is uncalled for in the traditional understanding of the philosophical activity. The mathematical substructure of a physical phenomenon may yield some physical laws, but this does not need to be the case with Philosophy, unless we uncritically make of it a Natural Science. Every species of Philosophy attempts to be the proper and adequate incarnation of the genus Philosophy. This third opinion would deny the fact that the difference between philosophical activity and any other human activity is that the question of its own identity belongs to itself and not to any other discipline. This general formal concept would yield, if successful, a non-philosophical conception of Philosophy.

I submit that the *fourth* of the abovementioned options is the most fruitful one, because the others suffer from the methodological flaw of the starting point: either too rigid or too murky. How, then, should we proceed? The fourth option is only partially convincing because, in my opinion, we have not yet elaborated an in-depth reflection on the very nature of this enterprise.

The *fifth* interpretation eliminates the problem altogether by not allowing any use of the plural, by virtue of the same *a priori* as the first attitude. In fact they coincide.

We would like to elaborate now the problematic of the fourth.

3. *The homeomorphic equivalents*

Indic culture does not have a univocal word for Philosophy. Western languages have no univocal word for *dharma*. Can we say that there is no Philosophy in classical India and no *dharma* in western culture? In a certain sense, yes. The respective cultures have not found the need to coin in one single word certain human experiences, most probably because they have not seen any common thread uniting those human activities in a meaningful way. Western languages have words for religion, duty, essence, element, rule, morals, energy. Yet it makes little sense to subsume all those names under any one generic word which could be the equivalent translation of *dharma*. It would appear artificial. There is no perceived need for it. And besides, each one of those names has a field of connotations which would not allow a common generic name. It would curtail the wider meaning of those single words. The parameters of understanding go a different way.

Something similar happens when the word Philosophy is applied to classical indic culture. There is no univocal translation. And yet there are scores of names which cover the field of what in the West is called a philosophical activity.

Two observations are pertinent here. The *first* one is that we are not at point zero of history, nor of culture. We cannot ignore the last two hundred years of mutual influence between India and the West, besides previous historical encounters. The very word 'philosophy', with a plurality of meanings, is already assimilated in an important sector of indic culture today. And there is no doubt that there are contemporary indian thinkers cultivating not only branches of western Philosophy but developing a proper intellectual activity inspired by ancient indic wisdom. They call this activity by the name of Philosophy. With what right are we going to say that 'it' is not Philosophy?[57].

The *second* observation concerns a more theoretical approach and also has practical consequences concerning the future, for there is no doubt that, in spite of the traditionally recognized 'philosophical' acumen of the indic mind, there has been no major contemporary contribution in the field of Philosophy coming from the indic subcontinent. Could it not be that the points of insertion between the two 'wisdoms' have not yet been found, so that no mutual fecundation has really taken place? If for instance, one dismisses the mîmâṁsâka-s as mere exegetes, and most of the vedantins as theologians, because Philosophy is, once and for all, what some modern philosophers dictate it to be, there is not much intercourse possible. This theoretical approach again poses the basic question of what we are looking for when we ask about Philosophy. Philosophy is a human activity before being an academic discipline or any kind of precise concept. If we speak in English we may ask what it means to philosophise. If within the indic world we were to ask what it means to strive consciously for liberation (*mokṣa*), would we call this pursuit Philosophy?[58]. Should we render 'philosophising' by *mumukṣutva*?

I have elaborated the notion of *homeomorphic equivalents* to denote those fundamental notions which, as third-degree analogates, perform an equivalent function in the respective systems[59]. A clear example may be God and Brahman. We cannot translate the one with the other. Belief in God cannot be equated with 'belief' in Brahman. Also the distinctive notes or attributes of the two differ radically. Not even their respective functions are the same. The one creates and is provident, loves and has a will, the other does not need to fulfill those functions. Both perform only an equivalent role. It is the role of being an ultimate point of reference, of being ultimate in the sense of not admitting anything beyond on any realm. For this they have to perform different functions. Brahman and God are homeomorphic equivalents. I enlarge metaphorically the use of the word homeomorphism which designates a definite function in geometry, in order to express a somewhat similar meaning in cross-cultural studies.

Within western culture, in spite of the fact that there are multiple views concerning 'philosophy', the question about the nature of Philosophy is both meaningful and significant. It is a question regarding that peculiar intellectual activity which generates a certain critical knowledge of ultimate situations, or attitudes.

If, however, we incorporate the insights of other cultures into western reflections, many of the western assumptions, accepted as a matter of course, do not apply. The problem regarding the nature of Philosophy is superseded since the problem itself is problematic. The correct method, therefore, is not one of seeking a more or less literal translation for a word, but rather to find those *homeomorphic equivalents* which describe the possible correlative activity or activities in other cultures.

In saying this we are already dismissing as inadequate the concern about univocal word-to-word translations. Each real word in any language opens up a whole field of meanings, with diverse connotations and denotations, which is practically impossible to cover with another single word of another linguistic universe. As there is no Man without a heart, there is no word without a soil. The original soil of the word Philosophy is greek and western. Is a 'pure' heart-transplant possible? Or should we also transplant some soil? Or?

Our question, then, is to ask for the homeomorphic equivalents of the western notion of Philosophy, and not just for the translation of the same concept. There is a biased start in asking whether the greek conception of Philosophy is present in the indic culture or not. The answer is irrelevant. It may well be that such a concept is present, but this would already limit the scope of the fundamental question, which is not to ask whether there is in India what the West understands by Philosophy, but to ask which is the ultimate and most important human conscious activity in indic culture, or that question which we may find most suitable regarding the very source of the human inquiry about the meaning of Reality or the sense of life. How can we think Being? asks the western mind since Parmenides. For it is assumed that "there is only being", and "Thinking and Being are the same"[60]. Without necessarily the Parmenidean identification, the Thinking/Being paradigm will remain paramount in western culture (Philosophy). "By what could one know that by which all this is known?" ask the Upaniṣads[61]. For it is assumed that to know is to become not only the object, but also the subject. How can we know the Knower?[62].

I repeat: the most important thing is the analysis of the scope of the question prior to answering it. *On the one hand*, what do we ask when we ask what is Philosophy in an open context? We have here what I have called a *pars pro toto* effect[63]. Our intentionality is directed to the *totum*, but until others make us realise it, we are not aware that by virtue of the *pars pro toto* effect we know the *totum* only *in parte* and *per partem*. We see the whole reflected in and refracted by our window. What do we do, within the indic culture, when we perform that activity which has impelled western culture to philosophise? What is it, at the source of the 'philosophical' question, which tallies with the corresponding indic activity? What is such an activity? We ask not only for the goal of the search (truth, liberation, happiness, power, just to know, etc.) but also for the impulse for such a search. This *svadhâ*, as the Nâsadîyasûkta, this ἦθός, as Heraclitus echoing the same Sanskrit root, would say, this *nixus* (*nisus*), as the european mediaevals wrote, or simply this urge, does not need to be a psychological motivation. The question is not just "Why do I — as individual — philosophise?". The question is whether the ultimate impulse to philosophise finds an homeomorphic equivalent in other cultures, and what forms it takes.

On the other hand, we cannot analyse any question if we do not delimit, at least implicitly, the range of the possible answers. Any question is about something, and the 'something' is already determined by our pre-understanding of it. Otherwise, we do not know what we are asking[64].

We have here another hermeneutical circle. It is similar to the classical argument of the christian Middle Ages and Vedânta that one would not search for God or for Brahman if one had not found 'him' already, or was inserted in 'it'[65]. This is why we cannot separate the first, historical, observation from the second, theoretical, one. Both belong together. Philosophy has existed for centuries as an indic category. Yet this mainly western and uncritically adopted category does not tally completely with the complex endeavour of classical India. The confluence of those two fields delimits our enquiry.

* * *

It is important to insert here a general consideration and apply it to our case. No human problem, I submit, can be adequately tackled if approached exclusively by means of what the western tradition has called the *logos* (word, intelligibility, rationality, reason). Any human situation is also dependent on the place and function of the *mythos* as well[66]. There is no *mythos* without logos, and this is why we can speak of the *mythos*. But there is equally no *logos* without *mythos*, and this is why we can never exhaust the sense and power of *logos*. Now, if the *logos* gives us a certain understanding, the *mythos* offers us the horizon, in any given space and time, in which we situate, and thus understand, the problematic (*qua* problematic first of all). The *mythos* offers us the *under* on which the under-standing *stands*. Now, our present writers, and those who will read them, both operate with certain tools of understanding, both see a definite set of problems, and envisage a possible and limited range of solutions. All this depends on the underlying *mythos* within which we live. It is the prevalent *mythos* of our contemporary situation. It is the situation of a certain group of humans which some call the 'intelligentsia'. Within this particular horizon we take some things for granted, accept certain common assumptions and use a certain language which emerges from our *mythos*. Within it we feel the legitimacy of the question for african and indic philosophies, without the need to abide by the rules of what until now has been called Philosophy in main-line western culture.

This consideration provides all our assertions with a factor of provisionality and relativity which we should not forget in this study. From an indic scholastic standpoint we may call it the *vyâvahârika* factor, and from christian scholasticism the *dissimilitudo* element. From a 'philosophical' perspective I have called this position one of *relativity* and distinguished it from relativism.[67]

* * *

Let us sum up. Heidegger may be right in delimiting with precision the nature of 'philosophy'. But our question is prior. We do not ask what is 'philosophy'? — and much less what is *that* which is 'philosophy'? We ask: what prompts the western or greek Man to do that activity which we call philosophise? This is not yet a cross-cultural question, but it is triggered by a cross-cultural concern, because it induces us to ask what homeomorphic urge shows indic culture. We may answer, re-echoing Aristotle: because all Men want to *know*[68]. We may also answer, recalling an even prior attitude, because everybody wants to be *happy* or lead a good life[69]. Man's immediate and universal concern is to overcome *duḥkha*[70]. We may go still further down the road of history and answer because the human being desires to be *saved*, to become *whole*[71]. What is there at the root of the ultimate human concern? Or rather, what is there at the root of the ultimate urge of the universe?

None of these statements is without presuppositions, but all of them also make sense within the indic world. Why is there Being rather than Non-Being? asks what we could call the acme of western Philosophy[72]. "At first was neither Being nor Nonbeing" wonders the Ṛg-vedic ṛṣi[73].

We may further add that, in order to know in a particular manner certain things, western Man philosophises, and, taking the whole range of Philosophy in the West, we may further add that Philosophy is cultivated in order to have a happy life, and to be saved or become whole, whereby each Philosophy will have a particular notion of all those key words.

In a very general way we may finally say that the business of Philosophy is that of acquiring a kind of *saving knowledge* or *gnôsis*, whereby the notions of salvation and knowledge are not specified — they may save us only from ignorance. Now, this is precisely what shall lead us to the homeomorphic equivalents of Philosophy in indic culture. What kind of human act is required or important in order to really *know*, be *happy*, *saved*, or simply to be? The Nyâya vocabulary is almost the same: *apavarga* (salvation, emancipation) comes only from *tattvajñâna* (true or proper knowledge of things). And this happens when we know the 16 *padârtha-s* (themes, categories) which are relevant for salvation. To have these 16 *padârtha-s* as *prameya* (objects of knowledge) is the purpose of Nyâya (which, of course also recognises other minor, secondary objects of knowledge).

With this in mind we may now proceed to look for the homeomorphic equivalents and give a number of Sanskrit expressions related to what Philosophy means in the western tradition. Each word discloses a new universe of discourse. There may be many more. We cite them by way of examples, and venture a certain typology.

II. The indic philosophemes

If western traditions are variegated and bountiful, the cultures of the indic subcontinent are no less polymorphous. We shall thus consider only some aspects of that bundle of traditions belonging to sanskrit India. It goes without saying that we shall have to be selective and incomplete regarding the notions we cite[74]. However, cursory as this study is, it may be sufficient to bring home our point. Our aim is not to describe "Indian Philosophy", but to figure out what Philosophy in classical India may mean.

There are three main *philosophemes* or activities of the human spirit which, in one way or another, could be homeomorphic equivalents to the 'philosophical' enterprise: *Kriyâ, icchâ, jñâna. Action. will, knowledge* could be the western pass-words.

In order to follow the indic way and not just walk in the wake of the western model we shall follow the most classical division of Vedic lore in *karma, bhakti,* and *jñâna*, well aware of its qualifications within the same indic tradition[75].

Although the Vedic scriptures do not contain the entire indic wisdom they are considered to contain the seeds of all that is to be done and known. *Trayî* i.e. the three (Vedas) is the consecrated name for the Vedic knowledge[76]. Veda means both the Veda, the crystallized knowledge of the *ṛṣis*, and *veda*, the vision, the science of any true seer. Tradition has considered the Saṁhitas and Brâhmaṇas as dealing with *karmakânḍa* or ritual, the Âraṇyakas with *upâsanâ-kâṇḍa* or meditation, and the Upaniṣads with the *jñâna-kâṇḍa* or knowledge[77]. The three are intrinsically related, and although later on, for many reasons, specialisation, and sometimes rupture, took place, the unity has never been completely lost[78]. *Karman* or ritual, in fact, does not mean a merely ritualistic action. It means that total human praxis by which we, human beings, reach the very goal of existence[79]. *Karman* is sacrifice in the deepest sense of the word[80]. *Upâsanâ* or *bhakti* does not mean just meditation or devotion. It conveys again that total human activity, concentrated in the heart, which fulfills not only the aspirations of Man but the very perfection of Reality. And finally *jñâna-kâṇḍa* is not directed to the satisfaction of mere intellectual curiosity, but to that total assimilation of/with truth, i.e. of/with Reality which can only be brought about by the transforming and saving *gnôsis*.

All in all, there is a human dynamism towards wholeness, salvation, happiness, peace, or whatever we may prefer to call it. This *svadhâ* would be the underlying urge prompting the philosophical activity. The three endeavours belong to what corresponds to the philosophical enterprise. It would amount to a disservice to Philosophy to reduce it exclusively to one part because the modern western notion has stressed only one side of the total human endeavour to reach the goal of existence. We shall consider the three separately.

Let us insist, to avoid misunderstandings, that we are not searching for translations of the word 'philosophy'. Guided by the respective homeomorphic equivalents we are trying to walk a two-way path: from the Mega-Hellas to the Mahâ-Bhârata, and from the latter to the former. What is that act or which are those activities to which Philosophy, among other notions, claims to be a case in point, thus offering a determined set of possible answers? Or, starting from a very restricted concept of Philosophy, we could argue as follows. We philosophise because we want to know. We want to know because we want to improve our existence, have more life, joy, power, control over things, . . . , because that knowledge orientates us in life — to go, of course, where we believe we want or have to go. In other words, whether we affirm that we do Philosophy out of sheer curiosity, for the sake of power, control, satisfaction, fulfillment, or salvation, ultimately it comes to this: Man has an innate or acquired capacity, or an urge, to steer one's own life towards something which may be called human fulfillment — be it also a cosmic and/or a divine affair. What is to be 'done' in order to live life to the 'full'? This is the locus of our question.

1) *Karma-Kâṇḍa. Action*

This first section may appear foreign or obsolete to a certain conception of Philosophy. All the more do we give due importance to it, because it may *enrich*, *deepen*, and *upset* the very notion of Philosophy. It may *enrich* the meaning of Philosophy by stressing the often forgotten praxis-factor in the philosophical endeavour. It may give to action a more predominant place than in most modern western philosophical systems. It may also *deepen* the notion of Philosophy, reminding the West of its own roots. There is in fact an entire western tradition which underscores that there is no rupture between theory and praxis, knowledge and action, holiness and insight, and the like. *Philosophiae desiderium* meant sanctity for long periods of the christian tradition. It may also *upset* a certain prevalent conception of Philosophy as an exclusively rational enterprise. The challenge here does not come from the irrational, which would simply revert us to a pre-human (and often anti-human) level. The challenge comes from the a- and supra-rational, i.e. from the reasonable assertion that reason is neither the only guide (to life) nor the exclusive criterion (for truth).

I insist, however, that the indic position does not merely run parallel to the western problematic. It has its own radical starting point from a differentiated but unbroken experience of Reality.

1. *Karman*

The first word here is *karman*[81]. The word has a bewildering variety of meanings, but we take it to mean what the *karmakâṇḍins* meant it to mean, i.e. the set of actions leading to the attainment of the goal(s) of life[82]. *Yajña*, sacrifice, is the vedic word for this integral and saving action[83].

It may be appropriate here to recall the fundamental sense of *karman*[84]. It means the working out of one own's salvation, thus contributing to the liberation of the entire universe. This work may require many generations and may imply many acts. The *karmamârga* of the *Gîtâ* may provide the most authoritative example[85]. *Karman* is then true Philosophy because it opens up for us the meaning of existence which is not just to possess a certain idea of what life is all about but actually to realise it. A recurrent theme of indic spirituality is expressed in the so-called *jñâna-karma-samuccaya-vâda*, or doctrine of the synthesis between knowledge and ritual action for the removal of human misery and attainment of salvation.

In a word, *karman* is not just doing or making (something) but doing and making oneself, viz. realization. The path is an existential one.

In recent times the study of the *karmakâṇḍin-s* has been practically neglected because modern cultivators of 'philosophy', have consciously or unconsciously used the word 'philosophy' in its mainly post-cartesian sense[86]. This is changing in our days[87]. Interestingly enough some who include *karmakâṇḍa* as Philosophy recall a verse of the *Gîtâ* saying that, "all actions with no exception culminate in knowledge"[88]. We should not split the two, since true action entails knowledge, and proper knowledge leads to action.

2. *Karmamîmâṁsâ*

Another word, which also has a long tradition, is *karmamîmâṁsâ. Mîmâṁsâ* means thinking inquiry. We shall return to this rich world again. It denotes the specific thinking effort. Now, here, thinking is directed to the most important human endeavour, that of providing for Man's survival, well-being, salvation, *mokṣa*... We should ponder that, in spite of priestcraft and all other abuses, Philosophy as *karma-mîmâṁsâ* stands for that undivided effort directed to the only thing that matters: human salvation. We should emphasise that "human salvation" stands here neither for mere salvation of the individual nor of humankind only, but for human cooperation in the salvation of the whole Reality.

The entire Pûrva-mîmâṁsâ then, and not only the so-called Uttara-mîmâṁsâ or Vedânta, can be said to be an homeomorphic equivalent for Philosophy. Those conversant with the primitive hellenic and christian traditions will recall that this same existential and holistic attitude was also called Philosophy[89]. And this in spite of the ambivalence introduced in christianity by the disparaging Pauline statements on a certain kind of hellenic wisdom[90].

3. *Dharma*

We are not concerned with the many opinions regarding the meaning of life. We are concerned with the enlightened praxis conducive to the fulfillment of the very goal of life — whatever life may be. The word which one would want to use here is *dharma*[91]. Philosophy is *dharma* in the active sense of the word. To perform properly one's own *dharma* requires a certain *consciousness* of the act[92]. This performance could be said to be the proper business of our equivalent to Philosophy.

We invert the priority. Philosophy is not primarily *knowledge* about the proper acts or sacrifices to perform. Philosophy is the actual *performing* of the proper acts; performing which entails an awareness of what one is doing — otherwise it is not a human performance.

Some may wonder whether to hold *dharma* as a homeomorphic equivalent of Philosophy is too far-fetched. If we recall that the root, according to most classical commentators, means to support, to bear, to keep together, and equally recall what the Hebrew Bible says about 'wisdom' along with the christian scholastic tradition, we may easily interpret the function of *dharma* as that of keeping together the universe, contributing to the sustenance of the world[93]. Philosophy in this sense is a cosmic function of collaborating in keeping the cosmos together, working towards its salvation. Indic, jewish and christian wisdoms concur.

I would venture to affirm that the passage from *ṛta*, or cosmic order (of the Vedas), to *dharman*, or active coherence of Reality (of later periods) represents one of the birth moments of what we may still call Philosophy. "Truth is *dharman*"[94]. In fact *ṛta* is cosmic harmony, but *dharman* is the knowledge and existential upholding of this cosmic harmony. It appertains to the Gods the maintenance of cosmic order: *ṛtasya dharman*. By knowingly accepting *dharman* and living in accordance with it the human being shares in the function and thus the destiny of the Gods. Man becomes conscious of human dignity and freedom. The sharing in *dharman* represents the passage from an objective and purely cosmic order to a subjective and also cognitive human order. Philosophy was born when humans shared in the destiny of the Gods.

4. *Sâdhana*

Another word which should be mentioned here is *sâdhana*. This word has been utilised mainly in *tantra* and from there has passed to practically all other systems and to the general neo-hinduistic world, and even beyond. It is differently translated as discipline, worship, practice. Coming directly from *sâdh-* (to go straight to any goal or aim, to attain an object and thus be successful) and indirectly from *sidh-* (to be accomplished, fulfilled, thus successful) it implies a triple element: effort, praxis and attainment[95]. Some have spoken of "practical philosophy". The indic reaction here would be that any Philosophy is practical, for an "unpractical philosophy" would not contribute to salvation and thus be no Philosophy at all. Philosophy here is interpreted as *karma-mârga*, way of praxis.

* * *

To sum up. The word Philosophy has developed a predominantly cognitive meaning. Yet the two greek components of the word *philo-sophia* suggest love and experience. The indic counterpart invites one to praxis, to a resolute conscious effort "to work out one's own salvation", to echo Buddha's alleged last word[96]. In this sense part of the buddhist tradition has understood buddhism to be Philosophy. This praxis element in no way also precludes a concommitant cognitive dimension. It only implies that praxis is paramount, and that the theory-praxis dichotomy is ultimately making theory short-sighted and praxis weak[97]. Both go together. The symbiosis is Philosophy[98].

2) *Upâsanâ-Kâṇḍa. Devotion*

For the English heading we have written *Devotion* instead of love. The latter word would have been equally legitimate, but its many meanings advise a great caution in using it. To be sure, *devotion*, in certain circles, may only convey sentimental attachment or a 'pious', 'religious' practice[99]. Here we should understand the word closer to its etymology — from the Latin *devovere* (to consecrate, dedicate oneself to something) which connotes the *votum*, the Sanskrit *vrata*, the vow[100]. Devotion here stands for the consecration of oneself and of the entire universe[101]. Devotion implies love and service, both to single beings and to the whole of Reality in a conscious way. It suggests the classical meaning of *studium* as dedication, consecration[102]. We may consider some words.

5. *Upâsanâ*

Since the very beginning of the indic lore *upâsanâ*, i.e. sitting near and thus waiting upon a master in an act of loving service, or proper adoration, worship and contemplation, has been considered the most effective and proper act of Man in order to fulfil the meaning of life[103].

Although we shall also propose *bhakti* as a legitimate homeomorphic equivalent for Philosophy, we have stressed the name of *upâsanâ* as more traditional and perhaps also clearer. *Upâsanâ* connotes an element of meditation, contemplation, loving gaze, cognitive surrender, conscious or 'superconscious' devotion bringing together, as it were, praxis with theory[104].

This notion is important because it stresses the unity of the 'philosophical' enterprise from a perspective which complements the usual western one. Philosophy, when it purports to be sacred or integral knowledge of the mystery of Reality, inasmuch as this is possible, will insist that this knowledge also integrates will and praxis. Here *upâsanâ* stands for an authentic consecration, i.e. a plenary dedication of our life to serve, contemplate or worship, the mystery of Reality. It also integrates an element of knowledge in it. It purports an awareness of what we are doing.

6. *Bhakti*

In spite of a certain modern western — and one-sided — image of India as the land of philosophical Vedânta, perhaps the foremost indic feature is that of *bhakti* in its manifold forms. Suffice to quote the *Bhagavadgîtâ* and her spirituality and the medieval saints, but more important still the entire life of the people. We will say only this much.

The consecrated word for it is *bhakti*. The root *bhaj-* suggests to participate, to share in. And affection appears as the most prominent way of sharing[105].

The conscious sharing in the *prâṇa* and *apâna* of the universe, in the throbing of the world, divine, cosmic and human, being present in awe and adoration, ready to serve because love is the ultimate constitution of Reality — this is the highest Philosophy. This love is not blind, nor is it lame. Mere reasoning may lead to clarity of ideas, but only contemplation leads to action and realisation. No other activity will deserve the name of Philosophy for a majority of the people of the indic subcontinent, even today. If we have to use the hellenic name as having a certain validity outside its post-cartesian field, we cannot exclude *bhakti* from being a homeomorphic equivalent to Philosophy.

Like the word love, *bhakti* has such a variety of meanings that some of them may not properly correspond to Philosophy. We do not have nor need to have word-for-word equivalents.

7. *Tantra*

Tantra is another name which should at least be mentioned in our search for homeomorphic equivalents. The commonly accepted root is *tan-* to stretch (the human knowledge beyond its limits). Other etymologies defend the root *tatri* or *tantri* with the meaning of origination (of knowledge).

Tantra is neither an intellectually blind nor an existentially shallow activity. It is "Anstrengung" not only of concepts, to paraphrase Hegel, but of life itself in order to live it to the full[106].

Tantra is a *sâdhana*, a discipline for attaining the final goal conscious of its aims and methods. There has been endless discussion of what kind of relation there is between the Vedas and the Tantras. Without entering into the discussion they certainly "correspond to the *upâsanâ-kâṇḍa* of the *śruti*"[107].

For our investigation it may suffice to say that *tantra* as Philosophy entails a sort of sacramental awareness of the universe in which theory is not divorced from practice, nor is the material from the spiritual. Reality is sacramental, that is, at the same time material, human, and divine. It is embodied in Language.

We underline that our study is not concerned with the contents of any Philosophy but with every possible effort at *stretching* the meaning of the word so as to legitimate its use when refering to equivalent experiences of other peoples.

An important element of *tantra* is the enlightened awareness that accompanies any of the actions leading to the liberation of the person. Person here stands as that portion of the universe about which we can exercise both more freedom and responsibility. The immediate purport of *tantra* is to raise our degree of awareness within the most ordinary activities of the human being. *Tantra* as a generic name should, of course, not be confused with tantrism as a particular religious tradition or with the *Śaiva-âgamas*. The generic name suggests both practical ritual and theoretical realization. *Tantra* connotes not so much *a* particular doctrine as an entire approach to Reality[108]. Is this not also Philosophy?

8. *Yoga*

Yoga is also another word which belongs to this section. Like *tantra* it means both a definite doctrinal system and an attitude, a way, implying both theory and praxis. The word already suggests a concern for bringing together, uniting, yoking the fragmented aspects of ourselves and of Reality[109]. The important thing here is connection in order both to know and thus attain the supreme goal[110].

Upâsanâ stresses consecrated service, contemplation, worship; *bhakt* union, loving realization; and *tantra* reunion of the disperse elements of the real, integration. *Yoga*, on the other hand, underlines a certain 'ascension' by means of a particular isolation which lets the superficial and inessential fall by the way. For this, particular attention has to be directed to the power and weakeness of the mind. The mental has to be overcome — certainly not neglected.

We are not referring exclusively to the Sâṁkhya-yoga background. But even the attempt to isolate *purûṣa* from *prakṛti*, to liberate the spirit by disentangling it from matter, implies a particular awareness of Reality, an urge to realisation leading to a definite attitude *vis-à-vis* the real.

* * *

To sum up: this second main and often neglected feature of 'philosophy' is also expressed in the very greek word. We need only to emphasise the *philein*, and even, as I have done elsewhere, translate philo-sophy more as the *wisdom of love* than as the love of wisdom. One thing emerges. Philosophy here does not connote the love of attachment (*sakti*), but a detached (*asakta*) love. It is not concupiscence, but an ἀγαπή which frees the human being from all those bonds of enslavement which prevent the full realisation of this unique 'portion', or rather mirror, of the universe which we call our-selves.

We should not forget that, just as there are manifold systems of Philosophy, these notions have an enormous variety of schools, opinions and directions. It is almost ironic that we have to stress that the names in this section are also legitimate equivalents for *philo-sophy*.

3) *Jñana-Kâṇḍa. Understanding*

We have been insisting that the equivalent notions of Philosophy in classical India cannot ignore the previous meanings which we have underlined so far. Yet it is obvious that the cognitive element of Philosophy should not be minimised. In fact, the indic tradition again shows here its typical and almost intoxicating richness[111].

We shall consider only a few of those names and divide them into somewhat homogeneous groups. We shall enumerate some notions which underscore: a) Knowledge, b) Science, c) Seeing, d) Reasoning, and e) Systematisation.

We chose the word *Understanding* for this section mindful of the very suggestion of the vocable: to under-stand invites us to stand-under the spell or power of the thing so understood. *Action* suggests activity and doing. *Devotion* implies service and surrender. Now *understanding* invites to passivity and obedience (listening).

a) *Knowledge*

Our first sub-section deals with the most basic function of understanding. To know, namely, is the eminently human way of both acquiring self-identity and entering in communion with the universe. Knowledge here does not primarily denote an epistemic activity. It denotes an existential and ontological act. It implies attaining that which is known by being it, it realises the union between the subject and the object, saying it in epistemic parlance. It symbolises the highest act of being—even without the need to totally identify *sat* with *cit*, Being with Knowledge, Reality with Omniscience.

9. *Jñâna*

This is a basic word of the indic culture in its hinduistic, jainistic and buddhistic manifestations, as well as in later traditions. It means, roughly, knowledge[112]. *Jñâna* can, of course, signify all kind of knowledge, but eventually it represents the highest form of being: that being which *is* by opening itself to the entire Reality — which potentially it is. To know is not only to be-come the thing known, but also to become oneself in as much as we fill up ourselves with Being, as it were: we actualize our potentiality, we become what we ultimately *are*.

Jñâna by itself designates the process (of knowing) and also the result of it (acquired knowledge). There is no severance made here between purely theoretical and more practical knowledge, between merely intellectual grasping and existential assimilation. Knowing we become the known, either by 'creating' (begetting) it, or by submitting ourselves to it. To know is the basic human act of ' incorporating' Reality and becoming it. We are here far from 'pure' and 'practical' reason. The prerequisite for *jñâna*, for example (and, to be sure, not only in Vedânta) entails renunciation (*vairagya*) of all that is not relevant, differentiation between the transient and non-transient (*nityânityavastuviveka*), the aspiration to salvation (*mumukṣutva*), self-control, and other virtues, like meditation, etc. Meditation is here the supreme act of being[113]. *Jñâna* is therefore more than the mere grasping of how a thing behaves or what features it shows; it is an incorporation of and an identification with that which one knows[114]. *Jñâna* leads to liberation[115].

It should be remembered that the traditional notion of *gnôsis* and of *intellectus* is somewhat similar to this indic insight[116].

The notion of Philosophy as *jñâna* represents a decisive step. It is now no longer a question of merely doing or loving, but of knowing. The ultimate goal involves knowledge. Henceforth this characteristic will remain as part of the self-understanding of the indic tradition.

Yet this knowledge is hard to get and is not wrung at once. This is sung impressively by the Ṛg Veda. One does not know what one is, or better, I do not know who I am, and yet I suspect that this knowledge alone will bring me salvation (*mokṣa*) and happiness (*ânanda*). Simultaneously, however, I sense that this knowledge is so overpowering and immasurable that it can only come to me as a gift:

> What thing I am I do not know.
> I wander secluded, burdened by my mind.
> When the Firstborn of Truth has come to me
> I receive a share in that self-same Word.[117]

Philosophy, here, is saving knowledge.

We may now consider a few related words.

10. *Vijñâna*

If *pra-* connotes horizontal and intensive dynamism (Cf. *prajñâ* later), *vi-* expresses a sort of extension and persuasion[118]. *Vijñâna* could suggest a detailed and even empirical knowledge up to analytical consciousness. *Vijñâna* could be a kind of discretionary knowledge, a *ratio discursiva*. Another way of rendering it would be critical knowledge. This is Philosophy. The prefix of the verb here has no analytical connotation, but implies critical and probably reflexive knowledge. This paradox of extolling knowledge up to its self-exhaustion is common in most mystical traditions[119].

The word becomes prominent in some buddhist schools with a rather positive and synthetic meaning. It probably goes back to that upanishadic *koan*:

"unknown by whom it is known
known by whom it is known"

or even shorter:

"unknown when known
known when unknown"[120].

This knowledge, *vijñâna* reaches its goal of identifying subject and object only when the knower does not know that it knows, for if the knower were to know (that it knows) then it would no longer be the knower but become the known. The text does not say further that it is known by those who know that they do not know. These already know something, i.e. their own ignorance, and thus they don't really know (*it*, *brahman*, reality, truth, realization, . . .). Reflexive knowledge may be the 'original sin', but pure knowledge it is not.

11. *Samjñâna*

The perfect yogi is *sarva-vit* "all-knowing" because the yogi posseses *samprajñâ*, omniscience. *Samjñâna* implies a certain intellectualism and could also be related to the *samyagdarśana* and *samyagdṛṣṭi* (the right wisdom) of many schools. This conception of Philosophy could be considered either as meaning the right system, right knowledge or, in an almost arrogant form, an omniscient Philosophy — whether this latter notion could still form part of our philosophical club or not, or perhaps only its goal. At any rate we also have to adduce this meaning of Philosophy as consummate wisdom.

12. *Brahmajñâna*

Related to the previous word, but with more a qualitative than a quantitative aspect is this important notion of indic wisdom. *Brahmajñâna* or knowledge of *brahman* is the *summum bonum* the *parama-puruṣârtha* or Vedânta, it is total enlightenment because the knowledge of *brahman* implies identification with it. It is obviously not a detailed 'know-how' of a nominalistic universe, but the holistic insight which remains the goal of the indic soul: the total realisation.

The well-known vedântic work by Maṇḍana Miśra is called *Brahmasiddhi*. *Brahma-vicâra* or the enquiry about *brahman*, *brahman-vâda* or doctrine about *brahman*, *brahma-vidyâ* or science on *brahman*, *brahma-mîmâṁsâ*, *brahma-yoga*, etc. are so many other current expressions in the 'philosophical' literature.

One point is here important to stress: the ideal. The indic purport does not fall short of the highest ambition. The indic effort is not contented with a certain dominion or control over things, prevision of events, or deciphering of intellectual riddles. It aims at the release, plenitude, joy, realisation, expressed by this word. Philosophy short of this ideal is not an appealing enterprise for which one consecrates one's own life.

If *brahmajñâna* remains the ideal, *brahmajijñâsa* appears as the self-reflective process towards the goal. In point of fact *Brahmajijñâsa* is a central word in the discourses of many schools in connection with the first *sûtra* of the *Brahmasûtra*[121]. It means to take a step towards that understanding by which the knower is known[122]. *Brahman* is the *prameya*, and the *śruti* the *pramâṇa*. Brahman is what is to be known and the Revelation (the Vedas) is the means to it. *Jijñâsa* is the inchoate knowledge which triggers the philosophical inquiry.

Some translators have rendered the *brahmajijñâsa* as inquiry about *brahman*. Madhva, among others may be more right in understanding it not as an external desire to know (*jñâna*) but as the very constitution of knowledge. *Jñâna* factually is *jijñâsa*, i.e. the internal dynamism towards knowing, precisely because *brahman* being infinite, attributeless, etc. is not properly an object and its knowledge never finishes. If knowledge stops, it ceases to be such. *Jijñâsa*, ultimately is the very call of *brahman*. We are pushed to know because we are known, and as such we tend to know[123].

Modern writers have sometimes translated the words *brahmajñâsa* and *brahma-jijñâsa* with 'theology'. This is done in order to emphasise an independent 'philosophy' which is then also supposed to exist in India, as if the western conflict between faith and reason were also an indic preoccupation. But the indic problem is of a different hue[124].

13. *Tattvajñâna*

This is knowledge of a thing just as it is. It is 'objective' reality. The *Nyâyakośa* circumscribes this word as *yathârthajñâna*, or true knowledge which, according to Vedânta, has "the highest Self as its object and is a means toward salvation"[125]. One usually translates *tattva* by 'essence', or rather 'truth'. Literally it is 'that-ness' or 'thus-ness': that thing which exists just as it is[126]. It is the knowledge of the 'thus-ness' of things, just as they are, which will lead us to the ultimate goal[127].

The notion of *tattva* has a particular importance for our study, both theoretical and practical. This latter is shown by the increasing number of contemporary works of Philosophy using this name[128].

The theoretical significance of this word lies in the fact that its connotations are both concrete and universal at the same time. The word has a long story which we cannot even insinuate here. It is the first word of the *parama-mahâ-vâkya*, the greatest of the upanishadic utterances, the *tat tvam asi*. The substantivised form of the demonstrative pronoun *tat* is *tattva*. We may recall the twenty-six or thirty-six *tattva-s* of tradition. For Râmânuja *tattva* is "the philosophical apprehension of Reality"[129], and in fact *tattva* designates Reality for most of the indic systems of thought. The current translation is both Reality and Truth. This polarity expresses both the ambition to concreteness and to universality of the indic 'philosophical' approach. *Tattvasaṅkhyâna* is the name of a work by Madhva. A seeker after truth is still called a *tattvajijñâsu*. *Tattva* could also be translated as principle[130].

We have already said that main words like *jñâna* and *vidyâ* allow for many compounds. But the relation is mutual. In point of fact we have expressions like *tattva-*

mîmâmsâ, *tattva-vidyâ*, *tattva-darśana*, etc., and even triple compounds like *tattva-vidyâ-śâstra*, and so on[131].

Expressions like *tattva-viveka*, *tattva-saṁgraha*, *tattva-bodha* and innumerable others are constantly found in 'philosophic' literature. *Tattvârtha-vicâra* is often used as meaning philosophical inquiry in a modern sense.

14. *Prajñâ*

This is usually translated as wisdom, although it has a plurality of meanings, according to different schools. The prefix *pra* denotes a kind of intentional movement, and the word is more common in buddhist than hindu texts. In the *Mahâbhârata* and elsewhere the verb *samprajnâ* (*samprajânâti*) is used with the meaning of discerning, distinguishing, recognizing and of accurately knowing. *Sampajanya* (*samprajâna*) means full consciousness, perfect knowledge. It can also mean *alertness*, attention[132].

b) *Science*

If the words of the first subsection stress a certain type of global and intuitive knowledge, this second group emphasises the type of a more detailed and rational knowledge. Yet these distinctions cannot be stretched beyond a certain point, since all these words are polysemic, used differently by different schools, and often employed almost indiscriminately as synonymous. Our division has a more heuristic that a strictly taxonomic character.

Many of the previous observations are also valid here. This will permit us to be even more succinct.

15. *Vidyâ*

The word means vision, knowledge, perception[133]. It also refers to a definite branch of knowledge, a science, doctrine or discipline[134]. The distinction of the Upaniṣads between *parâ* and *aparâ vidyâ*, "higher" and "lower" knowledge, has been roughly compared with the Aristotelian one between metaphysics (πρώτη φιλοσοφία) and physics, although the correspondence is only very approximate[135]. By means of *parâ vidyâ*, it is said, one knows *brahman*, and with *aparâ*, the transitory[136].

Vidyâ, properly speaking, is neither a science of objects, nor a mere cognition of subjects. It is a science of Reality. It is a science that pervades the whole being and, thus, which does not leave room for any action which is either contrary to or not transformed by cognition[137]. That somebody allegedly knows something and yet acts contrary to that knowledge would mean that the knowledge was not genuine. One cannot be an authentic scientist, *vidvas*, *vidyâdâtṛ*, a real teacher, and an immoral person at the same time. One cannot *know* that smoking is bad and go on smoking — distinctions notwithstanding.

This existential and not merely rational character of *vidyâ* is confirmed by the opposite meaning of the word. *Avidyâ* does not mean that we do not possess a certain rational 'know-how' of *something*, but rather that a fundamental ignorance seizes us. The result of *avidyâ* is not that we become unable to perceive objects or subjects, but rather that we cannot recognise who we really are and what everything around us actually is. We are ignorant in regard to the *âtman*. *Avidyâ* is sin. And *vidyâ* a very committed Philosophy indeed.

The embracing word for all types of knowledge is called generally *vidyâ*, as we shall see later.

We translate *vidyâ* as science; but we should not imagine that what has been traditionally understood as *scientia* (επιστήμη) is synonymous with science in the modern sense of the word[138]. We may repeat the scholastic definition of *scientia* as *certa rerum cognitio per causas* and add that modern science is basically measurement which allows prediction of the behaviour of phenomena. The claim of the former is to know things. The ambition of the latter to discover laws (regular patterns of behaviour).

Vidyâ is also used in a variety of combinations.

16. *Âtmavidyâ*

The science of Self is a frequently used name. That this Self has little to do with the 'psychological ego' is something that stands out clearly in the whole corpus of indic writings. Without this *âtmavidyâ* nothing is truly perceived; a fact stated in the Upaniṣads and repeated by many schools. *Âtmavidyâ* is that through which everything (else) becomes illuminated, known[139]. *Adhyâtmavidyâ* is also a common designation which is sometimes translated as 'metaphysics'[140]. At any rate it has an upanishadic flavour. The *âtman* is truly what should be known[141].

17. *Brahmavidyâ*

A related expression, *brahmavidyâ* and similar words are also used in this sense. The translation 'theology' is sometimes used. It has even found official recognition[142]. In the light of what has been said we need add nothing further[143].

On the other hand, it is necessary to make a note here. *Brahmavidyâ*, in certain circumstances, can designate that knowledge which has *brahman* as its object; this is not normally the case, however. *Brahman* is actually no object of knowledge[144]. This means, of course, the subjective genitive. *Brahmavidyâ* is the knowledge which brahman has (is). A *jîvanmukta* (one who is liberated from any type of (ultimate) *avidyâ*), a *sarvajñânin* (one who knows all), a *kevalajñânin* (possessor of absolute knowledge) is the one who has attained *brahmavidyâ*. It amounts to having known *brahman* and thus become *brahman*. The *vidyâ* here is not the science about *brahman*, but *brahman's* science, the knowledge which is *brahman*. When somebody can say *aham brahmâsmi*, "I am brahman", it is no longer the empirical *ego* that says this[145]. It is *brahman* expressing its identity in the perceiving subject which is (or has become) or has realised *brahman*. *Brahmavidyâ* is the consciousness of *brahman*; it is the same *cit* of the *sat-cit-ânanda* i.e., being, consciousness and happiness. That would correspond to the interpretation of theology as (λόγος τοῦ θεοῦ) in the sense of the Word of God itself and not our word about God[146].

18. *Râjavidyâ*

Is the name given by Lord Krishna to his doctrine of salvation in the *Bhagavadgîtâ*[147]. It is the royal, that is to say, the highest knowledge. *Philosophia* (or *theologia*) as *regina scientiarum* is a well-known case in the West, too.

* * *

Other combinations with *vidyâ* designate certain disciplines of knowledge, such as *tarkavidyâ* or *nyâyavidyâ*, which are usually translated as 'logic', although they usually presuppose a restricted sense of *logos*.

c) *Thinking*

If the first and second groups stress the knowing function of Philosophy and thus, in a certain sense, assume a holistic character, and therefore a saving power, this third set of words lays emphasis upon the very process of acquiring such a knowledge.

This process can stress either a general thinking aspect (*mati*) or a more hermeneutic function (*mîmâṁsâ*) or, finally, a more strict reasoning function of (*ânvîkṣikî*). We come nearer to the Hegelian "Anstrengung des Begriffes".

In general we detect here a meaning of Philosophy as a thinking activity. One hardly needs to underline the importance of this dimension of Philosophy. The somewhat mystical aspect of knowledge and science which seems to assume both a revelation or illumination from above and a holistic attitude and approach, is here balanced with the thinking effort of the human being. Philosophy is not just knowledge, not even a way towards wisdom. It is also a human discipline which demands, *above* and *beyond* the preparation of the heart and the readiness of the will, the work of the intellect, a critical stance, inquiry, checking, proving, probing,... in a word, thinking.

Now, this thinking effort can be directed in a twofold way: towards the nature of things and towards the very nature of thinking itself. It has often been repeated that "Indian Philosophy" is by and large not a critical Philosophy in the sense of a Kantian Philosophy and post-Kantian approach. This betrays the assumption that the critical problem can only be dealt with *à la* Kant, and secondly, it overlooks the entire indic reflection on *pramâ* or valid knowledge and on the *pramâṇa-s* or means for such legitimate knowledge, which is almost an obsession with some indic systems[148]. Buddhism, too, has reflected very deeply on the nature of the thought processes.

19. *Mati*

This word means thought, intention, opinion, conviction, belief, view, with an original connotation of prayer, veneration. The word also means understanding, intelligence, and is related to *mata*, thought, view, opinion, belief, doctrine. It can designate a complete system of thought or a specific view on a general factual situation in a more extended sense. However, the aspect of belief and veneration is consistently a part of the fundamental meaning[149].

The word *mati*, with mythical and cosmic meanings, appears in the *Pañcarâtra* and in the *Purânas*. Often it also has the meaning of ordinary knowledge (in the jaina vocabulary for instance: sensory knowledge).

20. *Mîmâṁsâ*

The word is formed from the desiderative of the same root which, before it was taken over to designate one or two systems (namely, Pûrva- and Uttara-mîmâṁsâ), had a more general sense. It had the meaning of profound thought, or, in actual fact, the intense desire to think, that is to say, reflection, scrutiny, and investigation. It can also mean theory. And so in one Upaniṣad investigation into happiness is termed *ânanda-mîmâṁsâ*[150]. It would be fascinating, and revolutionary, to relate this basic meaning of *mîmâṁsâ* with the related greek ἱστορία and have a cross-fertilization in the understanding of history[151].

This much we shall retain. *Mîmâṁsâ* could be rendered by hermeneutics in the modern sense of the word[152]. Philosophy as *mîmâṁsâ* corresponds to Philosophy as hermeneutics. Yet classical *mîmâṁsâ* (both *Pûrva* and *Uttara*) would emphasise two features. The first is this: not every text is equally worth investigating or interpreting. It has to be an 'Urtext' a primordially given text, an *apauruṣeyatva*[153]. This is to say that it has to be ultimate, with no agency behind to explain its meaning to us. Those ultimate symbols by which we grasp the meaning would then be the real words.

We should pay attention to this basic difference. Any text, the meaning of which may be deducible from another text, simply postpones ultimate interpretation. There has to be an end in this going back. Some meaning has to appear to us; it must be so transparent as not to require any further explanation, unfolding, interpretation whatsoever. These ultimate texts are the subject-matter of *mîmâṁsâ*. And this is the reason why the philosophical system with this name has been accused of atheism. The *mîmâṁsâ* is an ultimate investigation. It deals with the texts which have no author, *apauruṣeya*, because the relation of the word to its meaning, value, reality is reduced to zero. There is nothing else behind it. These are immediate because they are ultimate.

The second feature consists in the fact that the proper interpretation is not just a rational ex-planation but a spiritual in-sight. This intuition is, to be sure, more and not less than rational. This second feature is congruent with the first. In point of fact, no ultimate interpretation can be a merely rational one, a merely deductive procedure. It has to be immediate evidence or, better, existential realisation. After all, the proper objects of *mîmâṁsâ* are the *mantras*[154]. And the purport of the *mantras* is the liberation of Man.

21. *Anvîkṣikî*

Here we have a further word which may be very close to the main area of what in the West is termed 'philosophy'[155]. It designates a definite method which is applied in various sciences. The political scientist Kautilya referred to it as the "torch of all sciences, the tool for all activities, the foundation of all religious-social orders"[156].

The word is derived from *anu-* (after) and *îkṣ-*, (to see, look), and thus connotes to check, re-examine, investigate, reflect upon[157]. There is a critical side to it, an argumentative feature. We may note in passing that Latin *arguere* means to put in a clear light, to de-clare (all are also metaphors of the eye). *Ânvîkṣikî* means examination or investigation as a thought process[158]. *Ânvîkṣikî* is the science or method which above all else makes use of critical reflection[159]. The prefix *anu-* suggests that there is a seeing which precedes thought[160]. *Ânvîkṣikî* is always a secondary activity and proceeds from something which is previously given, and must also lead to something that comes after. According to the Upanishadic scheme, followed also in Vedânta, *śravana*, "the hearing of the Word", is what precedes thought[161]. This is elaborated in *manana*, reflection. It is the plane of *ânvîkṣikî*[162]. But this second plane leads to the third stage of *nididhyâsana*, meditation. Vâcaspatimiśra, who calls this *dhyâna* or *cintâ* (both words meaning 'meditation'), introduces yet a fourth level which he terms *âtmasâkṣatkâra*, the direct perception of self (also called *darśâna*, in the sense of direct vision)[163].

The system, which as a whole is called *ânvîkṣikî*, is Nyâya[164]. It seeks to differentiate itself from the spiritual science of the Upaniṣads (*adhyâtmavidyâ*) through its reasoning, reflective and logical character. Yet lest we lose sight of the saving—existential—character of *ânvîkṣikî* let us recall that Vâtsyâyana himself calls his doctrine, in another place, an *adhyâtmavidyâ* or "doctrine of self", because even the goal of logic is the attainment of salvation[165].

The close link between *ânvîkṣikî* and *âtmavidyâ* was wide-spread in the Middle Ages, and it is characteristic of the indic attitude, since a logical investigation without knowledge of self would be fruitless. Thus, Kâmandaki interprets *ânvîkṣikî* as "knowledge of self" (*âtma-vijñâna*)[166], and later, drawing upon etymology, he states that "The science of self might be an 'investigative science' because it examines desire and suffering. The Man who tests reality with it abandons both pleasurable and sorrowful excitement"[167].

The law book of Manu counts the "critical science of self" (*ânvîkṣikî-âtmavidyâ*) among those sciences which ought to be studied by kings[168]. This science differentiates itself from the vedic science (*trayî vidyâ*), which above all in its Upanishadic part is also an *âtmavidyâ*, by its critical method.

Whether we shall say that *ânvîkṣikî* is the proper name for Philosophy as Jacobi, and after him others, meant[169], or it represents only one of the *jñâna-kâṇḍas*, as others thought[170], the fact remains that this critical and deductive activity of the mind is an important constituent of that indic Philosophy we are looking for.

d) *Worldview*

This fourth group of words could also be brought together under the caption of *opinion* or even *system*. The main emphasis lies on being a more or less complete and systematic *vision* of Reality: an *intuition*. It may be an opinion or fruit of an experience, but at any rate it is a somewhat coherent and complete conception of the universe, a worldview or *Weltanschauung*, a cosmology in a certain sense.

We may insist for a moment on the importance of this set of words. They underline at least two features of the philosophical enquiry: the central place of seeing, of evidence, and the no less paramount position of cosmology.

Philosophy consists therefore in a way of seeing the world and of reaching evidence. The vision needs thus to be rational, coherent and, inasmuch as this is possible, complete.

Neither can Philosophy renounce the offering of a worldview in which the different ideas and notions may be inserted. Even more, it pertains to Philosophy to unravel the implicit picture of Reality where our human opinions move. This vision may have the danger of being too closed and even too rationalistically coherent. Nevertheless this is an important ingredient of the philosophical activity: to strive for rationality and coherence, other factors notwithstanding.

We allow ourselves to be here even more succinct than in our previous subsections, not because this one is less important, but only because it has been more studied than any other.

22. *Darśana*

This is the most frequently used word in the descriptions of ideological systems which are usually translated as 'philosophy'. *Darśana* means perception, idea[171]. *Darśana* is defined as that "by which one sees" (*dṛśyate 'nena iti*), and additionally means, among other things intellectual perception (*buddhi*), dream (*svapna*), mirror (*darpana*), and finally doctrinal system (*śâstra*)[172].

The *Nyâyakośa* offers the following meanings: looking with one's eyes (*câkṣuṣa-pratyakṣam*), cognition (*jñânam*), and a doctrinal system which is a means to the perception of things as they are (*tattva-jñâna-sâdhanam śâstram*). The more general sense encompasses physical or spiritual vision right up to mystical perception. The more strict sense, on the other hand, is that of a complete system leading to a true vision of Reality.

The traditional list of Brahmanic systems (*âstika*, the "affirmative" philosophy, as opposed to *nâstika*, the "negating" philosophy, such as buddhism, jainism, lokâyata) contains the six schools (usually grouped into three pairs) Nyâya-Vaiśeṣika, Sâṁkhya-Yoga, and Mîmâṁsâ-Vedânta. These intellectual systems, which are termed "canonical", have this in common: they were created "in order to reflect upon the Vedas and thus to achieve knowledge of Reality"[173]. Their 'orthodoxy' consists in the fact that they take the Vedas, the holy Revelation, as their point of departure, or, at least in a minimalist exegesis, do not contradict them. The *nâstika* systems would, on the contrary, not recognize the authority of the Vedas. We should keep in mind, however, that the division *âstika-nâstika*, as any orthodoxy-heterodoxy split, has more of a political than an intellectual nature. There is very little Vedic content in some of the 'orthodox' *darśanas*.

Words like *aupaniṣadam darśanam*, (the *darśanas*, vision of the Upaniṣads) *samyag-darśanam* (the right *darśanas*), *tattva-darśanas* (the correct vision of things), *âgama-darśana* (exegesis or interpretation of sacred Scriptures), etc., are constantly used in the literature[174]. Significantly enough the modern Philosophy of Religion is called *dharma-darśana* in contemporary India.

23. *Dṛṣṭi*

This word is related to the previous one, and often emphasises the exclusiveness of a system. It has acquired, in many intellectual circles, a sense of disparagement similar to the word 'dogmatism' in other places. It means a more or less exclusive 'philosophical' school which at least in principle claims to afford a complete vision of Reality. *Dṛṣṭi-vâda* is sometimes translated as "speculative philosophy"[175]. The word — even considering its etymology — could probably be translated with 'theory'[176]. The fundamental conception of these two designations rests upon the idea of vision, and ultimately, of evidence. There is a type of human behaviour which, should one follow it consistently, makes true Reality (*satyasya satya*) transparent. The various schools are precisely paths along which Man is led in order to attain the right vision of Reality, i.e. in order to realise oneself.

e) *Systematisation*

The classification of all forms of knowledge is, certainly, not an exclusive hellenic gift. It is a topic that is debated among indic authors. One of the most accepted classifications is that of Vâtsyâyana who divides all knowledge into *trayî* (the three Vedas), *ânvîkṣikî* (logic and metaphysics or 'philosophy'), *daṇḍa-nîtî* (governance), and *vârtta* (industries)[177].

Among the latter we have economics, commerce, medicine, agriculture and other practical arts and sciences. Manu adds a fifth science: *âtmavidyâ*[178]. Another popular division was that of 14 sciences: the four Vedas, the six Vedângas[179], plus *mîmâmsâ*, *nyâya*, *purâṇas* and *dharmaśâstras*[180].

If we add the four *upa-vedas* we have then 18 sciences, although the figures of 33 and even 64 have also been given.

What interests us here is to note the effort at systematisation and somehow the idea of an integration of all knowledge. From the time of the Upaniṣads there is tension in indic lore between the knowledge of the knower, knowing which all the rest is known, and the sum-total of knowledge, between Wisdom, we may say, and the particular Sciences. Orthodoxy will defend that all branches of knowledge are internally related and that they all, both individually and collectively, have but one single aim: *mokṣa*[181].

We shall write down only a few concepts.

24. *Mokṣaśâstra*

Human life, inasmuch as it is a manifestation of spatio-temporal existence, is perceived only as something transitory. That other state to which humans aspire is known, among other titles, as *brahman*, *nirvâṇa*, or even, at least provisionally, *svarga* (heaven). Hindu culture discusses with vehemence the nature of and ways to this goal: *mokṣa*, the liberation or salvation from every binding or transitory nature.

Mokṣa is, above all else, the goal, but also the threshold, which already belongs to the other shore[182].

The goal of this life is the attainment of *mokṣa*. The *śâstra-s* are those disciplines, books of instruction, or systems of life which contain the knowledge of the tradition. The ultimate and most central discipline is called, as a result, *mokṣaśâstra*; and precisely because the goal of life consists in attaining *mokṣa*, *mokṣaśâstra* is the system of salvation. It is not an end in itself but rather a means towards liberation. The name *śâstra* suggests 'regulation', 'direction' and, as a result, 'doctrine'. However, it is more a systematic training than a notional teaching[183]. The existential or practical aspect is emphasised just as much as the intellectual or theoretical. It is a question of a discipline of salvation practiced according to the dictates of tradition. That the *śâstras* later underlined the scholastic and intellectual aspect is readily understandable.

25. *Dharmaśâstra*

This word means both the general synthesis of all *dharma-s* with the corresponding knowledge thereof, and the series of treatises accumulated over the centuries expounding the different aspects of hindu lore[184].

Śâstram-ânvîkṣikam, *śâstra-vicâra*, and similar compounds also appear in the literature[185]. *Jñâna-śâstra*, *tattva-śâstra*, *padartha-śâstra*, *nyâya-śâstra* are equally common denominations. On a more popular vein the word *paṇḍitya*, or erudition (of the *paṇḍitas*) is also used.

26. *Hetuśâstra*

Hetuśâstra, or instruction on the causes, is another classical denomination[186]. The science of the logical reasons is certainly also connected with Philosophy. *Hetuvidyâ* is also a common name. *Hetu* means, of course, cause[187]. It is the *probans* in any argumentation, the reason on whose strength a conclusion is based. We could understand *hetuśâstra* as arguing Philosophy. From Nyâya to Vedânta the discourse on *hetu* is a prolific one.

27. *Tarka*

Tarka is obviously another name we have already mentioned. *Tarkaśâstra* is a common notion. It is generally translated as logic. *Tarka* means 'supposition', 'consideration', 'spiritual contemplation', or 'reflection', and emphasises the critical and 'logical' method, based on induction and deduction. Actually *tarka* means conjecture[188]. The current meaning is that of intellectual inquiry, scrutiny in the sense of reasoning, argument[189].

28. *Nyâya*

We have here another word with both a generic and a specific meaning. This latter is the name for one of the classical philosophical systems. The generic meaning is given by the word itself and its use in the general literature.

The word *ny-âya* (*nî-âya*), first of all contains the downwards and interior direction of the prefix *ni-* [190]. But in fact this is not precise enough, as the word comes from a verbal root, *nî*, which is the proper origin of the word. *Nyâya* means, in fact "that into which a thing goes back" (Monnier-Williams). In ancient times it was identified with *mîmâṁsâ*[191] and later on with *ânvîkṣikî*[192].

At any rate it has to do with the search for the archetypal, for the original, and thus for the rule, the standard. No wonder that the "linguistic turn" of the modern anglo-american tradition has been compared with the attention given to grammar in the fourth century BC in India[193]. Undoubtedly we have here a potent homeomorphic equivalent for Philosophy.

29. *Vâda*

This is also a generic name for discourse, speech[194]. It implies a more or less systematic doctrine about something. The word *vâda* has an extremely rich tradition. Vâtsyâyana speaks of three kind of *kathâ-s* (controversies): *vâda*, *jalpa*, and *vitaṇḍâ*, i.e. discussion, wrangling and cavil[195]. He follows, of course the *Nyâya-sûtras* which enumerate these as three of their sixteen categories[196]. Caraka adds *tarka* as an inner struggle to find truth within oneself before entering any debate[197]. At any rate *vâda* stands for a debate with the noble purpose of ascertaining truth and not for defeating an opponent or finding faults in others. *Vâdakathâ* would be the philosophical inquirer who stands in an open forum[198].

The normal use in philosophical parlance would roughly correspond to the suffix *-logy* in western languages, *âtmavâda*, *vivartavâda*, etc. are common expressions. In fact the word *vâda* means speech, discourse, talk, utterance.

The word *brahmavâdya* is related to *brahmanvâda* or discourse on *brahman*. *Brahmavâdya* meant rather discussion on *brahman* or brahmanical subjects i.e. rivalry in sacred knowledge[199]. In point of fact *brahmodya* seems to be the oldest expression (around IX century B.C.) to convey speculative discussions which we may call philosophical[200]. *Brahmopadeśa* would be instruction in sacred knowledge.

30. *Siddhânta*

Siddhânta, *samaya* and other words have also been utilised to indicate established doctrine, completion, perfection or arrangement, and collection, respectively.

* * *

To sum up: philosophy has to do with knowledge. It is wisdom. But this science (we use now the three as synonyms) is a practical science of salvation. We are summing up in western categories this main feature of the 'philosophical' activity of sanskrit India. Knowledge entails both theory and praxis in a non-dualistic relationship. The praxis does not follow the theoretical insights, nor is the theory the fruit of the praxis. The dichotomy is false. Knowledge is identification with the known. To reach the known is theory as well as praxis. And the conditions to reach it are theoretical as much as practical. One truly knows if one is on the way to knowing. And one is on the way if one knows. Philosophy, in this sense, is the liberating knowledge, not only because any authentic knowledge is liberating, but also because any authentic liberation is knowledge. Although many an indic system is similar to what is called gnosticism in the West, one should not identify this attitude we are trying to sum up with the *gnôsis* of Platonic origin. Abhinavagupta, to quote just one name, is not a gnostic and he would agree with our preceeding lines.

We are avering something more than *jñâna* is for the sake of *mokṣa*. We are affirming also more than salvation is knowledge (as some systems definitively assert). We are summing up this section by saying that the philosophical activity is the very reflexive (critical) awareness of our existential path, the intellectual side of the very human pilgrimage. And if by religion one were to understand the factual striving after the goal of life, Philosophy would then be the awareness of such a journey, the intellectual aspect of religion itself. It is not a vicious circle. It is the vital circle — of Reality.

III. ŚATAPATHAVEDA

1) *The wisdom of the hundred paths*

It is not proper to criticise other cultures without a thorough knowledge of the elements and the spirit of each tradition. We cannot correctly know any tradition without knowing it first from within. This demands a full plunge into that tradition. We have tried to show the complexity of the indic cultural world by adducing some key words concerning our problem. Yet, in the next moment, we need to criticise that tradition in the light of our gained cross-cultural understanding. And it is then, in a third moment, that a mutual fecundation, not an eclectical juxtaposition, can take place. We have only grazed the surface, yet we have gathered enough materials to proceed with caution and hope. An understanding among the cultures of the world may be today a necessary factor for human survival.

If there is any kind of result after the foregoing excursus it would like to be expressed in the sanskrit title. This title, which I have taken the freedom to forge, enshrines, I submit, the complexity of the indic world. *Śata-patha-prajñâ*: "Hundred-path-wisdom", hundred perspectives of the path, hundred ways to reach the vision, hundred visions to the one path, the hundredfold knowledge, or the wisdom of the hundred paths. The multivalence of the word is intended.

We prefer *prajñâ* to *jñâna*, in order to emphasize the dynamic and intentional aspect of our symbol, although we could also have used *vijñâna*, if interpreted as a critical and reflexive *jñâna*. Sometimes *jñâna* stands for a sort of living and discriminative consciousness, as the *Bhâgavata* suggests. We made clear enough that there is no possibility of univocal translations.

If we stress the significance of the prefix, we may say that *pra-jñâ* almost corresponds to *philo-sophia*. *Prajñâ* could mean, and actually this is explicit in hindu as well as in buddhist texts, a striving towards intuitive or saving knowledge, both practical and theoretical. *Pra-jñâtiḥ* is in fact the knowing of the right way. *Prajñâparâmitâ* is the supreme virtue in mahâyâna buddhism. A *sthita-prajñâ* is a *jîvan-mukta*, a person firmly rooted in knowledge, a liberated soul, says Vedânta.

For this reason we adopt this word as perhaps the most useful homeomorphic equivalent. But our title suggests something more: there are hundreds of 'wisdoms'!

Hundred, of course, stands here for indefinite, unlimited. There are as many ways to Freedom, Liberation, Wisdom, *mukti*, *nirvâṇa*, *jñâna*, *ânanda*,... as we humans are truly capable of finding out or marching upon. The paths of and to the Deity (Truth,...) are innumerable because the Divine (Truth,...) may manifest itself in infinite ways. The ways of Man are also infinite, and the visions are as many as there are seers. The vision is hundredfold, the perspectives innumerable, the Vedas are hundred-legged, the ramifications of Wisdom extend everywhere. *Philosophia* is *śatapathaprajñâ*. In another context we had almost dared to say: *śatapathadeva*. (Should I betray that *devaḥ* has also been a 'slip' of my pen?) Each truly human way to knowledge is also a way of and to the Gods. This does not deny that there may be false ways too. To be sure, truth lies hidden everywhere, no one has a monopoly on her—nor on wisdom for that matter—and she is of a pluralistic nature. But she lies hidden and the process of dis-covery, re-velation, ex-traction,... may eventually fail.

But we shall limit ourselves to the description of what an acceptable notion of Philosophy may entail.

2) *The meta-semantic problem*

Whether there is 'philosophy' in classical India depends first of all on what we postulate Philosophy to be. But the answer is not satisfactory unless we explain the basis of such a postulate.

In sum, the problem is first of all semantic. If we define Philosophy as a human inquiry done exclusively with the reasoning reason into the nature of that which precisely appears to our very reason, we shall have to conclude that only few of the candidates qualify for the bearing of that name.

But the problem is not exclusively semantic. The word has a meaning not just because it shows some intellectual content. It has a meaning also because it responds to a claim. This intentionality, in fact, is wider and deeper that its actual contents. If Philosophy is understood as a strict rationalistic enterprise, this is not because of an arbitrary postulate, but because those holding that position assume that only reason is the ultimate guide and the least unreliable agency for living a fully human life or for unravelling the riddle of the universe, as much as this is possible to us. This position does not exclude, of course, discrepancy regarding the nature of 'reason', the method to elucidate this very problem, and the function of 'reason' in life.

In other words, we may consider Philosophy to be that human activity which searchs for the over-all meaning of Reality by whatever means we have at our disposal. The word then is capable of encompassing many notions, according to what means we do believe we have at our disposal, and how we understand their nature.

We take a middle way between an exclusive concept of Philosophy, only proper to one single culture, and a merely atomistic notion of incommunicable human constructs. We defend the pluralistic stance of a *provisional but valid* notion of what we could still call Philosophy for our time and place.

Saying "provisional but valid" we would like to follow the middle way alluded to. There is, certainly, a sort of *philosophia perennis*, a kind of traditional wisdom incarnated in manifold ways in the human race. There is a valid Philosophy. We detect a somewhat common claim. We are capable of discovering homeomorphic equivalents. But it is all provisional. We are all thinking, philosophising, or even living, within a *mythos*. The entire enterprise shares in the contingency of everything that touches the *humanum*. Even if Wisdom were to descend from a higher Source it would need to unfold in time and space and be impregnated by humanness if it had to touch the humans at all. All our human achievements are but provisional — we do not (yet?) see face to face!

In other words, the answers to what Philosophy is, or should, or could be, are manifold, but the very question depends on a set of premises which we perceive now in a particular way without any guarantee that the problem may not vary radically. It is within this contingent context that we are able to operate.

Ours is far from being a semantic problem. It touches the kernel of the contemporary crisis. Post-modernity, post-structuralism, deconstructionism, end of metaphysics, "la nouvelle droite" and many other present day phenomena, even more visible in art and life-styles, all agree in the already old insight of the "end of the modern age". It seems as if the technocratic era is running, also spiritually and intellectually, out of petrol. Traditional Philosophy has been dethroned and theology has long abdicated. Humankind does not need to have a successor in the throne. This is the standing danger of revivalisms and totalitarianisms of any type. But Man cannot live in a vacuum. I surmise, and submit, that the insights of indic wisdom are here of crucial importance for the future of human life. Human cultures need each other. Our problem is not just an 'indological' issue, nor a question of sheer terms.

3) *The philosophical query*

Our subtitle contains an interrogation. "Should we speak of Philosophy in Classical India?". We may now attempt an answer. If by Philosophy we understand the *prevalent* western academic interpretation, classical India knew something of the kind, but did not pay much attention to it. It was not considered very important. That 'philosophy' was marginal, as it is today in the West. If, on the other hand, by Philosophy we understand what previous periods of western history took it to be, we will have to acknowledge intriguing parallelisms. But there is more. If, by Philosophy we understand that fundamental human enterprise seeking to make sense of Reality by all available means, we find a positive enrichment of Philosophy by the indic contribution.

Let us elaborate a little. That in India there is logic, metaphysical speculation, ethical reflections and the like, should be clear. That many of the philosophical activities of the western spirit have also been cultivated by the indic mind is without any doubt. But this, I submit, is not the question; it is not even the western philosophical question. We all speak, we all think, many of us write, and some even ride bicycles. The question about Philosophy is neither the abstract question of whether we all ride bicycles or know how to ride them, nor is it the general inquiry whether we all speak, think, or many of us write. The primordial question about Philosophy concerns the very nature of the 'philosophical' enterprise as the awareness (representation, actualisation,. . .) of the ultimate human act(s) by which Man reaches the fullness of what human beings are called upon, capable of, supposed to, engage themselves upon,. . . to be or to become. Philosophy has a claim to ultimacy which is certainly not objectifiable. Not whether we ride a bicycle or whether we speak or think is the philosophical question, but why ultimately do we ride a cycle, how is it that we speak, and what exactly is that which we call thinking; this is the philosophical problematic.

The question is not whether 'philosophy' also exists elsewhere. The question echoes rather the Augustinian existential query: *quaestio mihi factus sum*, or the Vedic interrogation: *yadi vedam asmi*? This fundamental question has two aspects: it asks about the nature and function of Reality ("what is it?") — generally through the conscious intermediary of the questioner (*ko'ham*, "who am I?") — and secondly, asks whether the very asking is the proper way of re-solving or perhaps solving, the very sting of the question.

In other words, the openness is double. On the one hand, the philosophical activity entails the readiness to practically and theoretically criticise the nature of the given and our opinions about Reality. On the other hand, it entails further the readiness to listen, whether the very question is not the pertinent one and/or not pertinent at all. There is a dimension of passivity, of obedience — of feminity I would like to say — in the philosophical attitude. To know and even to act imply at the same time to be known and acted upon. Put it differently. On the one hand, Philosophy needs openness towards transcendence, both on the horizontal level of cross-cultural voices from other human shores, and from the vertical level of a numinous cloud. On the other hand, it needs the closenness of immanence, the intimacy of introspection, critique and reflexion. Philosophy is constitutively both esoteric and exoteric, private and public, open and closed, meta-human and humane.

This attitude does not exclude negative answers to all those queries, interpreted as illusory projections of unfulfilled desires, or remnants of archaic ages. It only sets the stage.

We are well aware that some schools of Philosophy would prefer to be less ambitious and be satisfied with a critical analysis of our faculties of knowledge or our language. Yet, even then, either those schools leave the other task to another human endeavour — and then we only postpone the question — or they implicitly or explicitly acknowledge that the meaningful life is that which is satisfied with solving those analytical or linguistic riddles as much as possible — in which case they have already taken a metaphysical stance. But we are not directly concerned here with defending any particular idea of Philosophy. We are interested in delimiting the range of the word and in finding out whether it is possible, plausible, or even desirable, that we enlarge the scope of this word so as to cover, or at least to explicitely allow space for, other possible homeomorphic equivalents.

We have already mentioned the unavoidable political character of our problem. Our questions are inserted in and are a function of the political situation of the world, and our answers are, willy nilly, coloured by our perception of the present human situation. Even assuming that western Philosophy and/or Religion could furnish the solution to the actual human predicament we do not have the right to dismiss *a priori* the experiences, queries and expectations of other parts or other periods of the world. The best achievements of one human phylum do not make the history of other phyla obsolete, at least until they do not prove to be so. Unqualified evolutionism may lead to cultural genocides.

The concrete political option today is the following. Either we follow a certain linear evolutionary scheme and accept the modern-western worldview, and with it the corresponding conception of 'philosophy', as the central paradigm for the philosophical activity. (If there are heterogeneous activities they should not use the word 'philosophy'.) Or we may contest the dominant world-view and recognise the right of each tradition to follow its own internal dynamism. (Translators will inform us about the others.) Or, also, following a middle way, we may stretch the symbol Philosophy to the utmost so as to establish a relationship with the homeomorphic equivalents of other traditions.

We may recall now the three main traits of the indic tradition which we subsumed somewhat in the symbol *prajñâ*:

a) a *karmic* element containing a movement towards the realization of the goal of life, the praxis of salvation, the actual going into the path of liberation, wisdom, union, happiness, knowledge.
b) a *bhaktic* element of consecration, service, striving, desire, love to reach that goal, be it actually reachable or beyond every means, be it fulness or emptiness, the plenitude of existence or the way out of existence.
c) a *jñânic* element of an actual sharing, relishing, tasting, experiencing, knowing of Reality, be it perfectly or by approximations, in appearance or in actual fact.

Some may object that we have gone too far in including the first two groups in our 'philosophical' list. Not to have done it would have implied that we *a priori* decide that Philosophy is a merely theoretical approach to Reality discarding the possibility of more existential, concrete and even bodily, sentimental and practical approaches to the Mystery of Reality. In point of fact we find that the three traits share in that ultimate concern of leading us into, putting us in contact with, opening us up to, bringing us in communion with, letting us know of, orienting ourselves in, touching, reaching, loving,... or what, in each case, is 'considered' to be the main intentionality of that 'activity' which responds homeomorphically to the notion of Philosophy. We act with our minds as much as we love with our actions and know with our bodies.

Our middle way leads us further. While acknowledging and respecting the values of all traditions we should not freeze them, neither in their unpredictable processes, nor even in their starting points, let alone in their outcomes and possibly radical mutations. We mentioned the mutual fecundation among traditions. We intercalate only one comment in this respect.

Even a cursory study of the indic civilization shows her impassioned search for an Ultimate Wisdom which, at the same time, would not curtail the rightful existence of other aspects of life. We find a similar effort in the western adventure. The ultimate aim of life should in no way stifle life in her bountiful blossoming. Yet, we encounter in both cases the double temptation of neglecting the rights of the penultimate and more immediate spheres of experience for the sake of the ultimate, on the one hand; and the opposite reaction of renouncing the metaphysical for the sake of the physical. On the whole, the first temptation has been, by and large, more powerful in the indic tradition, and the second in the western one, especially the modern.

The challenging task for both traditions today consists in making room for the concrete without neglecting the ultimate, in discovering the *ontonomy* of regional ontologies without forgetting the unity of Being, and even of Knowledge. Science and Philosophy, Physics and Metaphysics are western words. *Jñâna* and *vijñâna*, echoing the Gîtâ, could be two sanskrit symbols. One may also speak of the unity of the four *puruṣârthas*.

The problem remains the glue, as it were, which may stick the *vyâvahârika* and the *pâramârthika* together. Here is where names like the Sacred, Transcendence, the Source, the Deity seem pertinent symbols. Perhaps a service of indic wisdom to western Philosophy is to help the latter to heal the split between what is being called philosophy and theology without for that matter falling into the pre-critical traps of which history has taught us so painfully. A service of western wisdom to indic Philosophy, may consist, on the other hand, in helping the latter to heal the split between the affairs of this world, the *vyâvahârika*, and those of the other shore, the *pâramârthika*, without falling into the desacralisation of the world which is threatening to destroy both human dignity and earthly existence. In both cases the resources are to be found in the interior of the respective traditions opening up to a cross-cultural fertilisation.

The word Philosophy may stand for the ultimate approaches to the Mystery of Reality as envisaged by the hellenic spirit, with all the subsequent modifications and reforms, up to embracing the similar enterprises of other cultures. Or it may stand as a symbol capable of receiving other symbolisations radically different from those just alluded to. This should not necessarily mean a Pandora's box of chaotic contents because of a third point we have still to make.

The openness of the symbol is not an amorphous open box in which anything is possible. The openness is limited to what at any given moment in space and time is considered to make sense, and in fact is seen as meaningful, albeit problematic. In other words, we are all living within a particular myth which situates our beliefs and visions. We reach existentially the mythical level when we do not question further. It is in this sense that we utilise the word myth.

The emerging myth of our times tends to make us believe, as a matter of course, that no single culture, tradition or religion, not even the actual ensemble of all of them, is capable of doing justice to the predicament of our world as seen through human eyes. What we can do is, first of all, listen to the human experience throughout the ages without despising any of them due to uncritical evolutionistic schemes. Secondly, to maintain an open dialogical dialogue with all that may appear in our horizon — not excluding the divine (to use just one expression).

In other words, we need some symbols standing for the human or superhuman effort at making sense of Reality. We need some myth which tells of our limited perspective, and, at the same time, does justice to our equally undeniable claim to ultimacy. We need symbols, human symbols, open but concrete. We have to be conscious of their limitations, but at the same time equally aware that we cannot renounce truth and validity. Philosophy, I suggest, could be *one* of those symbols.

NOTES

1. Because it comes from an unexpected source I quote from a theologian's writing just a century ago: "It would incur into the danger of self-deceit the person who would believe that mathematical truths are unanimously accepted because of their internal evidence. There is consensus because nobody is interested in rejecting them". Solowjew (Solovief) (1889: II, Ch. 6; my translation). Cf. for the modern problematic Habermas (1968).
2. Cf. Panikkar (1980a: 117–133).
3. Cf. what is perhaps Martin Heidegger's last poem (from March 1976) dedicated to the present author, which commences as follows:

   "Wann werden Wörter
   Wieder Wort?

   "When will words
   Again be word?

   Cf. Panikkar (1977b: 177). The english translation is by Sheehan (1976).
4. In books and essays I have striven to investigate the basic elements for a fruitful encounter between cultures and religions. This critical interpretation of other traditions I have called *diatopical hermeneutics*. Quoting oneself is almost always unpleasant; since, however, it is impossible to cover everything in a single essay, the reader is directed to other publications of mine. I could have plagued this study with ten times more endnotes, especially regarding the sanskrit sources, since I have been gathering materials for this work for the last twenty years. I have tried to reduce the notes to a minimum.
5. It is almost obligatory here to quote the beginning of the classical sanskrit work: "this *brahman*, without beginning and end, primordial word, imperishable sound". (*anâdinidhanam brahma śabdatattvaṁ yad akṣaram*). Bhartṛhari (1965 (1965: 1), p. 1) *Vâkyapadîya* (I: 1).
6. Here are two examples related to our case: "Yâska's greatness,.... lies in the fact that he is the first to claim *scientific* foundation ... for etymology" says Lakshman Sarup (1967: 64).
   "Pâṇini's *Aṣṭâdyâyî* represents the first attempt in the history of the world to describe and analyse the components of language on *scientific* lines "writes the Indian Ministry of Education (R. P. Naik) in the reprint of Pâṇini (1891; emphasis added).
7. Nowadays reflections on the problem of translation are numerous. Cf. Steiner (1975) (with 12 pages of bibliography); Elder (1976); Larson (1980); and García Yebra (1982).
8. The indic theory of language is here paramount. Besides the translation of the already quoted

Bhartṛhari by K. A. S. Iyer (1971 *sq.*) and his own study (1969); cf. Gaurinath Śastri (1959, 1980); Biardeau (1964); Coward (1976, 1980); etc.

9. Rudolf Otto (1930), represented a kind of scandal in Catholic circles. How far the scene has changed in the last fifty years can be seen in Vadakkekera (1981) in which mainly Catholic authors address the problem of grace in the hindu and christian traditions.
10. A further example: ... "we characterised faith as the substantial constituent of Christianity: Faith is rebirth". [Heidegger (1970: 29), lecture of 1927, revised, however, in 1969.] But faith as rebirth cannot be applied in a general way to other religions. Have we rather not to seek a notion that is also acceptable to those religions which are not based on historical consciousness? A more comprehensive conception of faith, would be one as the constitutive openness of human existence. Cf. Panikkar (1979a: 188–229). It is to be understood that Heidegger only wished to broach a christian argument. Cf. page 18 of his lecture.
11. We understand, of course, inter-cultural or cross-cultural values as values valid in more than one culture. Transcultural would be a value standing above all or some cultures, without specifically belonging to any of them. This is an important issue in Comparative Religion. Cf. Panikkar (1980b); Daya Krishna (1986); and Larson and Deutsch (1988).
12. "We believe that the essence of a global history of philosophy is not so much the history itself as the 'global perspective' which may result from studying such a history". Plott (1977: IX) 'Global perspective' means for Plott "absence of provincialism — geographical, cultural, temporal, dimensional" (*ibid.*). He differentiates, however, seven "types among man's theorising efforts towards understanding his place in the cosmos ...: Organic, Mechanistic, Panpsychic, Dialectical, Theogonic, Prophetic, and Revelational." (p. XIII) In fact what Plott suggests is a plurality of perspectives.
13. Cf. Panikkar (1985).
14. Cf. Gadamer (1972: 253); and Gadamer (1975: 238; english translation).
15. One of the best concise summaries I know of the nature of Philosophy is the corresponding entry in H. Krings et al. (1973). Written by the three editors themselves, it begins by saying that "Der Begriff der Philosophie ist selbst Gegenstand der Philosophie", and describes the "indefinite Diskussion" proper to philosophical reflection. Yet there is not a single reference to non-western Philosophy, nor to the possibility of another conception. This is the challenge of a cross-cultural approach: the possibility of a change of perspective, and with it the introduction of a new species and not only of another individual case within the same specific scheme.
16. For a radical critique of Philosophy in the West, within the Latin American situation, cf. for instance Dussel (1977, 1985). Asian reflection could profit from Dussel's approach as an example.
17. Cf. Dravid (1972).
18. An example will be more telling than many general statements regarding the difference of mentality. E. Roer (1850) translated the *Bhâṣâ-pariccheda* of Viśvanâtha Nyâya-pañcâna Bhaṭṭâchârya and in his introduction wrote thus:

> "You are compelled to think by reading the works of the Greeks, they introduce you into the process of their thoughts, and by this force you to accompany them with your own thoughts, until you arrive as it were by your own mind at the principles of their system, from which point it is easier either to look back upon the way you have made or to advance further. The Hindus, on the other hand, are dogmatical. They commence synthetically with a statement of their principles, yet do not condescend to unfold the train of thought which has led to them. Hence it is impossible for any one to understand their writings who has not previously, to a considerable degree, been practised in philosophical enquiries. The punishment however of this mystery and exclusion is the want of interest, felt in the study of their writings."

Almost a hundred years later Swâmi Mâdhavânanda (1940) wrote in the Preface of his new translation:

> "Students of Nyâya, however, should always remember that, no matter how good a translation is, they must be ready to do hard thinking for a proper understanding of the subject."

19. After writing this paper and collecting some 15 pages of footnotes Halbfass (1981) published his exhaustive study. The 550 pages of his book offer the proper historical perspective and furnish the pertinent data substantiating my more 'philosophical' approach. Cf. Halbfass (1988); cf. also a number of theoretical issues in most of the contributors to Larson and Deutsch (1988).
20. Cf. just at random from Asian scholars. Hiriyanna (1932, 1957); Radhakrishnan (1940, 1952b, 1960); Radhakrishnan and Moore (1957); Kunhan Raja (1960); C.D. Sharma (1960); S.N. Dasgupta (1969); Dravid (1972); and Nakamura (1983).
21. Cf. a typical statement: Vedânta "assigns to religious faith, theology, scholasticism, mysticism, art and science, their respective places in the one grand edifice of human knowledge, *as a whole.*" V.S. Iyer (1936) in his foreword to Nikhilânanda's (1949: p. v). The author goes on writing: "Philosophy, according to Gauḍapâda and Śankara, is an interpretation of the *totality* of human experience or of the *whole* of life from the standpoint of truth. Philosophy therefore, is the whole, of which Religion, Mysticism (*Yoga*), Theology, Scholasticism, Speculation, Art and Science are but parts"(pp. vi–vii)
22. A few examples: "It is a happy sign of the times that the facile and complacent statement that outside Greece no philosophy proper grew and thrived is much less repeated now with the passing of years", writes Haridas Bhattacharyya (1953) in his Foreword (p. XVI). And yet neither he nor S.N. Dasgupta (1953) in his "Introduction" nor T.R.V. Murti (1953) in the second introductory chapter seems to analyse the concept Philosophy in any kind of critical fashion.

    The same S. N. Dasgupta (1969) in his comprehensive work also fails to ask what Philosophy can, or rather should mean in India. In his introduction, after citing a History of Philosophy of 1941, in which it is stated that only the Greeks actually had philosophy and that everything else is only "mythological and ethical doctrines, and are not thoroughgoing systems of thought", he comments: "It is not necessary to say anything in order to refute these views, for what follows will I hope show the falsity of their beliefs" (Vol. I: 4). Although he is obviously right, he simply accepts that Philosophy also exists in India.

    "No one can today deny the supreme achievements of the Indian mind in the realms of metaphysics and philosophy", say the editors of a *History of Philosophy Eastern and Western* (sponsored by the Ministry of Education, Government of India), the chairman of whom was S. Radhakrishnan (1952: Vol. I, p. 5). And yet even here one fails to find a chapter on or consideration of the problem as to whether the word philosophy is an appropriate name.
23. "The expression 'Indian Philosophy' often conjures up certain images which most modern 'thought-minded' philosophers would be least disposed to consider," writes B.K. Matilal (1986: 2), having in mind a certain anglo-saxon slant.
24. "Indian Philosophy, religion and culture became stagnant and dead due to the Muslim and British rule for centuries" is the opening sentence of R.S. Srivastava (1984: 1).
25. Cf. for the contemporary problem of the study of Philosophy in India K. Satchidananda Murthy (1985). Klaus Klostermeier points out to me Zilberman (1988), which I have not been able to consult. A cross-cultural approach is tackled by Nakamura (1988), Panikkar (1988a) and in general in the Larson and Deutsch (1988) volume.
26. A cursory perusal of the *Indian Philosophical Quarterly* (and the former *The Philosophical Quarterly*) published by the University of Poona shows that the majority of articles deal with modern 'philosophical' queries of the anglo-saxon world. The highly interesting *Students' Supplement* even more so.
27. Cf. the proceedings of the *Indian Philosophical Congress*, which since 1925, offer copious evidence for what can only be touched upon here. For example, Kalidas Bhattacharyya (1955), and Rasvihary Das (1956) include no analysis of our problem. The contributions of the same Congress (1956) are of interest: G.R. Malkani (1956); J.N. Chubb (1956) and K.C. Varadachari (1956). Cf. also the symposium "Philosophy and Sâdhana", where J.N. Chubb (1949) and S.V. Dandekar (1949) posit the view that *sâdhana* possesses more of an existential character than philosophy, whereby both suggest that philosophy is more than purely a work of reason.
28. Cf. the two contributions offered under the same titles "Philosophy in India" by Humayun Kabir (1971) and Daya Krishna (1971) (both essays include extensive bibliographies) where once again the problem whether philosophy is the right word is not raised.

29. Cf. Panikkar (1971a). Ramachandra Trivedi (1965–1966) asks rhetorically "Can we then afford to identify philosophy with a rational or intellectual approach to reality?" (p. 118). He begins the article by stating: "Philosophy, according to Sri Aurobindo, should not be 'a seeking for abstract truths in a world of ideas and words', but 'it should be a discovery of the real reality of things by which the human existence can learn its law and aim and the principles of its perfection.'" (p. 111).
30. Cf. Halbfass (1976a). The same author (1976b) wrote an article in which he analyses the concept of substance and its applicability to the notion of *dravya*.
31. English, and above all North American, philosophical literature, when it is not specialised in ancient periods, uses words such as 'knowledge', 'memory', 'reason', 'object', 'thing', 'being', etc., in such a purely modern way that many of them are no longer appropriate equivalents for the otherwise corresponding names in Sanskrit. An encyclopaedia of basic words is a project that urgently needs to be undertaken. Cf. my project on which I worked for three years: *A Handbook of Fundamental Words of the Indian Tradition* (1968–1970) and which I subsequently had to put aside, since I did not seek for the necessary financial support. The *Institute for Cultural Cooperation* has taken it up and it may yield fruit some day. In the meantime there are other valuable projects under way.
32. "Personally I would have preferred the term 'religion' to 'philosophy' to indicate the thought currents in India", writes Kunhan Raja (1974) on the first page. "The three aspects of science, religion and philosophy formed a single current in India." (pp. 9–10). This author follows the opinion of K.C. Bhattacharyya (1956), who understands Philosophy "not only (as) an auxiliary discipline, but (as) an integral part of the religion and its characteristic self-expression" (Vol. I: 119). Something similar could be said of T. R. V. Murti.
33. Significant insights are provided by Betty Heimann (1964). Despite the fact that she hardly uses the word 'philosophy' concepts such as 'thought', 'psychology', 'metaphysics', are used as suitable categories, even if she is trying to reinterpret them. The converse endeavour — the arrangement of western terms according to indic classification, as far as I know, has not yet been undertaken. In the sphere of politics, too, this is an ominous state of affairs. What do such words as 'nation', 'democracy', 'individual votes', etc., mean in the indic linguistic consciousness?
34. Cf. Nakamura (1964, 1975).
35. Cf. for example, Keith (1925) where not a single mention of our problem is made. Even though Edgerton's (1965) is an anthology, the 30-page introduction remains silent on the subject of what Philosophy could be. Zimmer (1951: pp. 48–50, 56–66) a perceptive specialist in the field of indic wisdom, speaks of "Philosophy as way of life" and of "Philosophy as Power", yet our problem is scarcely mentioned. A pioneering book which attempts a Universal History of Philosophy is Misch (1926). He states in the expanded English edition: "Philosophy made its appearance at several points on the earth's surface: in India, China, and Greece (...) Despite this diversity, however, we can speak of the beginning of philosophy, using both words in the singular. Thus we approach the historical facts on the assumption that philosophy is a unity" (Hull, 1950 39). We must not forget Deussen (1894–1917) who simply assumes that there is an indic philosophy.
36. Cf. by way of examples, Gaurinath Śâstrî (1959: XXIV) writes of a "school of philosophy which is known by the name of Verbal monism, Śabdâdvaita or Śabdabrahmavâda". Or also: "The Supreme Reality in the philosophy of sanskrit Grammar has been called the Eternal Verbum". These quotations give us an idea of how the word is used. Cf. also Trivedi (1965–1966); P.T. Raju (1971), attempts with the assistance of indic philosophy, to reconcile Hegel and other idealistic philosophers with the existentialists.
37. Cf. Gonda (1950: 3): "It is, in general, advisable in books on Hindu subjects, to utilise Christian terms only with the utmost care, and, if possible, not to use them at all to explain Indian concepts." Nevertheless even Gonda uses other 'western' names as if they possessed a more universal significance. It is the merit of such journals as *Philosophy East and West*, and *Comparative Civilizations Review* to have promoted comparative studies in both directions: from West to East, and from East to West. Books such as E. W. Said (1978) have, of course, emphasised the political side of the issue.
38. We cannot evaluate the results of teaching philosophy and theology in the languages of India as it is being increasingly done nowadays. Significantly enough the Departments of Modern Sciences

and Technology prefer English. Why duplicate the effort if modern Science only uses terms and not words? Cf. the pathetic cry, which symbolises the almost tragic situation of most african 'philosophers' writing in French or English:

> Et ce désespoir à nul autre égal
> D'apprivoisier, avec des mots de France,
> Ce coeur qui m'est venu du Sénegal?

Laleau (1969). Cf. the sensible pages of Rahola (1988: 116–119) dedicated to this problem.

39. Cf. Panikkar (1971b).
40. Cf. Benz (1951); Darlquist (1962); Sedlar (1980); and the first chapter of Halbfass (1981: 13–36); etc.
41. The works of George Dumézil (1958, 1960, 1968) and Emile Benvéniste (1969) should be mentioned here.
42. "La filosofia dell'India é dunque in questo senso anche filosofia nostra e conoscerla non significa conoscere una parte di mondo separata senza rimedio da noi, (...) ma meglio e più intensamente conoscere noi stessi" affirms an authority as Raniero Gnoli (1960: 19).
43. The first issues of the *Recherches philosophiques africaines* are worth quoting: *La philosophie africaine* (1977) (Actes de la lère. Semaine Philosophique de Kinshasa); *Philosophie et libération* (1978); *Mélanges de philosophie africaine* (1978); *L'intersubjectivité dans les 'Proverbes Tetela'* (1979); *Langage et philosophie* (1981); etc. Our remark is that although the dependence of african Philosophy on the western one is recognized, a critical effort is made to reflect on the specificity of an african thinking.
44. Radin (1927) and Tempels (1945) were perhaps the first books to tackle this theme. Cf. also Kagame (1956, 1971); Hountondji (1971, 1976); both with ample bibliography; and Towa, (1979); etc.
45. Cf. Mbiti's (1970) title. Cf. Frazer (1974) in which the thesis of the title is put forward with reference to numerous examples. Cf. also Odera Oruka (1983); and Ruch and Anyanwu (1981). The african input is relatively unknown in the field of indology, and it could have a stimulating function for a contemporary reflection in India.
46. Niewöhner (1980) defends three theses:
    1. "Es gab früher eine 'jüdische Philosophie' — aber es gab den Begriff nicht" (p. 195).
    2. "Es gibt heute den Begriff 'jüdische Philosophie' — aber es gibt die Sache nicht mehr, für die der Begriff geprägt worden war" (p. 195). The author explains that for him "früher" means the Middle Ages and "heute" after 1948.

    His third thesis is that in the middle period (of some 500 years) there was a Jewish Philosophy in people like Spinoza and Cohen.
47. Cf. "La evolución del concepto de filosofía en occidente" in Panikkar (1971a: 68–73). The traditional use of the word 'philosophy' is not altogether lost in contemporary writings. Cf. a recent example in González Faus (1988), where the word "filosofia de la vida" is used as meaning the 'worldview' (of Jesus of Nazareth).
48. Heidegger (1956: 7), "Die gehörte Redeweise von der 'abendländischen Philosophie' ist in Wahrheit eine Tautologie. Warum? Weil die 'Philosophie' in ihrem Wesen griechisch ist". Heidegger is clear in this : "das Abendland und Europa, und nur sie, sind in ihrem innersten Geschichtsgang ursprünglich 'philosophisch'" (*ibid*) ("the West and Europe, and only they, are in the innermost course of their history, originally philosophical").
49. Cf. Halbfass (1979: 138): "Einen Namen und Begriff dessen, was in einem von den Griechen begründeten Sinn autonomen wissenwollends Philosophie heisst, haben die Inder zwar nicht entwickelt, die inhaltlichen und methodischen Entsprechungen sind jedoch so weitreichend, dass die von den Historikern der westlichen philosophischen Tradition nur zögernd akzeptierte Rede von 'indischer Philosophie' vollauf gerechtfertigt ist." And he adds "Jedoch hat der Gang der Forschung gelehrt, mit der Übertragung westlicher Begriffe und Denkmodelle vorsichtiger zu sein'.. ("Although the indians did not develop a name and concept of philosophy in the sense stablished by the greeks — that is to say, as an autonomous desire for knowledge, the correspon-

dences in both content and method are so far-reaching that it is entirely justified for historians of the western philosophical tradition to speak of what is only accepted with reluctance as "Indic Philosophy". And he adds: "The lesson to be learned from the course of investigations, however, is that one should exercise greater prudence in the transferring of western terms and patterns of thought..."

50. Heidegger (1956: 8 sq.), goes as far as seeing a direct relation between the modern natural sciences and the mode of approaching reality determined by (the greek) Philosophy. And it is this which explains the alleged universality of Modern Science.
51. Cf. The astonishing affirmation: "It is (...) because most of what is labelled *Eastern Philosophy* is not concerned (with argument...) that this book draws no materials from any source east of Suez" Flew (1971)—as quoted by Matilal (1986). This affirmation is all the more startling as the author, besides recognising that Philosophy "is characteristically argumentative" in the Preface of (Flew, 1979), asserts that Philosophy "usually embraces both value-commitments and beliefs about the general nature of things". Cf. the statement reported by Dasgupta in note 22.
52. The words of the editor of the latest *Historisches Wörterbuch der Philosophie* are characteristic: "Auf Anraten eines Kreises von Indologen, Sinologen und Japanologen werden ausser Begriffen der asiatischen Philosophie, die in der europäischen Philosophie schon Bürgerrecht geniessen, auch solche behandelt, die künftig in einer globalen Philosophie eine Rolle spielen könnten oder instruktive Parallelen zur europäischen Problem- oder Begriffsgeschichte bieten. Damit soll auf sinnfällige Weise deutlich werden, dass Philosophie nicht mehr nur als europäische Philosophie begriffen werden kann, auch wenn diese Mittel- und Schwerpunkt des Wörterbuches bleibt". "Following the recommendations of a circle of Indologists, Sinologists, and Japanologists, we have dealt not only with concepts in asian Philosophy which have already become naturalised in european Philosophy, but also those which could, in the future, play a role in a Global Philosophy or offer instructive parallels to the history of european problems and concepts. And so it ought to become clear in an obvious way that Philosophy can no longer be understood purely as european Philosophy, even if european Philosophy remains as the focus and emphasis of the Dictionary". (Richter, 1971: Vol. I: VII.)
53. This would correspond to the opinion of many contemporary investigators in Asia, as we have already shown. Nowadays we are beginning to witness the appearance of *Handbooks on World Philosophy*. Cf. a recent one by Burr (1980). The geographic division alone is characteristic if not suspicious: Part One ("Western Europe, Australia and Israel"); Two ("Eastern Europe"); Three ("The Americas"); Four ("Africa and the Republic of South Africa"); Five ("Islamic Countries"); Six ("Asia"). I am afraid that the present *Chronicles* may show a similar bias. We are improving, though. Cf. also Bahm (1979) where a "World Philosophy" is also spoken of; and Plott (1977); etc.
54. Heidegger (1956). Before that he gave the main argument: "Der Stil der gesamten abendländisch-europäischen Philosophie—es gibt keine andere, weder eine chinesische noch eine indische—ist von der Zwiefalt 'Seiendes-seiend' her bestimmt." (Heidegger, 1954: 136).
55. "Die griechische Sprache, *und sie allein*, ist λόγος!". Why? Because: "das in ihr Gesagte auf eine ausgezeichnete Weise zugleich das ist, was das Gesagte nennt." (Heidegger, 1956: 12; our emphasis).
56. We think of a great part of the indic reflections on the word from the Vedas to Bhartṛhari, etc. *Atha śabdânuśâsanam* begins Pâṇini. Much more recently K.C. Bhattacharyya (1956) wrote that "an explication of the concept of Philosophy appears to me more important than the discussion of any specific problem of Philosophy" (Vol. II: 100). It may be retorted that this original thinker of the last generation of British India was already influenced by greek thought (through Kant). But this would amount to saying that Plato was already influenced by the egyptians! The cited article was published in Radhakrishnan and Muirhead (1952a). Cf. also Mehta (1974). Cf. John E. Smith "Commentary" on Mehta's paper in the same issue of the Journal (pp. 89–93): Indian tradition may be 'alien', "but it need not on that account 'alienate' the tradition. "..." New strength comes from the discovery that a tradition can survive *outside* of its own milieu." (p. 90)
57. Cf. Panikkar (1988d).
58. It may be worth quoting the reflection of Karl Potter (1977), the editor of the ambitious project

Encyclopedia of Indian Philosophies in his preface : "that philosophers without extended training in Sanskrit and Indian studies are not in a very good position to appreciate the contributions made by classical Indian Philosophy toward the solution of perennial philosophical problems."

59. A homeomorphic equivalent is a kind of functional analogy of the third order. Being and *âtman*, for example, are neither analogous nor do they share the same function. These two designations nevertheless have a certain relation: a unique relation brought about by the respective corresponding function in the other system. Cf. Panikkar (1978: xxii): "Homeomorphism is not the same as analogy; it represents a functional equivalence discovered through a topological transformation."
60. Parmenides Fr. 3 : τὸ γαρ αυτὸ νοεῖν εστίν τε καὶ ειναι; Diels-Kranz (1960, Vol. I: 231).
61. BU II, 4,14. Cf. note 122.
62. Cf. Panikkar (1986b: 603).
63. Cf. Panikkar (1987: 134–140).
64. Cf. the answer of the Buddha to the monk Râdha when asked about the aim of *nirvâṇa*: "O, Râdha, this question goes too far: you can grasp no limit to this question." *Saṁyutta-nikâya* III, 189 and the analysis of the text in Panikkar (1979a: 268 sq.).
65. Cf. Io, VI, 44 and the many reflections on the subject by Macarius, Bernard, Cusanus, Pascal, etc. Cf. BS I, 1–2 and the many commentaries by Śàṅkara, Râmânuja, Madhva, Padmapâda, etc.
66. Cf. Panikkar (1979a: 100 *et passim*); Panikkar (1988c).
67. Cf. Rosemont (1988) for a cogent and spirited critique of relativism. As for relativity cf. Panikkar (1979a: 102, 316); etc.
68. Πάντες ἄνυρωποι τοῦ ειδέναι ορέγονται φύσει. ("Omnes homines natura scire desiderant") *Met*. I, 1 (1980a).
69. Cf. the outstanding work in five volumes, and unfortunately very little known, of Santiago Ramírez (1972). In spite of the subtitle *tractatus theologicus* the work is probably the most elaborated study from a christian-scholastic point of view. As for the 'theologicus' we should keep in mind the authur's saying: "velle Theologiam addiscere sine Philosophia est velle Theologiam *intelligere sine intellectu*." (I; xvii).
70. Cf. the Buddha's so-called fundational sermon of Varanasi, which gave birth to buddhism. *Saṁyutta-nikâya* V, 420 sq. Cf. translation and commentary in Panikkar (1989 : I, p. 3).
71. Cf. Panikkar (1986a), for the complex relationship between salvation and wholeness.
72. Cf. Heidegger's (1953 : 1) first line of his introduction to Metaphysics: "Warum ist überhaupt Seiendes und nicht vielmehr Nichts," echoing Leibniz and others.
73. So begins the famous *nâsadîya-sûkta* RV X, 129, 1.
74. Besides the quoted literature the reader may consult with profit general works on Indian Philosophy like: Hiriyanna (1932, 1957), Heimann (1937), Radhakrishnan (1940), Zimmer (1946, 1951), Frauwallner (1953/1956), Chandradhar Sharma (1960) and Raju (1985).
75. Yâska, to begin with, divides the Vegas only into two parts: *mantras* and *brâhmaṇas*, from which came the later interpretation of theory and praxis: what is to be thought and what is to be done.
76. The three are the Ṛg Veda, the Yajurveda, and the Sâmaveda, the Atharvaveda being a later addition. Jaimini's *Mîmâṁsâ-sûtras* (IInd century B.C.) ignore the Atharvaveda, probably because this 'fourth' Veda has little place for the sacrifice. Later, for Vedânta, the *prasthâna-trayî* or the threefold ground is considered to be formed by the *Upaniṣads*, the *Gîtâ* and the *Brahma-sûtra*.
77. "The doctrine of the two sections (*Kâṇḍa-dvaya*) had already been accepted by Śaṅkara and was adopted by his commentators (...) The *yoga-kâṇḍa* is set up in addition", says Nakamura (1983 : p. 413). And yet our tripartite division seems to be equally authoritative and traditional. Cf. Sharvananda (1958 : p. 183).
78. We may recall here the three *mârgas* (ways) of the *Bhagavadgîtâ* which consecrate our tripartite division.
79. Cf. a cross-cultural elaboration on the meaning of ritual in Panikkar (1979b).
80. Cf. The Vedic conception of sacrifice in Biardeau and Malamoud (1976); and Panikkar (1977a: 346–431) for the main texts and commentary. As for more textual commentaries, cf. the classical Levi (1898/1966); Thite (1975); and Vesci (1985).
81. The literature is proliferating. Cf. a selected one in Panikkar (1979a: 361–388). Cf. also O'Flaherty (1980); and Keyes and Daniel (1983).

82. *Yathâ-kari yathâ-cârî tathâ bhavati*, says BU IV, 4,5 "According to one's deeds, according to one's behaviour, so one becomes", and adds *yat karma kurute, tad abhisaṁpadyate*. "Whatever action one does, that one obtains".
83. Cf. Panikkar (1977a: 346–431).
84. Classical times seemed to feel already an analogous difficulty: "This course of *karman* is mysterious (*vicitra*, strange, manifold, wonderful) and difficult to discern (*durvijñânâ*)" says the YSBh II, 13. For karma spirituality, cf. Panikkar (1969: 466–474).
85. Cf. the main texts in Panikkar (1970: 54 *sq.*).
86. "The Pûrva Mîmâṁsâ", says no less an author than S. N. Dasgupta (1969) in spite of his own parenthesis "(from the root *man* to think – rational conclusions) cannot properly be spoken of as a system of Philosophy." (Vol. I: 68).
87. Cf. for instance Srimat Anirvan (1958) saying that the cleavage between knowledge and ritual did not exist in the vedic period: "knowledge is to be taken not in the sense of philosophical reason or intellectual co-ordination of experiences, but in the sense of spiritual realisation which aims at inner soul-experience of illuminated consciousness" (p. 314). He adds: "Realisation through rituals (*dravya-yajña*) and realisation through knowledge (*jñâna-yajña*) are not mutually contradictory". It is for us interesting to remark that the phrase "philosophical reason" implies already the acceptance of the post-cartesian pattern. Anirvan takes *tarka* as "philosophical reason" and distinguishes it from the "spiritual reason" (*mîmâmsâ*) of the intuitionist" (p. 317). Cf. also Anirvan (1983). "Karma, after all, is ritual" writes Staal (1988: 224) rightly underlying that Philosophy in India is mainly interpretation of ritual.
88. *Sarvam karm'âkhilam, jñane parisamâpyate*. "For knowledge is surely the culmination of all ritual works", BG IV, 33.
89. "True religion is Philosophy" says John Scotus Erigena. Cf. S. H. Nasr (1981 : p. 20), quoting some studies on Erigena. The whole of Nasr's book is an impassionate plea to recover the traditional meaning of Philosophy as sacred knowledge in the line of the *philosophia-perennis*. Cf. Panikkar (1971a: 51–82).
90. Cf. Col. II, 8; etc. Cf. the pertinent commentaries: "A quoi bon la sagesse humaine?... — Je le sais, mais pour y renoncer efficacement, il faut/d'abord la posséder.... il faut un minimum — et même un maximum – de sagesse pour comprendre que la sagesse ne vaut pas mieux que la folie." Thibon (1974: 73–74).
91. It is impossible to give even an approximate idea of this rich notion and to point out a selected bibliography. Cf. as examples on the general meaning of *dharma*: Zaehner's (1962) chapter on Dharma; and Hacker's (1965: 93–106) Innaugural Lecture to his Professorship in Münster. This does not mean I agree with everything said by Hacker. The author sent me an offprint with the dedication: "In freundschaftlicher Opposition"!
92. Cf. the notion of *svadharma* as described in the *Gîtâ* II, 31; II, 33; III, 35; and XVIII, 47.
93. The root is *dher-* (*dhera-*) "halten, festhalten, stützen" according to Pokorny (cf. the critical remarks of Mayrhofer). The biblical passage in Sap. I, 7 (τὸ συνέχον τὰ πάντα) "hoc quod continent omnia" and in general also Chapters VII to IX. Cf. the greek notion of σύνεσις (insight, intelligence), union (from συνίημι), coming from the Hebrew in Is. (XI, 3), etc. For the rich notion of *lokasamgraha*, cf. BG III, 20; III, 25.
94. Cf. *Satyam satsu sadâ dharmaḥ* (In the beings truth is always dharma) MB (*Śantiparva*, 162: 4) and comments in Panikkar (1956: 45; 1971a : 235). For *ṛta* cf. Miller (1985).
95. After Śaṅkara the third chapter of the BS is called *sâdhana*, implying the acquisition of all the requisites for the realisation of Brahman.
96. *Mahâparanibbâna-sutta*, VI. 10. Cf. Panikkar (1989) for related texts and commentary.
97. We must consider, however, that theory and praxis possess far deeper meanings in Greek and Latin. Here we witness an instance of the above-mentioned problem of translating.
98. *Theologia mentis et cordis* was a common expression in christian literature.
99. We have here some glaring examples of the erosion of the original meaning of words in modern western vernaculars. There is a world of distance between English 'pious' and Latin *pietas*, between to be a 'devout' person and a duly consecrated human being, between a vow that does not bind if too unconfortable and a *votum*, a *vrata* which by the very fact of having been pronounced

transforms the actor and the doer's actions – magic notwithstanding.

100. The root is *uer-* to speak (cf. Greek εἴρω, ῥήτωρ) and the meaning of vow is an oath. Monnier Williams adduces the verb *vṛ-* (*vṛṇoti*) to cover, screen, veil, conceal, obstruct, close. Mayrhofer does not accept the relations to *vṛṇoti* "he covers", to *vṛnîte* "he choses", and to *vartate*, 'he turns'. An interesting document which could trigger some pertinent reflections on Philosophy as a life-style could be J. W. Hauer (1927).
101. The idea of the *consecratio mundi* had a short-lived existence in the Catholic Church during and after Pius XII. Cf. Panikkar (1961).
102. Cf. the beautiful description: "animi assidua et vehemens ad aliquam rem applicate magna cum voluntate occupatio,"... Cicero, *De inventione* I, 25. Tellingly enough the first adduced example is *philosophia*.
103. Etymologically *upâsanâ*, (or *upâsana*) comes from *upa* (near, together) and *as*, (be): to be, sit or stand near, at hand, esteem, be devoted, attached to, etc.
104. The expressions *Îśvara-bhakti* and *Îśvaro-pâsanâ* are sometimes used in Modern India for 'religion'.
105. Cf. the *Gîtâ* for the pertinent passages. See Zaehner's (1969) commentary which shows clearly the place of *bhakti* as Philosophy. Extremely useful is the index of Bolle's translation (1979).
106. *Tanyate vistâryate jñânam anena iti tantram.* "The scripture by which knowledge is spread is tantra." (P. C. Bagchi, 1969: 211).
107. Cf. Gosh (1956).
108. The literature is immense. Just as references, cf. Chakravarti (1963); Panikkar (1969): 516–538); D. B. Patel (1978); and K. Mishra (1981).
109. "In Pâṇini's time the word yoga had attained its technical meaning, and he distinguished this root '*yuj samâdhau* (*yuj* in the sense of concentration) from '*yujir yoge*' (root *yujir* in the sense of connecting) "S. N. Dasgupta, 1969 : Vol. I, p. 226). He adds a pertinent remark: "In the *Bhagavadgîtâ*, we find that the word yoga has been used not only in confirmity with the root '*yuj-samâdhau*' but also with '*yujir yoge*'. This has been the source of some confusion"... (p. 227).
110. Cf. Eliade (1969); and Hauer (1958) which are probably the two best comprehensive works on the subject.
111. The Monnier-Williams (1851) already gives over 25 synonyms for Philosophy.
112. The root *jña* means to know, to understand. Pokorny distinguishes the two morphologically identical roots, *gen-gnô* to beget (*janati*, γενέσθαι, *gens*, *natio*, etc) and to know (*jânâti*, γιγνώσκω, (g)*noscor*, *notio*, etc.) Ultimately to know is to beget (the known – the Father begeting the Logos).
113. Cf. the introduction to Śaṅkara's BSBh and also his *Vivekacûḍâmaṇi*, 19. This discernment, together with *mumukṣutva* and *ihâmutra-phalabhoga-virâga* (renunciation of the enjoyment of one's own deeds), are the classical and accepted prerequisites for the study of indic wisdom. Cf. the sub-commentary of Vâcaspati Miśra (1933).
114. That is, in the final analysis, the primary meaning of the word in almost all traditional cultures.
115. Cf. BG IX, 1, for the relationship between *jñâna* and *vijñâna* and the interpretation of Śaṅkara and Râmânuja in the sense of intellectual experience or even perhaps of ritual adoration.
116. As late as his time Thomas Aquinas could still peacefully write that: "Manifestum est enim quod quanto aliquid magis intelligitur, tanto conceptio intellectualis est magis intima intelligenti, et magis unum: nam intellectus secundum hoc quod actu intelligit, secundum hoc fit unum cum intellecto." *Sum. theol*, I, q. 27, a. 1, ad 2. This is valid not only with reference to the Trinity.
117. RV I, 164, 37. Cf., purely as an item of curiousity, the following sentence from that philosopher of the word F. Ebner, (1963: Vol. II, p. 650) who wrote in his journal (entry for 12, September 1961): "Ich gehe durch das Leben als einer, der weiss, dass ihm das letzte Geheimnis des Lebens verschlossen ist." ("I walk through life as one who knows that the ultimate secret of existence is closed to him").
118. Heimann (1951: 7) interprets *vi-* as expressing "a kind of mechanical dissemination" and *pra* as "a more intentional, and thus more intensive but less extensive movement".
119. Cf. vg. *Tao te ching* 1; 56; 81; etc.
120. *avijñâtam-vijânatâm vijñâtam avijânatâm*. KenU II, 3. Cf. RV I, 164, 32; Dham V, 4 (63); *Tao te ching* 38; etc.

121. *Athato brahma jijñâsâ*, BS I, 1, 1: "Now the longing for the understanding of brahman", the aspiration to know *brahman*.
122. Cf. the famous passage: "How can He be known by whom all this is made known?" BU IV, 5, 15. "You cannot know the knower of knowing" has been said just before. Cf. note 61.
123. Cf. Panikkar (1981: 97 et seq.) for an analysis of the *jijñâsa* in BS (I: 1, 2).
124. Cf. by way of example, K.S. Murty (1959). Cf., in order to stress the difference, the study by Edwards (1972). But, "I must insist that all attempts to differentiate philosophy and religion in the cultural milieu of India must fail dismally... The differentiation of theology and philosophy is a Western phenomenon"... (in C. T. K. Chari 1967: 125).
125. *Idam tattva-jñânam ca paramâtma-viṣayakam niḥśreyasa-sâdhanam ceti bodhyam. Nyâyakośa*, sub voce *tattvajñâna*.
126. *tat* is a demostrative pronoun, and *tva* is an abstract suffix. *Tattva* could then be translated as reality, and some would also say being. It would be instructive to relate *tattva* with *thing* in the Heideggerian interpretation of *das Geviert* (earth, heaven, the divine and the mortal). Cf. Heidegger (1962, 1967).
127. *Tattva jñânânniḥśreyasâdhigamaḥ*, "through knowledge of truth (Reality) one achieves the highest salvation", NS(I: 1, 1). The text means, of course, the knowledge of all realities classifield under the Nyâya categories.
128. Halbfass (1981: 331) points out several works in Marathi and Gujarati, with titles of *tattvavidya* and *tattvajñâna*. Also in Tamil *tattuva* is being more and more used.
129. P. N. Srinivasachari (1953: 301).
130. Cf. the *tattvatraya* or three principles of many indic systems (*cit*, *acit* and *îśvara* in Râmânuja, for instance). Or the 28 or 26 *tattva-s* (principles) of many schools.
131. We could cite works like *Vedânta-tattva-dîpana-vyâkṣyâ* a commentary on commentary; *tattva-cintâmani-prakâśâ*, and many others. Cf. Dasgupta (1969) or Potter (1977).
132. Śantideva, *Bodhisattvâcharyâvatara* V. 23. The Tibetan word is *shes.bzhin*. The word is used together with *smṛti* (Tibetan *dran.pa*) and usually rendered as "mindfulness and alertness". Putting the two words together Philosophy would mean here the alertness of the mind mindful of all the wisdom which has preceded the actual philosophising. *Smṛti* here is the actual remembrance of all that one has received by tradition screened by the alert mind. *Smara* has also the connotation of love in the sense of longing for, nostalgia, remembrance of (the beloved).
133. The root is *vid-* Cf. Latin *videre*, English *vision*. The indo-european root is *u(e)di-* (Pokorny) or *ueid* (Mayrhofer): to see, and from there to perceive, know, and also to find, to discover, in the form *vindati*. Cf. Greek οἶδα (ειδώς) from the same root Fιδ–
134. Cf. the *vidyās* in the Upaniṣads. Cf. for example, Mukhopadhyaya (1960).
135. Cf. MundU I, 4–5, and Aristotle, *Met*. I, 2 (982 b 9) and VI, 1 (1026a 16), etc., although the *philosophia prôte* is also called theology (1026a 19).
136. *Nyâyakośa*, sub voce *vidyâ*. *Śaṅkara* (*Bhaṣya* to MundU I, 1,4) declares that *aparâ vidyâ* is the science which teaches the means and fruits of justice and injustice (*dharma* and *adharma*). And in BSBh I, 2, 21 precises that 'the distinction between lower and higher knowledge is made on account of the diversity of their results, the former leading to mere wordly existence (*abhyudaya-phalâ*), the latter to the *summum bonum* (*niḥśreyasa-phalâ*)".
137. Cf. the famous saying: *hi vai bhavati ya evam veda* "He who perceives *brahman* becomes *brahman*". BU IV, 4, 25. Cf. also *Nrsiṁhottaratâpinî Upaniṣad* 5 and 6.
138. Interesting the etymological suggestion of επιστήμη from επι and ἵστημι: to put, situate something in front (*Gegen-stand, ob-jectum*).
139. Cf. BU II, 4, 5–6.
140. Cf. Dasgupta (1969: Vol. I, p. 277) for instance.
141. There is a saying by Madhva which represents a common feature of indic Philosophy: *tame-vaikaṁ jânatha âtmanam / anya vaco vimuñcatha* ("Know only that which is âtman; abandon the talk of the other things") *Apud* Raghavendrachar (1953: 318). We notice that the verb here is *jâna* and not *vidyâ* showing the interchangeability of many such words.
142. The traditional Sanskrit Faculty of the Banaras Hindu University has as its official title: "Faculty of Oriental Learning and Theology".

143. Cf. Mundu I, 1, 1, *sq.* and Śaṅkara, BSBh I, 2, 21 besides the already quoted paragraph.
144. Cf. Śaṅkara, BGBh XII, 2, where it is stated that *vidyâ* is no characteristic (no *dharma*) of the perceiver.
145. Cf. Sureśvara, *Naiṣkarmyasiddhi* III, 1. Cf. the translation of Maximilien: "Lorsqu'un homme connait 'Je suis Brahman' a partir de 'Tu es Cela', etc. 'alors, ayant complètement detruit le Je et le Mien, il ne s'engage plus dans l'écoulement de la parole et de la penseée" (p. 68).
146. Cf. Panikkar (1963a: 91 *sq.*)
147. BG IX, 2.
148. The root is *mâ*, to measure. *Pra-mâ* would amount to well-measured. It may also connote the idea of basis, foundation. Right measure amounts to true knowledge, correct notion. *Pramaṇa* means a measure, scale, standard.
149. The extremely rich indo-european root *mê* (*m-e-t*) with the basic meaning of measuring is behind the word *mati*, which is also connected with the other root *men*, to think. The first root suggests to put stakes (in order to measure). The second root suggests to agitate and may be at the basis of the Latin saying *cogitare, id est coagitare*. For those two fundamental meanings of the thinking process: to measure and to moderate, cf. Panikkar (1986a: 27–28).
150. Cf. TU (II: 8, 1) (*ânandasya mîmâṁsâ*).
151. The greek word comes from ἱστορέω, to research, investigate, come to know, examine, observe, question. The root is Ϝιδ, to know, seen. Cf. Sanskrit *veda* Latin *videre*, Greek οιδα, etc. "Ιστωρ is the person who knows (also the jugde who knows the Law). History would then suggest not only the narrative of events, but also the investigation of Reality.
152. Cf. Panikkar (1963b: 343–370).
153. This central insight of Vedic orthodoxy should not be interpreted as if the Vedas had no author in the sense that nobody had chanted them or written them down. It means that the relationship between the word and its meaning has no further agency behind telling us what the words mean. Cf. Panikkar (1977a: 12–13). The word is both the revealer and the revealed, says Bhartṛhari, *Vâkyapadîya* (II: 32).
154. The root being the same as with *mîmâṁsâ* and *mati*. Cf. Gonda (1963).
155. Paul Hacker (1958: 81) closes his essay with the statement: "Es ist ein bezeichnender Zug der Hindukultur, dass sie zwar den Begriff der Philosophie kannte, aber kein Wort, das ihn ganz und ausschliesslich ausdrückte." ("It is a characteristic trait of hindu culture that the concept of Philosophy was known, whilst no word was found to express this in a definitive way"). Cf. the complementary and also critical remarks to Hacker's article in Halbfass (1981: 296–337).
156. *Arthaśâstra*, Introduction.
157. The root *îkṣ-* to see, could suggest we classify *ânvîkṣikî* as belonging to the following group of worldviews, but the traditional use of the word has always stressed its logical or rather *logos* character, often analytical.
158. Cf. Viśvanâtha's definition in NSBh, I, 1, 1: *śravaṇâd anu paścâd-ikṣa anvîkṣa, unnayanam;...*" das Nachsehen, das auf das Hören folgt, ist das Nachprüfen", das Folgerungziehen" The text is worth quoting in full: *tannirvâhikâ seyam ânvîkṣikî nyâya-tarkâdi-śabdair api vyavahriyate*. "die (Wissenchaft), die dasselbe ausfürht, ist *ânvîkṣikî*, sie wird auch mit den Worten *nyâya*, *tarka* usw. bezeichnet." Hacker's (1958:77) translation. Hacker (1958) p. 77.
159. Monnier Williams (1899) simply says: "logic, logical philosophy, metaphysics".
160. Cf. *anu-mâna*, deduction, inference. *Anu-* as a prefix to (mental) examination, clarification, seeing suggests an after-thought. Cf. the German *nachdenken* (reflexion).
161. Cf. the clear commentary by Radhakrishnan (1953: 133–137).
162. So, for instance Nîlakaṇtha explaining *Mahâbhârata* XII, 318, 28 (quoted by Hacker (1958: p.75). He seems to be the first to draw the equation: *manana-ânvîkṣikî*.
163. Cf. *Nyâya vârttika tâtparya tîkâ*, quoted in Hacker (1958: 76).
164. Vâtsyâyana calls it *nyâyavidyâ* in his NSBh I, 1, 1. Cf. Dasgupta (1969: I, p. 277 *sq.*) for the indological questions which we have to overlook here.
165. *iha tv adhyâtmavidyâyam âtmâdi tattvajñânam niḥśreyasâdhigamo' pavarga-prâptiḥ*, quoted by Hacker (1958: p. 73). Here we find *adhyâtmavidyâ*, *tattvajñâna* and the attainment of salvation closely linked.

166. *Nîtisâra* II, 7 apud Hacker (1958: 77).
167. *Nitisâra* II, 11; following Hacker's translation (1958: p. 74) (against Jacobi's rendering).
168. *Manu* VII, 43.
169. "*Anvîkṣikî* seems to signify philosophical knowledge, the *darśana* of later times" writes Handique (1962: 9). It is significant that more modern dictionaries seem to be reluctant to translate Philosophy with *darśana*, as remarks Halbfass (1981: p. 352).
170. Nîlakaṇṭha's commentary *apud* Handique (1962).
171. From the root *dṛś*: to see, view, consider, look into, perceive intuitively, contemplate, understand, notice, investigate, etc.
172. *Śabdakalpadruma*, sub voce.
173. ... *etâni tattva-jñânârtham vedân vicâryya* ... *kṛtâni*. *Śabdakalpadruma*, sub voce.
174. *Ṣaḍdarśanasamuccaya* is the name of a work by Haribhadra in the VIII century. The work *Sarva-siddhânta-sara-saṁgraha* and *Sarva-darśana-kaumudî* of the VIII-IX century is attributed to Śaṅkara. Râmânuja in the XIII/XIV century also wrote a *sarva-darśana-śiromani*. And often quoted is the work of Vidyâranya Madhva in the XIV century *sarvadarśanasaṁgraha* (where the expression *pâśu-pata-darśana* occurs). The expressions *vaiśeṣika-darśana*, *mîmâṁsaka-darśana* are also often mentioned.
175. Cf. T. R. V. Murti (1955: 140).
176. The Greek θεωρία is related to θεωρός, the spectator, and comes from οραω, to see. The root is θεά, in indo-european *dheia-*(*dhiâ-*, *dhî*) to see, to look. Cf. Pokorny (1959: p. 243). Cf. θαυμάζειν, to admire, astonish.
177. Cf. Vâtsyâyana, NSBh I, 1, 1. Kauṭilya in his Introduction to the *Arthaśâstra* also states: *ânvîkṣikî trayî vârttâ daṇḍanîtis ceti vidyâh*.
178. Manu VII, 43.
179. The six *Vedângas* are: *śikṣâ* — (phonetics), *kalpa* — (ritual), *vyâkaraṇa* — (grammar), *nirukta*-(etymology), *chanda* — (metrics), *jyotiṣa* — (astrology).
180. The fourteen *vidyâsthânâni* mentioned by Jayanta in the beginning of his *Nyâtamañjarî* are the four Vedas, the six Vedângas, the Dharma-śâstras, the purâṇas and mîmâṁsâ.
181. Cf. Klostermaier (1984).
182. *Mokṣ*- being a desiderative from the root *muc*, means to wish to free oneself. *Mukti* (past participle *mukta*) is release, liberation, salvation. Cf. the Greek απο-μυσσειν and the Latin *emungere* (to blow one's nose).
183. The verb *śastri* means to discipline, govern, order, and thus to instruct, teach.
184. Cf. the monumental work by P. V. Kane (1968).
185. Cf. Halbfass (1981: 342).
186. Manu II, 11.
187. The verb is *hinoti* with the original meaning of to impel, hurl, promote, assist, and also, stimulate. *Heti* has come to mean any weapon, missile, etc. It is interesting to note how from *hetu* impulse, it came to mean motive, cause, etc.
188. The root *tark-* means to turn, turn around, contemplate from all sides (Mayrhofer). From here it means to guess, suspect, infer, and thus also to reflect, speculate. It would be interesting to relate the root meaning of *tarka* (*tarkayati*, guesses, suspects), (Cf. greek ἄτρακτος spindle; Latin *torqueo*, turn), with the *ars conjecturarum* of Nicholas of Cusa. Cf. *De coniecturis* I, 2: "Quoniam autem [...] praecisionem veritatis inattingibilem intuitus es, consequens est omnem humanam veri positivam asssertionem esse coniecturam; non enim exhauribilis est adauctio apprehensionis veri". (As you have seen [...] that the exactness of truth is unattainable, it follows that every human affirmation about truth is only a conjecture; for the extension of the apprehension of truth is not exhaustible). Literally: "the increase (*adauctio*) in the apprehension of the truth is unexhaustible (we never reach the limit).
189. In the medical schools it is said: *tarko'pratyakṣa-jñânam*. "tarka (is) the knowledge (of things) which cannot be perceived". Cakrapâni apud Dasgupta (1969: Vol. II, p. 376).
190. Cf. Heimann (1951: 7).
191. Apastamba (before the third century BC) used the word *nyâya* in the sense of *mîmâṁsâ* (Dasgupta, 1969: I, p. 276).

192. We have already quoted Vâtsyâyana and Kautilya. The former defines *nyâya* as *pramânair-artha-parîkṣanam* ("to scrutinize an object by means of logical proof" — S. N. Dasgupta's translation who adduces Vâcaspati's *Nyâya-vâstika-tâtparyâ-tîkâ* I, 1, 1 as saying the same). Hacker translates: "das Nachprüfen eines Gegenstandes mit Erkenntnismitteln".
193. Cf. Potter (1977: Vol. II, p. 2).
194. From the root *vad*. The verb *vadati* means to speak, say, sing, raise one's voice, etc. and has numerous parallels in Greek, etc. Cf. Mayrhofer (1956/1980).
195. NBh I, 2, 1.
196. NS I, 1, 1.
197. Cf. Dasgupta (1969: II, p. 377).
198. Cf. Jogendranath Bagchi (1969).
199. Cf. SB XIII, 2, 6, 9 sq. as well as TB III, 9, 5, 1 sq. where *brahmodya* refers to apparently 'non theological' discussion, but for the sake of *brahman* splendour. Cf. SB XIII, 5, 2, 22: *sarvâptir vâ eṣâ vâcah yad brahmodyan*. (*Brahmodya* obtains all speech and obtains all).
200. Cf. K. S. Murthy (1985: 3).

Bibliography

Abhyankar and Limaye. Cf. Bartṛhari (1965).

Actas del V Congreso Eucaristico Internacional. Cf. Panikkar (1971b) and Raju (1971).

Anantakrishnan, S. V. (Ed.). Cf. Chari (1967).

Anirvan (1958). "Vedic Exegesis" in *(The) Cultural Heritage of India*, Vol. I, pp. 314–317.

—— (1983). *Buddhiyoga of Gîtâ and other Essays*, New Delhi (Biblia Impex), which is a reprint of already published papers.

Anyanwu, K. C. (1981). Cf. Ruch (1981).

Archiv für Begriffsgeschichte. Cf. Niewöhner (1980).

Bagchi, P. C. (1969). "Evolution of the Tantras" in *(The) Cultural Heritage of India*, Vol. IV, pp. 211–226.

Bagchi, Yogendranath (1969). "The Art of Philosophical Disputation" in *(The) Cultural Heritage of India*, Vol. III, pp. 562–580.

Bahm, A. J. (1979). *The Philosopher's Model*. Westport, Conn.: Greenwood Press.

Benz, E. (1951). *Indische Einflüsse auf die frühchristliche Theologie*. Wiesbaden: Steiner.

Benvéniste, Émile (1969). *Le vocabulaire des institutions indo-européenes*. Paris: Minuit. 2 vols.

Berchert, H. and Simson, G. V. (Eds.) (1979). *Einführung in die Indologie*. Darmstadt: Wissenschaftliche Buchgesellschaft.

Bhartṛhari (1965). *Vâkyapadîya of Bhartṛhari* (Sanskrit texts), edited by K. V. Abhyankar and V. P. Limaye. University of Poona Sanskrit and Prakrit Series — Vol. II. Poona.

Bhartṛhari. Cf. K. A. S. Iyer (1969) and (1971–1976); and Coward (1976)

Bhattacharyya, Haridas (Ed.) (1953). Cf. *(The) Cultural Heritage of India* (1953), 1956).

Bhattacharyya, Kalidas (1955). "The Business of Philosophy." In *Proceedings of the Thirtieth Indian Philosophical Congress*, Nagpur, pp. 175–187.

—— (1956). *Studies in Philosophy*. Calcutta: Progressive Publishers. 2 vols.

Bhattacharyya, Kalidas and Mazumdar, A. K. (Eds.) (1965/1966). *Indian Philosophical Congress*. West Bengal: Santiniketan.

Biardeau, Madeleine (1964). *Théorie de la connaisance et philosophie de la parole dans le brahmanisme classique*. Paris: Mouton.

Biardeau, M. and Malamoud, C. (1976). *Le sacrifice dans l'Inde ancienne*. Paris: P. U. F.

Bolle, K. (1979). *The Bhagavadgîtâ*. Berkeley: University of California Press.

Burr, R. (1980). *Handbook of World Philosophy: Contemporary Developments since 1945*. Westport, Conn.: Greenwood Press.

Chakravarti, Chintaran (1963). *Tantras: Studies on Their Religion and Literature*. Calcutta: Punthi Pustak.

Chari, C. T. K. (1967). "Tradition and Modernity in Recent Indian Philosophy". In S. V. Anantakrishnan

et al. (Eds.), *Indian Today: An Exploratory Discussion*, pp. 122–138. Madras: CLS (reprint 1968).
Chubb, J. N. (1949). "Philosophy and Sadhana" I (Symposium). In Proceedings of the XXIV Session of the *Indian Philosophical Congress*, Patna, pp. 1–12. Edited by N. V. Banerjee. Cf. Dandekar (1949).
—— (1956). "Does Indian Philosophy Need Re-Orientation?" II. In Proceedings of the XXXI *Indian Philosophical Congress*, Annamalainagar, pp. 79–97. Cf. Malkani (1956); Varadachari (1956).
Comparative Civilizations Review. Cf. note 37.
Coward, Harold (1976). *Bhartṛhari*. Boston: G. K. Hall.
—— (1980). *The Sphoṭa Theory of Language*. Delhi: Motilal Banarsidass.
(The) *Cultural Heritage of India*: Vol. I, "*The Early Phases*" (1958) (reprint 1970); Vol. II, "*Itihâsas, Purâṇas, Dharma and Other Śâstras* (1962) (reprint 1969); Vol. III, "*The Philosophies*" (1953) (reprint 1969); Vol. IV, "*The Religions*" (1956) (reprint 1969). (Five volumes, second Edition) 1958, *sq*. First edition 3 vols. 1937. Calcutta: Ramakrishna Mission Institute of Culture.
Dandekar, S. V. (1949). "Philosophy and Sadhana". In *Proceedings of the XXIV Session of the Indian Philosophical Congress*, Patna, pp. 13–21 Cf. Chubb (1949).
Daniel, E. Valentine. Cf. Keyes and Daniel (1983).
Darlquist, A. (1962). *Megasthenes and Indian Religion*. Stockholm: Almqvist & Wiksell.
Das, Rasvihary (1956). "What is Philosophy?" Presidential Address, *Proceedings of the Thirty-first Indian Philosophical Congress*, Amalnagar, pp. 3–16.
Dasgupta, S. N. (1953). Cf. (The) *Cultural Heritage of India* (1953), "Introduction", pp. 3–26.
—— (1969). *A History of Indian Philosophy*. Cambridge: Cambridge University Press. 5 vols. (1922–1955) reprint.
Deussen, Paul (1894/1917). *Allgemeine Geschichte der Philosophie*. Leipzig: F. A. Brockhaus. 6 vols.
Deutsch, E. Cf. Larson and Deutsch (1988).
Diels, H. and Kranz, W. (1960). *Die Fragmente der Vorsokratiker*. Berlin: Wiedmann. 3 vols. (several editions).
Dravid, Raja Ram (1972). *The Problem of Universals in Indian Philosophy*. Delhi: Motilal Banarsidass.
Dussel, E. (1977). *Filosofía de la Liberación*. Mexico: Edicol. English translation: *Philosophy of Liberation*. Maryknoll, N. Y.: Orbis, 1985.
Dumézil, G. (1958). *L'idéologie tripartite des Indo-Européêns*. Bruxelles. (Collection Latomus Nr. 31).
—— (1960). *Hommages à Georges Dumézil* (Bruxelles Collection Latomus, Vol. 45) (containing an almost complete list of his works).
—— (1968/71/73). *Mythe et epopée*. Paris: Gallimard. 3 vols.
Ebner, F. (1963). *Schriften*. München: Kösel. 3 vols.
Edgerton, F. (1965). *The Beginnings of Indian Philosophy*. London: Allen & Unwin.
Edwards, Rem B. (1972). *Reason and Religion*. New York: Harcourt, Brace, Jovanovich.
Elder, G. R. (1976). "Problems of Language in Buddhist Tantra". *History of Religions* XV/3: 231–250.
Eliade, Mircea (1969). *Yoga: Immortality and Freedom*. Princeton, N. J.: Princeton University Press, 2nd ed. Translation of the French (1954) edition.
Frauwallner, E. (1953/56). *Geschichte der indischen Philosophie*. Salzburg: O. Müller. 2 vols.
Frazer, D. (1974). *African Art as Philosophy*. New York: Interbook.
Flew, Antony (1971). *An Introduction to Western Philosophy*. London: Thames & Hudson.
—— (1979). *A Dictionary of Philosophy*. London: Pan Books.
Gadamer, H. G. (1972). *Wahrheit und Methode*. Tübingen: Mohr, 3rd. ed. (1st. ed. 1960).
—— (1975). *Truth and Method*. New York: Seabury. Translation of the second edition of 1965.
García Yebra, Valentín (1982). *Teoria y pràctica de la traducción*. Madrid: Gredos. 2 vols.
Gnoli, Raniero (1960). *Abhinavagupta: Essenza dei Tantra*. Torino: Boringhieri.
Gonda, J. (1950). *Notes on Brahman*. Utrecht: J. L. Beyers.
—— (1963). "The Indian Mantra." *Oriens* 16: 244–297.
González Faus, J. I. (1988). "La "filosofia de la vida' de Jesús de Natzaret". *Questions de Vida Cristiana* (Montserrat), Nr. 141: 68–80.
Gosh, A. B. (1956). "The Spirit and Culture of the Tantras", in *(The) Cultural Heritage of India*, Calcutta, Vol. IV, pp. 241–251.
Habermas, Jürgen (1968). *Erkenntnis und Interesse*. Frankfurt: Suhrkamp. English translation: *Knowledge and Human Interests*. Boston: Beacon, 1971.

Hacker, Paul (1958). "Anvîkṣikî". *Wiener Zeitschrift für die Kunde Südasiens und Archiv für indische Philosophie* II: 54–81.
____ 1965. "Dharma im Hinduismus". *Zeitschrift für Missionswissenschaft und Religionswissenschaft* 49/1: 93–106.
Halbfass, Wilhelm (1976a). "Indien und die Geschichtsschreibung der Philosophie". *Philosophische Rundschau* XXIII: 104–131.
____ (1976b). "Zum Begriff der Substanz (*dravya*) im Vaiśeṣika". *Wiener Zeitschrift für die Kunde Südasiens und Archiv für indische Philosophie* XX: 141–166.
____ (1979). "Philosophie" in Berchert and Simson (1979), pp. 138 –158.
____ (1981). *Indien und Europa: Perspektiven ihrer geistigen Begegnung*. Basel and Stuttgart: Schwabe.
____ (1988). *India and Europa: An Essay in Understanding*. New York: State University of New York Press. Revised and enlarged translation of (1981).
Handique, K. K. (1962). "Classical Sanskrit as a Vehicle of Indian Culture" in *(The) Classical Heritage of India*, Vol. II, p. 9.
Hauer, J. W. (1927). *Der Vrâtya: Untersuchungen über die nichtbrahmanische Religion Altindiens*. Stuttgart: Kohlhammer.
____ (1958). *Der Yoga: Ein indischer Weg zum Selbst*. Stuttgart: Kohlhammer.
Heidegger, M. (1953, 1966). *Einführung in die Metaphysik*. Tübingen: Niemeyer.
____ (1954). *Was heisst Denken*? Tübingen: Niemeyer.
____ (1956). *Was ist das—die Philosophie*? Pfüllingen: Neske.
____ (1962). *Die Frage nach dem Ding*. Tübingen: Niemeyer. 2nd. ed. 1975.
____ (1967). "Das Ding" in *Vorträge und Aufsätze*, Teil II, pp. 37–55. Pfüllingen: Neske.
____ (1970). *Phänomenologie und Theologie*. Frankfurt/Main: V. Klostermann.
Heimann, Betty (1937). *Indian and Western Philosophy: A Study in Contrast*. London: Allen & Unwin.
____ (1951). *The Signifiance of Prefixes in Sanskrit Philosophical Terminology*. Royal Asiatic Society Monographs, Vol XXV.
Hiriyanna, M. (1932). *Outlines of Indian Philosophy*. London: Allen & Unwin.
____ (1957). *Indian Philosophical Studies*. Mysore: Kavyalaya.
Hountondji, J. P. (1971). "Le probléme actuel de la philosophie africaine" in Klibansky (1971), Vol. IV, pp. 613 – 621.
____ (1976). *Sur la 'philosophie africaine'*. Paris: F. Maspero (with Bibliography).
Hull, R. F. C. (1950). *The Dawn of Philosophy*. London: Routledge and Kegan Paul. Translation and edition of Misch (1926).
Indian Philosophical Congress. Cf. notes 27 and 29.
Indian Philosophical Quarterly. Cf. note 26.
International Philosophical Quarterly. Cf. Odera (1983).
Iyer, K. A. Subramania (1969), *Bhartṛhari: A Study of the Vâkyapadîya in the Light of the Ancient Commentaries*. Poona: Deccan College.
____ (1971/74/76). *The Vâkyapadîya of Bhartṛhari* (several volumes published by Poona: Deccan College, and Delhi: Motilal Banarsidass.
Iyer, V. S. Cf. Nikhilânanda (1949).
Jano. Cf. Panikkar (1986a).
Jiménez Duque, Baldomero (Ed.). Cf. Panikkar (1969).
Kabir, Humayun (1971). "Philosophy in India" in Klibansky (1971), Vol. IV, pp. 549–563.
Kagame, A. (1956). *La philosophie bantu-rwandaise de l'être*. Bruxelles: Memoire de l'Academie Royale des Sciences d'Outre-Mer.
____ (1971). "L'ethno-philosophie des 'Bantu'" in Klibansky (1971), Vol. IV, pp. 589–612.
Kane, P. V. (1968). *History of Dharmasâstra*, 5 vols. (7 parts) Poona: Bhandarkar Oriental Research Institute, 2nd. ed. 1968. (1930 to 1962).
Keith, A. B. (1970/1925). *The Religion and Philosophy of the Veda and Upanishads* (Harvard Oriental series Vols. 31 and 32). Cambridge: Harvard University Press, 1925. Reprint Motilal Banarsidass, 1970.
Keyes, Charles and Daniel, E. Valentine (Eds.), (1983). *Karma: An Anthropological Inquiry*. Berkeley:

University of California Press.

Klibansky (Ed.) (1971). *Contemporary Philosophy: A Survey*. Firenze: La nuova Italia. 4 vols. (1968–1971).

Klostermaier, Klaus K. (1984). *Mythologies and Philosophies of Salvation in the Theistic Traditions of India*. Waterloo, Ontario: W. Laurier University Press.

Kranz, W. Cf. Diels (1960).

Krings, H., Baumgartner, H.M. and Wild, Ch. (Eds.) (1973). *Handbuch philosophischer Grundbegriffe*. München: Kösel. 3 vols.

Krishna, Daya (1971). "Philosophy in India" in Klibansky (1971), Vol. IV, pp. 564–577.

—— (1986). "Comparative Philosophy: What is it and What it Ought to be". In *Diogenes* 136 (Winter): 58–69. Reprinted in Larson and Deutsch (1988), pp. 71–83.

Laleau, León (1969). "Trahison". In *Anthologie de la nouvelle poésie négre*. Paris: P.U.F.

Larson, G.J. (1980). "The Format of Technical Philosophical Writing in Ancient India: Inadequacies of Conventional Translations". *Philosophy East* and *West* XXX/3: 375–380.

Larson, G.J. and Deutsch, E. (Eds.) (1988). *Interpreting Across Boundaries: New Essays in Comparative Philosophy*. Princeton University, N.J.: Princeton University Press.

Cf. Krishna (1986), Nakamura (1988), Panikkar (1988a), Rosemont (1988), Staal (1988).

Limaye, V.P. Cf. Bhartṛhari (1965).

Levi, Silvain (1898/1966). *La doctrine du sacrifice dans les Brâhmaṇas*, 2nd. edition. Paris: P. U. F.

Mâdhavânanda, Swami (Translator of) (1940). *Bhâṣâ-Pariccheda with Siddhânta-Muktâvalî* by Viśvanâtha Nyâya Pañcânana (Bhaṭṭacârya) 2nd. ed. 1954. Calcutta: Advaita Ashrama. Cf. Roer (1850).

Malamoud, C. (1976). Cf. Biardeau and Malamoud (1976).

Malkani, G.R. (1956). "Does Indian Philosophy Need Re-Orientation?" I. In Proceedings of the XXXI *Indian Philosophical Congress*, Annamalainagar, pp. 71–78. Cf. Chubb (1956); Varadachari (1956).

Matilal, B.K. (1986). *Perception: An Essay on Classical Indian Theories of Knowledge*. Oxford: Claredon.

Maximilien, G. Cf. note 145.

Mayrhofer, Manfred (1956/1980). *Kurzgefasstes Etymologisches Wörterbuch des Altindischen*. Heidelberg: Carl Winter. 4 vols.

Mazumdar, A.K. (1965/1966). Cf. Bhattacharyya and Mazumdar (1965/66).

Mbiti, J.S. (1970). *African Religions and Philosophy*. Garden City, N.Y.: Doubleday/Anchor Book (published in 1969 by Praeger Publishers).

Mehta, J.L. (1974). "The Problem of Philosophical Reconception in the Thought of K.C. Bhattacharyya". *Philosophy East and West* XXIV/1 (January): 59–70.

Mercier, A. Cf. Panikkar (1988b).

Miller, J. (1985). *The Vision of Cosmic Order in the Vedas*. London: Routledge and Kegan Paul.

Misch, Georg (1926). *Der Weg in die Philosophie*. Edited and translated into English by Hull (1950).

Mishra, Kamalakar (1981). *Significance of the Tantric Tradition*. Varanasi: Ardhanarîsvara Publications.

Miśra, Vâcaspati (1933). Cf. Śâstrî and Raja (1933).

Monnier Williams (1851). *English-Sanskrit Dictionary* (3rd. edition 1971). Delhi: Motilal Banarsidass.

—— (1899). *Sanskrit-English Dictionary*. Delhi: Motilal Banarsidass. (reprint 1974).

Moore, C. A. Cf. Radhakrishnan and Moore (1957).

Muirhead, J.M. and Radhakrishnan. (Eds.) (1952b). *Contemporary Indian Philosophy*. London: Allen & Unwin (2nd. impression).

Mukhopadhyaya, G.G. (1960). *Studies in the Upanishads*. Calcutta: Sanskrit College.

Murti, T.R.V. (1953). "Rise of the Philosophical Schools" in *(The) Cultural Heritage of India*, Vol. 3. pp. 27–40.

—— (1955). *The Central Philosophy of Buddhism*. London: Allen & Unwin.

Murty, K. Satchidananda (1959). *Revelation and Reason in Advaita Vedanta*. Waltair Andhra University.

—— (1985). *Philosophy in India: Tradition, Teaching and Research*. Delhi: Motilal Banarsidass.

Nakamura, Hajime (1964). *Ways of Thinking of Eastern Peoples* (revised English translation edited by P.P. Wiener). Honolulu: University of Hawaii Press. (reprint 1985).

____ (1975). *Parallel Developments: A Comparative History of Ideas.* Tokyo: Kodancha.

____ (1983). *A History of Early Vedânta Philosophy.* Delhi: Motilal Banarsidass.

____ (1988). "The Meaning of the Terms 'Philosophy' and 'Religion' in various Traditions" in Larson and Deutsch (1988), pp. 137–151.

Nasr, S. H. (1981). *Knowledge and the Sacred.* New York: Crossroad.

Neske, G. (Ed.) (1977). *Erinnerung an Martin Heidegger.* Pfüllingen: Neske Cf. Panikkar (1977b).

Niewöhner, Friedrich (1980). "Voriiberlegungen zu einem Stichwort, 'Philosophie jüdische'". *Archiv für Begriffsgeschichte* XXIV/2: 195–220.

Nikhilânanda, Swami (Translator and editor) (1949). *The Mândukyopanishad with Gauḍâpada's Kârika and Śaṅkara's Commentary.* Mysore: Ramakrishna Ashrama, 1st, ed. 1936. With a foreword by : V. S. Iyer.

Odera Oruka, H. (1983). "Sagacity in African Philosophy". *International Philosophical Quarterly*, Vol. XXII 4, No. 92 (December): 383–393.

O'Flaherty, Wendy D. (Ed.) (1980). *Karma: An Rebirth in Classical Indian Tradition.* Berkeley: University of California Press.

Oriens. Cf. Gonda (1963).

Otto, Rudolf (1930). *Die Gnadenreligionen Indiens und das Christentum.* Gotha: L. Klotz.

Panikkar, R. (1956). "Die existentielle Phänomenologie der Wahrheit". in *Philosophisches Jahrbuch der Görresgesellschaft* 64: 27–54. There is a Spanish revised version in Panikkar (1971a), 207–248.

____ (1961). "La Misa como *consecratio temporis*: la tempiternidad" in *Sanctum Sacrificium* (Actas del V Congreso Eucarístico Nacional) Zaragoza, pp. 75–93.

____ (1963a). *Humanismo y cruz.* Madrid: Rialp.

____ (1963b). "Sur l'herméneutique de la tradition dans l'hindouisme". *Archivio di filosofia* (E. Castelli, Ed.), pp. 343–370.

____ (1969). "Algunos aspectos de la espiritualidad hindú". In B. Jiménez Duque and L. Sala Balust (Eds.), *Historia de la espiritualidad*, Vol. IV, pp. 431–542. Barcelona: J. Flors.

____ (1970). *Le mystére du culte dans l'hindouisme et le christianisme.* Paris: Cerf.

____ (1971a). *Misterio y revelación.* Madrid: Marova. "Necesidad de una orientación de la filosofía india", pp. 51–82.

____ (1971b). "Die Philosophie in der geistigen Situation der Zeit". *Akten des XIV Internationalen Kongresses* für Philosophie (1968), Vol. VI, pp. 75–87. Wien: Herder.

____ (1977a). *The Vedic Experience.* Berkeley-Los Angeles: University of California Press.

____ (1977b). "Eine unvollendete Symphonie" in G. Neske (1977), pp. 173–178.

____ (1978[1984]). *The Intrareligious Dialogue.* New York: Paulist Press. New Edition, Bangalore: Asian Trading Corporation, 1984.

____ (1979a [1983]). *Myth, Faith, and Hermeneutics.* New York: Paulist Press. New Edition, Bangalore: Asian Trading Corporation, 1983.

____ (1979b [1973]). *Culto y secularización.* Madrid: Marova. Second edition of *Worship and Secular Man.* New York: Orbis, 1973.

____ (1980a). "Words and Terms" in *Esistenza, mito, ermeneutica*, M.M. Olivetti (Ed.), *Archivo di filosofia*, Roma, Vol. II, pp. 117–133.

____ (1980b). "Aporias in the Comparative Philosophy of Religion". *Man and World* XIII/3–4: 357–383.

____ (1981). *The Unknown Christ of Hinduism*, new edition. London: DLT and New York: Orbis.

____ (1985). "Global Perspectives: Spiritualities in Interaction". *Journal of Dharma* X, I (January–March): 6–17.

____ (1986a). "Medicina y religión", *Jano, Medicina y Humanidades* XXXI number 737 (July): pp. 12–48.

____ (1986b). "The Threefolfd Linguistic Intrasubjectivity". *Archivio di Filosofia* LIV number 1–3: pp. 593–606.

____ (1987). "The Invisible Harmony: A Universal Theory of Religion or a Cosmic Confidence in Reality?". In L. Swidler (Ed.), *Toward a Universal Theology of Religion*, pp. 118–153. Maryknoll, N. Y.: Orbis.

____ (1988a). "What Is Comparative Philosophy Comparing?" in Larson and Deutsch (1988),

pp. 116–136.
—— (1988b). "The Unknown Knower. Five Crosscultural Theses on Consciousness". In A. Mercier's Festschrift *Reflexions sur la libertè humaine*, pp. 133–159, edited by Maja Svilar, Bern: Lang, 1988.
—— (1988c). "Logomitía e pensiero occidentale". In P. Pellegrino (Ed.), *Rosmini e l'Illuminismo*, pp. 63 – 79. (Atti del XXI Corso della 'Cattedra Rosmini') (1987). Stresa: Sodalitas.
—— (1988d). "Indic and Western Approaches to Philosophy" read at the Session of *The Association of Christian Philosophers of India* (Madras) to be published shortly.
—— (1989). *The Silence of God: The Answer of the Buddha*. New York: Orbis.
Pânini (1981). *The Ashtâdhyâyî*, edited and translated into English by Srisa Chandra Vasu. Delhi: M. Banarsidas (reprint 1977). 2 vols.
Patel, Dadubhai N. (1978). *The Real Essence of Tantra*. Bombay: Yogi Divine Society.
Plott, John C. (1977). *Global History of Philosophy*, Vol I. Delhi: Motilal Banarsidass.
Pokorny, Julius (1959). *Indogermanisches etymologisches Wörterbuch*. Bern and München: Francke. 2 vols.
Potter, K. (Ed.) (1977). *Encyclopedia of Indian Philosophies. Indian Metaphysics and Epistemology: The Tradition of Nyâya-Vaisesika up to Gangesa* (Vol. II). Delhi: Motilal Banarsidass.
(The) Philosophical Quarterly. Cf. note 26.
Philosophische Rundschau. Cf. Halbfass (1976a).
Philosophisches Jahrbuch der Görresgesellschaft. Cf. Panikkar (1956).
Philosophy East and West. Cf. note 37.
Philosophy Today. Cf. Sheehan (1976).
Radhakrishnan, S. (1940). *Indian Philosophy*. London: Allen & Unwin. 2 vols.
—— (1952a). *History of Philosophy Eastern and Western* (Editor). London: Allen & Unwin. 2 vols.
—— (1952b). Cf. Muirhead and Radhakrishnan (1952).
—— (1953). *The Principal Upaniṣads*. London: Allen & Unwin (reprint 1968).
—— (1960). *The Brahma Sûtra: The Philosophy of Spiritual Life*. London: Allen & Unwin.
Radhakrishnan, S. and Moore, C.A. (eds.) (1957). *A Sourcebook in Indian Philosophy*. Princeton: Princeton University Press.
Radin, P. (1927). *Primitive Man as Philosopher* (first impression by D. Appleton). New edition New York: Dover, 1957.
Raghavendrachar, H.N. (1953). "Madhva's Brahma-Mîmâṁsâ" in *(The) Cultural Heritage of India*, Vol. 3, pp. 313 – 322.
Rahola, Pilar (1988). "Literatura africana negra". in *Revista de Catalunya* 24 (November): 111–130.
Raja, C. Kunhan (1933). Cf. Śâstrî and Raja (1933).
—— (1960 [1974]). *Some Fundamentals Problems in Indian Philosophy*. Delhi: Motilal Banarsidass, 2nd edition 1974.
Raju, P.T. (1971). "Man, Logos, and Philosophy" in the *Akten des XIV. Internationalen Kongresses für Philosophie* (2–8. November 1968) Vol. VI, pp. 103–116. Vienna: Herder.
—— 1985. *Structural Depths of Indian Thought*. New Delhi: South Asian Publishers.
Ramírez, J. M. (Santiago) (1972). *De hominis beatitudine*. Madrid: S.I.C., 2nd. ed. 5 vols.
Reserches Philosophiques Africaines. Cf. note 42.
Richter, J. (1971). *Historisches Wörterbuch der Philosophie*. Basel: Schwabe. 7 vols. until 1989.
Roer, E. (1850). *Division of the Categories of Nyâya Philosophy*, with a commentary by Visvanâtha Nyâya Pañchâna. Calcutta: Bibliotheca Indica. Cf. Mâdhavânanda (1940).
Rosemont, J.R. Henry (1988). "Against Relativism" in Larson and Deutsch (1988), pp. 36–70.
Ruch, E.A. and Anyanwu, K.C. (1981). *African Philosophy: An Introduction to the Main Philosophical Trends in Contemporary Africa*. Rome: Catholic Book Agency.
Said, E.W. (1978). *Orientalism*. New York: Random House.
Sala Balust, Luis (Ed.) (1969). Cf. Panikkar (1969).
Śantideva, (1979). *Bodhisattvachâryavatâra*, translated by Stephen Batchelor: *A Guide to the Bodhisattva's Way of Life*. Dharamsala: Library of Tibetan Works and Archives (reprint 1985).
Sarup, Laksham (Editor and translator) (1967). *The Nighantu and the Nirukta*. Delhi: Motilal Banarsidass (reprint).
Śâstrî, S.S.S. and Raja, C.K. (Eds.) (1933). *Bhâmatî*, sub-commentary of Vâcaspati Miśra. Madras:

Theosophical Publishing House.
Śâstri, Gaurinath (1959). *The Philosophy of Word an Meaning*. Calcutta: Sanskrit College.
____ (1980). *A Study in the Dialectics of Sphoṭa*. Delhi: Motilal Banarsidass.
Sedlar, J. W. (1980). *India and the Greek World*. Totowa, N. J.: Publisher.
Sharma, Chandradhar (1960). *A Critical Survey of Indian Philosophy*. Delhi: Motilal Banarsidass.
Sharvananda (1958). "The Vedas and Their Religious Teachings", in *(The) Cultural Heritage of India*, Vol. I., pp. 182–198.
Sheenan, Thomas (1976). *Philosophy Today* (Special issue after Heidegger's death).
Simson, G. von. Cf. Berchert and Simson (1979).
Smith, John E. (1974). Cf. Mehta (1974). "Commentary" on Mehta's paper, pp. 89–93.
Solowjew, Wladimir (1889). *La Russie et l'Eglise universelle*. Paris: (general modern edition).
Srinivanasachari, P. N. (1953). "The Visistâdvaita of Râmânuja" in *(The) Cultural Heritage of India*, Vol. III, pp. 300–312. (reprint 1969).
Srivastava, R. S. (1984). *Contemporary Indian Philosophy*. Ranchi: Sharda.
Staal, Frits (1988). "Is There Philosophy in Asia?" in Larson and Deutsch (1988), pp. 203–229.
Steiner, George (1975). *After Babel*. London: Oxford University Press.
Svilar, Maja. Cf. Panikkar (1988b).
Swidler, Leonard. Cf. Panikkar (1987).
Tempels, P. (1945). *La Philosophie bantue*. Paris: Présence Africaine (3rd. impression 1961).
Thibon, Gustave (1974). *L'ignorance étoilée*. Paris: Fayard.
Thite, G. U. (1975). *Sacrifice in the Brâhmana Texts*. Poona: Poona University Press.
Towa, M. (1979). *L'idée d'une philosophie négro-africaine*. Yaoundé, Cameroun: CLE.
Trivedi, Ramachandra (1965/1966). "Sri Aurobindo's Conception of Philosophy" in *Proceedings of the Indian Philosophical Congress*, pp. 111–123.
Vadakkekara, C. M. (Ed.) (1981). *Divine Grace and Human Response* (Studies in Christian and Hindu Spirituality — 2). Kumbalgud, Bangalore: Asirvanam.
Varadachari, K. C. (1956). "Does Indian Philosophy Need Re-Orientation?" III. In *Proceedings of the XXXI Indian Philosophical Congress*, Annamalainagar, pp. 99–103. Cf. Malkani (1956); Chubb (1956).
Vasu, S. Ch. Cf. Pânini (1891).
Vesci, U. M. (1985). *Heat and Sacrifice in the Vedas*. Delhi: Motilal Banarsidass.
Wiener Zeitschrift für die Kunde Südasiens und Archiv für indische Philosophie. Cf. Hacker (1958); Halbfass (1976b).
Zaehner, R. C. (1962). *Hinduism*. London: Oxford University Press.
____ (1969). *The Bhagavad Gîtâ*. Oxford: Clarendon.
Zilberman, David B. (1988). *The Birth of Meaning in Hindu Thought*. Boston: (Boston Studies in the Philosophy of Science).
Zimmer, Heinrich (1946). *Myths and Symbols in Indian Art and Civilization*. Washington, D.C. : Bollingen.
____ (1951). *Philosophies of India* (Bollingen Series XXVI). Princeton, N.J.: Princeton University Press.

Emerging new approaches in the study of classical Indian philosophy

DAYA KRISHNA

Indian philosophy by and large has been treated up to now in an antiquarian spirit, something belonging to past history which has no relevance to current concerns of philosophical thought in the world. Further, it has readily been assumed that the past traditions of philosophical thought had already reached a dead end and that there were no living representatives of that thought in modern times or that those who existed were concerned primarily with the preservation and safe-guarding of past knowledge in these domains. The intellectual and cultural domination of the west as well as the establishment of educational institutions on western models, along with the training of students in western universities, ensured that the paradigmatic model about philosophy and what was to count as "philosophical" was determined by the west. Also, as English was the language of both national and international intellectual life, it automatically ensured that those who did not know the language or could not write or converse in it would become intellectually non-existent even in their own country. This resulted in a situation where only those persons were paid attention to even in the field of Indian philosophy who wrote in English or in other major European languages. If one only knew Sanskrit and wrote in it, one had very little chance of being treated as a serious scholar or thinker even in those areas of philosophy which specifically pertained to that which was exclusively contained in the Sanskrit texts themselves. The scholar who could not read, write or speak in English thus gradually became invisible in the philosophical scene during the course of this century in India.

There were other reasons for the gradual disappearance of the visibility of the traditional Indian philosopher from the scene of Indian philosophy in India. The institutional segregation, the economic discrimination, the differences in the style of living and the radical diversity in the intellectual and cultural ethos of the two groups led to a situation where the privileged moderns were almost unaware of the existence of the deprived traditionalists, while the latter developed all the attitudes which any underprivileged minority which is discriminated against does. The deeper reason, however, was the attitude to all non-western traditions of knowledge sedulously developed by historians of thought in the west and internalised by intellectuals of most other countries which had not only modelled their educational system on those of the West, but also used West-european languages as their medium of instruction and taught through text-books written by western authors in those languages. It thus came to be an accepted belief, unquestionably held by all

Contemporary philosophy. A new survey. Vol. 7, 69–82.

educated persons, that not only was the west the source of all knowledge but also that only that would be regarded as genuine knowledge which was written in a western language, and that, too, by a western author. This alone may explain the strange fact that not only works written in various Indian languages are generally ignored by Indian intellectuals but even those which are written in English by Indians are hardly paid any attention by their colleagues in the same field of study. This has happened, surprisingly, even in such an area of study as Indian philosophy, where a large mass of significant material continues to be written in Sanskrit and other regional languages of the country.

The problem is further complicated by the fact that the model of what "philosophy" or any other cognitive discipline is supposed to be, is determined not only by the western paradigms but also by the west's own image of itself and that of other cultures which it would like everyone to accept unquestioningly. Thus, time and again, the question has been raised whether traditional Indian philosophy can be considered as "philosophy" in the western sense of the term. No one, of course, has asked if western philosophy can be regarded as "philosophy proper" that is, as "*Darśana*", or "*Ānavīkṣikī*", or "*Tattva-jñāna*", the Indian terms for what goes by the name of "philosophy" in the west. The very fact that such a question has not even occurred to minds engaged in the philosophical enterprise in modern India is a witness to the essential asymmetry in the cognitive status which is accorded to the two traditions not only in the west, but in India also.

The contrasting features ascribed to the two philosophical traditions usually are that, while the western philosophical tradition is primarily concerned with a rational consideration of things and believes in the primacy of reason, the Indian philosophical tradition believes in the primacy of the revealed word and the authority derived from it. It also sees philosophy as an ancillary activity to the achievement of a trans-philosophical end, that is, *mokṣa*. Besides these, there are a host of other alleged differences, such as being other-worldly, pessimistic, individualistic, etc.; but basically they are all derived from these. A questioning of these basic characterisations may be regarded as one of the chief features of the new way of looking at traditional Indian philosophy (Daya Krishna, 1965, 1966, 1984).[1] And, following from this, there has been the widespread feeling that in case this is true, classical Indian philosophical thought should have something to contribute to the discussion of contemporary philosophical problems as they are apprehended today.

One widely prevalent way of doing this formerly was to discover parallelisms in the Indian philosophical tradition to the problems and the patterns of their solution in the western tradition of philosophising. This, obviously, was not very intellectually exciting, either to the Indian or the Western philosopher, for the simple reason that if something had already been thought, it could hardly be of much interest to the active, independent thinker in the present, or even to a student of comparative philosophy. For, the only thing that could be of possible interest to a living thinker is the difference, either in what was apprehended as a problem or in the direction of its solution, rather than in the so-called similarities between the diverse traditions[2] (Daya Krishna, 1984b).

All cognitive traditions, whether in philosophy or in any other discipline, stake a claim to universal validity. The fact that they have arisen at a particular place or time or grown in the confines of a specific cultural tradition does not in any way stand in the way of the claim being there, as it is intrinsic to the very nature of the enterprise itself. True, there are realms of knowledge which seem to show a certain cumulativeness of growth, at least over certain periods of discernible time, and where the later phase in the growth of knowledge seems definitely to supersede the former. But philosophy is certainly not such a field of knowledge along with many other disciplines belonging to the social sciences and the humanities. In many of these, the past thinkers are not superseded by those in the later times. Rather they continue to be perennially relevant in a sense in which they can hardly be so in any of the natural sciences or even in some of the disciplines belonging to other realms. The fact is well known, but somehow most western thinkers tend to accept this only for the Greeks and those others who, in their opinion, belong to the western tradition of thought in their subjects. That is one reason why, though Plato and Aristotle are accorded full honours in any contemporary discussion of philosophical problems, no such consideration is extended to even the greatest of thinkers outside the western tradition. And this in spite of the heroic attempts at translation of some of the best philosophers outside the western tradition into languages that are easily accessible to the bibliography-conscious western scholar. Perhaps the segregation of Indological and Sinic studies in Research Institutions of the west may have been partly responsible for this, as the study of philosophical traditions of India and China was relegated to their care and treated as their exclusive concern. But, then, this segregation itself was the result of the mental attitude which regarded the intellectual contributions of all non-western civilizations as being of no relevance to contemporary cognitive concerns and hence only of antiquarian interest.

Yet, whatever the justifications for this attitude on the part of western scholars, there could be none for the non-western ones. However, surprisingly enough, most scholars belonging to cultures other than the western one accepted this as the natural state of affairs and considered their own rich intellectual heritages as completely irrelevant to their cognitive enterprise, excepting, of course, the Greek Masters whom the western scholarship has already accepted.

It is this unconscious acceptance of the western way of looking at their own intellectual traditions which has come under increasing scrutiny of a significant number of Indian thinkers in general, and those belonging to the field of philosophy in particular. It has increasingly begun to be asked why philosophers in contemporary India do not bring the classical Indian philosophical thought to bear on the discussion of contemporary philosophical issues. One example of this agonising concern is the special issue of the *Indian Philosophical Quarterly* (Poona; Oct.-Dec., 1984b) devoted to this very problem entitled "Swaraj in Ideas", containing the responses of a large number of Indian intellectuals to an address delivered under this very title and later published by K.C. Bhattacharyya, the most outstanding Indian philosopher of the first half of this century, as early as 1929. Also, if classical Indian philosophical thought was really vibrant and alive at one time and had the strength in it, even in contemporary times, to respond critically and creatively to philosophical issues in contemporary philosophy then why should it not be capable of independent growth and development on its own, even in modern times? Anything that is alive must be capable of modification, growth and development along new lines — a situation that can only be realised if the scholars who are deeply immersed in the traditional philosophy in the traditional way begin to look at it in a new way. And this can happen only if they begin to ask new questions, articulate their own dissatisfactions with the traditional philosophy as they apprehend it and suggest their own improvements in respect of the deficiencies they find in the traditional formulation of the problems or of their solutions.

But for all this to take place, there had to be a fairly large number of traditional practitioners of philosophy, adept in the classical way of philosophising in India. And assuming that they were there, they had to be located and discovered and some sort of contact established with them. Also, as most of them did not know English, the only language of serious intellectual discourse in India, the only way in which one could enter into a dialogue with them on philosophical or any other issues was to provide facilities for bilingual translation so that the participants in the philosophical interchange might express themselves freely in the language they could best express themselves in.

Professor M.P. Rege was perhaps the first person to realise the imperative necessity of providing such a facility to the participants if a meaningful dialogue between traditional and modern philosophers were to take place in contemporary India. He also was perhaps the first person to realise that if the dialogue was to be fruitful in the contemporary context, then it had to be about issues that were engaging the philosophical minds today. It was perhaps for this reason that he formulated some issues arising from Russell and Wittgenstein, got them translated into Sanskrit and circulated them to the traditional Pandits beforehand so that they may have some idea of the philosophical issues that were to be the subject of discussion between traditional and modern scholars in the field of philosophy later. He was also careful in choosing the scholars from both sides for the dialogue. The traditional Pandits invited for the dialogue were drawn not only from various disciplines, such as logic, grammar, linguistics, hermeneutics (Mīmāṁsā), which would be relevant to the subject under discussion, but also were of such outstanding authority in their respective fields that the very fact that they had consented to take part in such a dialogue made the experiment seem worthwhile to other Pandits. Professor Rege also met many of them personally and explained to them the purpose of the dialogue so that they had some idea of what was intended to be achieved through the dialogue and were favourably inclined towards it. Besides this, he took care to choose those amongst traditional scholars who had a relatively open mind about philosophical issues and were prepared to modify their positions in the light of counter-arguments or offer new arguments for holding to the old position. These qualities were preeminently embodied in such outstanding representatives of traditional scholarship as Pandit Badrinath Shukla, Pandit Sriniwas Shastri and Pandit Laxman Shastri Joshi who throughout guided the discussion for one full week during which time the dialogue was held.

Professor Rege was equally careful in choosing the University-based scholars trained primarily in the western tradition of philosophising for the dialogue. Most of them had a fair knowledge of Sanskrit, knew Indian philosophy — particularly *Nyāya* — quite well, and had a firm foundation in Western philosophy. Some of them had even studied Indian philosophical texts from the traditional pandits in the traditional way. A few had an equally fluent command of both the traditions and spoke Sanskrit as well as English equally fluently. These provided the expertise for translation from English into Sanskrit and *vice versa*, thus providing the bridge between the two traditions and making the dialogue possible. The week-long dialogue revealed two things — first, the necessity of continuing it; and secondly, the immense difficulties of translating from one conceptual structure into another. The experience was intellectually so rich and rewarding that there was no doubt, at least amongst those who were steeped in the western tradition of philosophising, that the whole thing should be put on a continuing basis so that the momentum generated might not be lost. On the other hand, it was equally clear that successful translation, which was the necessary prerequisite for the dialogue, was not such an easy job as many of us had earlier imagined. It requires not only a facility in both the languages, but also a grasp of the two philosophical traditions in their specificity. This, in a sense, had been anticipated to some extent. What was new and baffling to discover, was to find that languages are not conceptually neutral as we had thought them to be. The Sanskrit term or terms through which translation from English was being attempted, we discovered, had already been pre-empted to do a technical job in the interest of the conceptual structures which the Indian philosophical tradition has elaborated over millennia. There were, so to speak, no philosophically uncommitted or neutral terms in Sanskrit which could serve as even rough equivalents of what was sought to be conveyed through the English terms. Yet the difficulty, though disconcerting in the beginning, was gradually overcome, at least to the extent that the dialogue could be kept going and the participants felt that they were getting closer to understanding what the other side was trying to say regarding the issues under consideration.

The Rege experiment at Poona (11–12 July 1983) led to a number of startling discoveries on the part of those philosophy teachers in India who had lost all contact with the practitioners of the classical traditions of philosophising in the country. The fact that there were scores of living representatives steeped in the traditional ways of philosophising with whom it was almost a way of living itself came as a surprise to everyone who had thought that the classical traditions of Indian philosophy were dead and gone and hence a subject of only historical and antiquarian interest to be satisfied by rummaging among the dusty shelves filled with ancient texts, difficult to interpret and understand. To find it all alive and kicking and knocking at one's door was to realise that the first task was to fathom its depth and diversity and extent, so that one may also begin to relate to it in a living manner. The preparation of a *Who's Who* of traditional scholarship seemed the first priority, and getting in touch with those who were there, the next. Both were undertaken almost immediately. And, as everything living tends to proliferate endlessly, one thing led to another and the enterprise diversified itself and branched out in many directions.

One of these was the discovery that traditional scholarship in certain areas of philosophy such as *Nyāya*, *Mīmāṁsā* and Kashmir Shaivism was rapidly declining for a variety of reasons and that, if nothing was done to check the process, important texts would become unintelligible to modern students of the subject. It was decided, therefore, to convene meetings of outstanding scholars in these areas to allow them to interact with each other and discuss the future of their subject amongst themselves. The traditional scholars are not only scattered far and wide in different parts of the country, sometimes even in remote areas not usually associated with the presence of any man of learning there, but they have generally little occasion to meet each other in intellectual interchange. Many of them are old and retired and settled in their ancestral villages or in Maṭhas belonging to their own Saṁpradāya. To find their whereabouts and find funds for inviting them all to one place from all the four corners of India was a difficult and challenging enterprise indeed. But institutions such as the Indian Council of Philosophical Research, the Rashtriya Sanskrit Sansthan, The Tibetan Institute of Higher Studies helped to realise what had seemed a difficult, if not impossible, undertaking at one stage.

The first meeting of such a kind, i.e. of *Nyāya* pandits from all over India, was held at Benaras, the home of Sanskrit learning in India for millennia. The meeting, held for a full five days was such a resounding success and aroused so much enthusiasm that it was felt by everybody involved in its organisation that such a meeting must also be repeated with traditional scholars specializing in other schools of Indian philosophy. Steps were therefore taken later to organise a meeting of *Mīmāṁsā* scholars at Tirupati in Andhra Pradesh, and of scholars in Kashmir Shaivism at Srinagar in Kashmir. One of the constant features of these meetings was the challenging question posed before the scholars of these traditions as to how they would further develop the thinking in these schools in modern times. Also, what was the doctrine or the position which the adherents of those schools could not give up without giving up their identity as belonging to a recognisable school, distinctly different from others. The more sensitive question, perhaps, concerned their own dissatisfaction with the traditional formulations — with their defects, incompletenesses, weaknesses — as without these there can be no such thing as the living growth of a system.

This, in a sense, seemed to go against the seeming spirit of the Indian tradition where the original formulation was supposed to have said everything that could possibly be said about the system and the task that remained for subsequent thinkers was only to elucidate and explain away the seeming inconsistencies and contradictions in the text. Yet, once the issue was framed in that way and the challenge thrown to them, at least some of them took it up and suggested possible directions in which the old issues could be developed in new ways. One of the most outstanding amongst them, in fact, propounded in a public lecture delivered at the conference a most radical thesis concerning the dispensability of the notion of *Ātman* or soul for the understanding of experience within the *Nyāya* framework of analysis. The postulation of mind along with body were sufficient in his view to account for experience, and there was no necessity for positing any extra entity like *Ātman* when these sufficed for the purpose. This contention of Pandit Badrinath

Shukla was objected to by many traditional scholars who were present on the occasion and one whole session of the conference was devoted to a discussion of what he had said in his lecture. Later, it was suggested that the text of Pandit Badrinath Shukla's lecture be circulated to Pandits, who might then send their objections to what he had said in writing so that he might give his considered reply to them.

Some Nyāya Pandits at the Sarnath Conference at Benaras also showed an interest in modern logic during a discussion when the divergent approaches between the two traditions, the Indian and the western, were presented towards certain logical issues. Arising from this, the idea took shape of providing opportunities for traditional logicians, trained in the *Nyāya* tradition, to become aware of the western tradition in this field, and was implemented later on when the Ford Foundation showed interest in the whole enterprise and also funded it to some extent. However, as there seemed to be a complementary interest on the part of the modern scholars trained in western logic, the Summer/Winter schools were organised in such a way that both groups had the opportunity to learn from each other's specialised knowledge of logic as it had developed in the two traditions, Indian and Western.

The meetings of traditional scholars in the field of *Mīmāṁsā* and Kashmir Shaivism were later organised at Tirupati and Srinagar, respectively. Though these meetings did not result in any new, interesting contributions on the part of the Pandits in these recognised fields of traditional philosophising, they gave an opportunity to the traditional scholars to interact amongst themselves and respond to the new issues formulated with respect to these fields of studies in the light of formulations of problems relating to these fields in contemporary philosophy. Along with this, there was the other advantage for those primarily trained in the western philosophical tradition to become aware of the distinctiveness of these traditions — particularly of Kashmir Shaivism — and to engage in a dialogue on the subtle and complex issues relating to the problems of interpretation, especially those arising out of the *Mīmāṁsā* discussion on the subject.

The meetings and the dialogues gradually broke the ice between the representatives of the two philosophical traditions in India and established a rapport between the two groups, resulting not only in a profound intellectual respect for each other, but also in the exhilarating discovery of philosophically exciting differences in the way problems were perceived, questions posed, answers attempted and solutions sought. A contrasting analysis which leads in a different direction and gradually builds an imposing structure of philosophical construction over millennia is something whose possibility one can not even conceive of if one is confined to one philosophical tradition alone, which one inevitably treats as the only possible tradition and hence as universally paradigmatic in character.

However, the discovery of a living, alternative philosophical tradition cannot long remain confined to the field of philosophy alone. It inevitably raises the issue of alternative cognitive traditions in other fields of knowledge, too. As philosophy itself is not an isolated cognitive activity, it is intimately related to all other fields of knowledge and is both affected by them and affects them in a substantive manner over a period of time. But in case this is even marginally true, one has to seek and search for alternative cognitive traditions in other fields of knowledge in India and establish a dialogue with their living representatives, if any.

The story of this parallel search for alternative cognitive traditions in the field of political science, sociology and aesthetics has been told elsewhere (Daya Krishna, 1987).[4] But, as everyone knows, the search is unending. To grasp the cognitive traditions in different fields of knowledge that have had a long history of more than two thousand years behind them and conceptually articulate them in modern times and use them in the contemporary context for gaining insights into the processes as they are occurring today, is a task that can only be accomplished, if at all, by a large number of persons over a long period of time. But that would not be possible unless a beginning *were made* and the first step taken towards the fulfilment of that objective.

The branching out in other fields of the Rege experiment at Poona was attempted in the field of linguistics at Bhubaneswar in Orissa with the involvement and cooperation of the Central Institute of Indian Languages, Mysore, and the Central Institute of English and Foreign languages, Hyderabad. Current issues in linguistics were framed for discussion by these two institutions jointly and then translated into Sanskrit and circulated to traditional scholars beforehand. The dialogue proved as fruitful and promising as the earlier ones at Poona and Delhi and revealed the need for such dialogues in other fields, too.

The successive encounters with the Indian cognitive traditions in different fields of knowledge brought home the realisation that, in order to be really effective and meaningful, such an enterprise cannot be a one-time affair. Rather, it has to be planned on a continuing basis and pursued in depth in such a way that the intellectual issues which arise in each encounter are explored in depth at the next meeting, leading not only to a clarification of the issues concerned but also to an awareness of the problems challenging both the traditional and modern scholars alike to find viable solutions to them.

The search for in-depth exploration of one philosophical problem which has dominated the contemporary philosophical scene for some time past and which had also engaged the attention of classical Indian philosophers for a long time, viz. the problem of meaning and reference, was taken up by the Adyar Library and Research Centre for study and exploration in a series of successive seminars on the subject. The Adyar complex is a well known centre of Sanskrit learning in Madras and the seminars were planned and directed by Professor Kunjunni Raja, the present Director of the Library and Research Centre there. The Centre proposes publishing the results of the seminar as soon as the editing is completed.

All these activities taken together have generated at atmosphere in which philosophers trained in the classical tradition have ceased to be strangers to those trained in the western tradition in India. The lessening of intellectual distances between the representatives of the two traditions has also resulted in a feeling that philosophical activity, at least in India, is to be a joint activity undertaken through some sort of a cooperative, though dialectical, interaction between the two traditions where each questions the other and, by questioning, forces the thought to take new directions. In fact, a number of younger persons in both the traditions have started taking an interest in the other tradition after participating in these activities and the exposure to which they have been subjected in them.

Almost as a by-product of these encounters, philosophers educated in the western tradition in India have become aware not only of the traditional scholars in the field but also of the fact that many of them have been writing new philosophical works in the classical Indian tradition, unaffected by the western currents of which they have been completely unaware. This has not only completely demolished the widely held belief that philosophising in India stopped a long time ago, but also revealed the unbroken continuity of the philosophical tradition, as many of these scholars had vivid memories of their teachers and the intellectually exciting atmosphere in which they grew up.

The intellectual history of the philosophical tradition in India within the last few hundred years has hardly been paid any attention as it has been assumed, under the influence of the new system of education, that it was all dead and gone long ago, or that what remained of it was just a dull, dreary repetition of the past. Only a few works in the regional languages such as Hindi or Bengali had tried to convey the flavour of the living tradition in these fields, but they were primarily anecdotal and biographical, without giving the real philosophical contribution these thinkers had made to the past philosophical traditions of this country. Yet, as they give the list of their published and unpublished works, they provide an opportunity for their critical evaluation, provided a number of competent persons could be interested in doing so[15] (see Baldev Upadhyaya, 1983; see also Hemchandra Bhattacharya, 1977/1383). In fact, even the well-known philosophical works of the past have to be reassessed once again in the light of new questions that each generation tends to ask, if it is intellectually vibrant and alive.

The renewed interest in the classical schools of Indian philosophy, particularly *Nyāya*, is, to a certain extent, evidence of this. A number of scholars, both young and not so young, have begun to look at the classical texts from the *Sūtras* onwards with a new, questioning attitude which is not prepared to take the traditional way of looking at them for granted. They refuse to accept unreservedly the dictum that the founder philosophers of the system, or even those who were responsible for its later development, could not have been wrong or incoherent in their formulations. The very fact of their being so many *pūrvapakṣas* or counterpositions is an evidence, in their opinion, of there being a lively diversity of philosophical positions rather than the unquestioned finality of the one *siddhānta* or doctrine as propounded by some one person whose authority came to prevail in the tradition. The so-called self-image of the tradition as being uniform and unchanging from the *sūtradhāra* (author of the *sūtras*) onwards is, according to them, not only factually false but also almost fatal to the growth of Indian philosophy in modern times. The works of Pradyot Kumar Mukhopadhyaya (1984), Sukharanjan Saha (1987), Mrinal Kanti Gangopadhyaya (1982), B.K. Matilal (1986), Sibajiban Bhattacharyya (1987) amongst others, are an evidence of this. Professor Mukhopadhyaya, for example, writes "...I have in the main sought to reconstruct the views of the classical Indian realists" (Pradyot Kumar Mykhopadhyaya, 1984: p.iv). And, writes Sukharanjan Saha (1987: 11), "... the most sensible thing for us to do is to undertake a programme of reinterpretation of our own philosophy in the light of what we have been able to learn about Western philosopy." And, writes Bimal Krishna Matilal (1986: 2). "The writer on classical Indian Philosophy today is generally pulled in two different directions, towards the historical reconstruction of some classical views, and towards the critical examination of similar modern view." But whether it be "construction" or "reinterpretation", the idea is the same; that is, to look at traditional Indian philosophy from the perspectives of problems as posed in contemporary western philosophy.

The question has, of course, been raised why issues in contemporary philosophy alone should provide the basis for a dialogue with the tradition. Does it not create an asymmetry in the very foundations of the dialogue itself? Does it not prejudge in a certain way what shall be considered relevant or significant in the philosophical traditions of non-western cultures? And, can one really pick out arbitrarily certain aspects of the tradition and treat them alone as meaningful and significant, even when they are integrally related to other aspects which one is not prepared to accept.[1] And, at a still deeper level, is not the whole exercise of "reconstructing", "restructuring or "reinterpreting" to do violence to the past, which ought to be understood in terms of itself, and accepted or rejected on its own grounds?

These are all valid questions and need to be answered if the enterprise is to serve the larger cognitive purposes for which it is primarily intended. It is true that in order to become relevant to contemporary philosophical discussion, the issues and the problems posed to traditional scholars have been taken from that context, but they need not be confined to that context alone. In fact, as the discussion proceeds new problems emerge and they also get discussed. And, though it was inevitable that in the initial stages problems had to be taken from the contemporary western context, the situation need not remain asymmetrical for long and the pandits trained in the classical philosophical traditions of India may raise counter-questions of their own with respect to the western formulations. In fact, such questions become inevitable in the course of the dialogue as the intelligibility of the problems posed itself depends on assumptions which, though widely accepted within the western tradition, are hardly accepted within the Indian one. And, once the hidden assumptions are brought into the open they themselves become the subject of discussion. Further, as the traditional Pandits become aware of the western philosophical tradition, they are bound to raise questions regarding it and, to some extent, they have already started doing so. In one of the summer/winter schools in modern logic for traditional *Nyāya* Pandits, held on their own request to acquaint them with modern logic, this has occurred to some extent. In fact, these schools are experiments of a new kind in the sense that while traditional scholars seek to be acquainted with modern logic, the modern students of the subject are simultaneously taught traditional logic, that is *Nyāya*, by the Pandits. As both the groups had specialised knowledge of that tradition of logic which the other wanted to learn, there was no distinction between teacher and taught, for each was a teacher in one respect and a student in another.

The charge of doing injustice to the tradition by arbitrarily picking up certain aspects, features or issues to the deteriment or neglect of others and reinterpreting them to suit one's own purposes is, of course, far more serious and needs to be met adequately. The first thing to note in this connection is that there is no such thing as one uniform tradition, but rather diverse traditions in conflict with each other. Further, each of these traditions consists of multiple strands which are not exactly in harmony with each other, and which seem to pull it in different directions. One reason for this is that each tradition has had a long history during which it has undergone profound changes for both internal and external reasons. No system is ever a complete coherent whole and the inner inconsistencies always provide the internal pressure for change within it. Besides this, there is the continuous challenge from other systems as well as new needs arising from experience requiring to be accomodated within the system. These provide the external stimulus for change within the system.

The so-called tradition, then, was never, during the period of its growth and development, averse to change. Nor did it resist modifications, when they were needed. And this it did to what, at that time, was "tradition" for it. Why then, is it expected today that the classical tradition in Indian philosophy be preserved in its pristine purity and presented only in that form in which it would be acceptable to the past practitioners of the tradition? But the past practitioners are dead and, even if they were not, why should their judgment be preferred over ours — particularly when they themselves did not always accept what the past practitioners in their own field had held about matters they differed about?

The issue ultimately depends on how one looks at the past and what one wants to do with it. If one treats it as living and one wants to carry it forward and play some role in developing it further, then one's approach would be different than if one treats it as dead and closed, a matter of only textual exegesis and historical understanding. The attitude of the Indologists, it is obvious, cannot but be the latter. Whether it will be the same for those who by their very location happen to be the inheritors of the tradition is difficult to tell, particularly in face of the fact that there are literally hundreds, if not thousands, of living practitioners of the tradition which is treated by others as dead and buried in the catacombs of history.

True, most of the living practitioners of the classical philosophical traditions of India view themselves as the preservers of the tradition, rather than its developers in new directions. But as one of its most perceptive members said "In times of danger, what else can one do but try to preserve and see that something which was carefully developed over millennia is not entirely lost". Also, both the tradition of commentaries and the way the tradition has been understood by most Indologists and accepted, under their influence, by many of the Indian scholars themselves has helped in perpetuating the false image of the past and influenced thereby the traditionalist scholars' attitude to what was the proper thing to do in the present. But, in spite of this, there have been exceptions like Pandit Badrinath Shukla whom we have mentioned earlier. There may be others whose work is still unknown to us. There is so little knowledge about traditional scholars and their work that hardly anything can be said with confidence about it. However, the more we have come to know about them and their works during the last few years, the more suspicious we have become about the stereotypes built up over the past century-and-a-half under the leadership of western Indologists.

The problem of the creation of a culture's image by those who belong to a politically dominant group with an alien culture of its own, belonging to a totally different milieu and tradition, is too complex a problem to be pursued here. But this much may be said with fair plausibility: that, given the opportunities for continuous interaction amongst themselves, as well as with scholars from other intellectual traditions relating to their own field of knowledge, along with a challenge to the self-image of the tradition fostered over a long period of time, it is extremely unlikely that some new trends would not emerge amongst the traditional thinkers in the field of philosophy in this country. Not only this, one may also expect some new thought-currents to emerge amongst the younger group of western-oriented philosophers because of this continuous exposure to the challenge emanating from a dialogue with the living philosophical tradition as embodied in the flesh-and-blood pandits representing the classical tradition of philosophising in India.

This enterprise of discovering the classical traditions of India in the field of philosophy and entering into a dialogue with it on contemporary philosophical issues has recently taken a new turn with the discovery of what may be called the classical West Asian tradition of philosophising as it was continued and developed in India over centuries. This is usually known as Islamic philosophy, but this is a misleading expression, for even though most of its practitioners happen to be Muslims, that does not make their thought necessarily Islamic in any relevant sense of the term. Further, it appears that these thinkers themselves distinguish between theology, which not only accepts the Quran as the final revelatory authority on all matters but is also concerned with its coherent exegetics, and philosophy, which is concerned with logical, epistemological and metaphysical issues and deals with them through reason, involving argumentation and discussion.

The existence of Arabic philosophy is fairly well known, though it is usually treated either as an appendage to Greek philosophy or as a transmitting channel through which Greek philosophy was introduced to Western Europe. Its existence in its own right has seldom been recognised, nor its own intependent contribution to philosophical thought carefully assessed. Yet, even though it has never formed an integral part of the philosophical awareness of the Western countries or of those whose educational systems were substantially affected by their political and cultural domination, it still occupied some place in their awareness, however hazy and vague it might have been. The story of its transplantation into India and its independent development there is, however, hardly known to anybody. To have become aware of it, is itself an immense gain whose philosophical fruits can only be known in the future. But, like the discovery of the living tradition of the classical Indian philosophy amongst the pandits, it imposes on the philosophical intelligentia of contemporary India the same tasks and duties with respect to it. And this is to become aware of it and to establish a living dialogue with it, so that it may transform both itself and us in the process.

The dialogue with West Asian philosophy as it has developed over centuries in India presents problems which are in part similar and in part different from those encountered in the dialogues with the classical Indian philosophical tradition as it developed over millennia in this country. First, its practitioners seem far fewer in number than those representing the Sanskrit philosophical tradition in the country. Second, the language of the philosophical tradition, that is, Arabic, is known to far fewer persons at the present time in India than Sanskrit, and people who know both Arabic and English even far, far fewer than those who know both Sanskrit and English. Besides the linguistic difficulty, there is the additional problem of finding persons who are at home in both the philosophical traditions, as mere knowledge of languages hardly helps in translating thought from one language into another.

Yet, whatever the difficulties, one can generally get around them if one has the will to do so. And, at least these are not the sort of difficulties that can long stand in the way of a determined effort to get the dialogue going. One such effort has already been made under the auspices of the Indian Council of Philosophical Research by Professor Jamal Khwaja of the Aligarh Muslim University, and there is no reason why such attempts should not be continued in the future. In fact, the idea of a triangular dialogue between the Sanskrit Pandits, the Arabic Ulemas and the English-knowing philosophers of India around philosophical issues that are currently engaging the philosophical world (which, of course, is primarily western) or those which naturally arise from within the Sanskrit or the Arabic philosophical traditions is already engaging the minds of some people.

All this activity has emerged only within the last three or four years and pursued in a more or less systematic way. Each step has led to the apprehension of other relevant needs which demand to be fulfilled if the enterprise is to go forward. Up to now, there has been sufficient support, both financial and administrative, for the activity to go forward. The only substantive result, besides the new and radical formulation of the *Nyāya* position with regard to *Ātman* in Pandit Badrinath Shukla's lecture, and the publication of *India's Intellectual Traditions*, giving a tentative preliminary account of the attempt at conceptual reconstructions in the field of sociology, political science, aesthetics, and law, has been the change of awareness and atmosphere with regard to the urgency of establishing critical and creative linkages with the cognitive traditions of India. Whether any more substantive results will follow, only time can tell. But, at least, the first steps have been taken.

NOTE

1. This point was forcefully argued by Professor A.K. Saran, the well-known sociologist, in a recent discussion of the issue at Lucknow.

BIBLIOGRAPHY

Bhattacharya, Hemchandra, *Bangiya Sanskrit Adhyapaka Jivani*. Calcutta: Navin Prakashan, 1977 (Bengali Year 1383).

Bhattacharyya, Sibajiban, *Doubt, Belief and Knowledge*. Delhi: Indian Council of Philosophical Research, 1987. Distributed by Allied Publishers, Delhi.

Gangopadhyaya, Mrinal Kanti, *Nyāya-Sūtra with Vātsyāyana's commentary* (Indian Philosophy in Its Sources Series). Calcutta 1982.

Krishna, Daya, Three Conceptions of Indian Philosophy. *Philosophy—East and West*, USA (January 1965).

——, Three Myths about Indian Philosophy. *Diogenes*, Paris, Vol. 5 (July – September, 1966).

——, Indian Philosophy and Mokṣa—Revisiting an Old Controversy. *Journal of Indian Council of Philosophical Research*, Vol. II, No. 1 (1984a).

——, Comparative Philosophy: What it is and what it ought to be? Paper read at the Conference on "Comparative Philosophy," Hawaii, 1984b.

——, (Ed.), *India's Intellectual Traditions: Attempts at Conceptual Reconstructions*. Delhi: Motilal Banarsidass, 1987.

——, *Indian Philosophy – A Counter-Perspective*. *Delhi*: Oxford University Press, 1991.

Matilal, B.K., *Perception*. Oxford: Clarendon Press, 1986.

Mukhopadhyay, Pradyot Kumar, *Indian Realism: A Rigorous Descriptive Metaphysics*. Calcutta: K.P. Bagchi & Co., 1984.

Saha, Sukharanjan, *Perspectives on Nyaya Logic and Epistemology*. Calcutta: K.P. Bagchi & Co., 1987.

Saṁvāda: A dialogue between two philosophical traditions, Indian and Western. Ed. Daya Krishna, M.P. Rege, R.C. Dwivedi, Mukund Lath, Delhi, Indian Council of Philosophical Research in association with Motilal Banarsidass, 1991.

Upadhyaya, Baldev, *Kashi Ki Panditya Parampara* (Hindi). Benaras Vishvavidyalaya Prakashan, 1983.

India

Indian metaphysics

MARGARET CHATTERJEE
Delhi University

Introduction

Any distinction made between metaphysics and speculative thought in an Indian context needs pondering over. Kant's distinction between a metaphysic of experience and the kind of thinking which beats its wings in empty space is too well known to need comment. On such a view speculative thinking remains unscientific, 'natural', but seductive. A more recent distinction, that between descriptive and revisionary metaphysics, takes its stand on linguistic considerations, incidentally, in the view of the present author, underestimating the subtlety of the texture of descriptive language. A further distinction, that between the phenomenological and the ontological, makes yet another type of demarcation, one which might dissolve if the phenomenological is itself found to have ontological weight.

If the metaphysical be traditionally regarded as contrasted with the experiential we need to concede that classical Indian philosophical thinking made no such distinction. The higher reaches of speculative thought were regarded as based on experience, no doubt experience of a kind. This aside has some importance. Long before phenomenologists mapped various 'modes' of experience, Indian philosophers took it for granted that experience was not all of a piece. The Carvakas apart, no one was inclined to regard sense experience as in any way rock-bottom, let alone paradigmatic. In Indian thinking, speculation, being based on 'privileged' experience, so far from being distanced from *dṛsti* (seeing) provided its own characteristic mode of seeing.

The dismay felt in certain quarters in India when the verification principle eventually made an impact (this happened after a time lag of several decades), understandably made no impact on traditionalists. It affected those who saw fit to have second thoughts about the 'real' world, but made no impression on those who took the distinction between phenomenal and noumenal (*dṛsta* and *adṛsta*) for granted and whose 'speculative thinking' derived no less from a claim to experiential validity than from the verdict of texts deemed to be canonical or near-canonical. In fact there had traditionally been three levers in Indian philosophical thought — experiential warrant, "authority", and logic. In the past, considerable dexterity was shown in weaving a web in which these very diverse threads might very well pull in contrary ways.

Modern thinking is characteristic in no longer sensing an *obligation* to juggle with all three. Crises of conscience are felt most by those who still feel such an obligation and yet wish to evidence an awareness of what is *de rigueur* on the

Contemporary philosophy. A new survey. Vol. 7, 83–105.

world philosophical scene. Those who would regard it as a compliment to be called metaphysicians often look at a classical thinker through the spectacles of a modern one (cf. looking at Shankara through the eyes of Husserl) or proceed in the opposite direction, i.e. looking at Husserl through the eyes of Shankara. Such procedures can give rise to innovative scholarly approaches (not always genuinely 'comparative') but tend to inhibit *original* work. The tendency to write footnotes to a major thinker is perhaps a common phenomenon today, regardless of which country we may live in. The sharpening of dialectical tools has given rise not to "speculative thought" but to analytical expertise directed to fields which lead away from metaphysics. Those who 'do' analytical philosophy in India do so, not so much because they have fallen under the verificationist spell, but because they are keen to detach the "genuinely" philosophical from the metaphysico-religious. All in all it is not at all easy to identify those who can be said to have made an original impression in the field of metaphysics during the last twenty to twenty-five years excluding, as I shall, exegetic work within classical systems or on particular philosophers.

METAPHYSICS AS SEEN BY SOME CONTEMPORARY INDIAN PHILOSOPHERS

The period surveyed in these pages includes the years reviewed in an article published by J.N. Mohanty[1] in September, 1974. This covered "Philosophy in India 1967–73". The present focus, however, is centred on metaphysics during the last twenty years. The "anti-metaphysical thrust of contemporary philosophy" referred to in Mohanty's article has lost some of its bite. There are certainly many philosophers in India engaged in what they would like to describe as analytic philosophy. But this is less from any anti-metaphysical bias than from an interest in patterns of argument and, in certain quarters, a re-discovery that similar strategies have been pursued for centuries by "classical Indian philosophers". There are signs that former stereotypes about Indian philosophy are giving way to a rather different image, the nature of which we shall shortly try to determine. The predominant earlier stereotype about Indian philosophy has been characterised as follows by B. Matilal:[2]

> "Indian philosophy has unfortunately come to denote a group of ancient religious cults, a system of dogmas and an odd assortment of spirituality, mysticism and imprecise thinking, concerned exclusively with spiritual liberation".

It would perhaps be not too wide of the mark to say that a large percentage of overseas scholars who come to India to study Indian philosophy are interested in *darśana*, *brahmavidyā*, *ātmajñana* and the like. That is to say, they are often 'seekers' rather than critical investigators. Looking for good reasons for what is believed on instinct has always been the mark of a certain kind of philosopher. It is only fair to mention, however, that there are far more overseas scholars than there are Indians who investigate the set of phenomena mentioned by Matilal from the standpoint of the anthropologist and the sociologist. The distancing needed for scientific enquiries can perhaps best be offered through an objective perspective.

But what remains if cults and mysteries become the preserve of the social scientist and strategies of argument are taken over by the logician? Is there still a metaphysical terrain which lies in between? The classical metaphysician maintains that the study of Being is "identical with the I-am (I-consciousness, *ahaṁdhi*)...". P. T. Raju,[3] whose dictum this is, goes on to say that "This is not speculative metaphysics, for speculation lacks restfulness and stands suspended as the construction of a possible hypothesis". Metaphysics for him, it is clear from the above, is by no means an "in between" affair but a discipline which is grounded and grounding. A Vedantic turn is also found in Debabrata Sinha's[4] understanding of metaphysics, albeit seasoned with Kantian and phenomenological overtones. He speaks of:-

> ".....metaphysics (that is, metaphysics of experience) as the 'transcendental' enquiry into the presuppositions — those of ordinary experience, knowledge and language, and so on — taking 'transcendental' in a broadly 'phenomenological' sense".

Deeper reflexion reveals *tattvas*, things as they are. "Transcendental reflexion" moves "from the experienced continuum of phenomena to the over-empirical essence or essences in which they are grounded". Metaphysical thinking here seems to be given the task of a progressive refinement of the content of experience, closing in on the data, as it were, using penetrative and intuitive powers rather than discursive ones.

An interesting comment about metaphysics was made by K.K. Banerjee,[5] who was probably India's most gifted Nyayaika in recent years. He said that: "Which metaphysical system could be acceptable to which individual or society of individuals depends on the individual or the society of individuals, and the condition of acceptability is not logical, but that of growth....."

This comment reveals the insight of a historian of philosophy who perceived how standpoints come into existence thanks to a multiplicity of factors. Different metaphysical systems, he advised, can be subjected to reflective consideration and this can have the useful effect of protecting us against ideology. What he had in mind was perhaps a caveat *vis-à-vis* the unexamined, and even more, the doctrinaire.

Several modern Indian metaphysical thinkers find entry to a *metabasis eis allo genos* within stances of reflection. Such reflective shifts seem to me to take the place of the classical distinction between the *paramārthika* and the *vyavahārika*. To the phenomenologically inclined they take the place of the turn from the naturalistic standpoint to the post-epoché stance of *ratio intuitiva*. However, it is usually maintained that a purged mode of reflection, that is, one which has left behind the unreflective, can provide insight into the nature of what is. But since these insights are so various, philosophers like K.K. Banerjee and Kalidas Bhattacharyya concede how diverse our standpoints can be. Herein lies both their modesty and, I would say, their realism. Rather than give a condensed survey of several metaphysicians, which would mean scarcely doing justice to any of them, I have chosen to present three thinkers in depth, all of whom belong to the post-K.C. Bhattacharyya era, turning first to a philosopher who over a period of years has come to formulate a distinctive system of his own.

Jehangir N. Chubb

Jehangir N. Chubb (1910–) has wrestled with metaphysical questions over a period of many decades, his early work centring on logic and the theory of perception. In 1937 he took a D. Phil. from Oxford where he worked on the problem of error and illusion. His contact with Oxford philosophy stretches over the timespan from Collingwood to Ryle and Strawson and he has acknowledged his debt to all of these. The two chief Indian influences on his work are K.C. Bhattacharyya, whom he knew personally, although he was not a pupil, and Sri Aurobindo. He has been through the mill of linguistic analysis and had things of his own to say about knowledge and belief and similar puzzles which occupy the epistemologically oriented. But this has not stood in the way of close examination of the ontological argument and the classical theme of metaphysics in the grand manner, that is, the theme of God and the Absolute. All this goes along with an underlying Indianness of concern with the self, freedom and the ultimate destiny of man.

His paper on "Critical Philosophy"[6] opens up what "philosophy becoming self-conscious" involves, granted that it is important for a philosopher to be aware of what philosophising involves. Such an awareness comes about through reflection on incompatible points of view. Whereas this reflection set Ryle to work on the nature of philosophical *arguments* in his inaugural lecture at Oxford, Chubb advocates[7] "cultivating at the level of philosophical thinking what in the spiritual discipline of the Samkhya and the Vedanta is called the *Saksin* or the Witness attitude". Chubb obviously agrees with H.H. Price's celebrated maxim "Clarity is not enough". But, whereas Price advised a combination of analytic and synthetic approaches, Chubb wants us to go in for "the activity of contemplating or witnessing thought". Critical philosophy practised in this way is elsewhere[8] described by him as a form of "epistemological introspection". It is reckoned to result in a clear awareness of "the origin of our thought, i.e. the points of orientation and on what they rest".[9] It seems to me that, in so far as the activity advocated involves something very like intellectual biography, the 'findings', in the nature of things, must necessarily be different. However, Chubb says they are not exempt from criticism. This criticism, I would suggest, might very well be self-criticism, something that there is evidence of when a philosopher adopts a radical change of standpoint.

The plot thickens when Chubb goes on to contrast critical philosophy with what he calls 'constructive' thinking. The latter is described a "the process of expanding and maturing a point of view into a philosophical system".[10] It is necessarily "biased thinking", the adjective 'biased' not being used in a pejorative sense. And here he makes an interesting point — "in philosophy we do not *arrive* at a conclusion, but merely explicate and mature *foregone* conclusions".[11] "The philosopher is committed to a point of view and his reasoning is interior to it. Errors, whenever they arise, do so not because of biased thinking but because of "imperfect observation or insight". Philosophical arguments cannot prove or disprove anything and so, once we see this, the bite and sting can be removed from philosophical discussion. We can exchange ideas in a civilised manner and presumably (although he does not say this) thereby understand other people better.

All this reminds one of H.D. Lewis' ultimates or "ways of looking", and, as far as discussion is concerned, the reader may well recall Rorty's comments on conversation. But Chubb does not seem merely to be advocating a tolerant acceptance of pluralism for at the end of the paper under discussion he seems to plump for "*direct verification* of philosophical propositions in an ultimate experience".[12] This sounds like moving away from the stance that it is misguided to think that "the truth" can be possessed and making a triumphant claim to precisely that. Moreover, since we are beings who live in time, what meaning can be given to the expression "ultimate experience"? Are we not open to further and further experiences, some of which may not be 'unspeakable' in a felicific sense but, in this century of ours, catastrophically such.

Equally serious seems to be the question how one can conceive the relation between 'reflecting' till one reaches 'the origin', contemplating it, then going on to extrapolate it in a 'constructive' way in the form of a system. Epistemologists will fasten on Chubb's apparent ruling out of philosophical disagreement. But even if one has little faith in the productivity of philosophical interchanges, the activity of philosophising surely involves a strenuous grappling with ideas where we are urged on by the desire to ferret out the true in so far as we can perceive it.

The task of intellectual autobiography (this is how I describe Chubb's initial move towards 'critical' thinking) is usually embarked on in the process of self-explanation rather than self-discovery although the two are surely interrelated. However there are further difficulties. The terminus of "epistemological introspection" is reckoned to be detached contemplation of the roots of our thought. But such a terminus would seem to amount to a heightened awareness of bias and therefore to rule out the possibility of contemplation. One might compare with this, say, the culminating point of a session of psycho-analysis where the successive work of unveiling and unmasking results in a self-awareness which is deeply disturbing, if not world-shattering.

As far as imaginative entry into the viewpoints of others is concerned, something which I think Chubb would concede as a possible outcome of an exchange of ideas, this too would hardly give rise to 'contemplation'. It could give rise to what H. H. Price used to call 'entertaining' rather than affirming or denying, or we might invoke Ryle's term 'heeding' in order to refer to such a state of affairs. More importantly, unless we make up our minds beforehand to be quite impervious to what others say, we may very well *modify* our standpoint in the light of discussion. Others, on the other hand, may modify *their* stance in the light of what *we* say. In other words, standpoints, dogmatic ones apart, tend not to be the rockbottom sorts of thing that Chubb presents them as being.

But we must move on, for his own standpoint is worked out further in his later writings. Chubb's book *Assertion and Fact*[13] was published five years after the article just discussed. Dedicated to K.C. Bhattacharyya and Gilbert Ryle, it deals with "the categories of self-conscious thinking" which he details as concepts like true, false, meaning, knowledge etc. It may be recalled that one of Ryle's early writings was on categories. In sharp contrast to Ryle, however, Chubb regards categories as "indirectly ontological". If this expression smacks of ambiguity we can compare with it Ryle's odd use of the expression "mongrel categorical" which occasioned similar qualms when it first appeared. Such concepts, Chubb says, "have a nisus towards being and find their fulfilment in it".[14] Ontological concepts (object, quality, relation, event and situation) are descriptive, whereas the concept of fact is partly evaluative.[15] A fact "discloses itself as what was always on the other side of the epistemological gap". This gap obtains between what is asserted and "what is out there". At this point Chubb joins issue with Austin and Strawson regarding the nature of reference. Facts are not directly ontological because they are not "independent of all reference to the subjective attitude of the person who uses the concept of fact."[16] This is the way Chubb takes up a position *vis-à-vis* the old controversy between idealism and realism. Heedful of standpoint and attitude as he is, he cannot afford to neglect the subjective pole of all epistemic relations. But that facts have a kind of intermediate status is clarified in the following:[17] "To state a fact is, in part, to stake a claim or to recommend something for acceptance". Facts are by no means pseudo entities which exist in the world. This is how Chubb steers a course between the correspondence theory of truth and the idealist treatment of judgment. Fact language, he thinks, is always deployed in a context where doubt and belief cannot be completely eliminated. Talking about facts involves making a kind of assessment (my language) and this is perhaps why he brings in the notion of evaluation which we have otherwise been taught to regard as far from fact talk as you can get.

Chubb's analysis in this difficult but very meaty book never remains at the purely linguistic level. He is confident that the mind can transcend itself and "confront reality" and he approves of idealism for its metaphysical principle that there is "nothing outside the Real". Towards the end of the work[18] he maintains that ".... meaning has a nisus towards being through the intermediate term 'assertion'". Earlier on[19] he had spoken of *concepts* as having a nisus towards being. Now Chubb is not a phenomenologist and it would be incorrect, I believe, to read a theory of intentionality into his oeuvre. The background of this particular book is the treatment of categories from Aristotle, through Kant up to Ryle, and at the same time it takes into account twentieth century treatment of the problem of reference. What he does is to reintroduce the subjective pole of assertion left out by the linguistic analysts and recognise that in asserting there is an element of striving, of balancing pros and cons which, for all that, does not mean that we are *exiled* from the real.

So far there is scarcely a clue that Chubb's sights are set on the Absolute. But between 1977 and the present day it is this which he seeks to express, making explicit his penchant for Sri Aurobindo's way of philosophising. In 1983 Chubb published a book called *Faith Possesses Understanding*[20] which tries to rethink the issue posed in St. Anselm's ontological argument, and seeks to establish that faith not only *seeks* understanding but attains it. The work is certainly more overtly metaphysical than *Assertion and Fact*, but there is a distinct link between the two in so far as Chubb has already set out in the earlier book his reasons for advising us that talk about the real is based on what is believed to be the case. Chubb is certainly concerned with the possibility of "a new direction in rational theology", but we shall fasten on the metaphysical implications. He recognises the extent of the use of analogical predication in classical natural theology but holds that we must go further, into "direct realization".[21] How this is compatible with vindicating the autonomy of reason is elucidated by the comment that "What is not rationally *grounded* can still be rational, at the very least in the sense that it conforms to the laws of thought".[22] What do we do when we use argument in rational theology? It is a process of "*exhibiting* the self-enclosed rationality of religious belief". Reason is not limited to make room for faith but "faith contains its rationality within itself".[23] The response of the whole being "takes up into itself the tacit but total consent of our rational nature".[24]

Metaphysics centres round the concept of the Infinite as the "self-existent self-luminous course of all that exists". Now rationalist discussions of the idea of God made much of the problem of the adequacy of our concepts, given that man is finite. To Chubb, however, our concept of God can become progressively more adequate as religious consciousness develops. He even thinks that religious consciousness need not "fall short of perfection."[25] The bearing that the *mode* of consciousness has on a particular concept is brought out in his comment that the concept of God is "impregnated with its characteristic mode of experience". Apart from the point that there is no one characteristic mode of religious experience, which could be mentioned in criticism, it must be noted that Chubb is saying something important here regarding the relation of concepts and experiences. In the sphere of religious consciousness concepts are neither instruments which penetrate, not nets which seek to capture, but rather ideas which bear the mark of the experience that gives them birth. We have obviously moved away from concepts which enable or which serve in a regulative manner (Kant's two kinds) to concepts which are necessarily *imbued* in a certain manner. This certainly throws some light on perhaps a leading characteristic of authentic religious discourse.

There is more to come in an as yet unpublished paper on "The Hindu metaphysical tradition on the Meaning of the Absolute".[26] In this Chubb stimulates reflection on whether there is a difference between the metaphysical and the religious import of absolutism, a difference which is not conceded by the Advaitin. The metaphysical quest, according to Chubb, is a quest for unity. This quest finds its culmination in "a Mystery continuous with logic though transcending it, and erected on the pedestal of reason". In this article Chubb identifies his standpoint as that of Sri Aurobindo whom he regards as 'fulfilling' the Indian tradition. The crown and summit is not a system but an "integral vision and experience". Methodologically we have to leave analogy behind and go in for parable and poetry. But the latter are not means of reaching the goal so much as evidence of having reached it. They are the modes of expression proper to the *sādhaka*.

There are several difficulties here, barring those attending matters of Aurobindean exigesis which lie beyond the scope of this study. I only mention a few. While metaphysics can readily be identified with the quest of Being in the sense that we know what is *meant* by identifying it in this way, it is hard to see how metaphysics can be regarded as a search for an *experience*. Secondly, the language of realization seems to put paid to the possibility of further and further developments of consciousness, something which Chubb advocates in his earlier work and which Sri Aurobindo also seemed to encourage. Growth in the biological sphere proceeds to the stage of maturity, beyond which there is a decline. Quite possibly things are different once the biological sphere is left behind. Finally, the appeal to an ineffable experience does not go along too well with a fairly consistent insistence on rationality, something which Chubb has been at pains to exhibit over many decades. Is it the destiny of metaphysics to transcend itself? Many who talk in terms of spirit perhaps take this for granted. At all events Chubb's latest thinking involves an explicit conflation of metaphysics and religion, and although this is typical of the Advaitin position, some of Chubb's contemporaries, by contrast, are able to maintain a metaphysical stance which keeps clear of religion, as we shall see shortly.

Kalidas Bhattacharyya

Kalidas Bhattacharyya (1911–1984), as Chubb does, belonged to a generation of Indian philosophers who were equally familiar with Indian and Western thought. The key to Bhattacharyya's mode of philosophising, however, lies not in this so much as in the oeuvre of his father K.C. Bhattacharyya. His life work, in restrospect, appears almost like a sustained meditation on the work of his senior. This makes the task of presenting his own viewpoint extremely difficult. I would neither describe him as a metaphysician in the grand manner, nor as a phenomenologist, both of which descriptions have been variously applied to him by his contemporaries. My ground for the former assessment is that he did not propound a *system* of metaphysics but insisted, rather like Heidegger, that philosophical thinking is always *am Weg*. Those who describe him as a phenomenologist take their cue from Marvin Farber's letter to his father, describing the latter's *The Subject as Freedom* as phenomenological. But Kalidas's method starts from the empirical in order to penetrate it further. This seems to me as different from the Kantian contrast between the phenomenal and the noumenal, as it is from Husserl's distinction between the unbracketed and the bracketed, or indeed from the Shankarite *vyvahārika* and *paramārthika*. K.C. Bhattacharyya had spoken of the 'demands of experience' and Kalidas too finds possibilities rather than 'stoppers' in experience. We shall come across the word 'demand' again in N.V. Banerjee's thinking, although in a different sense. What is behind the various meanings of 'demand' presented by these three thinkers is a challenging consideration, namely whether the main spur or leverage of metaphysics is the desire to *know* or the *desire to be*, 'desire' of course being here used in the sense of metaphysical impulse (cf. Kant's 'disposition').

Drawing on my own recollection of his mode of philosophizing during the course of discussion, I regard his distinctive contribution as being a 'showing' or exhibiting of metaphysical thinking in process. Whereas Chubb's position is rounded off and crystallised in a definite standpoint of his own, Bhattacharyya's work remains strangely incomplete. Looking back, in 1965, he wrote:-

> "My endeavour has always been to pave the way for a metaphysics that cannot be assailed by positivism."[27]

This confession is itself remarkable, for most Indian metaphysicians take the stand that their discipline is based on experience and is therefore impervious to the logical positivist critique. It may be useful at this point to look at some of Bhattacharyya's pronouncements on the nature of metaphysics. The programmatic character of the metaphysical claim is referred to in the following quotation from the same book:-

> ". . . . metaphysics proposes to arrive at a deeper structure primarily through non-sensuous intuition but proximately, and for listeners, though arguments and analysis and, therefore, referring so far to a large extent to empirical details."[28]

Herein he pinpoints what has certainly been true of the classical metaphysical quest, a thrust to penetrate beneath the phenomenal, a tendency to be seduced by the possibility of intuitive knowledge of a special kind, and the employment of a discourse which strives to convince through argument. The last part of the quotation recognises the way in which we need to proceed from what we know, using examples wherever we can. All this is *proposed*. Whether it is attained or not is a different matter.

Nearly a decade later he reconciled the idea of the search for a deeper structure with the need to start with the empirical through the concept of the trans-natural. "Metaphysic . . . . studies . . . facts . . . of a trans-natural world."[29] The use of the word 'fact' evidences his belief that metaphysics has cognitive import. In the following year he wrote:-

> "Metaphysics is a sort of analytical study of some particular types of presuppositions of science and common life, the distinction between science and common life being only of degree of clarity, precision and order."[30]

The examples given of such presuppositions include "concepts like space, time, substance, matter, cause, being, non-being, self and the like" . . . . , a list which combines the sort of thing treated from Kant to Russell, along with earlier Aristotelian concepts plus those more associated with Indian standpoints. At the same time other things are maintained as well; metaphysics is predominantly non-logical (even though it resorts to arguments); it involves the 'practice' of transcendence. I see this as a frank recognition of the intrinsic problematic of metaphysics and of its combination of analysis with claims to insight. But by now we need to know how Bhattacharyya started his own metaphysical activity.

Bhattacharyya's first published work, *Alternative Standpoints in Philosophy*, presented his "logic of alternation", showing how philosophy inevitably presents alternatives no one of which is theoretically stronger than another. We can detect in this echoes both of Kant's treatment of antinomies of reason and of the Jaina insistence on plurality of viewpoint. But it provides a daunting initial insight, admitting as it does that whatever a particular philosopher might maintain there will always be others who quite legitimately will maintain something very different. I believe this starting point of his derives both from logical considerations and from his father's frequent references to indetermination. Whether this indetermination was ontological or whether it was based on human cognitive failings remained to the end a baffling problem for him.

His father, K. C. Bhattacharyya, put forward a theory of alternative forms of the absolute in an article entitled "The Concept of Philosophy".[31] These alternative forms, he said, are truth, freedom and value, each of these being the proper target (my language) of knowing, willing and feeling respectively. If knowing discloses *cit*, willing discovers *sat* and feeling fulfils itself in *ānanda*. What Kalidas does, and in this interpretation I remain most tentative, is to detach all this from 'classical' overtones and start afresh, using epistemology and logic as tools. To do this is to take up the concept of disjunction, since this is what alternation is about. The matter is introduced in his book *Alternative Standpoints in Philosophy* as follows:-

> "In the disjunctive judgment there is a disjunctive unity of alternatives. So here also knowledge and object, or the subjective and the objective attitudes, have to be taken as alternatives forming a disjunctive unity. Each attitude is valid, but alternatively. So we cannot have both knowledge and object at the same time...... Actual philosophy.... is either from the subjective or from the objective point of view, not from both...... Each alternative fulfils itself in a system.... Idealism and objectivism are such alternative systems... From each to both there is no passage in philosophy the ultimate logic of which is alternation. This view may be called 'Alternative Absolutism'— each is absolute but alternatively."[32]

This theoretical position is then applied to an analysis of experience (rather than an analysis of the knowledge-situation). Psychical processes can concern cognition, feeling or conation, and he finds there is a "correspondence" between each of these and "the subjective, the objective and the dialectical attitudes on the other".[33] This is worked out in a highly original fashion.

In the cognitive group of processes, sensation, perception, memory, and thought, he finds a characteristic withdrawal from the object and a movement in the direction of pure subjectivity. But, in the sense that all cognitive activities *refer* to something, presumably there is always an accusative, an 'object' of some kind or other. In the case of feeling, Bhattacharyya takes the position that this leads us to objectivity. While at first sight this sounds odd, what he has in mind is probably the way in which affective states of mind are, as it were, swamped by their content. The *difference* between the objective contents of experiences as diverse as pain, the pity felt at the sight of an accident, or the intense delight experienced in hearing music which appeals to us, brings out what Bhattacharyya means by the dominance of the object in feeling states. We are 'taken out of ourselves' or 'lose ourselves' in such experiences. The third attitude is identified by him as conative and dialectical. This point recalls Kierkegaard's Either/Or, his dialectic of choice, except that for Kierkegaard this dialectic can never yield up an Absolute as Bhattacharyya thinks it can.

In this early stage of his work Bhattacharyya's position reflects the sense that truth can be approached through various experiential stances and that there can be no theoretical criterion for regarding one as more valid than another. That he should move from the logic of judgment to the structure of experience no doubt shows a certain Kantian bias. His early thinking, moreover, bears a resemblance to Kant's in respect of its delineation of different faculties in the mind, a delineation on which he did not insist so much in his later years. But whereas Kant's architectonic sees a once-for-all structure in experience Bhattacharyya's insight centres round a proliferation of possibilities.

A pressing problem poses itself as to whether alternative standpoints are a feature of our human endowment or whether reality itself incorporates alternation. Reflecting on his father's treatment of the alternative absolutes of truth, value and freedom, Gopinath, Kalidas' brother, wrote that "alternation appears to be constitutive of the absolute".[34] No doubt in terms of ontology this is a rather idiosyncratic position to take, and Kalidas does not go so far as to opt for it. During the seminar in 1981 at which he contributed replies to the participants who wrote papers on his philosophy he commented: "Probably there is not much of significant difference between these two positions — between the absolute itself as alternating and alternative formulations of the absolute".[35] This may give us pause. To say that there is not much difference between the two would be rather like saying there was scarcely and difference between speaking of God as "He who is" and taking the line of the process theologian. Furthermore, there is surely a real difference between regarding reality as indeterminate and regarding it as susceptible of infinite determination. The way Bhattacharyya wrestles with these problems at least makes us aware once more of the main metaphysical problem, the limitations of the subject and the transcendence of the object. In his final reflections he seemed to recognise the extent to which we *oscillate* between alternatives rather than choose between them. He also mentioned the possibility of being neutral *vis-à-vis* alternatives. This, however, would surely be yet another 'alternative'. At all events Bhattacharyya was "modern" enough to exclude the alternative of a perspective *sub specie aeternitatis*, a perspective which virtually marks out other alternatives as radically wanting.

But we must move on to his other contributions. His treatment of reflection was initiated in *Alternative Standpoints in Philosophy* in a distinction drawn between consciousness "which is non-distinct from the awareness of itself"[36] and "reflective awareness" where consciousness grows more inward, and is "enjoyed". In reflective knowledge of an object, the object, "though already known, has now revealed characters which it had not revealed before ...".[37] In unreflective experience these characters are 'fused' or objectively implicit. This recalls both the traditional Indian distinction between the *dṛsta* and *adṛsta* and the advocacy of discrimination as a proper task for the philosopher. What reflection reveals to the metaphysician, he says, can be described as essences. These are seen as 'proper' and 'distinct', possessing 'autonomy' and also interrelation. This sounds very much like a claim to intellectual intuition of the sort disallowed by Kant. J.N. Mohanty compares Bhattacharyya's treatment of essence with that of Husserl[38] and Bhattacharyya was certainly in later life greatly interested in phenomenology, largely thanks to Mohanty's own influence. I believe, however, that Bhattacharyya's concept of 'fusion' was worked out by him considerably before he had much acquaintance with Husserl's thought and that his treatment grew out of reflection on the nature of erroneous perception, again a typically classical Indian epistemological concern.

But is there any kind of reflective intuition which reveals 'essences' or 'idealities' such as "space, time, self, God etc."? And how would one apply the concept of the 'proper' to any of them? Is the 'proper' analogous to what the scholastics called 'adequate'? We are not given answers to such questions. There is a further aspect of Bhattacharyya's treatment of essences to be mentioned too. In *Philosophy, Logic and Language*[39] he maintains that they are "intuited through the use of language", that they "cannot be known except as spoken". This enables him to avoid a problem which otherwise might have presented itself, the adequacy of language to express the findings of intuition. A nagging worry persists, however, as to whether talk of intuition is compatible with talk of alternatives. There may be a way out in that to bring in language is to recognise that there are *alternative languages*. But, if we concede this it is difficult to retain the notion of what is 'proper'. Bhattacharyya seems to have recourse to metaphysical faith and this fits in with his remarkable freedom from any hint of a sceptical spirit. Now his faith in the viability of metaphysical intuition is connected with what he has to say about transcendence and freedom and to this we turn next.

According to Bhattacharyya, man has the capacity to stand apart from nature and it is really this capacity that makes him able to think metaphysically:

> "If man is thus, as a matter of fact, sometimes outside Nature, it means that he has a capacity to remain so, for it is not by accident that he has slipped from Nature; through sustained effort he can acquire a more or less permanent attitude of so standing aside and wherever he stands aside he feels, rightly or wrongly, that he is more in himself".[40]

The linking of metaphysical intuition (or reflection) with the capacity to be detached is not stated by Bhattacharyya himself, but I think there is sufficient warrant in his writings for such a linking. Detachment involves withdrawal from nature, including in nature the determinants from within. In a paper presented at the International Society for Metaphysics Conference held at Visva-Bharati in 1976 he uses a phrase strikingly reminiscent of Sartre, detachment is a vacuum, a hole in Nature. William James' critique of the block-universe concept likewise implies fissures in being, gaps which allow room for free movement, free action. Gap and hole language suggest a lack, and to this extent smacks of negation. The positive aspect, however, says Bhattacharyya, is the way in which, through this fissure, nature and the trans-natural meet. At the level of bodily movement man asserts his freedom in manipulating nature according to his desires and interests. But as subjectivity deepens (my language) man's freedom shows itself in manifold ways. He writes "every man's authentic awareness of himself as free posits transcendence in the most evident and convincing form".[41] What he is suggesting here is that man's capacity to dissociate himself from his own body and from nature is a symptom of self-consciousness, of subjectivity, of "pure consciousness", of transcendence. This freedom of self-transcending shows itself in cognitive, emotive and conative forms. In his later writings he strove to show that each of these forms passes through different stages, terminating in the alternative absolutes pertaining to each. This combination of the concept of alternation with that of stages was implied even earlier in so far as an alternation can only be identified from a supervenient perspective (my language).

Among the different stages of freedom understood positively, reason has an important place for Bhattacharyya since it makes communication between selves possible. We have already noted how he associates the intuition of essences with the use of language. Reason operates as a universalizing factor whether as discoverer of laws of nature or as the source of moral imperatives. He believes, however, that reflective consciousness can go even further than reason, and here his thinking closely approaches that of his father in his work entitled *The Subject as Freedom*.[42] If the freedom of man shows itself par excellence in the attitude of detachment from objects (including under this rubric all possible types of accusatives) the culminating point will be a state where consciousness is turned in on itself. However Kalidas's own stress on alternation keeps him from any dogmatic pronouncement regarding the end term of the metaphysical quest. In fact he ventures an innovative use of the *līla* (sportive play) idea traditionally applied to the absolute, saying that it is the *philosopher* who almost in a playful way approaches the absolute through various alternative paths. This image which he left with us in his latter days (when arguments often ended in laughter on his part) is as different from that of the detached observer as it is from the stance of those who claim to have "realized" Himalayan heights. Along each path, and at every rung of the scale of ascent, that is to say in multiform ways, Bhattacharyya finds evidence of what he calls the 'spiritual'. He was never attracted by a synthetic attitude that combined disparate approaches. He was, moreover, sufficiently interested in the relations that selves bear to each other to write critically about the concept of *mukti* (liberation). It is almost in the spirit of a critique of culture that he wrote:-

"It is not without reason, therefore, that the Advaita doctrine of inaction — a necessary consequence of cognitive *mukti* — has found favour with a large section of educated Indians."[43] This dilemma which inevitably attends the cognitive path, we shall find, is a central theme in the metaphysics of N. V. Banerjee to whom we shall turn next.

Nikunja Vihari Banerjee

Nikunja Vihari Banerjee (1897–1982), unlike the two philosophers already discussed, was not troubled by the diversity of philosophical viewpoints. He saw the philosopher's task as that of working out a distinct philosophical position of his own regardless of what others might have said. As an Indian philosopher, he inherited a tradition of which a large part concerned itself with the problem of human liberation, this problem being understood as a metaphysical one. The philosophical scene was vastly different from the Indian one in the London of the early thirties when he went there to undertake his doctoral studies. Epistemological realism was very much in the air and it is not surprising that this became an important strand in Banerjee's own thinking throughout his life. After being a pupil of Brajendranath Seal, Hiralal Haldar and S.K. Maitra in Calcutta University, he was now to have John Macmurray as his supervisor, a fortunate circumstance, for Macmurray's interests were as wide as those of his pupil. Without insisting on residence in London, Macmurray packed him off to a village in Sussex where he was to work on his own on Kant's philosophy of the self. Both Chubb and Banerjee worked out their philosophies for the most part on their own, unlike Bhattacharyya who, as we saw, always remained under the strong influence of his father.

In the round I might anticipate what follows by saying that, in my view, Banerjee fills up many notable lacunae as far as Indian philosophy is concerned. This might sound a tall claim and quite surely he would not have claimed this for himself. But consider the following. Liberation had been viewed in a transcendental manner, and, the *jivanmukti* idea apart, for the most part in other-worldly terms. Cognition and morality had been kept in separate baskets. The undeniable plurality of selves had only too often been swept under metaphysical carpets or, less philosophically, relegated to the sphere of a hierarchical society. Temporality had scarcely been taken into serious metaphysical account other than by the Buddhists. It had been largely assumed that egoity should be overcome, but the remedies prescribed had been drastic, sometimes going so far as to throw out the baby with the bathwater. The suffering to which man is heir had been bemoaned by all, but it had only too often been assumed that the way out involved a turning one's back on all that is human and all that could be described as "joy of living". Here are but a few of the issues on which Banerjee has refreshingly new things to say.

As in the case of any other philosopher, the reader can often detect influences even within the structure of innovative stances. The influence of realism and neo-realism has already been mentioned. To both, Banerjee's own response is striking. We find this chiefly in *Concerning Human Understanding*.[44] The sense datum theory, he maintains, is none other than a spurious attempt to give a status to the erroneous object. Perspectival and non-veridical aspects of things cannot be put under the same umbrella. The status of an illusory aspect is indeterminate and indeterminable. The novelty of this criticism can be appreciated if it is set aside the usual contexts of discussion including the problem of causality, the relation of sense data to scientific objects, or more technical questions regarding how a categorical proposition ("there is one such") can be derived from a series of hypotheticals. Moreover, he was alive to the epistemological puzzles involved in perceiving a common world at a time considerably before this was generally granted to be a problem in theory of perception. Banerjee would later say that error entered not only into our understanding of the perceptual world but into our own ill-conceived sense of insularity.

Throughout his thinking, his sustained dialogue with Kant on the one hand, and Shankara on the other, shows itself. Of the two, his relation to Kant is the more important. In *Kant's Philosophy of the Self*[45] he holds that "the analysis of moral consciousness reveals the fact that the self has a moral ideal and not that it is, as Kant holds, the source of moral laws to which it is itself subject, and in virtue of its subjection to which, it is free as a self-determined being."[46] He then goes on to say that "personality does not belong to the self as such but the body as associated with the self or to the self as associated with its body". This negatives the possibility of immortality, nor is the latter regarded by him as in any way a necessary implicate of morality. As far as Kant's mention of the human body is concerned we have to look at his pre-critical treatise, "Concerning the Ultimate Foundation of the Differentiations of Regions in Space"[47] in which the ultimate foundation is treated precisely in terms of the body. Even so there is no doubt that in the first *Critique* Kant shows more interest in time than in space and in this Banerjee follows him. Banerjee's treatment of the body is confined to whatever is tied up with "biological birth" (his language) and this in turn is associated with a "gamut" of emotions all of which concern human bondage.

He regards Kant's discovery that "the self is other than Nature" as a valuable one but concludes from this that it is without attributes. As far as the relation between selves is concerned, this concerns *praxis*, not *theoria*. For example, it does not make sense to ask whether "denken überhaupt" is singular or plural, or how 'we' is constituted. Banerjee's line here is to eschew formalism in favour of concreteness. He will later offer an ontological understanding of the relation between selves and also link this up with what then is to be done. The ground theme is set in the work on Kant and in his hunch that in Kant the relation between the transcendental 'I', the empirical self, and the noumenal self is not convincingly elucidated.

Whereas Banerjee's first published work dealt with epistemology, in subsequent years the focus of his attention turned to an analysis of the human situation. When he met Heidegger in 1957 the two philosophers found they shared a common conviction that cognition was not enough. Both concentrated on the position of man in the world and saw this as an existentially ontological matter. Both, following Kant, sought a key in the nature of human endowments. Nonetheless both arrived at anthropocentric matters through very different routes. The preface to Banerjee's work *Language, Meaning and Persons*[48] contained the important 'declaration' that "in philosophical investigations, linguistic analysis, however necessary and useful it may be, must culminate in the surrender of the all-importance of language, yielding place to philosophic vision".[49] The preface also contains the viewpoint that philosophy has undergone misadventures, not through mock battles, but through having had an "age-old nomadic career". He had first thought of calling his last work "Adventures of the Human Mind", the idea being that these would yield a proper path for the philosopher. However, he eventually gave the work, which appeared posthumously, a different title. To regard doing philosophy as an adventure seems to me to offer a neat reply to those who think that we must choose between 'descriptive' and 'revisionary' paths, the modern form of the dichotomy between 'critical' and 'speculative' philosophies. Philosophizing is not a linguistic exercise. It involves a grappling with the human predicament and an attempt to find a way out of the situation of bondage. All this is tied up with in-depth awareness of who and what we are and a reaching out which to some might seem speculative, but which Banerjee used to refer to through words like demand/vista/destiny. He was sufficiently appreciative of the ancient Indian insight that we cannot become other than what we are to see the central 'adventure' in the discovery of a realm from which, through our own ignorance, we have been in a state of exile. This is the "realm of persons".

Language, Meaning and Persons sets the stage for Banerjee's most original contribution to metaphysics, his view that selves are "essential to one another". Our intelligence, he says, has been bewitched not by language but by the belief that men are mere individuals. This is not ignorance simpliciter if ignorance be taken to be a cognitive deprivation. The bewitchment comes out of the gulf between knowledge and performance which pertains to our ordinary actions (the *vyavahārika* level, if you like, except that for Banerjee I suspect there is no meaning in speaking of a *paramārthika* level). Human activity can be understood either as 'knowledge-with-a-view-to performance' which he calls 'Imagination', or 'performance-with-a-view-to knowledge' which he calls 'Understanding'. In addition we have the human passions which are possessed of power of a blind and reckless kind and with which Understanding is often allied. The examples given of the gap between knowledge and performance include eating while others go hungry, and motoring while others go on foot. This brings out the point that Banerjee has in mind activities which impinge on the lives of others, either positively or by default, and the way in which non-fulfilment of social norms of conduct is so often characteristic of our day to day existence.

He now introduces what he calls the temporal triad, before-now-after. Knowledge is related to the 'now' in the case of perception and to the 'before' in the case of memory. What remains is the 'after' and this he associates with the activity of Imagination and 'the plan for action'. It is the role of Imagination to *envisage* 'the plan for action'. Banerjee is not talking here about different particularist strategies for the future. Imagination has its source in the 'we', that is the realm of the personal. This gives it a certain authority. But it is nonetheless vulnerable. The Butlerian problem about power and authority *vis-à-vis* conscience is transferred by Banerjee to Imagination. Imagination seems all set to be in a position to redeem the human condition, but we find that its freedom and authority are fragile, being under threat from the power of the passions. The question then arises not of removing human passions, for that is impossible, but of shorning them of their dehumanising power.

The analysis next shifts to the problem of time. In the posthumous work entitled *Knowledge, Reason and Human Autonomy*[50] time is referred to as "the form of Imagination". Time is not to be identified with the temporal triad before-now-after, for 'before' and 'now' are but "temporalised spatial determinations".[51] Time is essentially the 'after' conceived as "ceaseless and endless". As for the content of Imagination, Banerjee identifies this with "the one universal plan for action" which points to the "remedy for the deprivation with which our ordinary actions are infected".[52] This gives the important clue that the central human flaw is not ignorance, the traditional source of human bondage according to Indian thinkers, but is to do with some defect in performance. The basis for this diagnosis is set out in an analysis of different types of otherness.

At the level of basic otherness there is absence of inter-communication and this is the relation that obtains between ourselves and material objects. In the case of mere otherness there is at least behaviourial inter-communication and we are able to relate to animals in this way. But both of these fall far short of mutual essentiality which is proper at the human level and wherein otherness is conquered. This, however, remains an ideal, for human relations more often than not fall into other paths, including those of indifference and hostility. Banerjee distrusts intersubjective stances based on feeling, finding these transient and unreliable (cf. Kant's distrust of *Schwärmerei*).

What then can take us beyond knowledge, that is to say, beyond the stance of 'I' versus the other. In approaching this question Banerjee makes an excursion into the scope of art, religion and morality which are all spheres of practical reason, and which in his view are all attempts to conquer otherness in one way or another. Art is based on the will to create and works of art *invite* appreciation. Religion is founded on the will to believe and arouses the sense of surrender. Moral experience witnesses to the will to adjust, and moreover, the will to adjust to a life of bondage. One would scarcely expect to find in the symbolising activities of the theoretical disciplines, mathematics and logic, a means of going beyond knowledge. Technology can only too easily serve the interests of individual and collective egoity (my phrase). Linguistic communication can always fan hostility and therefore bears no necessary relation to the actualisation of selves in their essentiality to each other. As for society, this at best can offer a life of "I *and* others". All available alternatives, then, seem to bear little or no relation to the fundamental predicament of the human being, his illusory sense of insular individuality *vis à vis* his fellows, and his consequent sense of alienation.

Banerjee's next move is to find in the Kantian term 'task' a way out of the existential impasse, maintaining that "man's being with his fellows is a *task* just as his being held in the relation of mutual otherness with his fellows is an illusory supposal".[53] The humanising task is to be embarked on through the discovery of certain fundamental practical principles. This is tied up with the preceding stages of his analysis as follows:[54]

> "... it is *time as after*, the form of Imagination, which provides the concept of being-with — the concept which, as we have already seen, is the most suitable means of the understanding of the peculiarity of human existence, and which, moreover, proves to be the starting point of an analytical process which ultimately yields the idea of the principles of strictly human conduct".

These principles, which are regulative, are the following:- overcoming ego-consciousness, conquering the dread of death and its compensatory counterpart, the desire for immortality, and bearing the cross of disvalue. The pursuit of these principles, Banerjee thinks, would bring about liberation from bondage. Such a liberation would involve a sense of expansion, communion, love, blessedness, and joy of living.

Not to be missed as a background to all this is Banerjee's perceptive criticism of Advaita Vedanta, especially in his *The Spirit of Indian Philosophy*,[55] on the ground that it involves an unwarranted inflation of selfhood. The empirical world is not to be down-graded through *māyavāda*, nor is *ānanda* postponed to a putative future state but rather regarded as a characteristic quality of *living* in the realisation of a sense of essentiality to one's fellows. His work on Buddhism[56] inclined him to believe that there could be a remedy for suffering and convinced him of the importance of *karma* and of rules of conduct, factors which could be transformative of life in spite of the causal stringency which prevails in the order of nature. In *The Future of Education* (1976)[57] the three regulative principles are elucidated with reference to the three paradigmatic figures of Socrates, Christ and Buddha. In Socrates Banerjee finds the nobility of the self-critical *examined* life and lack of concern for the hereafter. The phrase "bearing the cross of disvalue" seems to synthesise the self-sacrificial love of Christ and the compassion of the Buddha. Everything points to a linking up of the notions of persons, communion and love. But talk of principles of conduct does not quite amount to a metaphysic of love. The principles of conduct, he once told me, are to be regarded as the *manifestations* of love. Now the condition of love between man and man is not a fact but a *task*. When pressed on whether he holds a theory of potentiality or not, Banerjee treads warily, and here I draw on conversations with him.

Banerjee was suffiently steeped in Indian traditions to prefer a metaphysic of being to one of becoming. And yet he was firm on the point that there is clearly something *to be done* if man is to live in full mutuality of relationship with his fellows. He speaks in terms of the actualisation of what man is *destined* to be. Another way of putting this is to say that man bears within himself 'the promise' of a life lived in a new way. The idea of promise seems to me to involve both ought and can. In any case it offers a treatment of being which makes room for plurality and for change. Since man is a temporal being he is subject to change. But this change needs to be directed in accordance with his 'nature'. What is unique about Banerjee's working out of this is how he is able to avoid the classical overtones either of an *eschaton* or of *moksha*, both of which, to him, smack of other-worldliness. He advocates a thoroughly this-worldly metaphysic, and it is in fact very refreshing to find an Indian thinker envisaging such a concrete and available-to-every-man type of quality of life.

A further clue to the status of the principles of conduct can be found in his K. C. Bhattacharyya Memorial Lectures on *The Concept of Philosophy* delivered at the University of Calcutta in 1964. Reference has already been made to the influence of K. C. Bhattacharyya on the other two metaphysicians discussed in this survey. It was in the course of his Calcutta lectures that Banerjee introduced the notion of 'heuristic principles'. He indicated his agreement with K.C. Bhattacharyya on two points (1) I am unaccountably embodied (2) I am in personal relationship to other selves. However he disagreed with the tenet that "I am in communion, or in the relation of identity-in-difference with the over-personal self".[58] He adds, "What I am insisting on is simply that philosophy has no means of building a road from the assertion 'I am' to the assertion 'I am in communion with the over-personal self'".[59] He could detect an inbuilt narcissism in all talk of an over-personal self and with this he could have no truck. Banerjee wanted to steer clear both of absolutism and of insular subjectivity and finds an alternative in 'I with others'. However this is "not the content of any mode of our consciousness" but is "transcendental"[60] and there is a demand for its actualisation. It is to be noted here how Banerjee has shifted from a Kantian view of the transcendental (as condition of the possibility of experience) to what is set as a task and how he formulates the ensuing problem in language which has strong Kantian overtones, "...the problem, not as to whether, but as to *how* this demand can be fulfilled". Also to be noted is his insistence that humanising principles of conduct are not to be confused with ethical norms, for the latter are concerned with adjustment to a life of bondage whereas the former deal with its transformation.

Banerjee's metaphysic is a humanistic one and concerns man as an agent. His exposition of the concept of essentiality provides a twentieth-century version of the life of the *jivanmukta* thought through in inter-personal terms. The *Leitmotif* of love provides the key to the humanising principles which would give shape to life lived in the light of an awareness of that essentiality. In the context of contemporary Indian philosophy he kept a noticeable distance from those who tried to legitimate Indian philosophy on the world scene by claiming that Indian logic and epistemology had anticipated much that was now fashionable in the West. He detected that the motivation of those who "did Indian philosophy" in this way was largely anti-metaphysical. Taking refuge in technicalities, he cannily saw, was only too often a disguise for poverty of original thinking. His own work showed it was possible to shift philosophical attention from ultimate reality and yet be metaphysical. This could be done by a rigorous analysis of "the realm of persons", that is to say, an analysis of the total human situation.

Banerjee's final position evidences engagement in a lifelong cross-cultural dialogue with thinkers from many traditions. Brajendranath Seal, one of Banerjee's 'guru' figures, had written of the synthetic spirit of the ancient Hindus. Banerjee was bold enough to think out his own synthesis, but only after he subjected available standpoints to rigorous analysis and criticism. The themes which gripped his attention chime in with much that occupies philosophers in diverse parts of the world today, and among these themes the critique of objectivism is second to none. He provided a distinctively Indian style of humanism, building it moreover without the usual historic pillars of rationalism, science, or individualism and yet, nonetheless, giving it a firm metaphysical base.

Both Chubb and Bhattacharyya devoted their philosophical attention to the whence and whither of thought. Banerjee does this too, but the main thrust of his metaphysical investigations was with the whence and whither of man.

NOTES

1. "Philosophy in India, 1967–73", by J.N. Mohanty, *The Review of Metaphysics*, 28. 1 (September 1974): Issue no. 109.
2. Vide *Epistemology, Logic and Grammar in Indian Philosophical Analysis*, (The Hague: Mouton, 1971), Preface, p. 10.
3. Vide *Indian Philosophy Past and Future*, ed. by S.S. Rama Rao Pappu and R. Puligandla, (Delhi: Motilal Banarsidass, South Asia Books, 1982), p. 94.
4. *The Idealist Standpoint : A Study in the Vedantic Metaphysic of Experience* (Santiniketan, 1965), p. 282.
5. Presidential address entitled "Metaphysical Systems", given at the Fifty second Session of the Indian Philosophical Congress at Gauhati. Printed in *The Indian Philosophical Quarterly*, V 11 (January 1978). The comment about ideology needs to be understood in the context of events in the Calcutta of his day.
6. In *Current Trends in Indian Philosophy*, eds. K.S. Murty and K.R. Rao (Waltair; Andhra University Press, 1972), pp. 107–121.
7. *Ibid.*, p. 114.
8. Vide his paper on "The Concept of Philosophy" published in *The Proceedings of the All India Philosophical Congress*, 1960.
9. *Current Trends*, p. 115.
10. *Ibid.*, p. 115.
11. *Ibid.*, p. 116.
12. *Ibid.*, p. 121.
13. Published by Somaiya Publications Pvt. Ltd., Bombay, 1977.
14. *Assertion and Fact*, p. 13.
15. *Ibid.*, p. 21.
16. *Ibid.*, p. 22.
17. *Ibid.*, p. 25.
18. *Ibid.*, p. 221.
19. *Ibid.*, p. 13.
20. Published by Concept Publishing Company, New Delhi, 1983. This and the previous book mentioned were both written while he was a visiting faculty member at the University of North Carolina.
21. *Faith Possesses Understanding*, p. 60.
22. *Ibid.*, p. 68.
23. *Ibid.*, p. 91.
24. *Ibid.*, p. 92.
25. *Ibid.*, p. 187.
26. *Person and God*, eds. G.M. McLean and Hugo Meynell (International Society for Metaphysics; and Washington: The University Press of America, 1988).
27. *Philosophy, Logic and Language* (Delhi: Allied Publishers Private Ltd., 1965), Preface.
28. *Ibid.*, p. 114.
29. In "Presuppositions of Science and Philosophy", *Presuppositions of Science and Philosophy and Other Essays* (Visva-Bharati, 1974), Chapter 1.
30. "Metaphysics — A Genuine Cognitive Pursuit", in *Indian Philosophy Today*, ed. by N.K. Devaraja (Delhi: The Macmillan Company of India Limited, 1975) p. 19.
31. *Contemporary Indian Philosophy* (London: Allen & Unwin, 1936; second edition, 1952), pp. 105–125.
32. This was his Ph.D. thesis submitted in 1945. It was published by Das Gupta & Co., Calcutta, 1953 (out of print), p. 153.
33. *Ibid.*, p. 208.
34. G.N. Bhattacharyya, *Studies in Philosophy*, ed. of Vol. II (Calcutta: Progressive Publishers 1957), Editor's Introduction, p. 31.

35. *The Philosophy of Kalidas Bhattacharyya*, ed. by Daya Krishna (Poona; I.P.Q. Publications, 1985), p. 151.
36. *Alternative Standpoints in Philosophy*, p. 66.
37. "The Nature of Reflection in Metaphysics", in *Current Trends in Indian Philosophy*, eds. K. S. Murty and K. R. Rao (Waltair: Andhra University Press, 1972), p. 37.
38. *See* "Philosophy in India, 1967–73", by J.N. Mohanty, *Review of Metaphysics* (September, 1974).
39. Allied Publishers Private Ltd., 1965, p. 41.
40. *Philosophy, Logic and Language* (Delhi: Allied Publishers Private Ltd., 1965), p. 155.
41. In "My Reaction", *The Philosophy of Kalidas Bhattacharyya*, ed. by Daya Krishna, p. 182.
42. Included in G. N. Bhattacharyya, ed. of I and II of *Studies in Philosophy* (Calcutta: Progressive Publishers, 1957).
43. *Philosophical Papers*, ed. by Kalidas Bhattacharyya (Visva-Bharati, Santiniketan: Centre of Advanced Study in Philosophy, 1969), p. 142.
44. Allen & Unwin, London, 1958.
45. Completed in 1932 but not published till 1974 by Arnold-Heinemann Publishers (India) Private Limited, Delhi.
46. *Ibid.*, p. 213.
47. *See* Immanuel Kant, *Selected Pre-Critical Writings*, trans. by G. H. Keyford and D. H. Walford (New York: Barnes and Noble, 1968), p. 38.
48. Allen & Unwin, London, 1963.
49. *Ibid.*, p. 10.
50. Somaiya Publications Pvt. Ltd., Bombay, 1985.
51. *Knowledge, Reason and Human Autonomy*, p. 135.
52. *Ibid.*, p. 144.
53. *Ibid.*, p. 240.
54. *Ibid.*, p. 243.
55. Arnold-Heinemann, New Delhi, 1974.
56. *Buddhism and Marxism* (New Delhi: Orient Longman, 1978).
57. Progressive Publishers (Calcutta, 1976).
58. *The Concept of Philosophy* (Calcutta University, 1968), p. 73.
59. *Ibid.*, p. 74.
60. *Ibid.*, p. 82.

SELECTED BIBLIOGRAPHY

General

Books

Contemporary Indian Philosophy (Second Series) ed. Margaret Chatterjee. First edition Allen & Unwin, London 1974; second edition Motilal Banarsidass, Delhi 1988.

Self, Knowledge and Freedom, eds. J.N. Mohanty and S.P. Banerjee. The World Press Private Limited, Calcutta, 1978.

Indian Philosophy Today, ed. N.K. Devaraja. The Macmillan Company of India Limited, Delhi, 1975.

Indian Philosophy: Past and Future, eds. S.S. Rama Rao Pappu and R. Puligandla. South Asia Books, Motilal Banarsidass, Delhi, 1982.

Articles

"Philosophy in India, 1967–73", by J.N. Mohanty, *The Review of Metaphysics*, 28/1, (September, 1974); Issue no. 109.

Jehangir N. Chubb

Books

1. *Assertion and Fact* (The Categories of Self-Conscious Thinking). Somaiya Publications, Bombay, 1977, p. 228.
2. *Faith Possesses Understanding*, Concept Publishing Company, Delhi, 1983. p. 225.

Articles

1. "Critical Philosophy", In K.S. Murty and K.R. Rao (eds.), *Current Trends in Indian Philosophy*, pp. 107–121. Waltair, Andhra University Press, 1972.
2. "The Hindu Metaphysical Tradition on the Meaning of the Absolute". In G.M. McLean and Hugo Meynell (eds.), *Person and God*. International Society for Metaphysics and the University Press of America, Washington, 1988.

Kalidas Bhattacharyya

Books

1. *Alternative Standpoints in Philosophy*. Dasgupta & Co., Calcutta, 1953 (the mss. were completed as early as 1942, now out of print).
2. *Object, Content and Relation*. Dasgupta & Co., Calcutta, undated. 160 p. (now out of print). This was his P.R.S thesis, 1951.
3. *Philosophy, Language and Logic*. Allied Publishers, Bombay, 1965. 280 p. (now out of print).
4. *Presuppositions of Science and Philosophy and Other Essays. Centre* of Advanced Study in Philosophy, Visva-Bharati, Santiniketan, 1974.

Articles

1. "Metaphysics — A Genuine Cognitive Pursuit". *Indian Philosophy Today*, ed. N.K. Devaraja. Macmillan, Delhi, 1975.
2. "Nature and Freedom", *Man and Nature*, ed. George MacLean. Oxford University Press, Calcutta, 1978.

Nikunja Vihari Banerjee

Books

1. *Concerning Human Understanding*. Allen & Unwin, London, 1958. 333 p.
2. *Language, Meaning and Persons*. Allen & Unwin, London, 1963. 173 p.
3. *The Concept of Philosophy*. Lectures delivered in the University of Calcutta, 1964, and published by that University in 1968. 86 p.
4. *Kant's Philosophy of the Self*. Arnold-Heinemann Publishers (India) Private Limited, Delhi, 1974. 220 p.
5. *Knowledge, Reason and Human Autonomy* (An Essay in Philosophical Reconstruction). Somaiya Publications Pvt. Ltd., Bombay, 1985. 265 p.

The Pandit World with special reference to problems of logic

S. SANKARANARAYANAN

INTRODUCTION

Indian Logic is vast. Its problems are many. The Pandit World is also quite big. The erudition of pandits is well known and their mastery of many difficult original texts of the *śāstra* is amazing. Many pandits have suggested many solutions in their speeches and in their writings. Their speeches have not been recorded. Many of their writings are also not available; yet a great number of them are available in Sanskrit. The available material itself looks like an ocean indeed! But how much of it have I drunk and digested? Hence to write under the above heading is a precarious job. Only a fool could venture to contain this ocean-like subject within a peppercorn-like paper of some fifteen pages. Inescapably, therefore, one has to choose one or two convenient problems only for the paper. Let me proceed. I am also a pandit, in a way.

1. PROBLEMS OF AMALGAMATION OF THE NYĀYA AND THE VAIŚEṢIKA SYSTEMS[1]

1.1 *Independent development*

The two systems of Nyāya and Vaiśeṣika developed quite independently to begin with, as their respective *sūtra*-s indicate. Of these two, the Nyāya system specialised in epistemology and methodology of investigation and it was not very much interested in metaphysics and ontological categories. But the case seems to have been in the reverse order with the early Vaiśeṣika system. The *NS* commences with the very expression *pramāṇa* 'means of cognition' and ends with a *sūtra* on *hetvābhāsa* 'fallacious reason', a *pramāṇābhāsa* 'pseudo-*pramāṇa*' and it discusses over and again *pramāṇa* in the middle, too. All these features (*upakrama-upasaṁhāra-abhyāsa-parāmarśāḥ*) connote that the main concern of the N. system is epistemology and methodology (See below). Further, among the sixteen items enunciated and listed in the very first *NS*, as the subject matter to be taught in the N. system fifteen are concerned with epistemology and methodology of investigation. The remaining one, i.e. the second in the list, A. designates as *prameyam*. This term is used here in the specialised sense 'objects to be rightly cognised for gaining liberation' [*NB*, 32, 183] and not in the general sense 'objects of right cognition', in

Contemporary philosophy. A new survey. Vol. 7, 107–120.

which sense it is used in the Vaiśeṣika system to denote collectively all the six or seven categories (*padārthāḥ*) of the system, i.e. all that exists in the universe, including the universe itself. A. employs the word *padārtha* only once [*NB*, 2.2.66], that too in the singular, to denote 'word-meaning'. Early authors like Vāt., U. etc., use the word only to refer to the sixteen, listed in *NS*, 1.1.1 — which they designate *sat* 'existents' [*NB*, *NV*, 28 etc.] — as subject matter, dealt with in the system; and they do not seem to use the word in the V. sense of the term. A sort of non-recognition of this difference in the N. concept of *prameya* and *padārtha*, as against the V. concept of the same, has driven some pandits to find fault with the logic and propriety in A.'s classification of 'existents' as sixteen [H. Narain, 56–57; Varadacarya, iv]. In fact in the *NS*, 1.1.1, A. seems to intend to convey the idea that 'liberation is attained by the right cognition of the *prameya* (No.2, in the list) by means of the rightly cognised *pramāṇas* (No.1 in the list) by following a methodology of investigation indicated by *saṁśaya* etc., (the rest of the list) rightly cognised; and hence a right cognition of all the sixteen leads to liberation either directly (in the case of No.2) or indirectly (in the case of the rest)' [*NB*, *NV*, 32, 33].

On the other hand, unlike A., K. does not classify or define the means of right cognition (*pramāṇāni*) required for his system, even though he touches them here and there incidentally. He uses the word *pramāṇa* perhaps only once, [*VS*, 9.2.4] and that too as a synonymn of *hetu* 'probans'. At the same time, in his entire V. *sūtra*, K. enunciates, defines and examines what he clearly designates (1.1.4) *padārthāḥ*,[2] i.e. the six ontological categories. Thus it is certain that these two systems are mainly concerned with two different, but not mutually opposing, subjects, each system claiming liberation (*niḥśreyasam*) as its final goal. Hence the tradition calls A.'s system *pramāṇśāstram*, 'science of means of right cognitions', and K.'s system *prameyaśāstram* 'science of objects of right cognition, i.e., 'categories'. That is why the post-K. works of the V. system are mainly named after categories as a whole — such as *Padārthadharmasaṅgraha*, *Daśapadārthasāra*, *Saptapadārthī*, *Padārthatattvanirṇaya*, etc. Even the *Tarkasaṅgraha*, mainly a V. work, is, in fact, *Vaiśeṣika-padārthasaṅgraha* only.[3]

At the same time the treatises of writers of heterodox systems, mainly intended to demolish the N. epistemology, are found, named after the means of right cognitions as a whole—such as the *Pramāṇasamuccaya*, *Pramāṇavārttika* etc. And the N. treatises of subsequent ages, viz., *TC*, *NSM*, *TS* etc., have their subdivisions named after the names of individual means of cognitions, i.e., *Pratyakṣam* (perception), *Anumānam* (inference), *Upamānam* (comparison) and *Śabdaḥ* (verbal testimony). Pandits have written a good deal on points of differences between the N. and V. systems.[4] But the above basic differences are not always fully recognised.

1.2 *Process of amalgamation*

We saw above that the N. and V. systems were specialising in *pramāṇa* (means of cognitions) and *prameya* (objects of cognition) respectively. So it is nothing but natural that they started, in course of time, attracting one another, as their subject matters are interdependent.

In his *NS* [1.1.29] A. defines *samānatantrasiddhānta*, 'the tenet peculiar to one system'. His *samānatantra* signifies only 'one individual system' as opposed to two or more systems, and it does not denote 'allied or sister systems', as it is often understood by pandits [See e.g. *Primer*, xlil; etc.]. Under this *sūtra*, Vāt. cites *asatkāryavāda* (the theory that the effect, before its birth, not being inherent in its cause) as a tenet peculiar to the Yogas (not the Naiyāyikas), i.e., the Vaiśeṣikas, as some great pandits hold [*LD*, 19; *GJ*, 60].[5] But, with no demur, Vāt. speaks of the internal organ (*manaḥ*) as a super sense-organ, as accepted by the V. system, in spite of the fact that this has been not included in the list of five sense-organs given by A. [*NS*, 1.1.12]. Again, he clearly recognises the existence of all the six V. categories, not listed in the *NS* [*NB*, 185]. Thus it would appear that, even in early days, earlier than Vāt., the Naiyāyikas had started freely adopting some tenets and theories of the V. system; but they seem to have preferred to keep that system at a safe distance.

U. takes a big leap forward. He speaks of the V. system as complementary to the N. system. He declares in clear terms a general dictum: A doctrine of the other (*paramatam*), i.e., the V. system, when not spoken against, is to be treated as a doctrine accepted by us, viz., the Naiyāyikas [*NV*, 125. Cf. *Ency*, 12]. Because U. calls this dictum 'a methodological device' (*tantrayuktiḥ*) and because this dictum had already been ridiculed by Dignāga [*TT*, 125], it could not have been an innovation by U. The fact that U. refers to the V. system as a system of others (*paramatam*) indicates that the Naiyāyikas continued to treat the V. as an outsider, even during the time of U. and Vāc. Yet U. defends with all force, and more ably than P., the much controverted V. category, inherence (*samavāyaḥ*) and the important V. theory of 'the composite whole being separate from the component parts (*avayavāvayavinor bhedaḥ*)'. And he refutes more vehementaly the theory of 'identity of cause and effect (*kārya-kāraṇayos tādātmyam*)' [*NV*, 158 ff., 471 ff., 491 f.]. Perhaps the *NV* of U. marks an important stage in the process of the N. and V. amalgamation a process that is found developed in the *Saptapadārthī* of Śivāditya (12th century) a V. work incorporating into it the threefold inferences, the concepts of five limbs of syllogism and of fallacies (six, i.e., one, in addition to those of N. system) of probans (*SP*, 28–30) as in the N. system and unlike in *PB*. But *SP* does not favour the N. scheme of four means of cognition.

In fact, it is not totally improbable that Vāt, has used *Yōgāḥ* (See above) in a specialised sense 'allies'[6] signifying the Naiyāyikas and the Vaiśeṣikas as a single group or as one system. For, in the same context U. illustrates the *samānatantrasiddhānta* by citing the theory of sense-organs being made of the elements (*yathā, bhautikāni indriyāṇi iti yogānām*) [*NV*, 264]. And we know that this theory is a basic tenet common to both the Naiyāyikas and the Vaiśeṣikas and it does not belong to anyone else. [*NS*, 1.1.12–13; *PB*, 13–17, 26]. If this suggestion is accepted, then it would go a long way to prove that the process of alliance and amalgamation of the N. and V. systems had started earlier than the time of Vāt.

While commenting on the twelve *prameyas*, 'objects to be rightly cognised' [*NS*,

1.1.9], Vāt. clearly states that these items are given here because the right cognition of these alone leads to liberation (*apavargaḥ*) and their wrong cognition to the birth-death cycle (*saṁsāraḥ*). And, in this connection, Vāt. recognises the existence of the six V. categories and their innumerable subdivisions — all being *prameya*, and all falling beyond the twelve-*prameya*-scheme of A. Here, in this context, we find no reference to any other categories of any other system, like those of the Sāṅkhyas. From that time onwards, the Naiyāyikas have been using the V. categories for establishing, or rather illustrating, their own epistemological and methodological theories. Were the Naiyāyikas, down the centuries, conscious of the inadequacy of their own system in not having suitable categories and hence they were depending more and more on the V. categories? It is to be noted that all the time, neither P. nor his V. successors are found to depend so much on the N. system. This may be one of the reasons why a syncretization of the two systems could not be effected successfully. (See below).

1.3 *Anomalous results*

The lack, or impossibility (?), of complete syncretization and at the same time the desire on the part of some later writers to work out a sort of amalgamation, have led to certain anomalies: We know that the sixteen 'existents' of the N. system are only an insignificant fraction of the all-inclusive six or seven V. categories. Keśava Miśra (c.13th century) also knows this fact. He also knows that the 'objects' (*arthāḥ*, the fourth in the list of A.'s *prameyas*) are not more than the five sense-objects, like smell, taste, colour, touch and sound [*NS*, 1.1.14], and that they are only five among the twenty-four (or fewer) qualities of the V. system. Yet the author tries, in his *TB* [60 ff], basically an N. work, to contain in this *artha* all the V. categories. Evidently, this treatment is acceptable neither to the Naiyāyikas no to the Vaiśeṣikas.

Gaṅgeśa's famous *TC* is mainly an N. work with a definite purpose: to teach mainly A.'s four means of right cognition. Hence the treatise contains four parts under the headings perception, inference, etc., (*pratyakṣakhaṇḍaḥ*, *anumānakhaṇḍaḥ* etc.) as we saw above, all the four incidentally dealing with the V. categories. And there is also a fifth section, *guṇakhaṇḍaḥ* on the V. qualities , which the author could not adequately deal with in his earlier four parts. Similarly, the later works, like *NSM*, *TS*, *TD*, that are chiefly and avowedly V. works and are meant to teach the seven V. categories, discuss only incidentally, (*prasaṅgāt*) [*NSM*, 412; *TD1*, 323] but lengthily, the four N. means of right cognition, either as instruments of cognition, which is one of the qualities of the soul, or as means of knowledge (*jñānam*), which is one among the twenty-four V. qualities. Yet these works have sections with the same headings as those of the *TC*. It is strange that these authors have to name the sections of their works after those matters which, they openly say, they deal with only accidently and casually .

Again, this sort of incidental treatment of the very important subject, viz. the means of cognition of the N. system and a sort of disregard for the fact that the sixteen 'existents' of the N. system, are not 'categories' in the V. sense, seem to have compelled Mahādeva and Annambhaṭṭa etc., to have the exercise of explaining at length how those 'existents' could be contained in the V. scheme of seven categories. [e.g. *DK*, 50–54; *TD*2, 396–409]. All these look rather odd. But the pandit logicians continue to approve of it. Obviously, they have become more and more Vaiśeṣikas and less and less Naiyāyikas as far as their study of the categories is concerned, and *vice versa*, with regard to their views on the means of cognitions. Evidently, they are satisfied that these authors have worked out some apparent synthesis of the two systems — whether logical or otherwise.

Furthermore: the logicians have added the seventh category 'negation' (*abhāvaḥ*) to K.'s list of six categories and have maintained the same against all opposition. But they have not thought of providing any suitable relation for it to exist on its locus, just as K. has provided the sixth category 'inherence', as a common relation for all the other five categories (*pañcapadārthavṛttiḥ*) [*NV*, 59]. The new entrant, negation, had to wait centuries for a relation, until R., with shrewdness and boldness, added for the purpose another category, viz. 'the relation of being qualified (*vaiśiṣṭyam*)[7] to K.'s list'. To the same list pandits have also recently added some more items 'Qualificandness' (*viśeṣyatā*), 'Qualifierness' (*prakāratā*) etc., that may be designated as 'categories of technical nature' (*pāribhāṣikapadārthāḥ*).[8] Similarly to K.'s list of seventeen qualities [*guṇāḥ*, *VS*,1.1.6], P. had added seven more [*PB*, 3].

Thus while K.'s scheme of six categories has got additions, A.'s scheme of sixteen 'existents' has shrunk to a considerable extent. In fact, the pandits are now conversant only with eight or nine items[9] of the sixteen 'existents' of A.'s list. They have ignored the other items, obviously because these do not serve any important purpose in the new way of investigation that has replaced the old one. Modern pandits are themselves unconscious of these changes.

Further, R. has boldly chipped off two (out of the three) inferences, viz. the one, by the purely positive concomitance, and the other, by the exclusively negative one. These two inferences had influenced and controlled the logicians' thought, and their mode of investigation, at least since the time of U. But at the same time R. has also added one more means of right cognition, viz., 'presumption' (*arthāpattiḥ*) to A.'s list of four means of cognition [SS2]. Pandits, however, prefer to set aside and ignore these pertinent suggestions of R., obviously because they think that his views go counter to the ultimate authority, viz. the *NS* and the *VS*. Evidently they do not care for the fact that in the field of logic there is no such thing as ultimate authority and they have not inherited the broadmindedness of the Ṛgvedic seers who could pray, 'May good thoughts come to us from all quarters!' [*Ṛgveda*, 1.89.1].

1.4 *The N. and V. amalgamation as viewed by others*

In general all philosophical systems receive light from Logic (*ānvīkṣikī*) as stated by Kauṭilya [*KA*, 1.4] and by Vāt [*NB* 64]. Yet the writers of those systems do not hesitate to refer to, and to refute, certain specific theories of the N. and V. systems. This gives us an idea of how these authors viewed the N. and V. alliance. Let us see here the points of view of the Advaitins. On studying their views one gets the idea that they are not recognising such an alliance. Bādarāyaṇa assigns exclusively two sections, consisting of as many as seven *sūtra*-s [*BS*, 2.2. 11–17] in refuting the atomic theory, the theory of inherence and the theory of effect not inherent in its cause even before its birth (*asatkāryavādaḥ*)—all distinctly belonging to the V. system only. And he does not question, even once, any of the specific views of the Naiyāyikas. At the same time, many of *BS* are in the form of probans (*hetuḥ*), i.e. the middle term in syllogism (*nyāyaprayogaḥ*)—a feature indicating that the entire system of Vedānta is modelled on the lines of *Nyāya*. The Vedāntins seem to take pride in calling their system a *tarka*, a *nyāya*, reasoning, logic etc. [*BSSB* and *Bhāmatī*, 83]. Maybe A.'s *NS* is later than the *BS*. But, as indicated by Jayanta Bhaṭṭa [*NM*, I, 5], the principles enshrined in the latter must have been of much earlier date.

Śaṅkarācārya refers to A. reverentially (without even mentioning his name) by the honorific title *ācārya* 'master' and admires the logic contained in *NS*, 1.1.2 [*BSSB*, 121–122]. But, with great contempt, he condemns the abovementioned V. theories of K. and blacklists them as semi-nihilistic (*ardhavaināśikāh*). He also says that, on this account, V. theories are not accepted by the Vedists [*BSSB*, 523]. Perhaps this attitude of the Vedic orthodoxy explains why Vāc., a staunch Vedist, did not think of contributing any work to the V. system. But his voluminous *TT* is a very valuable and epoch-making contribution to the N. system. His contributions to other orthodox systems of philosophy are also very significant.

Further, while exposing the non-finality and baselessness of pure logic (*tarkāpratiṣṭhānam*), Ś. brackets Kaṇāda with Kapila and describes both of them as eminent logicians[10] and as founders of philosophical systems (*tīrthakarāḥ*—*BSSB*, 448). Evidently Ś. views these two teachers as experts in pure logic. In this context, the name of A., the author of *NS* and the 'founder of the N. school', is conspicuously absent.

The celebrated Advaita commentator Brahmānanda Sarasvatī (17th century) censures V. as the worst and vile among the philosophers, i.e. orthodox philosophers (*vaiśeṣiko hi dārśanikānām adhamaḥ*), because the latter, it is alleged, does not recognise the validity of verbal testimony as a means of right cognition. But he is not so harsh with the *Tārkika*, i.e. the Naiyāyika. He simply belittles the latter by calling him an inferior philosopher (*nyūnaḥ*), on the ground that the latter accepts the validity of verbal testimony only partially [*Nyāyaratnāvali*, 49]. On the abovementioned charge, Pt. MM. Anantakrishna Sastri too calls V. an atheist (*nāstikaḥ*) [*BSSB*, Introd.1].[11] Whether the charge of non-recognition of the validity of verbal testimony levelled against K. is justified or not, we will see later. But this much one may say now: these Advaitins perhaps view the N. and V. combination as an alliance between the holy and the unholy.

Further Brahmānanda Sarasvatī quotes profusely in his *Laghucandrikā* [276–80, 331] from Udayana's *Bauddhādhikāra* i.e., *Ātmatattvaviveka* [220–230] and tries to establish that the Nyāyācārya is a zealous admirer of Advaita-Vedānta.[12]

Years back, Pt. Lakshmana Sastri Drāviḍa, a great scholar in Nyāya and Advaita Vedānta, put forward this thesis: "It is accepted on all hands that there are six indicators for ascertaining the real purport of any given treatise (*ṣaṭ tātparyaliṅgāni*), such as the opening and concluding passages; repetition in the middle; and so on (See above). We have not even one of them indicating that *NS* really intends to teach philosophy (*adhyātmavidyā*). Rather, they seem to suggest that the N. system is interested only in teaching the science of reasoning and investigating (*nyānaḥ, ānvīkṣikī*), a science that helps every branch of learning, just as grammar does.[13] So, whenever *NS*, *NB*, etc., deal with the V. theories, they do so, only by way of illustrating their own N. methodology of investigation and not by way of establishing and proving the correctness of those V. theories. The Naiyāyikas, in course of time, have lost sight of this fact and have become themselves semi-Vaiśeṣikas [*LD*, 19 ff]". Thus, according to this panditji, there is no such thing as a special alliance of the N. with the V. system; for the former is an ally of each and every branch of learning.

Recently Pandit V.S.V. Guruswami Sastri has built up an altogether different thesis.[14] He affirms: The self-knowledge taught in N. system is the same as that which is imparted in the Advaita Vedānta. So, Vāc. betrays his ignorance of this "fact" when he says, in the *Bhāmatī* (*BSSB*, 121), that they are different. Accordingly, in this pandit's view, the well-known N. and V. amalgamation would be an alliance between an Advaitin and a Dvaitin — one abhoring the other! We need not examine here, how far the above views of these pandits are tenable.

Let these writers on philosophy remain on one side. In ancient India there had been at least a few monarchs and scholars who seem to have viewed the N. and V. systems as a single united system of logic. Some two copper-plate grants from Andhra Pradesh belonging to the reigns of the Cāḷukya emperor Pulakeśi II and his nephew Jayasiṁha (both of the 7th century), describe some of their recipients as scholars, well versed in what is called *aśīti-tarka* 'A logic of eighty'. This expression is probably used to denote the N. and V. systems, as one single unit. We know that the N. system deals with sixteen 'existents' which are subdivided again broadly into forty-five by A. in his *NS*. Likewise, the V. system deals with six categories which are also sub divided into thirty-five items by K. in his *VS*. (More items originated in the post-K. age. See above). So *aśīti-tarka* may signify the combined N. and V. systems, collectively teaching eighty subjects (45 + 35 = 80). A parallel instance mat be found in the expression *ṣaṣṭi-tantra* 'A. philosophy of sixty' [*SK*, 78] that signifies the Sāṅkhya system teaching sixty subjects.[15] These grants perhaps indicate that a sort of syncretization of the N. and V. systems was visualised sometime round about the seventh century A.D., at least in the Krishna-Godavari delta.

However, it is a fact that the Naiyāyikas and Vaiśeṣikas generally preferred to remain separate down the ages, even though they — perhaps the Naiyāyikas alone, as we saw above — were prepared to have some mutual adjustments. Why could not they work up to a total and enduring amalgamation? What could have been the factors standing in their way to such an achievement? Answers to these questions must, of course, be quite interesting.

2. Problems of means of right cognition

No other system of Indian Philosophy attaches so much importance to the study of the means of right cognition as the N. system. Hence it is rightly called *Pramāṇaśāstra*, 'Science of means of cognition' (see above). Since the time of A., logicians recognise four such means of cognitions, viz. Perception, Inference, Comparison and Verbal Testimony. And they do not take seriously P.'s contention that the first two alone are the means of cognition and these two include the other two. There are scores of problems posed by each of these means. But we shall touch here only on a few of them.

Before we proceed, it may be good to notice an interesting result of the above stand of P. We saw above that A. had not given a good scheme of categories to the N. system, nor K. a well defined scheme of means of cognition to the V. system; and that both these systems started attracting one another as their respective subjects happen to be mutually complementary. But now P. has introduced in the V. system a scheme of two such means and has also shown how to accommodate all the other means of cognition in these two. Naturally the Vaiśeṣikas of the post-P. age must have felt no need to look up to the N. system for means of cognition. The spirit of interdependence, a necessary condition for syncretization, must have started waning in the minds of Vaiśeṣikas. Naturally this stood in the way of a successful amalgamation of the two systems. This suggestion of ours, of course, goes counter to the view that since the time of P. the Vaiśeṣikas showed a tendency to syncretism [*SP*, Introd. x].

2.1 *Perception*

A.'s basic definition of perception is: Perception is a cognition, produced by sense-object-contact [*NS*, 1.1.4]. This has its parallel in the definitions of the same in other systems too [*Jaimini-sūtra*, 1.1.4; *SK*, 5]. The Naiyāyikas first classify perception broadly into two: indeterminate and determinate perceptions (*nirvikalpakam* and *savikalpakam*).

2.1.1 *Indeterminate and determinate perceptions*

Surprisingly, pandits have discussed the origin of the concept of the first one. They hold different views on the subject. Pt. Jagannāthaopādhyāya believes that it came to the N. system from the Buddhist sources. Dr. Varadachary opines that its origin cannot be traced to any particular original text of Indian Philosophy. However the generally accepted view[16] is that, besides defining perception, the *NS*, 1.1.4 itself, classifies it into two, mentioned above; and in the *sūtra* these two are mentioned by the words *avyapadeśyam* and and *vyavasāyātmakam* [*TT*, 110 ff., 114 ff.; *VV*, 120].

But Pandit Srinivasa Sastri has recently shown in detail that the concept of indeterminate perception can be traced to *VS*, 8.1.9; it was devéloped by P. [*PB*,

94–95], by the Buddhists and by Kumārilla Bhaṭṭa and his followers; Vāc., being influenced by them all, and also by his N. teacher Trilocana, has introduced it afresh into N. system, suitably interpreting in *TT*, the N. *sūtra* concerned. For, Sastri also points out, in his earlier work *Nyāyakaṇikā* the same Vāc. had an occasion to comment on *NS*, 1.1.4; but there his interpretation of the *sūtra* reveals that at that time he himself had been unaware of those concepts as being contained in the *NS*.

2.1.2 *Influence of N. on the Vedānta system*

Jayanta Bhaṭṭa refers to the views of earlier commentators according to whom the word *avyapadeśya* is to ward off from the realm of perceptual cognition those cognitions that stem from verbal authority, even when there is the sense-object-contact [*NM*, I, 73 ff]. Here one may note the following interesting controversy, well known in the Advaita school. In Vedānta, Maṇḍana Miśra and Vāc. build up the theory that the verbal authority, viz. the Upaniṣadic 'Great sentences' (*mahāvākyāni*) like 'Thou art That' produce only indirect cognition (*parokṣajñānam*, i.e., *Śābdabodhaḥ*) which, when augmented by meditation etc., leads to direct knowledge (*aparokṣajñānam*, 'perception') of the total identity of the Jīva with the Brahman. This theory is refuted by Sureśvara and other Vedāntins of the *Vivaraṇa* school who on the basis of *Śruti* accept *śābdāparokṣa*, 'perceptual cognition stemming also from the verbal authority'.[17] It appears not improbable that Maṇḍana Miśra and Vāc. had received inspiration from the above thesis of the early Naiyāyikas referred to by Jayanta Bhaṭṭa. It may also be worthwhile to investigate whether Maṇḍana and Vāc, could also have thought along the following line of the Naiyāyikas: maybe there are some instances where the effective causes of both the perceptual and verbal cognitions of one and the same object coexist. In those instances what is born must be perceptual cognition only, no doubt; but it cannot be said to be the result of the verbal authority, the cause of verbal cognition. For the latter cause is not present in other instances of perception and hence it must be a dispensable antecedent (*anyathāsiddham*) as far as the result, perception, is concerned. [cf. *TD*, 232 f.]. And so *śābdāparokṣam* is a self-contradictory expression (*vyāghātavacanam*), logician pandits would say. To this problem the *Vivaraṇa* school has its own suitable solution, which, being not N. like, need not be detailed here.

2.1.3 *Cognition of negation*

The Naiyāyikas classify into six the sense-object-contacts (*indriyārthasaṁnikarṣāḥ*) responsible for perception. The sixth one, viz. the contact called 'qualifierness', is introduced to obtain the perceptual cognition of negation (*abhāvapratyakṣe viśeṣaṇatāsaṁnikarṣaḥ*) (U. also accepts the same contact for the perception of inherence). U. is probably the first known writer to list these six contacts in clear terms for specific purposes [*NV*, 94] and also to view the cognition of negation as a perceptual one. It may be noted that A. and Vāt. take the negation-cognition to be inferential cognition only [*NS*, 2.2.2; *NB*, 576]. There, in that context, while elucidating the *NS* and *NB*, the author of *NV* falls in line with them and he does not voice his personal view, aired earlier. P., too, evidently under the influence of A. and Vāt., brings negation-cognition under inference only [*PB*, 111].[18] Thus U. is most probably the first philosopher to view negation-cognition as perceptual. That he speaks in two voices on this point seems to signify that he still remains undecided on this question.

The Naiyāyikas of the subsequent ages, however, follow only the former personal view of U. [*TC*.—Bib. Ind. Ed-I, 669; *TB*, 51 f.; *NSM*, 438 ff.; *TS*2, 43 f.; *TD*2, 239 ff.). Pandits do not generally take note of all these differences of opinions among those early authors. Similarly they also seem to ignore another important fact: the cognition of pot-negation (*ghaṭābhāvaḥ*) is possible only if the cogniser in question has a knowledge of pot, the counterpositive relatum of the given negation. Thus the cognition of negation arises from a knowledge of counterpositive relatum, just as the cognition, inference arises from the knowledge called *parāmarśa*, 'subsumptive reflection.' Hence, it is fundamentally different from the perceptual cognitions of pot etc., which do not result from any other knowledge. The Navya Naiyāyikas too define perceptual cognition as 'a cognition, the efficient cause of which is not a knowledge'.

2.2 *Inference and its varieties*

A.'s *sūtra* on inference [*NS*, 1.1.5] is variously interpreted in *NB* itself. This has recently led Professor S. D. Joshi to conclude that there are three layers of thought, even in this limited part of the *NB*; and that Vāt. has only compiled the different interpretations of earlier authors available to him. U., too, adds many more interpretations to the *sūtra*; and, most probably borrowing from the Sāṅkhya and Buddhist sources, he has introduced in the N. system the concepts of two additional inferences, i.e. the one by invariable concomitance of purely negative nature, and the other by that of exclusively positive nature. However, these two are rejected by R. (See above).

2.2.1 *Inference and presumption*

R. finds his own reason for accepting, in the place of the inference by exclusively negative concomitance, which he rejects, another means of cognition called *arthāpatti*, 'presumption', and here he goes against *NS* [2.2.2], *NB* [576] etc. (See above).

The writers of yore and pandits of recent times invariably consider the following two instances of presumption in order to deny it the status of being means of cognition [*NV*, 576 f; *TT*, 576, 581 f.; *TB*, 50; *NSM*, 796 ff.; *TD*2, 353 ff.]; (1) The reasoning from one's perception that 'Mr. X, who is alive, is not in the house', leads one to the conclusion (presumption) that Mr. X is outside (a case of *dṛṣṭārthāpattiḥ*). (2) One's hearing that 'Mr. X, who is fat, does not eat during daytime', leads one to the conclusion (presumption) that 'Mr. X eats at night' (*śrutārthāpattiḥ*). But the view of R. (see above) perhaps provides a stronger case in favour of presumption.

Similarly, one learns from 'the scriptures that "the performance of Vedic sacrifices leads one to heaven, (i.e. after one's death)". At the same time, in reality one sees the act of performing the sacrifices disappears soon, even when one is alive. A logical solution to this anomaly leads one to conclude (to presume) that there is an unseen agent (*adṛṣṭam, dharmaḥ*) resulting from the performance of sacrifice and it exists in the Self of the sacrificer and leads him to heaven. Likewise, one hears in the Upaniṣad, that "Here there are not many (i.e. not more than one)". But one actually observes many here in the world. By way of solving this anomaly one concludes that "the many" that appear are unreal (*mithyā*). These results cannot be easily achieved by inference alone, the Mīmāṁsakas and Advaitins would say. These cases of presumption are not taken into account by the logician pandits.

2.2.3–2.2.4 *Comparison and verbal testimony*

The growth of N. literature connected with verbal testimony has been phenomenal since the time of appearance of Gaṅgeśa's *Śabdakhaṇḍa*, the fourth part of *TC*. Many famous authors including R., Jagadīśa, Gadādhara and others are responsible for this growth. Of the recent additions to this branch of N. literature, the *Gūḍhārthatattvāloka* of the versatile scholar, Pt. Dharmadattasūri and the *Śābdataraṅgiṇī* of Paṇḍitarāja V. S. Subrahmanya Sastri are noteworthy.

Dignāga accepts neither comparison nor verbal testimony as separate means of cognitions. So does P. also [*NV* and *TT* 172, 177 ff; *PB*, 106–110]. But U. quotes and refutes only the former and not the latter. Why? Was he unaware of *PB*? Similarly Ś.'s elucidation of the V. atomic theory [*BSSB*, 505] seems also to betray that that *ācārya* too was unaware of *PB*. Researchers, however, assign P. to a period much earlier than that of U. and Ś. [*Ency*, 282]. Further, we earlier referred to the later Vedāntins, condeming V. as an infidel on the charge that the latter does not recognise verbal authority (of the Vedas) as a means of cognition. But Ś. does not level that a serious charge against V. Why?.

In fact, K. seems to describe twice — once in the beginning and once at the end of his work [*VS*, 1.1.3 and 10.2.9][19] *āmnāya* (the Veda) as an authority in teaching right conduct (*dharmaḥ*). Authorities on the V. system speak of the existence of a not a very popular branch of Vaiśeṣikas recognising three means of cognition, viz. perception, inference and verbal testimony.[20] *Thus it would appear*: (1) K. really recognises verbal testimony as a means of cognition and he accepts the Vedas as sole authority in matters of Dharma. (2) It is only P., who, under the spell of Buddhism, has tried, as far as he could, to deny verbal testimony the status of being an independent means of cognition, and to accommodate it under inference. The

Buddhistic leaning of P. is well known [*Ency*. 282]. (3) By this act of dethroning the verbal testimony in the above fashion, P. has earned from the authors of other schools for the V. system the condemnation of being infidel. (4) This unorthodox and also, perhaps, pro-Buddhist attitude of P. (and of those who followed him in the system), may be another factor which would explain why the out-and-out anti-Buddhist N. system could not happily be syncretized with V. system. For, as Jayanta Bhaṭṭa feels [*NM*, I,5], one of the main aims of the Naiyāyikas and of their unique system is to defend the verbal authority of the Vedas against the mighty forces of the unorthodox systems and to bring home that authority to those who entertain doubts or misconceptions in that regard.

Lets us hope that the Pandit World will soon take up and investigate these and many other similar problems of logic for the benefit of the scholarly world. Let us bear in mind that in the world of logic —

> "Not all views are correct, simply because they are old; and no view is wrong, merely because it happens to be new. Men of wisdom accept either one by discerning it. But the dullard is led astray by another's opinion."[21]

NOTES AND REFERENCES

1. This is one of the most fundamental problems of the N. and V. systems. *Primer* (xxiii. ff.) and *Ency*. (12 f.) give us an idea of it. We have similar questions in other systems too. E.g. scholars often discuss the question whether the Mīmāṁsā and the Vedānta systems make one syncretized whole or not.
2. So also does *PB* (2). Cf. *Ency*. (693. n.7).
3. *DP*, (52) and SS1 thereunder.
4. See e.g. *Primer*, xx ff.; MM. Kālipada Tarkācārya, *Nyāyavaiśeṣikadarśanayoḥ pṛthagdarśanatvam* in the *Nyāyadarśanabinduḥ*, Varanasi Sanskrit University Spl. Lecture Series, No.4 (Saṁ.2002); Pt. Sukhesvara Jha: *Nyāyavaiśeṣikayor bhedakatattvāni*: *Sāra*, XV, 311–18; *Paṇḍitarāja* V. Subrahmanya Sastri, *Nyāya-Vaiśeṣika-tantrasaṅgrahaḥ*: Tiruchirapalli, 1973; and so on. See also *Ency*, p.687, n.9.
5. The argument that *yoga* is a synonymn of *yukti* 'reasoning' and hence the word may signify both the N. and V. systems (*Primer*, xvi) is hardly convincing as that sense does not seem to fit in the context.
6. Cf. the *Bhagavadgītā*, 2.39 where *Sāṅkhya* and *Yoga* are used in unusual senses. See the glosses of Śaṅkara, Madhusūdana Sarasvatī etc., there on. Even the pandits who would identify Vāt.'s *yogāḥ* with Vaiśeṣikas or that of U., with Naiyāyikas have to take the word in unusual sense only. See above.
7. *Padārthatattvanirṇaya*, 76–77. The necessity of this *vaiśiṣṭyam* seems to have been felt by some earlier Vaiśeṣikas also. But, Śivāditya rejects it and tries to manage the contingency with what is called *viśeṣaṇa-viśeṣya-bhāva-sambandha* 'the relation of being attribute and attributive' [*SP*, 39].
8. Kālipada Tarkācārya, *Navyanyāyapāribhāṣikapadārthāḥ*: *Sāra*, XIX (No.4): 263–96.
9. See No. 1–4, 8, 10, 13, 16. For the names of these 'existents', see *Ency*, 222.
10. It may be of interest to note that Kauṭilya (3rd century B.C.) also clubs together the Sāṅkhya and Yoga (i.e., Vaiśeṣika. See above) systems, side by side with the system of atheist (?) Lokāyata and designates all the three as *ānvīkṣikī* — *KA*, 1.4.
11. Dr. Anantalāl Thākur and Pt. Vāsudeva Sāstri have partly refuted this contention in their respective Sanskrit papers: *Kaṇādo nāstiko na vā* and *Kaṇādaḥ āstiko vā nāstiko vā*, both in *Sāra*, Year 15 (V.S.2017), 297–306 and 307–09.

12. See also *Ātmatattvaviveka*, Pt. Ḍhuṇḍhirāja Śāstri's Introd. (Skt).
13. Cf. the popular hemistich: *Kāṇādaṁ Pāṇinīyaṁ ca sarva-śāstropakārakam*. "The Logic and the Grammar Help All Sciences".
14. See Sāstri's Introd. to his ed. of the *Bhāmatī tilaka* of Allālasūri. Madras 1984.
15. For further details, see SS3, 145, 151.
16. See, e.g., N.S. Rāmānuja Tātācārya, *Nirvikalpakajñānavicārah*: Chārudeva Śāstri Felicitation Vol. (1974), 150–160; *Ency*. 148. For more Sanskrit research papers by different pandits dealing with the matter in the conventional manner, see *Sāra*, years 2, 19–21, 30; etc.
17. For details, see *Brahmasiddhi* by Maṇḍana Miśra with commentary: Ed. MM. T.S. Kuppuswami Sastri Madras 1937; the editor's learned Introd., pp. xxviii ff.; *Pañcapādikā* of Padmapāda with commentaries: Madras Govt. Ori. Series, No. CLV (1958); Ed. S. Rāma Sāstri and S. R. Krishnamūrti Sāstri, the editor's scholarly Introd. (Skt), pp. 118 ff.
18. In his *Kiraṇāvali*, Udayanācārya twists the relevant passage of *PB*, to derive a meaning that the negation cognition may be either inferential or perceptual. Here, obviously, Udayana is trying to strike a compromise between U. and P.
19. The traditional interpretations of these two *sūtra*-s, having identical wording, are hardly convincing. In fact these two *sūtra*-s seem to stand for *tad-vacanam āmnāyasya prāmāṇyāt*.
20. See Anantalal Ṭhākur, *Sāra*. XV, 304–05.
21. Cf. Kālidāsa, *Mālavikāgnimitram*, Prologue, verse 2.

Bibliography with Abbreviations

Advaitasiddhi of Madhusūdana Sarasvatī with commentaries including *Laghucandrikā* by Brahmānanda Sarasvatī. Ed. Anantakṛishṇa Sāstri Bombay 1937.

Ātmatattvaviveka of Udayanācārya with commentaries. Ed. Ḍhuṇḍhiraja Sāstri. Benares 1940.

Brahmasūtra-Śāṅkarabhāṣya (*BSSB*) by Śaṅkarācārya (Ś.) with *Brahmasūtra* (*BS*) of Bādarāyaṇa (B) and *Bhāmatī* by Vacaspati Miśra and other commentaries. Ed. Anantakṛishna Sāstri etc. Bombay 1917.

Dharmadattasūri (Bacca Jhā), *Gūḍhārthatattvāloka*: A commentary on Gadādhara's *Vyutpattivāda*. Bombay 1911.

Dīpikāprakāśa (*DP*) with *Tarkasaṅgraha* (*TS1*) *Tarkasaṅgraha-dīpikā* (*TD1*), of Annambhaṭṭa, *Śaktisañjivinī* commentary and S. Sankaranarayana's *Ṭippaṇī* (*SS1*). Tiruppur 1971.

Gaṅgānātha Jhā (GJ), Gautama's *Nyāyasūtras* with *Vātsyāyana-Bhāṣya*. Eng. translation. Poona 1939.

Harsh Narain — *Evolution of Nyāya-Vaiśeṣika Categoriology*, Vol. I: *Early Nyāya Vaiśeṣika Categoriology*. Varanasi 1976.

Jagannāthopādhyāya, *Nirvikalpakajñāna-tajñeyayoḥ Svarūpavikāsaḥ*: *Sāra*, y.14; No.3 (1959): 91–100.

Joshi, S.D, The Text of the *Nyāyabhāṣya* on *Nyāyasūtra* 1.1.5. *Mysore Orientalist* XIV (n.d.): 21–27.

Kārikāvalī with the commentaries *Nyāyasiddhāntamuktāvalī* (*NSM*), *Prabhā*, *Mañjusha*, *Dinakarīyam* (*DK*), *Rāmarudrīyam* and *Gaṅgārāma jaṭīyam* Ed. C. Śaṅkararāma Śāstri. Madras 1923.

Kauṭilīya Arthaśāstra (*KA*). Ed. R.P. Kāṅgle. University of Bombay. Vol.I (Text), 1960; Vol.II (Eng. translation), 1963.

Kuppuswāmi Śāstri, S.MM., *A Primer of Indian Logic* (*Primer*). 3rd. Ed., Madras 1961.

Lakshmaṇa Śāstri Dravida (LD), Sanskrit Introduction to his edition of *Khaṇḍanakhaṇḍakhādya* of Śrī Harsha, with commentaries. Benares 1914.

N = Nyāya, Naiyāyika.

Nyāyadarśanam: *Nyāyasūtra* (*NS*) of Akṣapāda (A) with the *Nyāyabhāṣya* (*NB*) of Vātsyāyana (Vāt), *Nyāyavārttika* (*NV*) of Uddyotakara (U) *Tātparyaṭīka* (*TT*) of Vācaspati Miśra (Vāc.) and *Viśvanātha Vṛtti* (*VV*). Calcutta Sanskrit Series, Vol.I, No. XVIII (1936). Ed. Tārānatha Nyaya-Tarkatirtha and Amarendramohan Tarkatīrtha; Vol. II (1944), Ed. Amarendra Mohan Tarkatīrtha and Hemantakumār Tarkatīrtha.

Nyāyamañjarī of Jayanta Bhaṭṭa: Part I (1936). Ed. Suryanarayana Sukla; Pt. II (1934), Ed. Ḍhuṇḍhirāja Śāstri Benares.

Padārthatattvanirṇaya by Raghunatha Śiromaṇi (R), with commentary. Ed. Pt. V.P. Dwivedi. Reprint from the *Paṇḍit* Kāsi 1915.

Potter, Karl. H, *Encyclopedia of Indian Philosophies*, Vol. II (*Ency.*). Delhi 1977.

Śaṅkaranārāyanan, S (SS2), Threefold Inference of the Naiyāyikas: A Historical Study. *Adyar Library Bulletin*, Vols. 45–46 (1980–81): 107–119.

Sāṅkhyakārikā (*SK*) of Īśvarakṛṣṇa with *Sāṅkhyatattvakaumudī* by Vācaspati Miśra with commentary. Benares 1937.

Sankaranarayanan S. (SS3) *The Vishnukuṇḍis and Their Times* (An Epigraphical Study), Delhi 1977.

Saptapadārthī (*SP*) of Śivāditya with commentaries. Ed. Amarendra Mohan Tarkatīrtha and Narendrachandra Vedāntatīrtha. Calcutta Sanskrit Series, No.VIII (1934).

Śāra = *Sārasvatisuṣamā*.

Siddhāntabindu of Madhusūdana Sarasvatī with commentary *Nyāyaratnāvali* by Brahmānanda Sarasvatī. Ed. Panditaraja S. Subrahmanya Sastri. Varanasi 1978.

Śrīmad Bhagavadgītā with *Śāṅkarabhāṣya* and seven other commentaries. Ed. Wasudeva Laxman Sastri Panśīkar (Rpt). New Delhi 1978.

Śrīnivāsa Śāstri, T.S.: *Savikalpaka-Nirvikalpakavicāraḥ*. Extension Lecture: Kendrīya Sanskrit Vidyāpeetha, Tirupati, (December 1984); not published; xerox copy of the lecture given to me by the lecturer.

Subrahmaṇya Śāstri, V.S, Paṇḍitarāja, *Śābdataraṅgiṇī*. Madras 1969.

Tarkabhāṣā (*TB*) of Keśava Miśra with comm. *Tarkabhāṣā-prakāśikā* of Cinnambhaṭṭa. Ed. D.R. Bhandarkar and Pt. Kedārnāth. Bombay Sanskrit Series, No. LXXXIV (1939).

Tarkasaṅgraha (*TS2*) of Annambhaṭṭa with nine commentaries including *Tarkasaṅgrahadīpikā* (*TD2*) of the same. Ed. S. Chandraśekhara Śāstri. Madras 1920.

V. = Vaiśeṣika

Vaiśeṣikadarśanam: the *Praśastapādabhāṣya* (*PB*) of Praśastapāda (P) and *Vaiśeṣikasūtra* (*VS*) by Kaṇāda (K) with the commentary *Upaskāra* by Śaṅkara Miśra. Ed. Pt. Dhuṇḍhirāja Śāstri. Benares 1923.

Varadāchāri, V., A Note on the *Nirvikalpaka* and *Savikalpaka* perceptions in Indian Philosophy. *Srī Venkateśvara University Oriental Journae*. Vol. VI (1963), 11–21.

Varadāchārya, K.S, Introduction (Skt) to his edition of *Nyāyamañjarī* of, Journal Vol. VIJayantabhaṭṭa, Vol. II. Mysore 1983.

The implicit philosophy and worldview of Indian tribes

BAIDYANATH SARASWATI

1. Introduction

There is nothing in Indian thought structure, nor in the traditional organisation of Indian society, that may correspond to the Western evolutionary notion of tribe as a vestige of savagery and barbarism. Yet, tribe exists in India as a political category created by the British Raj, intellectualised by the colonial anthropologists, and sanctified by the Constitution of modern India. These so-called precultural groups, properly speaking, are the people of Oral Culture.[1] This essay attempts to explore their implicit philosophy and worldview, their vision of the cosmos. In so doing, it examines the myths of the North-East India tribes.

Myth[2] may be defined as a body of revelatory knowledge of the unseen reality that flows eternally in time and space. As a self-reflective and self-validating statement of the true nature of the phenomenal world, it is founded in the faith that supports all knowledge, and formed in the style of a narrative that makes all intricate theory understandable. It is usually stated in the past tense; it describes the reality that has already been seen. But, then, it is not history, the event of the past; it is the truth of life unfolding itself everywhere and every moment of the existence. It is not an ideational construct; it is an experiential reality. It is not an empirical description of the natural phenomena; it is an expression of the experience of the mystery of existence. Myth is not a moral code; it is a treatise on the divine cosmology. For a traditional man, it is the wisdom of the ancients that remain all-time new.

Since myth is concerned with matters of transcendental reality, its linguistic construction is translucent. As a primary presentation of the unseen and unmanifest, it is articulated in a poetic, or metaphorical, form with multiple possibilities of interpretation and reformulation. But, viewed comparatively, all such successive presentations are regressive in character. This is what the tribal myths speak of paradigmatically: at first there was a common language for the gods, men and animals. God wrote all wisdom on pieces of skin and gave one to the tribe, but the hungry tribesmen roasted that piece of skin and ate it up.[3]

This discourse confesses its limitations arising from the author's inexperience of the myth that he deals with. As stated, myth is an expression of a revelatory experience. Though experience and its expression are not unrelated,[4] the latter is only a partial manifestation of the former. Moreover, unless myth is lived from

Contemporary philosophy. A new survey. Vol. 7, 121–136.

inside, one cannot comprehend it fully. The tribals follow their myths without interpretation; the purity of a revelatory knowledge is lost the moment it is manipulated. To them the true meaning, or value, of a myth lies in its intuitive realisation through ritual action rather than in its intellectualisation. The present analysis may not, therefore, convey the kind of meaning which the tribes themselves attach to their myths.

2. The supreme principle

The texts of the tribal myth are varied and complex because of its primordial simplicity. Its translation into another language, or mode of transmission, is problematic. The only way to have some understanding of it is to follow the text faithfully. To explore its implicit philosophy we may consider the creation myths under three moments:[5] (1) solitude, (2) sacrifice, and (3) integration.

2.1 *Solitude*

The tribal cosmogenic myths[6] agree that at first there was nothing, nothing at all; neither earth nor sky. Most of the myths state that in the beginning all was water. The Singpho myth[7] describes that there were only clouds and mist. The Hrusso Aka myth[8] tells us of two great eggs which appeared like gold. In the Panggi belief,[9] the world was wrapped around by darkness. All these imply radical solitude. The primeval ocean, cloud and mist, the golden egg, and darkness are concrete representations of the unspeakable mystery that shrouds the cosmos.

The mystery deepens further as 'life' emerges and the original unity is broken. The earliest form of life comes from things incomprehensible: woman in the form of cloud; moving soft rocks; revolving soft golden eggs; and the like. The woman cloud gave birth to a pair of male and female snow. From the rocks a female rock was born; from her another female rock which united with a male rock. The eggs broke open; from one came the Earth, from the other the Sky, her husband.

The first principle of the cosmos is, then, 'One-two-and-many'.

2.2 *Sacrifice*

There are myths of creation by sacrifice, the divine transformation. The Apa Tani myth[10] reveals that Kujum-Chantu (the Mother Earth) died of her own accord for the sake of the world; and every part of her body became the part of the world, her eyes turned into the Sun and Moon. According to the Bori,[11] at first there were two spirits, one of whom had the form of a *mithun* (*bes frontalis*) and the other the form of an elephant, who fought and killed each other, and from their flesh and bones the world was formed. The Hill Miris[12] speak of the original tree, which grew out of the sea and fell to the ground; its bark became the skin of the world, the trunk turned into rocks and the branches into hills.

The sacrifice of the Mother Earth was the indicator of the primordial love that binds all the creatures of the world. What had provoked the spirits to sacrifice themselves is again a pointer to beings' worldly attachment. The primordial energy becomes manifest with the sacrifice of the tree and other great personage.

Thus, what sets the cosmos into motion is the sacrifice.

2.3 *Integration*

The myths describe how the creatures played an important part in the construction of the world. The Gallongs[13] say that at first, when everything was water the prawn collected a great pile of rubbish on the surface, the crab dug a pit below to drain off the water, and thus the earth emerged. The Hill Miris[14] describe how an enormous tree grew out of the sea and worms ate the wood until, by the constant falling of the dust into the water, the world was formed. In Minyong tradition[15] a great *mithun* dug a pit into which the waters poured and allowed the dry earth to appear. The Sherdukpens[16] believe that two heavenly brothers threw a lotus down onto the surface of the ocean, whereupon it was covered with flowers, over which the four winds blew clouds of dust, until the earth was formed. A Taraon Mishmi tale[17] describes how there was a lot of mud below the waters and a god created in it a pillar, up which a swarm of white ants climbed with their mouths filled with earth. The Kaman Mishmis tell[18] of two powerful spirits who do not actually create the world but come down from the sky and make the hills and mountains.

A creature born miraculously[19] becomes the creator of another creature. The Bugun Khowa myth[20] says that two sons of Zongma (a god) were formless. They made the Earth and the Sky. The Earth and the Sky made love and every kinds of creature came into being. In a Bugun story[21], Apuphulwa, an elderly man, married a hideous creature. From their union the first rock was born and from it flowed the first rivers. Then she bore a son who went into the Sky and made the thunder and lightning, and another who went below the Earth to cause the earthquake. A fourth son was a rain-maker, the fifth originated the four winds and the sixth caused the rainbow. Other children were snakes, poisonous insects and demons of disease. Last born was the pathetic child, Death, who unknowingly killed his own parents and then, in searching for them, he goes on killing everyone he meets.

In these myths the creation of the world is not attributed to a Supreme Being. The distinction between the creature and the creator is unclear. The early phase of creation is characterized by total integration of all that came into being. Everything that has been created has its own life, and hence no ontological difference between the man and the mountain.

Thus, what makes the cosmos the One undifferentiated reality is the supreme principle of nonduality.

3. The unity of all experience

The truth of the unity of all experience is expressed in the tribal eschatological belief.[22]

> The Apa Tanis believe that the soul, Yalo, of all those who die a natural death go to the Neli, the place of the dead, which looks like an Apa Tani village with long rows of houses. At the entrance of the Neli, they are met by Nelkiri, the guardian spirit who questions all newcomers about the exploits of their earthly life. He inquires how many enemies and wild animals they have slain, how many slaves they have bought and how much land they have owned. As Apa Tani lived on this earth so will he live in Neli: a rich man will find the cattle he has sacrificed during his lifetime, but those animals which has passed to his heirs are for ever lost to him. Every woman returns to her first husband, but those who died unmarried may there marry and beget children. Life in Neli is similar to life on this earth: people cultivate and work, and ultimately they die once more and go another Land of the Dead.

The Khasis also believe in a continued existence of man's life in blissful condition.[23] Nothing is so pleasant and entertaining to a Khasi as 'betel-nut eating' in his earthly career. Hence, after this life on the earth when he goes back to the House of God, it is believed, he would be eating betel-nuts in the company of his dear-and-near ones, who have all assembled there.

The belief in the continued existence is demonstrated in practice.[24]

> When a Wancho Chief dies, his body is bathed, wrapped in cloth, and laid on a mat. A special kind of basket is placed by his head. His relatives and subjects come one by one and each puts a little bark, betel and tobacco in it saying, 'This is for you, to save you from trouble'. Sometimes a friend may add a piece of cloth and says, 'This is for you; this is for you to wear'. All the dead man's possession, his *dao*, gun, spear, ornaments, bags are put by him, and later are hung on the tomb, which quickly becomes a most desirable museum. For the deceased must take these things below; in the Land of the Dead he will have to build a house, so he will need his *dao*; he will go head-hunting so he will need his gun; there will be swiddens to cultivate. If these things are not given, the ghost says, 'I have no cloth, I shiver with cold; I have no gun, how can I hunt?' I have no *dao*, how can I clear the path? They won't give me anything, so I will not give them good crops'. If the dead man had owned a tree, a little of the bark or a twig is brought, for otherwise he might worry about it. Then the mourners kill an animal, a pig or fowl, and place its head by the corpse saying, 'We give you this so that you will not be hungry'. The head is later hung by the tomb.

The belief in continued existence is a support to the tribal perception of death as beneficient. According to the Hill Miri[25], before the death came, the world was so full of people that there was not enough land to give them food, and they were hungry and unhappy. Most miserable of all were the old: they became dry as sticks, there was no blood in them, they had nothing to do. For they got older and older and did not die. Death was then brought to men by Abo-Tani for their benefit. When life in earthly existence loses the savour of youth and becomes a burden, death comes to man as a boon and takes him on to the heavenly existence. A Wancho tale[26] describes how at first men did not die but changed their form, just as the caterpillar and silkworm do, with the result that the world got too full of human beings who gave a lot of trouble to the birds and animals. Death came to the world, not in this case as an act of mercy to men (who did not grow old and miserable, but constantly renewed their youth) but out of pity for the animals and birds.

The ontological experience of life and death is the same. For death brings to man only a new existence. The errant self moves on in the cosmos between the earth and the sky through an unending cycle of birth and death on this earth, also in the other world, from one abode to another.

4. The harmony of all existence

The tribal myths deny the uniqueness of man insofar as his origin is concerned. The Dhammai Miji[27] myth describes how from the union of the Earth and Sky a male and a female frog were born first, and when they mated they had human children. In another type of myth, the first human beings are described as emerging from various material things. The Moklums[28] describe how the daughters of Rang, the Lord of the Sky, gave birth to a lump of flesh which was neither dead nor living. She dried it by the fire and presently it burst open and bits of it scattered about the world and turned into human beings. A Khampti story[29] tells us that in the beginning there was a great tree, from the berry of which there grew a flower; out of this flower came a pair of human beings, the first man and woman. In one Singpho story[30] the great spirits who live at the dawn of the world leave three children, one of whom is human, the second a rock and the third a gourd. As the children are playing together the rock-child breaks open the gourd and many little people came out of it: these are the first human beings. A Taraon Mishmi story[31] also describes how many of the first men and women came out of the magic tusk of an elephant.

Man and animal are close kins. There are many stories of marriage of human beings with gods, spirits, animals (real animals, not human beings in disguise, or under enchantment) as well as leaves, trees, and even fire. A Sherdukpen tale[32] describes how a woman gave birth to a human child, and then to snake, monkey, tiger, cow and dog. Monkeys, in a Nocte story[33], were originally men. In a Hrusso tale,[34] a woman turned into an animal to avoid incest with her brother to whom she had been forcibly married. A man turns into a tiger, a woman gives birth to twins, of whom one is human and the other a tiger.[35] There is a Minyong story[36] of two brothers, one of whom turns into a cat and the other into a tiger. A Wancho story[37] describes how men and tigers exchange their teeth. Animals talk and also often behave like men.[38] Indeed, in those early days, there does not seem to have any real distinction between man, animal and spirit.

Man is not unique even in the possession of knowledge. Primordial knowledge came to him from birds and animals. In a Hill Miri story[39] the god sends two birds to the first man and woman (who were innocent of any knowledge about the facts of life) how to people the world. In a Bugun story,[40] after the first couple have come into being, a god excites them to intercourse. In another version[41], a boy and a girl are fertilized almost as flowers are fertilized by bees. There is an excellent Singpho story[42] describing how the first men learnt how to make the pillars of a house from the legs of an elephant, the poles from the body of a snake, the roof from the skeleton of the buffalo, and the thatch from a fish's scales. Another tribal myth[43] describes how at a time when all men lived naked, the growing of cotton and the art of weaving was taught to men by the gods, in a dream. The Singpho story[44] tells how from the spider, a girl learnt how to weave. According to Wanchos,[45] from the first rat who fell down from the sky man learnt how to cultivate. The Singphos[46] believe that man's knowledge of iron is a gift from goddess Lepchan. Men got fire from various animals.[47] A bird brings water to mankind.[48] Men learnt from birds how to weep.[49]

Like men, the gods and spirits are also not unique; the myths about them follow the same pattern. As stated, from the union of an elderly man and a hideous creature were born the god of death, the god of water, the god of rain, the god of lightning, and the god of earthquake. At first there was only one man in the whole world. He united with the god Yang's daughter; from her many children were born. Of the two brothers, one became the ancestor of men and the other the ancestor of the spirits. Following a fight between the two brothers, the sphere of influence of men and spirits were divided. When the land was divided between man and spirits, the fathers of mankind got the best land from the ancestors of the spirits.[50] The Khasis[51] believe that in the beginning god was walking hand in hand with man. At that time there was a tree which served as a ladder to the original sixteen families of human beings for their communication between heaven and earth.

The statement that neither man nor god is unique implies that every creature is an organic part of the cosmos. Nobody reigns supreme; one is only different from the other. Every creature performs the same paradigmatic act of (1) creation, (2) preservation, and (3) destruction. The cosmic order is maintained by the harmonious functioning of each one of them. The tribal myths present an example of the early age.

4.1 *Creation*

As stated, a creature becomes the creator of another creature. It is through this process that the intrinsic difference between man, animal, and god is totally eliminated. The ontological meaning and significance of creation is conveyed by the myth of incest. At the beginning of the world the first brother and sister came together to produce the human race. In an Idu Mishmi story[52] the parents of the first boy and girl decide that since there are no other people in the world, they will have to marry each other. The girl tries to escape, but in the end she is compelled to marry her brother and in due course, gives birth to a tiger and a god in human form. The shadow of incest lies across many of the stories of creation. But all these myths have only one message: that the act of creation is paramount, higher than human morals.

4.2 *Preservation*

Every creature, howsoever small and insignificant it may appear, plays the role of the preserver. In a Nocte tradition[53] the cock claims to be the Chief of men and animals because even the Sun and the Moon come into the sky at his command. In a Singpho tale[54] the cock crows to tell men when the first waters have subsided and they can come to occupy the earth. In a Kaman Mishmi tale,[55] a bird seeing great gods levelling the mountains, becomes frightened that his nest will be destroyed, and brings them a message that their wives are sick and their parents dead, thus saving the hills and mountains for the world. In a Singpho tale[56] a parrot warns men to escape from the power of a dangerous demon. In a Tagin tradition[57] a little bird persuades men and animals to act in a way that saves them.

4.3 *Destruction*

As mentioned, death is brought for the benefit of the world. Before it came life in the world was miserable. Death eventually serves as a leaven for rejuvenation and preservation of life. The eschatological beliefs of the tribes show that, beneath this tragic world, the eternal life flows; the soul changes its body to acquire new body for a better existence. According to the Wanchos,[58] death came to human beings for the preservation of the birds and animals. For they not only killed them but took away their food. One day a crane caught a fish and was about to eat it, when a man came by and snatched it from its beak and ate it instead. The crane lost its temper and cursed him, saying that he would soon die. Another day a crow got a mouthful of meat and was about to eat it when a man snatched it from its beak and ate it instead. The crow exclaimed: 'If men live for ever, we birds will surely die'. The god Tatchak Namlong heard the crow and declared that men should henceforth begin to die. In a Singpho story[59] seven suns destroy mankind for its inequity. The first men were destroyed by fire and flood, and only one man and woman were saved: from them modern humanity has descended. In a Wancho tale[60] the world is destroyed by flood and the tribe is descended from a girl who was impregnated by the wind. In the same way, many traditions hold that the first race of men was destroyed and a new race created.

5. Man as a ritual being

As the cosmos was constructed by a primordial sacrifice, all creatures in essence, become ritual being. Man is not unique in respect of ritual, he is only more expressive than the other creatures. As a ritual being he demonstrates his paradigmatic role, communicates with others, protect himself from the unseen forces of destruction, and becomes the beneficiary of the benevolent powers.

5.1 *Priestcraft*

The Dafla myth[61] reveals that ritual is the ontological need of every being, and hence there are priests of men, priests of the earth, priests of gods and spirits, and priests of tigers and all animals. Tarngum Sung-Sung was the ancestor of all the priests in the world. The Sherdukpen myth[62] describes how the great god Lopung-Chungba created the Jiji priest and sent him to teach mankind to sacrifice to the gods and spirits so as to be saved from distress. In the Singpho tradition[63] when sorcery began and as a result people fell sick and died god's help was sought. Four gods help the Tumsawa priest who fights against evil spirits. One god watches over them from the sky while three other gods go about the world making men priests, and teaching them what to do.

The priest brings the desired harmony between the creatures confronting inequality of 'power'. The celestial creatures are described as more powerful than the terrestrial creatures, and even within each of these categories all are not equally powerful. Then, there are also benevolent and malevolent forces. To maintain equilibrium in the world-order these forces have to be harnessed. In the tribal tradition neither spirits nor gods are omniscent and omnipotent. All beings and all powers subsit in sacrifice. Hence the priests were born to transmit the ritual knowledge not only to mankind but to all creatures. Man's priestcraft is, however, more manifest and far more complex than that of the other creatures.

5.2 *Recreating the cosmos*

As the cosmos origins from the sacrifice, ritual becomes the strongest ligament that binds together the temporal and the eternal, the creature and the creator. Aware of this reality, the tribesmen perform ritual not only for their personal wellbeing, or private destiny, but also and pre-eminently, for the maintenance of the cosmic order. Fertility rites and the sorcery rites of destruction are known to mankind. The cosmic drama is enacted in a grand ritual where the priests of man performs his paradigmatic role in the cosmic creation. The entire community celebrates this occasion of man's greatest performance as creator of the cosmos. The Lai Haroba ritual dance of the Meiteis is an apt example. Vatsyayan[64] has interpreted its symbolic meaning and significance with remarkable clarity.

> The *Lai Haroba* is a collective ritual which all members of the (Meitei) community participate. It is conducted by hereditary priest (maiba) and priestess (maibi). The maibas and maibis are selected for life..... When a person shows special signs, i.e. of any kind of supernatural powers, whether a

child or an adult, male or female, he or she can be identified to be a maiba or a maibi.

The *Lai Haroba* is a recreation of the cosmos. It lasts like the yajna for ten to fifteen days. As elsewhere the universe is ritually created and ritually destroyed. The festival begins with an important ritual invoking the deities which lie in the waters. A procession is led by a maiba and a maibi from the centre of the village to a stream outside the village. Dressed in immaculate white, an impressive austere procession arrives at the stream. Now the nine gods, who are believed to have brought earth from heaven and the seven goddess who took the earth and threw into the waters are invoked. This is done through music and dance. Ritually two leaves are placed in the waters — one facing upward and the other downwards. These two symbolically represent the female and the male principle. The seed is symbolically within the leaves. The first part is called Lai Ekauba and the second Lai Themba. These leaves are placed on two water pots which have also been carried to the stream empty. Now they are full of the energies of creation. The procession solemnly walks back to a temporary shrine which has been constructed for the festival. Two earthen pitchers are placed near the shrine. The deities in the shrine can be made of cane, wood or mud; in short, all ephemeral material which will be destroyed at the end of the festival. Once the earthen pitchers are placed near the shrine symbolically and ritually, creation takes place, the human being is born. As in the case of other festivals in other parts of India for this time and duration, a new universe has been created. It is essential therefore also to invoke the deities of the four directions. These are Wangbaren, the deity of the South, Thangjing, the Lord of Moirang tribe, who protects the South-West. Marjing, the Lord of the horses, protects the North and Koubru, the mountain guardian of the North-west. Having invoked the directions, the stage is ready for recreating the life-cycle of the human being. First there is the birth of a child. This sequence is called Laipou. Now a procession moves in two rows and the whole growth of a human being in various stages is recounted. The maibi touches different parts of her body to indicate the making of the human being as the procession moves, one can almost see how a human being is being formed. Thereafter is the sequence of the human being needing shelter and thus the construction of a house. There is the need of human being for sowing and for harvest. Each function of the life-cycle is recreated day after day in a beautiful moving procession and movements accompanied by equally fascinating strains of music to the sole accompaniment of that unique string instrument called pena. The Pam Yanba and the Phi Shaba recreate the different phases of cultivation, harvesting of lashing (cotton plant) and the weaving of textiles. Movements of the loom called Khuthek are typical of these sequences. In each phase the choreographical pattern is that of an interwined serpent. There are other beautiful and vibrant sequences......

On the concluding day of the *Lai Haroba* festival, many martial dances are performed. They are known by the generic terms *Thang Haiba* and *Takshaw Saiba* or *Ta-Khoushaba*. The mood of quite austerity is replaced by a virile fierceness: daggers, spears, swords, shields, maces are held. Movements are quick, swift and sharp. Crisp hops and extensive open movements with spear attacking the ground are dominant in some phases: the art of attack and defence strongly reminiscent of fencing techniques is characteristics of others . . . .

As has been mentioned above, the whole festival is a recreation of the cosmos — the movement of the universe from the beginning, i.e. Shrishti (creation) to Samhar (destruction) — is presented. It moves from the most esoteric to the exoteric.

5.3 *Communication with divinity*

Through ritual man communicates with the divinity. In this act he is helped by other creatures. There are myths concerning the creation of birds for sacrifice. In a Minyong myth,[65] the cock is one of the animals who goes to find the Sun who has hidden in fear and sorrow. He acts as a messenger to tell mankind what must be done to persuade her to return and crows triumphantly when she appears. Fowls are specially suitable for sacrifice to the gods. An Idu myth[66] describes in detail the process of transforming a small round stone into a fowl for this purpose. A Hrusso myth[67] explains the dilemma of a man who is told to sacrifice a domestic fowl and pig for the relief of his sick children, and how he manages to create them.

Ritual serves as a bridge between man and god. During the performance of a ritual the Meitei[68] priest and priestess usually fall into a trance and speak in unintelligible syllables which are interpreted as the answer of the god. The Khasis[69] make a divine consultation through the medium of an egg, or a fowl, or any other animal judging the divine response from the signs that he ask for. When, for instance, the sacrifice for a sick person is done with a cock-sacrificing ritual, after ripping open the intestines to find out the answer, the performer asks the god to state the cause of sickness.[70] From the position of the intestine the signs are read and the cause is found. This having been done, god's grace for the recovery of the patient is asked and a word of satisfaction from god comes forth.[71]

Such methods of communication with the divine being are adopted in all important events in tribal life.

6. The foundation of the tribal thought

Evolutionary interpretations[72] of the tribal mind may be summarised in such epithetic terms as prelogical, prescientific, irrational, savage mind. A more sympathetic interpretation would allow logical but not scientific understanding to the 'savage'. For instance, the Khasi method of finding out the cause of sickness by cock-sacrificing ritual may be considered logical but unscientific. Such considerations have led many a scholar to conclude that the savages are natural philosophers[73], that the savage mind puts the philosophy of the finite into practice,[74] or that the savages, too, are rational creatures, but their mode of rationality is different from our mode of rationality.[75] No anthropologists, nor even a philosopher, would, however, grant conceptual and abstract thinking to the tribal mind.

Without contesting all these racial prejudices, or prejudgements, we may straightaway approach the foundation of the tribal thought.

In the tribal worldview, myth is entry into unthought and unthinkable; ritual is an act that leads into direct contact with the reality. Myth and ritual are, therefore, inseperable in tribal cultures. Myth, as an expression of the ultimate experience, stands for the reality, but in itself is not the reality. It represents a symbolic reality, the indirect message of the transcendence; but the transcendental reality is not the symbolic reality. The knowledge of ritual is revelatory. The priests of all creatures were born at the beginning of the creation. Man is not unique in the possession of knowledge; primordial knowledge came to him from birds and animals. The implication of this belief is that the universal knowability lies with the transcendence, the unthought, the unthinkable, the unknowable, the unmanifest, and not with man, god, or the other creatures.

In such a perception of human faculty, it is no wonder why the intellectual construct of knowledge has been given no place of pride in the tribal cultures. It is also true that modern scientific thinking — rational, logical, conceptual and abstract thinking — as cognitive part of man, cannot decrease the space of the unknown, the infinite. For, man's faculty is limited and regressive. The only tool of grasping the unknowable, which is available to man, is the ritual. Hence, the Khasi cock-sacrificing ritual is not only logical but also the most perfect way of knowing the unknowable cause of sickness. The only language in which the unthinkable, and the unknowable, can be thought and spoken of, even partly, is the symbolic language of the sacred myth. In the tribal thought structure, myth is more intelligible and acceptable than any other mode of knowing the reality. Hence, it places itself above philosophy.

The fact that the thought comes from the unthought is borne by the tribal cosmogonic myths. We have seen how in the primordial time, or no-time, the manifest comes from the unmanifest. The unmanifest remains unnamed, unqualified, expressed only symbolically in terms of primeval ocean, darkness, or the self-effulgent light. From water comes the lotus, the symbol of the manifest. The manifest is named and qualified. Hence all named gods, spirits, men and animals are creatures and creators of the second order. In the tribal perception, none of them is omniscent and omnipotent. The first principle of the universe is: one-and-many. Hence, the tribesmen think of nonduality at all levels of existence. This is contrary to the modern anthropological interpretation of human understanding in terms of binary opposition.[76]

Obviously, one's understanding of the universe depends on how he interprets the reality, or what he calls the reality. Modern man's formulation of existence is based on his personal aspirations and interest in the temporal world; for him that alone is real, the ultimate. Traditional man's vision of existence and well-being is based on a much wider horizon, the transcendence, where his true and complete identity is undifferentiated from the other beings.

Such reflection of human understanding is found not only in the tribal thought structure but also in the traditional organisation of the tribal life and culture.

APPENDIX

The following myths have been recorded by Elwin (1968).

1. At first Kujum-Chantu, the earth, was like a human being; she had a head, and arms and legs, and an enormous fat belly. The original human beings lived on the surface of her belly.

   One day it occured to Kujum-Chantu that if she ever got up and walked about, everyone would fall off and be killed, so she herself died of her own accord. Her head became the snow-covered mountains; the bones of her back turned into smaller hills. Her chest was the valley where the Apa Tanis live. From her neck came the north country of the Tagins. Her buttocks turned into the Assam plain. For just as the buttocks are full of fat, Assam has fat rich soil. Kujum-Chantu's eyes became the Sun and Moon. From her mouth was born Kujum-Popi who sent the Sun and Moon to shine in the sky. (Apa Tani).
2. At first there was nothing but water. There were two brothers, Poling the elder and Tering the younger. They lived separately in the water and neither of them knew that the other existed. Terring had the form of a mithun and Poling had the form of an elephant. Each thought he was the only creature living in the world.

   But after many years had passed, as they moved about the lonely ancient waters, they met each other, and Poling said, 'Who are you and where do you come from'? Terring asked, 'Who are *you* and where *you* come from'? Though there were only these two in all the world, they did not make friends; each accused the other of invading his territory, they quarrelled and in the fight that followed they killed each other.

   From the flesh of Tering, the earth was formed, from his bones came the trees and from his hair the grass. From Poling's bones came the rocks and from his flesh the mountains. (Bori).
3. In the sky were the old man Puk and his woman Yuk. At that time the earth was covered with water. The prawn and the crab tried to dry it up. The prawn used his antennae to collect the rubbish, leaves, grass and bits of wood that were floating on the surface and the crab dug a deep pit down below. The water poured into the pit and the earth appeared. The prawn's pile of rubbish turned into mountains.

   But the earth was very soft at first. Then the Sun and the Moon came up and hardened it by their heat. (Gallong).
4. Everything was water, water as far as the eye could see. But above the water rose the tree Teri-Ramula. As time passed a worm was born in the tree and it began to eat the wood. The dust fell into the water, year after year, until slowly the world was formed.

   And then at last the tree fell to the ground. The bark on the lower side of the trunk became the skin of the world; the bark of the upper side became the skin of the sky. The trunk itself turned into rock. The branches became the hills. (Hill Miri).
5. Long ago, before the world was made, the Wiyu Chungum-Irum was living with his wife Chingum-Erum. One day when this woman was lying asleep, a little water fell and flowed into her. She woke up and felt bitterly cold inside. She looked everywhere to see what had made her so cold, but there was nothing. She told her husband about it, how she was shivering with the cold thing that had entered her body but he too could not make out what it was.

   Twenty days later, Chingum-Erum laid an egg. Husband and wife looked at it in wonder and put it carefully aside. After six months it broke open of its own accord, and out of it came Sichi, the earth, Togle the great mountains and Togji the little hill. (Hill Miri).
6. At first there was no earth and sky; there were only two great eggs. But they were not ordinary eggs, for they were soft and shone like gold. They did not stay in one place, but went round and round. At last, as they went round, they collided and both the eggs broke open. From one came the Earth, from

the other the Sky, her husband.

Now the Earth was too big for the Sky to hold in his arms and he said, 'Though you are my wife you are greater than I and I cannot take you. Make yourself smaller'.

The Earth accordingly made herself pliable and the mountains and valleys were formed, and she became small and the Sky was able to go to her in love.

When the Sky made love to the Earth, every kind of tree and all living creatures came into being. (Hrusso Aka).

7. Long long ago there were two powerful spirits named Draku and Supaidang. They were strong and powerful but rather lazy and spent their days in a free and happy way.

   One day Draku met Supaideng on the bank of a river and said, 'Look brother, let us do some real work which will remain on earth in token of our strength and power'. Supaideng agreed and after some discussion they decided to level down the hills and make the surface of the earth equal. They each went to opposite corners of the earth and crushed the hills and mountains to sand and dust. (Kaman Mishmi).

8. At first there was nothing, nothing at all, but rocks and water. The first living beings were the rocks, but they were not as rocks are now: they were soft and they could move about: From the rocks, a female rock Eling-Litung-Tune was born and from her another female rock Peddong-Nane was born. When Peddong-Nane grew up she was ashamed, for she was naked. She married another rock, Eling-Limuk-Muktum, and her first child was the fish; it went to live in a mountain stream. Then she gave birth to the big frog and then to Etut-Tigling, the little frog, and then to the land-frog. After that she gave birth to the insect Kungung-Pangam, which lives in water, and then to another fish.

   But now Peddong-Nane left her husband and went to the sky village Engo-Takar among the stars, where she married Karba-Bojong. By him she had many children — Ninur-Botte, Lomang-Botte, Doini-Nibu and Darro-Rabbo — and when she had borne them all she died. Doini-Nibu and Daro-Rabbo prepared rice beer for her funeral. When the millet was ready and they poured water over it, a great cloud arose, and from the cloud was born Polung-Sabbo, the mithun.

   They held a Kebang to discuss how to make an earth. 'There is nothing but water. Who will dry it up and let the earth appear?' 'I can do it', said Polung-Sabbo and forthwith he dug a great pit with his horns and when the water had poured into the pit, the dry land appeared. After this the rocks became hard as they are today. (Minyong).

9. It was night. There was no day. There were Wiyus in the world, but not yet men. (Panggi).

10. Before the earth was made everything was water. There were two brothers who were supreme in the sky. The elder was Lopong-Rimbuche and the younger was Chom-Dande. One day they said to each other, 'When men are created, how will they live if there is nothing but water in the world?' There was lotus flower growing in the sky. The brothers threw this down and immediately the water was covered with flowers. Then they called the winds from the four quarters. The east wind brought white dust and scattered it on the flowers. The west wind blew yellow dust, the south wind red dust and the north wind black dust. The wind blew the dust round and round and mixed it up together until the earth was formed. This is why the earth is of different colours. (Sherdukpen).

11. At first there was no earth nor sky, but only cloud and mist. From it a woman called Khupning-Knam was born, and since she came from the mist she was a sort of cloud. In time, she gave birth to a girl called Ningon-Chinum and a boy called Tung-Kam-Waisun. They had the appearance of snow. When they grew up they married each other and from them were born a girl called Inga (Earth) and a son called Mu (Sky). Inga was mud and Mu, a cloud. These two also married and had a boy called Imbung, the Wind. When he was born he blew so strongly that he raised the cloud, his father, into the sky and dried up his mother, the mud. In this way heaven and earth were made. (Singpho).

12. At first there was nothing but water in the world. Below the water there lived the god Techimdum. He came up to the surface and saw nothing but water.

    Now below the water there was a lot of mud and in this Techimdum erected a great pillar. White ants swarmed round and crawled up to the top with their mouths full of earth, which they spread over it. When it was dry, the wind blew the earth all over the surface of the water.

    Techimdum said to the ants, 'I will put up four more pillars and you must help to bring the earth up from the bottom of the ocean'. He put one pillar in each corner of the world and the ants became very busy carrying earth up above the surface of the water. Along with the earth they brought the seeds of flowers and trees. In some places they put too much earth and there were hills and mountains. (Taraon Mishmi).

NOTES

1. In 1970 I had suggested a classification of cultures in terms of three principal modes of transmission of knowledge: oral, textual and transcendental. This was further developed in 1986. From the evolutionists' point of view writing is the hallmark of civilisation. Hence the preliterate tribals are a precultural group. My contention is that writing is another mode of transmission of knowledge and in no way superior to the oral (Saraswati, 1988).
2. Raimundo Panikkar, one of the leading religious thinkers of our time, has set standard for hermeneutic interpretation of myth. My own initial understanding of myth has come largely from his works: but he is in no way responsible for any unclarity that one may find in the present formulation, which I have essayed independently.
3. This story, in various versions, has been recorded from Gallongs, Noctes, Panggi Adis, Taraon Mishmi and Wanchos, and is known to many other tribes (Elwin, 1968:107).
4. For a comprehensive philosophical analysis of human experience, see Panikkar (1983:292–308).
5. Panikkar has analysed the Hindu myth of Prajapati under these three levels (*Ibid.* 70–78). The present interpretation follows this format but not the contents.
6. See Appendix.
7. *Ibid.*
8. *Ibid.* The Vedic Risis had revealed the creative energy of Hranyagarbha, the Golden Germ, which caused heaven and earth, waters and all that lives (Rig Veda X).
9. *Ibid.*
10. *Ibid.* This idea of creation by sacrifice echoes the famous *Purus Sukta* of the Rig Veda (X). For philosophical interpretation, see Panikkar (*op. cit.* 70–86).
11. *Ibid.*
12. *Ibid.*
13. *Ibid.*
14. *Ibid.*
15. *Ibid.*
16. *Ibid.*
17. *Ibid.*
18. *Ibid.*
19. *Ibid.*
20. Elwin, *op. cit.* 9–10.
21. *Ibid.* 10–14.
22. Furer-Haimendorf (1953), cf. *ibid.* 299–300. The truth of unity of all experience was visualized by the Upanisadic Risis:

    Whatever is here, that is there,
    What is there, that again is here.
    He who sees any difference here
    Goes from death to death. (*Katha Upanisada* II.1.10)

23. See Mawrie (1981:29–32).
24. Elwin, *op. cit.* 308–309.
25. *Ibid.* 317–318.
26. *Ibid.* 329.
27. *Ibid.* 103.
28. *Ibid.* 104.

29. *Ibid.* 105.
30. *Ibid.*
31. *Ibid.*
32. *Ibid.* 390.
33. *Ibid.* 393. The myth reverses the modern ideas of evolution.
34. *Ibid.* 390.
35. *Ibid.* 104.
36. *Ibid.* 394.
37. *Ibid.*
38. *Ibid.* 103.
39. *Ibid.* 105.
40. *Ibid.*
41. *Ibid.*
42. *Ibid.* 213.
43. *Ibid.*
44. *Ibid.* 219.
45. *Ibid.*
46. *Ibid.* 227
47. *Ibid.* 251.
48. *Ibid.* 377.
49. *Ibid.* 319.
50. *Ibid.* 154.
51. Mawrie, *op. cit.* 33.
52. Elwin, *op. cit.* 106.
53. *Ibid.* 376.
54. *Ibid.*
55. *Ibid.* 376.
56. *Ibid.*
57. *Ibid.*
58. *Ibid.* 329.
59. *Ibid.* 105.
60. *Ibid.*
61. *Ibid.* 76.
62. *Ibid.* 269.
63. *Ibid.* 282.
64. Vatsyayan (1987:132–35), the internationally known aesthetist, has made such excellent analysis of the tribal ritual dance from different parts of India.
65. Elwin, *op. cit.* 376.
66. *Ibid.*
67. *Ibid.*
68. Parratt (1981:57).
69. Mawrie, *op. cit.* 28.
70. *Ibid.* 31.
71. *Ibid.* 35.
72. For instance, Levi-Bruhl's and Sartre's interpretations. See Chandra (1988).
73. Evans-Pritchard (1935).
74. Levi-Strauss (1966).
75. For a critique of anthropological and philosophical interpretation of the savage mind, see Chandra (1988).
76. Durkheim (1965), and many of his followers. See Saraswati (1984:3–7).

References

Chandra, Suresh (1988). 'Evans-Pritchard on Persons and Their Cattle-clocks: A Note on the Anthropological Account of Man'. Paper presented at the National Seminar on Philosophy, Anthropology and Tribal Societies. North-Eastern Hill University, Shillong, 7–8 June 1988.

Durkheim, Emile (1965; 1912). *The Elementary Forms of the Religious Life*. New York.

Elwin, Verrier (1968). *Myths of the North-east Frontier of India*. Shillong.

Evans-Pritchard, E.E. (1935). *Science and Sentiment*, University of Egypt. Quoted by Peter Winch in *Ethics and Action*, p.11. London, 1972.

Furer-Haimendorf, C. von (1953). 'The After-life in Indian Belief'. *Journal of the Royal Anthropological Institute* LXXIII: 37.

Levi-Strauss, Claude (1966). *The Savage Mind*. Chicago.

Mawrie, H.O. (1981). *The Khasi Milieu*. New Delhi.

Panikkar, R. (1983). *Myth, Faith and Hermeneutics: Cross-Cultural Studies*. Bangalore.

Parratt, Saroj Nalini (1980). *The Religion of Manipur*. Calcutta.

Saraswati, Baidyanath (1984). *The Spectrum of the Sacred: Essays on the Religious Traditions of India*. New Delhi.

—— (1986). 'Thinking about Tradition'. *Man in India* 66(3).

—— (1988). 'Textualizing Tribal Tradition'. Paper presented at the National Seminar on Philosophy, Anthropology and Tribal Societies, North-Eastern Hill University, Shillong, 7–8 June 1988.

Vatsyayan, Kapila (1987). *Traditions of Indian Folk-dance*. New Delhi.

Pramāṇa epistemology: Some recent developments

PURUṢOTTAMA BILIMORIA*
Deakin University, Australia

Preamble

Indian phlilosophy — if the qualification 'Indian' is at all warranted here — has in modern times suffered from a particular handicap, due largely perhaps to the narrow interests promoted by 18–19th-century Orientalists, in that its concerns have been associated with what is generally classified as religion in the West, and in some measure with an explication of a particular worldview, or metaphysics. The more technical concerns in logic, epistemology and problems of language and meaning have been set aside or have remained largely unknown and unappraised in the West: if these, too, have not also been dismissed as being of little interest and sophistication, from Locke and Hume to Flew *et al.* (Matilal, 1986:2–5).

But for a modern thinker, who might be versed equally in Western and Asian traditions of philosophy, this scenario raises a fundamental issue, as follow.

Should not problems of epistemology or theory of knowledge be settled before the questions of metaphysics are dealt with, so that metaphysics — if this lofty goal is ever to be attained — may be established on firmer foundations? Could the ancients have been blind to this problematic? The early history of Indian philosophy in the modern era appears largely to have overlooked this challenge. This is evidenced in the plethora of works generated in the earlier part of this century which attempted to link or compare, say, the metaphysics of Śaṅkara or Rāmānuja, or of the Buddhists, variously with the metaphysical thinking of Kant, Hegel, Spinoza, Bradley, Bosanquet, Royce, Whitehead, amongst others. But for a handful of more enlightened philosophers, in particular, Krishna Chandra Bhattacharyya (writing in the 1920s), this trend appeared to have been well entrenched in the Indian philosophical establishment. It peaked in the 1930s. Sarvepalli Radhakrishnan, to give an instance, whose name became almost synonymous with Indian philosophy, even as he ascended to the Spalding Chair in Oxford, perpetuated this bias and gave a distinctive religious flavour to Indian philosophy. Although it must be to the credit of Radhakrishnan for having made available to a largely Western audience much material, argumentation and theories of the

* I would like to express gratitude to the late Professor Bimal K. Matilal (All Souls College, Oxford) for reading and commenting on an earlier draft of this essay.

Contemporary philosophy. A new survey. Vol. 7, 137–154.

classical Indian thinkers (Hindu, Buddhist, Jaina, and the materialists), without which we would be even more impoverished today. And, of course, some commentarial and derivative work had been carried out all along. Still, the overwhelming importance given to metaphysics at the expense of epistemology was not seriously questioned, it might be fair to say, until around the 1950s and after, whence a new generation of Indian graduates had begun to acquaint itself with the contemporary analytic and phenomenological developments in the West.

A considerable amount of groundwork, entailing a closer reading of the ancient texts, had to be performed before any serious or novel response and developments were to be forthcoming in the area of Indian epistemology (*cf. EIP*, 1977, 1978). Hence the recent developments that we wish to speak of here have a relatively short history. For the purposes of this chapter we will not be able to survey the developments in any great detail and breadth, partly in view of the (terminological) technicalties involved in such a philosophic enterprise and in view of the space available. What we hope to do is to highlight, by way of examples and pointers to current research on some of the salient features of the topic under discussion.

1.

Notwithstanding the doubt cast about the wisdom of the ancients in our preamble, if the Sanskrit dictum: *pramāṇādhīnā prameya-sthitiḥ* (*ibid.* 8) has any particular merit, it is clear that in the minds of the ancients the discussion of logic and epistemology preceded that of metaphysics. In classical Indian philosophy this concern is expressed in terms of the wide-ranging discussion and debate on what has been commonly referred to as *pramāṇa*, which roughly translates as 'sources or means of knowledge'. In the main, *pramāṇa* deals with the question of the possibility and grounds of the valid means of knowledge. Questions seen to be relevant for any theory of knowledge to be viable are raised within the ambience of the branch of thought that came to be known as *pramāṇa-śāstra*, whose closest equivalent in the West is the term 'epistemology'.

Mysore Hiriyanna (1973:178), who was amongst the first modern Indian philosopers to attempt to make the discourse of *pramāṇa* intelligible to a non-Indian audience, explicated *pramāṇa* in terms of its three basic functions: first, *pramāṇa* as a *karaṇa* or means or 'instrument' that becomes effective as a source of knowledge; secondly, *pramāṇa* as a means of scrutinising, criticising, evaluating through ratiocination the understanding or knowledge-claim derived through the source; thirdly, as *prāmāṇya*, the 'measurement' in terms of the criteria for 'truth' or 'falsity', which is characterised as the problem of validity. Together these constitute the grounds for the possibility of knowing. To elaborate just a little, *pramāṇa* is first defined in terms of the instrument and the concatenation of conditions and factors efficacious in bringing about a desired result, in this instance, a genuine piece of knowledge. The criteria for discerning the truth or otherwise of the knowledge-episode is also given alongside the definition. However, the definition, the generating conditions of both the knowledge and its truth, the criteria for determining the latter, as well as the number of *pramāṇas* accepted and the relative emphasis each receives in respect of its capacity to lead to significant knowledge, have all been matters of much dispute across the different schools of Indian philosophy. Still, there is no gainsaying that these issues received prime attention in classical Indian philosophy. Modern scholars have tried to resolve some of these issues by drawing upon parallel developments in modern Western philosophy.

In the discussion that follows we shall delineate three major areas within *pramāṇa-śāstra* which have engaged modern minds, and describe ways in which these have been interpreted and their ramifications explored or extended in the light of contemporary critical reflections. These comprise *jñāna* or 'awareness-episode', *prāmāṇya* or the notion of validity or 'truth', and the *pramāṇas* of perception and *śabda* or linguistic utterance.

2.

Jñāna

One of the concepts central to the *pramāṇa* theories is *jñāna*, which at a general level is cognate with the Western term 'cognition'; only after much qualification could it be rendered as 'knowledge', as we shall see. Not all schools of Indian philosophy, however, were agreed on the exact epistemological status of *jñāna*. It was ambiguously used for experience which is neutral as to its truth or falsity and for cognition with a truth-value. Thus there could be *jñāna* that is neither true nor false, and there could be *jñāna* that could be true, as there could be *jñāna* that could be false or in error. This convention led to much ambivalence, evidenced in the univocal translation of *jñāna* as 'knowledge', from the days of Max Müller to even recent times (Mohanty, 1966). There are further differences of views, (a) on the causal nexus in the generation of *jñāna*, (b) as to whether it involves identifiable mental or psychic processes, and if so, what are their relative stages and significance, (c) as to whether *jñāna* must have content, and (d) on the question of its truth-value: i.e. is *jñāna* intrinsically true, or are 'truth' and 'falsity' predicable of *jñāna*, etc.? These questions may have been asked in the interest of different theories of truth and not merely as a matter of definition.

Recent authors have discussed these issues at length in the attempt to arrive at some consensus on these matters. The rendering of *jñāna* as 'knowledge' has, by and large, been rejected, since the term *pramā* seems better suited. But, again, a theory may circumscribe its conditions and requirements in such a way that *jñāna* is identified with *pramā*. We shall return to this shortly. Very simply, *jñāna* may be taken to be a mental or psychical act or 'episode', and in this respect may denote 'ideation', 'remembering', 'doubting', 'wondering', 'guessing', 'dreaming', 'inferring', 'understanding', and so on, as modes of experience (*anubhava*). More fundamentally, however, *jñāna* has the nuance of 'cognition' or, more correctly, 'awareness'. Potter (1984) has argued that *jñāna* should be rendered as 'an act of awareness' in the sense of a cognitive or mental episode or occurrent, rather than as a disposition, which, say, 'belief' implies. In short, *jnana* is an awareness, and only such *jñāna* would amount to knowledge (*pramā*) as would yield truth. *Jñāna* as awareness, then, is the starting point in *pramāṇa* theory, for it provides the 'subjective' or phenomenological basis of 'objective' knowledge that is the supposed goal of epistemology (Matilal, 1986:23; 105–107).

At the outset two things should to be said about *jñāna*: first, that it is direct and immediate (at least in the case of sensory awareness); and, secondly, that it has a content (or 'contentness', *viṣayatā*), that is to say, an awareness is intentional for it is *of* or *about* or *directed* towards an 'object' (either externally or internally). Now Matilal (1986), elaborating on the Navya-nyāya (literally, 'New Logic') phenomenology of awareness shows that it is possible to move from 'appearances' to 'raw' experience wherein is discerned the primitive or pre-existent and indeterminate awareness-events, before the awareness is apprehended in its structured or 'propositional' form. Matilal links this observation with Strawson's notion of a pre-theoretic scheme or view of the world reflected in, and presupposed by, all our mature perceptual judgements. (1986:11) It is to be noted that the indeterminate or primary stage of awareness (*nirvikalpajñāna*) is not reduced to an opaque sensation: it is only that the qualificative structure underlying each mature judgement (*savikalpajñāna*) has not been consciously disclosed. It follows that each *jñāna* has a relational-qualificative structure, which may be stated in terms of the two following stages: (1) Whenever an object x features or 'floats' in my awareness, it features there as something, i.e. under some guise or mode, distinguished in some way or other (this distinguishing element is called (*viśeṣaṇa*)); (2) The guise or *mode* (*prakāra*) under which the object is presented is the purported property or qualifier of the object x, the qualificand. This qualifier – qualified structure is regarded as implicit in the content of any cognitive episode that has a claim to be regarded as knowledge: $F(a)$ or a as qualified by or distinguished by *F-ness*. Although there is causal element involved, this by no means amounts to a representational theory of cognition (Matilal, 1985:374ff.).

A further observation is made that the structure of a qualificative awareness-episode can — but need not necessarily — be represented linguistically. The qualification is meant to restrict the stronger view supported by some classical Indian philosophers, such as Bhartṛhari (*c*. 400–500 C.E.), that all knowledge is, as it were, shot through with language, i.e. that the phenomenological datum is itself essentially linguistic. Some earlier Nyāya writers seemed also to have maintained that the penetration of conceptual and verbal elements is necessary for an immediate but unformed experience to be presented as a well-formed judgement to consciousness. For most others cognition and language are two distinct events (Sibajiban Bhattacharyya, 1977).

The merit of the insight, perhaps, is that, much like language, cognition has a *mode* and a *content* and that underlying all cognition is a deeper structure and structuring process; each part is in relation to the other, and each cognition (like a 'sign') is integrally related to a preceding string of cognitions, wherein its fuller meaning is achieved. The appeal to language is more that of a 'model' than its coincidence with consciousness — very much as anthropologists, notably Lévi-Strauss, appealed to structural linguistics as a way of understanding the complex of covert or beneath-the-surface structures of a culture, its habits, practices, kinship relations, myths, and so on. Or it has resonances with Searle's (1983) recognition of the fundamental and 'intrinsic' intentionality of mental states *and* its linguisticality, i.e. resemblances or sharing the structure of propositions.

However, modern authors take the fact of the expressibility or propositional formulation (in the Western sense) of *jñāna* as an interesting heuristic postulate, as this allows the qualificative cognition to be discursively formulated as a declarative sentence (Matilal, 1968:12). And this proves useful for analyses of the constituents and the relation of the mode of presentation to the purported property of the object apprehended, that is to say, the qualifier – qualified relation as between the distinguishing feature of the cognition and the object of which it is a cognition. The laws that govern this relation can then also be unearthed, and its 'meaning', in terms of an undergirding logic interpreted. There is here a clear separation between psychology and logic, as Mohanty has tried to demonstrate in defending Navya-nyāya against the charge of psychologism (Mohanty, 1985). Whether this applies to indeterminate or the primary stage of awareness is a difficult question. Nevertheless, one implication of the foregoing analyses is that all seeing may be *seeing-as*, that the object does not shine in its nakedness, untinged with any qualifier (Matilal: 1985, 374). This, however, is not universally accepted.

The ramifications of these insights for perception have been admirably articulated by Matilal (1986), which will be discussed a little later. Matilal believes that the insights from the *pramāṇa* theory, particularly perception, can feed into and enrich the age-old controversy between phenomenalism and realism, or immaterialism and materialism (Naïve Realism), and the problem of universals and particulars, all of which continue to command significant attention in Anglo-American analytic philosophy.

3.

Prāmāṇya

Prāmāṇya is the other important category that we have chosen to discuss. *Prāmāṇya* has been rendered as 'truth' by Mohanty (1966:77), and hence *prāmāṇya* has to do with a theory of truth and criteria for evaluation of truth and falsity of *jñāna*. But the theory is complicated by the fact that there are two wide divisions within Indian epistemology on the very notion of 'truth', and how truth is constituted. The two divisions fall roughly into what are called the '*svataḥ*' ('intrinsic') and '*prataḥ*' ('extrinsic') theories or theses of truth. Either the conditions that generate the *jñāna* also generate its truth (i.e. a *jñāna* is self-validating or is self-evidentially true), or the two conditions are different such that the 'truth' of the cognition requires certain additional or conditions *ab extra*. (We shall not concern ourselves with a related issue of whether the truth, in either case, is also self-illumining or stands in need of another cognition for its awareness.) *Prima facie*, the *svataḥ* thesis seems to analytically rule out the possibility of error, for if the conditions that generate a cognition are those that yield its truth then no cognition would arise without its truth tagged on to it, so to speak. The *parataḥ* thesis would seem to be more in keeping with common-sense recognition of the possibility of error in every instance of awareness-episode. Mohanty (1966:70–80), like a few other modern thinkers, tried to wrestle with the chasm between the two views, and came to the opinion that perhaps these views were responding to rather different sets of problems in the interests of quite different definitions and theories of truth. They might even be incommensurable (*Ibid.* 1–2).

Potter has tried to accommodate or reconcile the opposition between the two views by urging a radically different way of conceiving the relation between *jñāna* and its content (which it must have). This relation is characterised by Potter (1984) in terms of purposive fulfillment, or satisfaction, taken as a mark of 'fit' for the content so 'measured out' by the awareness-episode in question. Potter construes the classical Sanskrit term that characterises the relation of 'fit', namely, *yathārtha*, to mean "in accordance with the purpose that motivated J[*jñāna*]" (*ibid.* 312). The play obviously is on the troublesome term '*artha*', which could mean 'object', 'purpose', 'goal', 'system of ends', and so on. The question, however, is whether any of these senses of '*artha*' are epistemically-weighted as its appears to be in *prāmāṇya* discourse. Potter juxtaposes the thus-derived 'purposive' structure on the more familiar classical *prāmāṇya* theories, viz.: There is *prāmāṇya* where the qualificandum (i.e. the subject term) of a *jñāna* (J) possesses a certain property (*viśeṣaṇa*), which property is the chief qualifier (predicate term) of the *jñāna* (J). Potter argues that the notion of 'right' awareness, implied in later Navya-nyāya definitions, as in certain Buddhist accounts, invariably relates itself to "previously identified purpose or purposive object" (1984:313). But Potter has overstepped the strictures that the *prāmāṇya* concept is normally goaded with, and interpolates this in terms of another, albeit connected concept, namely, *sāmarthya* (or *saṃyaktva*), which has to do with satisfaction or purpose or desired goal. This enables Potter to argue, against the usual reading of *prāmāṇya* as 'truth', for a more or less pragmatic or 'workability' criterion in terms of the capacity of *jñāna* to effect satisfaction in action. If a 'correspondence to reality' is entailed in such a theory, for Potter this is to be interpreted in terms of the 'fit', or the 'frustration', of satisfaction that results from *jñāna*. Thus, 'correspondence' or *yāthārthya* (literally, 'such-as it actually is') is subsumed under the satisfaction or *sāmarthya* principle (which might encompass ethical, normative and aesthetic descriptions). In short, Potter opts for a reworking of the *prāmāṇya* concept towards a 'workability' theory of truth, wherein the chasm between 'fact' and 'value' is virtually collapsed (in the interest also of a noncognitivists approach to value theory).

A number of scholars have taken up Potter's challenge in different ways, but virtually all are agreed that Potter's proposal does not seem to square easily with the spirit of the classical theories. Mohanty, whose earlier ascription of a Justified True Belief theory of knowledge to the Indian philosophers was called into question by Potter (in the light of Gettier's paradox), shifts his strategy and questions in turn the viability of the 'workability' interpretation of *prāmāṇya*. Mohanty (1984) first clarifies the two distinct understandings of *prāmāṇya* in the two opposing (Indian) theories of truth, in the following terms. It is true that the *svataḥ* theorists identify *prāmāṇya* with *jñānatva* ('knowingness') in such a way that the generating conditions of the two coincide. This *a fortiori* rules out error, which is opposed to *pramā* (true judgement), but by this definition *jñāna* is also opposed to error, considering also that *jñāna* reveals its object as having had no unknown existence prior to being apprehended while error does not. Thus, to have a *jñāna* (in the stronger sense) is to know something to be true. Either *apramā* or error and falsified judgements are not regarded as apecies of *jñāna* (excepting mistaken instances); or, alternatively, one might say with the Prābhākaras that all *jñāna* are *pramā*, there being no such thing as erroneous awarenesses or *apramā*. The Advaita Vedānta school opts for the former, while Nyāya maintains that even error or *apramā* is a species of *jñāna*, so that the generating conditions of truth (and falsity) do not coincide entirely with those of *jñāna*. Thus, to know (in the weaker sense) is tantamount to a 'belief' or simply a judgement that awaits confirmation or falsification. That is not to say, according to the Nyāya writers, that the truth of the cognition cannot arise simultaneously with the cognition; Nyāya says that their generating conditions are different. Hence, the judgement that it is raining outside might arise confirmed or with due certainty (*niścaya*), and without any doubt as to its truth. If a doubt subsequently arises, one might have to examine the special feature or 'mark of excellence' (*guṇa*) that gave the judgement its particular truth-value. There might have been a defect (*doṣa*) or a vitiating condition that gave rise to the error in the judgement, in which case the *jñāna* has to be rejected as being false.

As to the suggestion of *prāmāṇya* as 'workability', Mohanty and others have taken issue with Potter, arguing that: (a) it is based on a rather stretched interpolation of *artha* as purposive 'object-value' or *satisfaction*; (b) it involves testing in a practical action, when in fact *jñāna* need not as such entail in any action; (c) it equivocates on the dicta: p works because p is true, P is true because p works; and (d) practical outcome may well be significant for corroboration of the 'truth-claim', but is not essential to the determination of truth, which, according to the *svataḥ* thesis at least, is taken to be self-evidentially given ('I *know* that a is F; it works, so I am quite sure that a is F'). If falsification comes about as a result of doubt and subsequent tests or scrutiny, then the *jñāna* is repudiated and indeed its very claim to being a *jñāna* is thereby undermined. Advaita would argue therefore that a *jñāna* is not a piece of knowledge which is not 'true' knowledge, and that 'false' knowledge is no knowledge at all: it is rather the opposite of knowledge, however mistaken one might have been about its truth-value. Clearly, an epistemological requirement is built into the the Advaita definition of *jñāna*, and which therefore, under these circumstances, is quite appropriately translated as 'knowledge'. (Mohanty, 1984:332).

Most recent authors are agreed that the first component of a *prāmāṇya* analysis involves claims about 'correspondence' (*yāthārtya*), though not in the sense in which a standard Correspondence Theory of Truth is understood. The 'correspondence' has to do with the 'equiformity' or 'sameness' of the property that qualifies the awareness with the propertry that belongs to the object of which it is an awareness. Thus, the perception of silver, to be veridical, must be such that 'silverness' qualifies both the object and the percept. Potter, it is claimed, has reduced the *definiendum* to the *definiens* (the elements w, x, z, in terms of which T is defined or explicated), and it is here (in the *definiens*) that 'workability' makes a significant occurrence as one of the more important discriminatory marks of the *definiendum*, so that:

if $T = Df.\ w, x \ldots z$; then one cannot say $w \rightarrow T$ (say, in the absence of $x \ldots z$)

(nor $-w \rightarrow -T$, if $x \ldots z$ are present)

Truth, it is argued, is not an ontologically-neutral concept (as between realism and idealism) in the *prāmāṇya* theories. *Artha*, as for instance in *padārtha* ('categories') in Nyāya metaphysics, does answer to certain ontological descriptions, and it is a function of the *prāmāṇya* enterprise to find out, for any knowledge-episode, whether the description 'fits', or it does not fit, the reality being apprehended. Hence the notion of 'objectivity' need not be dispensed with, in the interests of values, 'facts' (of the Wittgenstenian sort), or whatever.

Nevertheless, the importance attached to the 'workability' principle in classical *prāmāṇya* theories is not to be underestimated. Earlier Nyāya writers, such as Udayana (*c.* 1050 C.E.) and Keśava Miśra (*c.* 1400 C.E.), had proposed the setting up of an hypothesis, which utilises 'workability' inasmuch as it enables one to make a tentative inference that, if what presents itself as *niścayajñāna* (*Jn*) or 'certainty-awareness' does not work then *Jn* could not be true (based on the *vyatireki* or negative generalisation that what is not true does not yied *samartha*, i.e. does not work) (Chakrabarti, 1984; Bilimoria, 1988:267). This at least motivates the subject to proceed with the tests she may indeed decide upon. If the test does not lead to 'satisfaction' (in terms of the projected successful productive activity (*saphala*)), she has immediate grounds to seriously doubt the presumed truth of the judgement. Where satisfaction is forthcoming, this need not be taken as a direct confirmation of the truth of the judgement, but rather indirectly on the grounds of the 'familiarity' (*tajjātīya*) this judgement (*B*) presents with an earlier confirmed judgement (*A*). The truth of judgement *A* functions transparently as a second-order judgement (the middle term) for inferring the truth of *B* (the major term), on the strength of the analogy. Similar inferential moves are made, which enable the subject to, at least, bracket a possible falsification of judgement *B* as a consequence, say, of unabashed doubts. While she may not be confident in saying that *B* is true, she could be confident in saying that it is not false. Or, alternatively, she might feel that her behavioural disposition, indeed, 'belief', is justified on the grounds of the secondary evidential support, and she has a right to be sure about this. It has been argued that this is the minimal that a *svataḥ* thesis should require to get by with, and was possibly very nearly achieved in later Advaita epistemology (Bilimoria: 1988: 248; 256–272). For what is self-evident—*svataḥ*—, according to the Advaita theorists, is the intuition, which arises with the *jñāna*, (*Jn*), that *Jn* which has *yāthārthya* or 'correspondence' in the sense explained earlier, possesses the following characteristics: (a) something new (previously unknown existence of the object) is revealed (*anadhigata*); (b) it is not falsified or contradicted by knowledge pertaining to the same state of affairs (*abādhita*); (c) there is felt certainty or tenacity in the awareness, which dispels possible doubt (*niścaya*); (d) it is congruent or coherent with the larger background body of knowledge (*samvāda*); (e) it is capable of leading to successful result or satisfaction (*pravṛttisāmarthya*).

While a *parataḥ* theorist might accept some only of the above as the characteristics of truth, he is likely to protest that these are not 'given' in and through the judgement itself (for reasons stated earlier), but that these may be what one tests for *ab extra* or extraneously to the judgement. In other words, their confirmation occurs after the awareness has arisen. But such a process, it could be objected, may go on *ad infinitum*, and the rules that connect the substantial evidence may well be arbitrary; unless, of course, one is prepared to accede to the self-evidentiality of a *jñāna* at some point in the chain of invoked confirmatory *jñāna* (which is what the *tajjātīya* principle almost comes to).

Bilimoria (*ibid.*) has argued for a modified or 'syncretic' *prāmāṇya* thesis which aims at combining the *svatah* and *parataḥ* principles using the criterion of falsification. He argues that the classical formulation of *tatprakārakatvam*, which in a broader sense requires the conformity of the features of the judgement with the qualifying features of the 'object' or state of affairs being apprehended, should be construed as involving a 'soft' epistemological requirement. For instance, if one has the awareness of a typewriter, the object of that awareness should have all the features of the typewriter, and not, say, of a cash register placed nearby. Is it, however, ever self-evident that the perceived state of affairs actually obtains in reality? Protagonists of the *svataḥ* thesis have no hesitation in giving a positive response to this question, with the proviso that the *modus operandi* here is an accredited *prāmāṇa*. But even if something *x* appears self-evidentially to be *y*, does it necessarily follow that it might not turn out to be false? As Śrīharṣa and Citsukha (11th century Vedānta philosophers) had argued, we can never be sure that a judgement taken to be true now will not be falsified at a later time (Granoff, 1978:37). This critique is itself a sufficient recognition, in classical thought, of the viability of the falsificationist criterion and the attendant methodology, which was not developed to the extent to which it could have been. Two insights are drawn upon here. First, that of Gaṅgeśa (a leading proponent of the 'New Logic' school), that not all experience (even 'awareness') is *jñāna*, nor even necessarily a candidate for *jñāna*, and that a cognition would be false if its qualifier (*y*) is not actually related in the same way as it is to the qualificand (*x*) *qua* the object (*ayathārthas tu arthavyabhicārī apramāṇajaḥ* (Mohanty, 1966:64;206ff.)). Secondly, from the *svataḥ* thesis, that a *jñāna* generated via an accredited *pramāṇa*, when it arises, arises as being either true or false: truth is not something, a character or a mark, that is constructed or constituted *ab extra*, but is there, as it were, *ab initio*. It is then argued that any *jñāna* should be assumed at the outset to be not-false, whose truth however awaits to be made '*firm*' (*con-firmare*, affirm, re-affirm), or simply corroborated (in the strongest possible sense). The characteristics (a–d) mentioned earlier under the *svataḥ* thesis, should function as the criteria for tests. If there is success one could say that the truth of the *jñāna* has been 'recognised'. One of the tests obviously could invoke the *sāmarthya* or 'workability' principle, in the following move: either the cognition is fit to lead to satisfaction or it is not; if it is not then it is not true; if it is, then doubts about its not been true are dispelled (i.e. its truth is corroborated). One could continue to try and refute it through the other criteria. To formalise the schema with the help of an illustration, suppose one looks across and reports seeing water at some distance. His judgement bears the structure $F(a)$ (where a is *that something* and F is *waterness*; and b will be 'wetness'; q 'quenching of thirst'; G, *guṇa* or 'mark of excellence' in respect of these; s 'soothing effect', on t 'dryness of mouth'; while f is 'fire'). The following hypothesis may be set up:

$a \rightarrow b$

where a judgement $Jn = F(a)$ because G is b

assume Jn to be false, $= -F(a)$ because $G = -b$

but $-b$ *iff* $-(q + t)$

q has s but not f, *etc.*

Supposing the test fails to show f, but shows s instead and its locus is q. There is independent evidence that q, together with t, does obtain. But this is opposed to $-(q + t)$, which in turn repudiates $-b$. Thus $-b$ has not been confirmed, thus a is also not negated (from $a \rightarrow b$). It follows that Jn has not been shown to be false. One need not expect 'indefectible' certainty for every piece of knowledge-episode that arises in this way. Such a move rescues the *parataḥ* thesis from the possibility of *regressus*, as it also forces the *svataḥ* thesis to be more open-ended in respect of the future status of the *jñāna* presently taken to be true. In terms of this 'syncretic' theory outlined here, then, a judgement which is not known to be false, and has been corroborated against doubts etc., could be taken to be, if not incontrovertibly true, at least, *in all probabality true*—until it is shown to be otherwise. (*cf.* K.C. Bhattacharyya, 1956:83. Bilimoria, 1988: 246: 256–259)

4.

Perception

There are several theories of perception in the Indian tradition, and the age-old controversies between idealism – realism – phenomenalism surface here as much as elsewhere. While the variety of phenomenalism, represented largely by the Buddhists (e.g. Diṅnāga, fourth century), has been expounded in earlier works (see Hattori), the overt realism in Indian theories of perception, represented largely by Nyāya-Vaiśeṣika, has not been analysed, nor contrasted with the 'realisms' of Anglo-American variety, until very recently. On this Matilal has perhaps made the most advancement, and the ensuing discussion will draw on this particular work of his (*Perception*, 1986).

Bearing in mind the analysis in the foregoing sections, especially as regards the episodic ("event-based") interpretation of cognitive awareness, the general observation is made that all perceptual experience, including those that we call illusory, imaginary, erroneous, even hallucinatory, have objects which they grasp, and that a causal explanation, in principle, is possible for each object grasped in perception. However, the precise ontological status of the preceptual object and the kind of causality involved are matters of intense dispute among the rival theories. While for the Buddhist, illusion is a 'revelation of the non-existent', for the Nyāya the principal object in illusion is a material object which, however, may not be present in the locus of perception but rather presented by memory, recollection, or sensory-stimulation, and wrongly connected with the experience, resulting in an overall, but not a totally, non-existent object. But the question arises: what is it that one "sees" first which triggers off this event? The answer to this has implications for the further analysis of non-illusory perception.

It might be argued that in the case of perceptual illusion, nothing is "seen", that there is really no *perceiving* going on, only a belief or an inclination to believe that we are *immediately* perceiving something (Armstrong, in Matilal, 1986:218). The occurrence is said to be 'perceptual' only in a very loose sense of the term. But the Nyāya view, as explicated by Matilal, is that one does have a strong visual, tactical, ect. experience in such an illusion, and in this respect it is perceptual and so calls for explanation.

Alternatively, a sense-data theorist might say that what is immediately presented is a raw baggage of sense-data (even *unsensed* sense-data), which is mind-dependent, and that it is this that is causally responsible for our 'mediate' perception of objects. He might further state that in the case of illusion a false inference is made of the existence of an actual object where none is present. Of course, much hinges on the coherence of the assumption, typified in most representationalist views, that our perception is based upon a causal inference from what is presented through the senses. Could it not be possible, as the phenomenalist maintain, that what we perceive is, in the final analysis, *constructed* out of what is presented through the senses? Let us say that for both the representationalist and phenomenalist views the *immediate* object of awareness is sense-impression, or a *sense-data*, or a sensible quality (thus the given in our sensory experience). The difference is in the way each links the sense-impressions to the external objects, regressively or progressively.

The Nyāya brings several objections to these two major positions, as well as against a variation of the phenomenalist view which considers the sense-data or percepts ("sensibilia") not to be mind-dependent but to be as *real* as the perception itself or the very object by which the awareness is also designated. Basically, it is argued that these views make a great deal of the distinction between immediate/direct and mediate/indirect, and that they posit entities (mental or physical), which are presented "up front" as the properties or constituent parts and from which the perceptual object is (causally) derived or constructed piecemeal.

For one, the part-whole distinction seems to be spurious. (1986:273) There are no parts without a whole: to speak of parts is to assume a whole of which they are said to be parts, even if they are now scattered apart. Do I see a "part" of a chair (even if I do not see its rear legs), or do I see the whole chair, and possibly the parts in virtue of seeing the whole chair? Do I see a tomato in virture of the red-patch that appears in front of me — for if I did, would I not be seeing two things, albeit in sequence, viz. the red-patch and the tomato? Do we ever see properties, as it were, hanging in mid-air, independent of the substance which they qualify? Even those Buddhists who dismissed the existence of substantive entities, were moved to grant nominal existence to the composite entity made up of the sensibilia or sensas of colour, shape, tacticle feel, etc., which in themselves might be imperceptible or "unsensed" (1986:246).

Secondly, although it is true that our normal perceptual judgement 'goes beyond' the sensory experience which give rise to them, the 'gap' need not be as large as it has often been assumed; and while in many instances we do make inferences from what is immediately presented, a theory of perception need not devolve around these factors and these factors alone. Again, the Nyāya argument in favour of its own more radical theory of, what Matilal has christened 'naïve' or 'direct realism', dwells on the inappropriateness of the part-whole distinction underlying most 'indirect' theories of perception (whether realist or idealist). Nyāya gets round the obviously common-sense assumption that the relation between parts and whole (e.g. branch and tree) is contingent by suggesting that the relation that obtains is in fact one of inherence (*samavāya*), such that one could say that the tree is as much in the branch as the branch is in the tree! Thus when I see a tree and someone asks me later, 'Did you see that drooping branch?' I might well reply, 'I can't quite recall... but I did notice the tree being uneven.' I may have to go back and take a second look to 'see' the branch (a part) apart from the tree (the whole). Thus Nyāya argues in favour of whole-part distinction at the ontological level; while talking about perception Nyāya claims that we see the *whole* first in virtue of its having parts.

Arguing in such a manner, Matilal takes the Nyāya theory a stage further in contending that the traditional distinction between mediate and immediate objects of perception does not quite hold (1986:267). And this provides grounds for casting serious doubts about the representationalist theory of perception, in respect of both its long-suspected assumption of sense-data and its formidable view of the causal relation between the sense-data and the object — apart from being locked into the Lockeian dualism of secondary and primary qualities, one of which is imperceptible and therefore has to be inferred on the basis of the other (presumably perceptible).

To be sure, the troublesome term is not 'causal', for Nyāya has its own way of making the part 'causally responsible' for the perception of the body-whole. The following quote from Matilal clarifies this point: "Nyāya direct realism eliminates the third entity, sense-datum, entirely from the discussion of perception. By giving a different *causal* explanation of perceptual illusion and hallucination Nyāya aims at characterising the causal factors in a naive or non-specialist conception of perception. Obviously, then, the consideration for the so-called causal relation between sense-datum and the material object (which has been the cornerstone of modern representationalist and causal theorists) is conspicuous by its absence in Nyāya. It is not contended here that a person claims to have a perception of a familiar body in virtue of the occurrence of some sense-impression." (1986:268).

On the contrary, Matilal goes on to argue, using the earlier example of the tomato (and switching to substance-property-language), that the visual perception is to be regarded as the material substance (tomato) not the property, or that the *thing* tomato is the substratum of a particular instance of the property red, which in its turn, instantiates the property-universal *red*. In short, one sees the tomato *not because* one sees the red colour-patch (appearance) one first encounters, but that one sees the tomato because, among other things, it is coloured, and one sees the red colour because the colour-particular resides (inheres) in a composite substance and it has a specific nature (1986:283).

Putting the view advanced in another way, the 'in virtue of' relation is moved from the 'sense-data' representationalist account, after denuding it of the sense-datum assumption, and placed differently. This is succintly expressed in the following remarks, which may well be regarded as the '*motto*' of Nyāya-Vaiśeṣika (direct) realism: 'One does not see the tomato *in virtue* of seeing a coloured shape. Rather one sees the tomato *in virtue of* its having a coloured shape.' (1986:285)

5.

Śabdapramāṇa: 'Testimony as knowledge'

The other *pramāṇa* or category of knowing in Indian epistemology that has begun to receive considerable attention is *śabdapramāṇa* or (verbal) knowledge derived from linguistic utterance or testimony. The classical material on this problem is extensive and varied; what we present will only be a gist of the thinking. The basic element of *śabdapramāṇa* is said to be *śābdabodha* or linguistic understanding; and the distinctive 'instrument' or *karaṇa* for *śābdabodha* is said to be the 'word' or *śabda*, more specifically a *vākya* or sentence utterance. *Śābdabodha*, like any cognition, has a relational-qualificative structure, constituted of 'meanings' or *arthas*, their relations (*saṃsarga*), and the intentionality relevant to the particular speech-act or utterance. Obviously, the 'hearer's side' is more of a concern in this analysis than the 'speaker's side', provided there are no defects in the communication process. (Kalidas Bhattacharyya, 1985). There is here a recognition of the 'power' of words, the efficacy of verbal communication, as well as the essential linguistic structure of verbally-generated understanding, and last but not least, the epistemic character of the understanding — hence also, its implications for a theory of knowledge.

Much debate in classical and modern literature on *śabdapramāṇa*, however, has revolved around the question of whether *śābdabodha* is *eo ipso* a linguistic knowledge, i.e. (*śabda*)*pramā*, or does *śābdabodha* amount to knowledge only when certain specifiable conditions, in addition to its generating conditions, are satisfied (Mohanty, 1988). Very generally, the semantic and phenomenological conditions that are responsible for generating *śābdabodha* are four, viz. *ākāṅkṣā*, or the mutual expectancy that a word rouses for another in the absence of one of them; *yogyatā*, or the semantic fitness or compatibility that words share if they are to be meaningfully related so as to yield a coherent *śābdabodha*; *āsatti*, or the contiguity of the words, and maybe the word-meanings, in their juxtapositioning in an utterance; and, *tātparyajñāna*, or the awareness of the intentionality of the utterance (so that there is a clear distinction between the literal and the otherwise intended, or metaphorical, suggestive, poetic, etc., senses to be grasped).

One point of view argues that *śābdabodha* is *pramā*, i.e. true, so that a false (*apramā*) *śābdabodha* would be a contradiction in terms. The argument turns on the kind of theory of 'linguistic meaning' presupposed. The protagonists of this view are prepared to settle for, namely, a purely referential theory, which suggests that, a word *qua* sentence refers "directly to the thing, expresses the thing, *touches* it in a sense", and "[T]he sentence at once refers to an objective relation" (K.C. Bhattacharyya, 1956:83). In a stricter sense, this is what the requirement under *yogyatā* could be said to amount to, so that not only sentences such as "He is sprinkling with fire" and "Idleness is green" are ruled out, but also sentences such as "There is jar on the floor" when there is actually no jar on the floor in the context of the utterance. It follows that there can be no *śābdabodha* of a false sentence. The reasoning offered is that, a sentence '*a is F*' is characterised by fitness (*yogyatā*) only if *a* is *F*. But this evidently makes *yogyatā* collapse with 'truth'. Only true sentences are fit and so can generate *śābdabodha* (Chakrabarty, 1986). Truth is, more or less, made a condition of sentence-understanding, in the absence of a more interesting theory of *linguistic* understanding, as Mohanty points out in questioning this view (Mohanty, 1988:viii). Mohanty is concerned that this appears to be counter-intuitive, *qua* common sense, for one does *understand* false sentences, more than recollect from memory the unrelated bits of referents of the component words. Secondly, we could say that, in the case of a true sentence, the undestanding becomes a component of knowledge, while in the case of a false sentence, the knowledge-claim is revoked, but the understanding stands all by itself (*ibid*). Obviously, this line of interpretation requires a theory of linguistic meaning which is not a purely referential theory. Bilimoria (1988) demonstrates that there *is* such a theory of linguistic understanding, which is not constrained by the "direct object-relation" theory, particularly in the Mīmāṃsā-Vedānta theory of meaning, and he utilises the previously outlined 'syncretic' *prāmāṇya* theory (with the addition of *āptabhāva* or reliability of the authority for the testimony) to show how truth becomes a component of *śabdapramā*, i.e. verbal knowledge. The view is taken that, (a) meaning and reference are not to be collapsed, (b) the signification (i.e. the propositional understanding) of a sentence is independent of the question of truth or falsity, and (c) a claim to knowledge on the basis of this understanding would require knowing what would be true *if* it were true. As Strawson, in another context, had put it: "To understand a sentence is to know what thought it expresses or is capable, in given contextual circumstances, of expressing; and to know this is to know what we would be believing if we took that thought to be *true*." (Strawson,1976:274). And *mutatis mutandis*, if we took the thought to be *false*. Both false or, better, 'mock' understanding and 'false knowledge' are distinct possibilities within this analysis; neither amounts to significant knowledge, but that is not the whole issue here. We 'know' a lot of things about the world, some of which is true, some dubious, and some perhaps not very interesting, or plainly boring (Bilimoria,1988:268–272).

There are, however, other, perhaps even more challenging problems with the doctrine of *śabdapramāṇa* that we are not able to pursue in this short compass. For example, is all verbal knowledge of the relational, qualificative, kind? What of the 'identity' statements that Vedānta is seen to champion, such as 'You are That', 'I am I'? And, in what sense is a linguistic understanding a grasping of the relational ontological structure (or of a properly qualified individual)? Suppose the same object or state of affairs is perceived by one person and heard of or about by another person, do both their awarenesses share the same structure of propositions, or is one more directly related to the 'objective' reality than the other, or might one be more 'linguistic' than the other, and would they be interchangeable without compromising the significance and truth-claim of one against the other (i.e. is there verisimilitude)? If there is difference, is it in respect of the intentionality of each, or is it fundamentally in the way each is phenomenologically constituted and related to its 'objective reference'? Further, what are the ramifications of this doctrine for a hermeneutics of the *text*, particularly of scriptures, and moreover, of that class of scriptures thought of, in the Hindu philosophic tradition, as embodying 'authorless revelation' (*apauruṣeya-śruti*)? These and other questions, only partly attended to in current research, stand in need of more thorough investigation.

SELECTIVE BIBLIOGRAPHY

Banerjee, S.P. and Mohanty, J.N. (eds.) (1978). *Self, Knowledge and Freedom: Essays for Kalidas Bhattacharyya*. Calcutta: The World Press.

Bhattacharyya, Kalidas (1954). 'Indian Concepts of Knowledge and Self'. Calcutta: *Our Heritage*, II and seriatum.

—— (1985). 'Some Problems Concerning Meaning'. In B.K. Matilal, and J.L. Shaw (eds.) (1985), pp. 173 –188.

Bhattacharyya, Krishna Chandra (K.C.) (1956). *Studies in Philosophy*, 2 parts. Calcutta: Progressive Publishers.

Bhattacharyya, Sibajiban (1977). 'Some Principles and Concepts of Navya-Nyāya Logic and Ontology'. Calcutta: *Our Heritage* XXIV and XXV.

—— (1984). 'Some Indian Theories of Meaning'. Calcutta: *The Journal of the Indian Academy of Philosophy* XXIII/2: 25–41.

—— (1985). 'Abstraction, Analysis and Universals: The Nyāya Theory'. In Matilal and Shaw (Eds.) (1985), pp. 189–202

Bilimoria, Puruṣottama (1988). *Śabdapramāṇa: Word and Knowledge: A Doctrine in Mīmāṃsā-Nyāya Philosophy (with reference to Advaita Vedānta-paribhāṣā 'Agama')*. Dordrecht/Boston/London: (D. Reidel) Kluwer Academic Publishers.

—— (1989). 'On the Idea of "Authorless Revelation" (*Apauruṣeya*'. In Roy Perrett (ed.), *Indian*

Philosophy of Religion. The Hague: Martinus Nijhoff 1989, pp. 143–166.

Chakrabarti, Kisor Kumar (1984). 'Some Remarks on Indian Theories of Truth'. Dordrecht: *Journal of Indian Philosophy* XII/4 (December): 339–355.

Chakrabarty, Arindam (1986). 'Understanding Falsehoods: A Note on the Nyāya Concept of Yogyatā'. Calcutta: *The Journal of Asiatic Society* XXVIII: 10–20.

Chatterjee, Margaret (ed.) (1974). *Contemporary Indian Philosophy*. London: Allen & Unwin. (Vol I. and Vol II, forthcoming; Delhi: Motilal Banarsidass.)

Chinchore, Mangala R. (1988). *Vādanyāya: A Glimpse of Nyāya-Buddhist Controversy*. Delhi: Sri Satguru Publications.

Datta, D.M. (1972). *The Six Ways of Knowing*. Calcutta: University of Calcutta.

Deutsch, Eliot and van Buitenen, J.A.B. (eds.) (1971). *A Sourcebook of Advaita-Vedānta*. Honolulu: University of Hawaii Press.

D'Sa, S.J., Francis X. (1980). *Śabdaprāmāṇyam in Śabara and Kumārila*. Vienna: de Nobili Research Library.

EIP = *Encyclopedia of Indian Philosophies* (Karl H. Potter, General Editor). (1977). *Nyāya-Vaiśeṣika: Indian Metaphysics and Epistemology. The Tradition of Nyāya-Vaiśeṣika up to Gaṅgeśa* (part 1); (1978). *Advaita Vedānta up to Śaṅkara and His Pupils* (part 1). Princeton: Princeton University Press. (Delhi: Motilal Banarsidass; Vols. II & III respectively.)

Hattori, M. (1968). *Digṅāga: On Perception*. Cambridge Mass.: Harvard University Press.

Hayes, Richard P. (1987). *Digṅāga on the Interpretation of Signs*. Dordrecht/Boston/London: (D. Reidel) Kluwer Academic Publishers.

Herzberger, Radhika (1986). *Bhartṛhari and the Buddhists*. Dordrecht/Boston: D. Reidel. & Co.

Hiriyanna, Mysore (1973). *Outline of Indian Philosophy*. Bombay: George Allen & Unwin.

Gandhi, Ramchandra (1974). *Presuppositions of Human Communications*. Oxford: Oxford University Press.

Granoff, Phyllis (1978). *Philosophy and Argument in Late Vedānta*. Dordrecht: D. Reidel.

Inada, K. (1970). *Nāgārjuna: A Translation of His Mūlamādhyamakakārikā*. Tokyo: Hokuseido Press.

Ingalls, D.H.H. (1951). *Materials for the Study of Navya-nyāya Logic*. Cambridge, Mass.: Harvard University Press.

Larson, Gerald. J. (19??). *Classical Samkhya*. Delhi: Motilal Banarsidass.

Matilal, Bimal K. (1968). *The Navya-nyāya Doctrine of Negation*. Cambridge Mass.: Harvard University Press.

—— (1971). *Epistemology, Logic and Grammar in Indian Philosophical Analysis*. The Hague and Paris: Mouton.

—— (1985). *Logic, Language and Reality: An Introduction to Indian Philosophical Studies*. Delhi: Motilal Banarsidass.

—— (1986). *Perception: An Essay on Classical Indian Theories of Knowledge*. Oxford: Oxford University Press.

—— (1985). 'Awareness and Meaning in Navya-Nyāya'. In Matilal and Shaw (eds.), pp. 373–391.

—— and Shaw, J.L. (eds.) (1985). *Analytical Philosophy in Comparative Perspective: Exploratory Essays in Current Theories and Classical Indian Theories of Meaning and Reference*. Dordrecht: D. Reidel.

—— and Evans, D. (eds.) (1986). *Buddhist Logic and Epistemology: Studies in Buddhist Analysis of Inference and Language*. Dordrecht/Boston: D. Reidel.

Miśra, G. (1975). 'Analytical Studies in Indian Philosopical Problems: Śaṅkara's Doctrine of Adhyāsa'. Poona: *Indian Philosophical Quarterly* II/3.

Mohanty, J.N. (1966). *Gaṅgeśa's Theory of Truth*. Visvabharati, Santiniketan: Centre of Advanced Study in Philosophy. (Revised edition 1989. Delhi: Motilal Banarsidass.)

—— (1970). 'Nyāya Theory of Doubt'. In his, *Phenomenology and Ontology*. The Hague: Martinus Nijhoff.

—— (1980). 'Indian Theories of Truth: Thoughts on their framework'. Honolulu: *Philosophy East and West* XXX/4 (October): 439–451.

—— (1982). 'Indian Philosophy Between Tradition and Modernity'. In Rama Rao Pappu and R. Puligandla (eds.). *Indian Philosophy: Past and Present*. Delhi: Motilal Banarsidass.

—— (1984). 'Prāmāṇya and Workability: response to Potter'. Dordrecht: *Journal of Indian Philosophy* XII/4 (December): 329–338.
—— (1985) 'Psychologism in Indian Logical Theory'. In Matilal and Shaw (eds.) (1985), pp. 203–211.
—— (1988). Foreword, to Bilimoria (1988) (*qv*). vii– xi.
—— (1992) *Reason in Indian Thought*. Oxford (forthcoming)
—— *Essays in Indian Philosophy: Traditional and Modern* (1992). Ed. P. Bilimoria. Delhi: Oxford University Press.
Murti, T.R.V. (1960). *The Central Philosophy of Buddhism*. London: Allen and Unwin.
Murty, Satchidananda K. (1985). *Philosophy in India: Traditions, Teaching and Research*. Delhi: Indian Council of Philosophical Research/Motilal Banarsidass.
Nagatomi, Masatoshi (1979). '*Mānasa-Pratyaksa*: A Conundrum in the Buddhist *Pramāṇa* System'. In N. Nagatomi, B. K. Matilal *et al.* (eds.). *Sanskrit and Indian Studies*. Dordrecht: D. Reidel.
Nakamura, Hajime (1983). *A History of Early Vedānta Philosophy*. Delhi: Motilal Banarsidass.
Pandey, R.C. (1963). *The Problem of Meaning in Indian Philosophy*. Delhi: Motilal Banarsidass.
Pajin, Dushan (1987). 'The Legitimacy of the Term "Philosophy" in an Asian Context: The Beginnings of Indian Philosophy'. Dordrecht: *Journal of Indian Philosophy* XV/4 (December): 349–3627.
Potter, Karl H. (1963). *Presuppositions of India's Philosophies*. Englewood Cliff, N.J.: Prentice-Hall.
—— (1970). 'Realism Speech-Acts and Truth-Gaps in Indian and Western Philosophy'. Dordrecht: *Journal of Indian Philosophy* 1.
—— (1975). 'Some Thoughts on Nyāya Concept of Meaning'. Dordrecht: *Journal of Indian Philosophy* III: pp. 209–216.
—— (1978). 'Towards a Conceptual Scheme for Indian Epistemologies'. In Banerjee and Mohanty (eds.) (1978), pp. 17–30.
—— (1984). 'Does Indian Epistemology Concern Justified True Belief?' Dordrecht: *Journal of Indian Philosophy* XII/4 (December): pp. 307–382.
Radhakrishna, S. and Muirhead. (eds.) (1966). *Contemporary Indian Philosophy*. London: Allen & Unwin.
Searle, John R. (1983). *Intentionality: An Essay in the Philosophy of Mind*. Cambridge: Cambridge University Press.
Shastri, D.N. (1964) *Critique of Indian Realism*. Delhi: Motilal Banarsidass.
Shastri, Indian Chandra (1986). 'Jain Theory of Knowledge'. Delhi: *Journal of Indian Council of Philosophical Research* IV/1 (Autumn): pp. 73–86.
Sinha, Debabrata (1978). 'Cognitive Language in Vedānta'. In Banerjee and Mohanty (eds.), pp. 213–228.
Smart, Ninian (1969). *Doctrine and Argument in Indian Philosophy*. London: Allen Unwin. (Reprint: E. J. Brill, Leiden, 1992.)
Sprung, Mervyn (ed.) (1973). *The Problem of Two Truth in Buddhism and Vedānta*. Dordrecht/Boston: D. Reidel.
Staal. Frits J. (1960). 'Contraposition in Indian Logic'. In. E. Nagel. P. Suppes and A. Tarski (eds.) (1962). *Logic, Methodology and Philosophy of Science: Proceedings of the 1960 International Congress*. California: Stanford University Press.
—— (1962). 'Negation and the Law of Contradiction in Indian Thought'. London: *Bulletin of the School of Oriental and African Studies* XV.
—— (1969). 'Sanskrit Philosophy of Language'. In Thomas A. Sebeok (ed.), *Current Trends in Linguistics*, Vol. 5 of Linguistics in South Asia. The Hague and Paris: Mouton.
Strawson, P.F. (1976). 'Knowledge and Truth'. Poona: *Indian Philosophical Quarterly* III (April): pp. 273–282.
—— (1974). 'Meaning and Truth'. In his, *Logico-Linguistic Papers*, pp. 170–189. London and New York: Methuen.
Streng, F. (1967). *Emptiness: A Study in Religion and Meaning. New York: Abingdon Press.*
Taber, John (1981). 'Reason, Revelation and Idealim in Śaṅkara'. Dordrecht: *Journal of Indian Philosophy* IX/3: pp. 283–307.
Thakur, Anantalal (ed.) (1967). *Nyāyadarśana — with Bhāṣya of Vātsyāyana, Vārttika of Uddyotkara,*

Tātparyaṭīka of Vācaspati Miśra and Pariśuddhi of Udayana. Mithila: Mithila Institute Series, No. 20.

Vācaspati Miśra. (edited 1917). *Tattvabindu (śābdabodhakaraṇa)*. E.J. Lazarus and Co. *La Tattvabindu* (1979) (French translation by) Mde Biardeau. Pondicherry: Institut français d'indologie.

Philosophy of language in India

K. KUNJUNNI RAJA

INTRODUCTION

The potentialities and limitations of language have attracted the attention of Indian thinkers from ancient times. The poet-philosophers of the *Ṛgveda* were greatly concerned with the powers and limitations of language as a means of communicating their mystic, personal experiences of an ecstatic nature to their fellow brethren and consciously tried to stretch the power of language to the realm of the inexpressible, almost to the fringe of silence. Two approaches to language are seen, one praising it as a powerful and benign deity (*Vāc*) ever ready to bestow favours on her devotees, and the other complaining about the inadequacy of language for communicating the intimate, personal experiences, and for dealing with the ultimate reality. Does language act as a barrier or as a bridge? Is it a veil on the face of Reality which we must tear asunder before we can see her face to face? Or, is it a sort of coloured lens which helps us in seeing Reality, but at the same time distorts our vision?

The grammarian philosopher Bhartṛhari (5th century A.D.) pointed out in his *Vākyapadīya* that the whole phenomenon of material existence is only an illusory appearance (*vivarta*) of the Speech Principle of which symbols and meanings are only two aspects. It is the same speech-essence which appears in the form of the various ideas or meanings on the one hand and their symbols or expressions on the other hand, and thus constitutes the entire phenomenal world.

The Indian approach towards the study of language has been characterised both by analysis and synthesis. *Vyākaraṇa*, the term for grammar in Sanskrit, literally means 'linguistic analysis'; and Pāṇini's monumental work on Sanskrit grammar is mainly concerned with the analysis of words into the verbal roots, prefixes and suffixes, and verbal and nominal terminations. At the same time grammarians like Bhartṛhari advocated the theory that the real unit of language is the complete utterance, and that the analysis into words and further into roots, prefixes, suffixes, etc. is only an unreal, but necessary means for understanding.

The analytical method was older and more popular. The speech utterance was analysed into sentences and words, stems and suffixes, morphemes and phonemes. The verbal root was considered as the core element to which pre-verbs, primary and secondary suffixes, as well as nominal or verbal terminations were added to evolve the word. Rules of coalescence (*Saṃdhi*) between these various elements and between words in a compound word or sentence were also studied carefully and

Contemporary philosophy. A new survey. Vol. 7, 155–164.

attempts were made to identify the cementing factors helping to form an integral unit.

Language studies were carried out in India for several centuries by the various schools of thought as a result of which insights into the working of language have been brought forth.

They had to deal with the totality of language behaviour. The mystics and philosophers, as well as the inspired poets were interested in extending the frontiers of language for communicating intimate personal experiences and emotions. Like Bergson they believed that 'though language may be incapable of grasping Reality directly, it can bring the hearer to a point where he himself may transcend language and pass to incommunicable thought'.

The Nyāya school of logicians, mainly interested in the theory of knowledge and the truth or falsity of judgements, had to be concerned with the theory of meaning since understanding the meaning of a proposition was a primary requirement for making any significant study about it. They believed that what cannot be expressed clearly need not be discussed at all.

The literary critics, who were concerned with the understanding and appreciation of literature, were very much interested in the stylistic analysis of language and in finding out its deviation from ordinary language in order to see how far the poets have been able to communicate their vision of beauty and aesthetic experience through the medium of language. The theory of suggestion (Dhvani) and the solution of the problem of communication of emotional factors are their main contributions.

The Mīmāṃsā school, called *Vākya-Śāstra*, 'the science of sentences', was concerned with the methodology of textual interpretation with special reference to the scriptural texts. Their basic assumption was that the scriptural texts are authoritative and faultless, and have to be interpreted to explain away any apparent absurdities or contradictions. They developed the theory of metaphor and *arthavāda* (that the exaggerations have to be taken not in their literal sense, but as a sort of advertisement in praise or condemnation of certain actions).

According to the analytical method adopted by the schools of Mīmāṃsā and Nyāya, the basic unit is the word. If each word in a sentence gave only its own isolated meaning, of a universal nature, how can the hearer get a connected, unified sense as the sentence meaning? This problem was discussed in India from ancient times and three main factors have been pointed out as affecting the unity of the sentence meaning — mutual expectancy, consistency and contiguity. Even Pāṇini seems to have hinted at some of these points by the term *Sāmarthya* (capacity) prescribed for all operations between words — between the members of a compound word, and among the words in a sentence. This term has been explained as mutual expectancy (*vyapekṣā*) and unity of sense or unity of purpose (*ekārthībhāva*). These two are mentioned in the Mīmāṃsā definition of a sentence *arthaikatvād ekam vākyaṃ sākāṅkṣam ced vibhāge syāt (Mīmāṃsāsūtra* II. 1.46). The Mīmāṃsā explains the term *arthaikatva* as unity of purpose, but Bhartṛhari has pointed out that it can also be explained as unity of sense.

Language is used to communicate information to a hearer or a group of hearers.

The Indian theories on communication generally view the meaning of utterances from the hearer's point of view. This is especially so in the case of the Navya-Nyāya theories, though they also take into consideration the intention of the speaker. The hearer has to understand the intention of the speaker on the basis of the utterance heard; but it is on the basis of the well known meanings of the words and their syntactic relation that the hearer gets at the intention. The speaker cannot create new meanings to the words he uses, he can only pick and choose the appropriate words in the context. The so-called intention of the speaker is not absolute, it depends on the meanings of the words used in the particular context. Jagadīśa seems to feel that there is no need to take into consideration the intention of the speakers; contextual factors control the meanings of words, especially in the case of ambiguous expressions, and the hearer's understanding of the speaker's intention is on the basis of the contextual facts.

One of the basic problems discussed by all schools of thought in India is the relation between the linguistic expressions and their meanings. A linguistic sign becomes a sign only when it is meaning-bearing. What is the unit of language? Is it the phoneme or the word or the sentence? According to the Sphoṭa theory of Bhartṛhari, it is the complete utterance or the sentence that forms the unit, though at the analytic level the word can be taken as the unit. Those who know the language very well think and speak in units of sentences, and also hear whole sentences. It is only those who do not know the language intimately that hear words or phonemes and have to struggle with them to get the connected sentence meaning.

THE SPHOṬA THEORY

The Sphoṭa theory, fully promulgated by Bhartṛhari in the fifth century A.D., is one of the most important contributions of India to the problem of meaning. According to this the fundamental linguistic fact is the complete utterance or the sentence. Just as a letter or phoneme has no parts, so also the word and the sentence are to be taken as complete integral units, not made up of smaller elements. Linguistic analysis may be helpful in learning a language, but it has no reality of its own. In a normal speech situation, communication is always through complete sentences. The fact that expression is always through a temporal or spatial series of phonemes or words does not warrant our considering it as made up of parts. A painter normally conceives a picture in his mind and paints it on the canvas. He may use various colours and make various strokes; still the picture is integral and we see the picture as a unit, not as a conglomeration of colours and strokes. Similarly the meaning is also understood in an instantaneous flash of insight, (*Pratibhā*).

According to the grammarians it is the *Vākya Sphoṭa* or the sentence as an integral sign that is uttered and heard. The words and the phonemes are the lower units that make up the integral sentence-Sphoṭa. The cognition of the Sphoṭa is not sudden and spontaneous, but gradual, as in the case of evaluating a gem. Successive observations of the gem, through their accumulated impression, help in its clear evaluation. In the same way the successive phonemes, together with their order of sequence, help in the identification of the sentence Sphoṭa. Each phoneme suggests the full Sphoṭa, the first one in a vague manner, the succeeding phonemes helping in the clearer identification of the Sphoṭa from vagueness to clarity. The order of sequence is also necessary for the cognition of the identity of the *Sphoṭa*.

The *Sphoṭa* is grasped through perception. It is the sentence-Sphoṭa that is uttered and heard. To those who know the language well, the integral sentence-Sphoṭa is perceived.

It may be noted that according to the theory of the integral and indivisible sentence-Sphoṭa as the real unit of language, syntax and semantics will have no reality. Syntax means the interlinking of the different words in their proper order; if the sentence-Sphoṭa alone is real, the words have no real existence as such, and the order of sequence of the words is also unreal. Syntax is possible only in the articulate speech where words have reality. Semantics is possible in the articulate speech, taking word-meanings as units. Grammar demands syntax and semantics. So according to the Sphoṭa theory, grammatical analysis has no reality.

Again, since it is on the binary relationship between expressions and their meanings that linguistic communication is possible, the hearers are expected to know the meaning of the expressions. If only the sentence is real, the hearer can understand only those sentences which he has already heard previously. But in all living languages new sentences are being used, even by children. In the case of long sentences, too, it is difficult to get the meaning without coming down to the level of the words. The Sphoṭa theory may be all right as far as short sentences used in everyday life are concerned, as in sentences like 'How do you do!', 'Good morning', 'Excuse me', 'Beg your pardon' etc. The grammarians' reply would be that, in all linguistic discourses, one has to come down to the level of words and word-meanings for understanding, though at the absolute level the Sentence-Sphoṭa is the reality.

THE MĪMĀṂSĀ SCHOOL

The Mīmāṃsā philosophy of language was mainly concerned with the methodology of textual interpretation in order to give a cogent explanation of the scriptural texts of a prescriptive nature. It had to deal with apparent absurdities, inconsistencies and self-contradictions, besides ambiguities, and evolved rules of interpretation which were later accepted generally by all schools of thought and used freely in legal practice and commentarial literature. The Mīmāṃsā gave a semantic definition of a sentence, evolving the concepts of mutual expectancy (*ākāṅkṣā*), consistency (*yogyatā*), and contiguity (*āsatti*) as factors necessary for a sentence. In order to explain away the apparent absurdities, it evolved the theory of metaphors.

Ākāṅkṣā, or mutual expectancy, consists in a word not being able to convey a complete sense in the absence of another word or other words. Literally it is the desire on the part of the listener to know the other words in the sentence to complete the sense. The Mīmāṃsāka-s were more interested in the psychological expectancy than in the syntactic expectancy. The logicians and rhetoricians were stressing the need for syntactic expectancy. A noun in the nominative case has an expectancy towards a verb, and a verb will have expectancy towards a noun in the nominative case. For all practical purposes the grammarians also accepted syntactic expectancy, though according to the Sphoṭa theory, grammatical analysis and expectancy etc. have no absolute reality.

The next condition for a valid sentence is taken to be *yogyatā* or semantic consistency. The meaning of every word can be subjected to componential analysis. If there is conflict between the semantic items of terms in a sentence, there is obstruction to the understanding of the sentence meaning. The term 'bachelor' implies human being, male and unmarried person. The term 'mother' refers to the semantic items human being, woman, having one or more children. The phrase 'bachelor mother' is semantically self-contradictory since one term implies a male and the second term a female. A Sanskrit example is *vandhyā-putra*, 'son of a barren woman'. There is difference of opinion as to whether *yogyatā* should be taken as a positive condition or not. If the inconsistency is apparent and can be explained away by recourse to metaphoric transfer, the sentence can be accepted as valid; if that is not possible, it will be a non-sentence.

Grammarians did not emphasise the importance of *yogyatā*, for to them it was enough for a sentence to give a syntactically connected meaning. Its veracity is not a condition. A sentence is a sentence whether it is true or not. From the grammarians' point of view even metaphoric transfer (*lakṣaṇā*) is not of interest. 'He is a boy' and 'He is an ass' are equally valid sentences. Even empty phrases like 'the child of a barren woman' are linguistically valid, since the grammarians are not concerned with the external existence of the thing meant by the expression. The doubt as to how a barren woman can have a child shows that the hearer has understood the literal meaning of the sentence. Kumārila bhaṭṭa, the Mīmāṃsaka, has declared that sentences convey the idea of even non-existent things; it is only the truth value that is affected.

If the lack of *yogyatā* can be explained away by resorting to the metaphoric meaning of one of the words in the sentence, there is no difficulty in understanding the meaning.

The third condition for a sentence is *āsatti* or contiguity, which refers to the uninterrupted utterance or the unbroken apprehension of the words in the sentence. In the case of elliptical sentences, some of the missing words are understood from the context and help in the understanding of the sentence meaning.

To these three conditions some add a fourth one, namely *tātparya* which refers to the intention of the speaker, or to the general purport of the sentence as understood from contextual factors. The logical school generally lays stress on the intention of the speaker, while the Mīmāṃsā school says that the general purport can be understood from contextual factors.

Sentence meaning

The Sphoṭa theory of the grammarians takes the sentence as the unit (*Vākya-Sphoṭa*) and its meaning also an integral unit, an instantaneous flash of insight (*pratibhā*). According to the analytical school followed by the Mīmāṃsā and Nyāya schools, there are two main theories called *anvitābhidhāna* ('expression of the related') and *abhihitānvaya* ('relation of the expressed'). The sentence meaning is something more than the sum total of the word meanings; it consists of the individual word meanings and their mutual syntactic relationship. The *abhihitānvaya* theory followed by the Bhāṭṭa School of Mīmāṃsā considers that each word in a sentence first gives out its individual unrelated meaning and stops with that; these word-meanings then get mutually related on the basis of mutual expectancy etc. and the contextual factors, through the function of a metaphoric transfer. Speech is purposive in nature, and it is assumed that the co-utterance of the various words in the sentence is to convey a connected meaning: and by resorting to metaphoric transfer (*lakṣaṇā*), the syntactically related sentence meaning is obtained. According to the Prābhākara School of Mīmāṃsā, however, the words in the sentence themselves give out both their individual meanings and their syntactic relationship. Thus the entire sentence-meaning is conveyed by the words themselves. According to the *abhihitānvaya* theory, the sentence-meaning is grasped in a single comprehension, after getting the meanings of all the words in the sentence; while according to the other theory the comprehension of the sentence meaning is gradual, step by step. The logicians generally believe that the sentence meaning is an association of the word-meanings, termed *saṃsarga-maryādā*. In a phrase, 'the blue lotus', the term 'blue' refers to a thing qualified by the universal 'blueness', and the term 'lotus' refers to any particular lotus flower qualified by 'lotus-ness'; the meaning of the phrase is the association of the two.

The Buddhists like Dignāga held that the meaning of any term is a concept in the mind, not directly connected with the thing-meant, and that this concept (*vikalpa*) has no reality of its own, its function being the preclusion of everything else, (*apoha*). Though the word-meanings are negative in nature, their sum-total or the sentence-meaning is of the nature of *Pratibhā* or a flash of insight.

Word meaning

The binary, permanent relationship between a word and its meaning is generally accepted by all schools of thought in India. This implies that every word has a meaning of its own, and that every object can be designated by a single word. Ambiguity will arise if this rule is not observed. In actual practice, however, this rule is not rigidly observed, and the possible ambiguity is to be solved with the help of contextual factors, both syntactic and situational. Polysemy can be of two kinds, homophones and homonyms — two words having the same form and one word having two meanings. Yāska's (500 B.C.) discussion about the principle of derivation of words sheds considerable light on the origin of polysemy. Nouns are normally derived from verbal roots denoting action; if this principle is strictly followed, every object will have as many names as the actions with which it is associated, and each noun could be applied to as many objects as are associated with that action indicated by the verbal root. Yāska's answer is that there are restrictions; in designating objects language uses only one of the many activities associated with the object. Hence there can be some sort of permanent relation between a word and its meaning.

It is also accepted that the primary meaning of a word is not definitely circumscribed. The boundaries of the meaning often change on the basis of contextual factors, even in ordinary words. 'Man is mortal' does not mean that 'woman is immortal'. When there is a conflict between the etymological meaning and the popular usage, the meaning current in popular usage among the educated elite is to be accepted.

Metaphoric transfer (Lakṣaṇā)

Sometimes a word is used to denote a referent other than its normal one. Such metaphoric usage is common in all languages. If we take the word as denoting its normal primary meaning, the sentence may become nonsensical in the context. This produces a 'psychic resistance' in the listener and there is a sort of break in the flow. It excites attention and requires interpretation for understanding the purpose. The actual referent of the word has to be taken as different from its normal one, but in some way connected with it, either through similarity or through some other relation. This function of the word, denoting a referent different from its normal and primary one, but somehow related to it is called *lakṣaṇā* or metaphoric transfer.

Even though every word is accepted as having a primary, stable meaning core, in actual practice, however, shifts in meaning, metaphoric transfers and secondary usages are quite common. There are three conditions considered necessary for resorting to the secondary meaning. (1) Inconsistency or incongruency, if the words are taken in their literal senses. 'He is an ass' or 'He is a firebrand' cannot be taken in their literal senses, because the human being who is described in these sentences cannot be an animal or an inanimate firebrand. Hence, to remove the discrepancy and arrive at an acceptable sense in the context, the sentence has to be explained taking the unsuitable words as referring to the qualities — 'ass' referring to 'foolishness' since the ass is notorious for its foolishness according to fables, and 'firebrand' referring to brilliance and hot-temper. In the sentence 'The village is on the Ganges', since the village cannot be on the current of the river, it has to be taken to mean that the village is on the banks of the Ganges. (2) The second condition is that there should be some relation between the primary meaning of the word and the actual meaning in the context. This relation can be a common quality or similarity; it can be any other relation, like proximity, cause and effect, contrariety, etc. (3) The third condition for resorting to the secondary significance is either sanction by popular usage as in the case of faded metaphors and shifts in meaning, or a special purpose for which it is resorted to, as in the case of intentional metaphors. The inconsistency of the primary meaning can be the impossibility of syntactic connection from the point of view of meaning or it can mean inconsistency in the context. In the sentences, 'Let us move; look at the sun', the latter part implies that it is getting late. A sentence like 'See that crows do not spoil the food' implies that the food should not be allowed to be spoiled even by dogs etc. The Nyāya school accepts metaphoric transfer only for a particular word in a sentence, but the Mīmāṃsā school says that it can also mean that the whole sentence implies a metaphoric sense. Metaphor plays an important role in poetry, especially symbolic poetry.

Lakṣaṇā is subdivided into two classes: that where the relation between the primary meaning and the actual meaning is on the basis of similarity or common quality, and that where the relation is based on something other than similarity. There is another division into three classes on the basis of the degree of intimacy between the primary and the actual meanings. (1) where the primary meaning is completely rejected, (2) where the primary meaning is partly retained and (3) where it is fully understood.

In the sentence, 'the village is on the Ganges', the primary meaning has to be completely rejected; in the sentence, 'the village is burnt', only part is rejected, for only part of the village is burnt. In the sentence, 'people with umbrellas are coming', the literal meaning is not rejected; it implies that many of the people in the group have umbrellas.

Sentences of the type A=B come under the *lakṣaṇā* where part of the primary meaning is rejected and part retained. 'This is that Devadatta'. Here the identification is only for the person named Devadatta; the man seen here, now, is identified with the same man seen previously at some other place. The famous statements in the Upaniṣads like '*Tat tvam asi*' (That thou art) belong to this type. Brahman or the Absolute Reality is identified with the self in man. If the identity is accepted literally it becomes nonsense and consequently there is a feeling of resistance. When the machinery of interpretation is brought into play, there is a feeling of richness of meaning. The word 'Thou' in the sentence does not mean Uddālaka, son of Śvetaketu, with limited intelligence, and the word 'That' does not mean the omniscient Brahman. Stripped of all individual attributes, the sentence implies the identity of pure consciousness in the individual with that in the Universal Soul.

This may be compared to Frege's famous example 'the morning star = the evening star'. Frege gives it to explain his distinction between sense and reference (*Sinn* and *Bedeutung*), and says that the identity is only for the reference and not for the meaning. Here the identity is only for the substantives, not the substantives as qualified by the attributes. This is a good example of the type of *lakṣaṇā*. or metaphoric transfer, where the primary meaning of both the terms are partially retained.

Sentences of the type A=A, normally rejected as tautology by the logicians in India as well as in the West, can be brought under one type of *lakṣaṇā*. This is called 'pregnant use' by Empson. His example is 'An explosive is an *explosive*, it must be handled carefully.' The Indian example generally given is, 'When the Spring season arrives, the crow is the *crow*, and the cuckoo is the *cuckoo*'. Here the second term means not only the primary meaning but also the primary meaning with all the suggestive (good or bad) qualities. This kind of metaphor is common in poetry. 'The lotus flowers become *lotus flowers* (in the true sense of the term) only when blessed by the rays of the sun'.

Lakṣaṇā is based on the apparent inconsistency of the literal meaning of one of the terms in the sentence in the context. How can the metaphoric sentence be accepted as a sentence at all, since one of the conditions for an acceptable sentence is *yogyatā* or semantic congruity? How can we distinguish between a non-sentence (due to the lack of *yogyatā*) and a metaphoric sentence (where also semantic inconsistency is present?) The answer seems to be that if the inconsistency in the literal sense can be explained away by resorting to the metaphoric transfer, the sentence is acceptable as a metaphor, and if it cannot be explained away, it is a non-sentence. According to the Mīmāṃsā school all Vedic sentences are authoritative, and there is no possibility of its being unacceptable. Any apparent absurdity or lack of consistency has to be explained away somehow. In the case of other sentences there is a possibility for mistakes, and the utterance can be a non-sentence.

In poetry metaphor plays an important role. A sentence like 'Her face is the full moon', occurring in the description of a girl, does not obviously mean the identity of the two terms seriously. From the context it is obvious that it is the description of the face of the girl, and hence the expression has to be taken as metaphoric and the meaning is that her face is beautiful like the moon. Thus as in the case of the scriptural statements, in poetic description also we have to assume that a cogent meaning is intended and that necessary interpretation, as metaphor, is to be resorted to. 'The sacrifical pole is the sun' found in the scripture has to be interpreted to mean that the 'pole is to be as bright as the sun'.

Vyañjanā (Suggestion)

The theory of literal and metaphorical meaning developed by the Mīmāṃsā and Nyāya schools in ancient India, was further extended by Ānandavardhana (c. 850 A.D.) to include emotive and other associative meanings also under linguistic meaning. He confined his attention to the field of poetic meaning. He accepted the literal and metaphoric meanings, but in addition postulated a third potency of language, namely suggestion (*vyañjanā*), 'the capacity to suggest a meaning other than its literal meaning'. Emotive meaning and social-cultural meaning cannot be brought under the literal or even the metaphorical meanings, but can be explained satisfactorily only by the assumption of the suggestive power of language. Logicians are not interested in including these under meaning; but for literary critics these are the most important aspects of poetic meaning.

The suggestive meaning cannot be included in inference, because there is no invariable concomitance between the expression and the suggested meaning. The suggested meaning is often vague and depends on contextual conditions and the mental make-up of the hearer: and the same expression may suggest different ideas to different hearers. Moreover, emotional factors which are the most important poetic meaning cannot be conveyed by merely using the terms referring to them. It has to be suggested by describing the situation, the characters involved, their actions and reactions. The motive element in intentional metaphors is not part of the function of metaphoric transfer; that is conveyed through suggestion.The suggested meaning understood depends on the cultural background of the hearer and is not constant.

Indian philosophers envisage four different stages in the manifestation of language – *parā, paśyantī, madhyamā* and *vaikharī*. Of these, *vaikharī* is the last stage and refers to the actual sounds uttered by the speaker and heard by the listener. The *madhyamā* is the linguistically relevant form comprehended by the intellect, with the order of sequence in a latent form. *Paśyantī* is the mentally conceived sentence, the sentence-sphoṭa and the meaning in the form of *pratibhā* (a flash of insight) mingled into one. It may be taken as the principle of consciousness. The final stage *parā* is the Absolute Reality, beyond the reach of the mind, about which nothing can be said definitely.

SELECT BIBLIOGRAPHY

Aiyar, K. A. Subrahmanya, *Bhartṛhari*. Pune: Deccan College, 1969.

Biardeau, M., *Theorie de la Connaissance et Philosophie de la Parole dans le Brahmanisme Classique*. Paris 1964.

Raja, K. Kunjunni, *Indian Theories of Meaning*. Madras: Adyar Library, 1963.

Sastri, Gaurinath, *The Philosophy of Word and Meaning*. Calcutta 1959.

La pensée rationnelle en Chine ancienne

JEAN-PAUL REDING

1. Introduction

Le lecteur occidental ne manquera pas d'aborder le sujet proposé avec une certaine appréhension. Pourquoi s'attarder, dira-t-il, sur un aspect mineur de la pensée chinoise, au détriment des moments éthiques et contemplatifs de cette pensée, seuls domaines où la mentalité occidentale pourrait tirer profit de la profondeur d'esprit orientale? A quoi bon se pencher sur les balbutiements de la logique chinoise, dont nous ne pouvons plus rien apprendre?

En assumant la tâche de présenter les vestiges qu'a laissés la pensée rationnelle en Chine ancienne, je m'engagerai donc à démasquer ces objections comme autant de préjugés mal fondés. Je me proposerai de montrer qu'il y a, en Chine ancienne, une pensée logique non seulement originale — c'est-à-dire qui ne répète pas un ou des stades de développement de la pensée logique occidentale ou même indienne — mais encore digne de notre intérêt et d'un niveau élevé.

Auparavant, il me faudra brièvement évoquer les raisons pour lesquelles la phase rationnelle qu'a traversée la pensée chinoise est si mal connue, non seulement en Occident, mais en Chine elle-même.

Tout d'abord, les premiers philosophes qui se sont intéressés à la pensée chinoise n'étaient guère disposés à y rechercher l'élément rationnel; la pensée chinoise avait alors la fonction de servir d'antidote aux formes tentaculaires du rationalisme occidental. Ensuite, les textes qui auraient pu révéler à ses premiers explorateurs les aspects rationnels de la pensée chinoise étaient mal — ou pas du tout — édités, dépourvus de commentaires et donc difficiles d'accès, même aux spécialistes. Enfin, la tradition chinoise elle-même n'a jamais "réintégré" ces textes au sein de ses classiques et les a longtemps ignorés. En effet, le confucianisme s'est très tôt imposé comme la pensée officielle et a éclipsé de manière durable tous les autres grands courants de pensée de la Chine ancienne, à savoir le taoïsme, le légisme, le moïsme etc.

HU SHI et FENG YOULAN, deux historiens chinois formés aux modes de penser occidentaux avaient, les premiers, attiré l'attention sur le potentiel rationnel de la pensée chinoise (cf. bibliographie ; HU [1], FENG [2]). Toutefois, il pesait sur leur tentative le soupçon de vouloir préparer la civilisation chinoise à son envahissement imminent par la mentalité occidentale technocrate, ce qui les priva malheureusement d'une crédibilité scientifique pourtant méritée, autant par le

Contemporary philosophy. A new survey. Vol. 7, 165–193.

sérieux de leur travail philologique que par leur capacité de discernement philosophique. Nous devons la renaissance des études contemporaines dans le domaine de la logique chinoise à l'éminent sinologue anglais A.C. Graham, éditeur notamment du texte logique chinois le plus important, le *Canon moïste* (*Mojing*).

2. L'ÉPOQUE, LES HOMMES, LES TEXTES

L'époque qui nous intéresse est celle dite des "Royaumes Combattants" (-481 à -221). Elle va, en gros, de l'éclatement du royaume féodal chinois à sa réunification au début du IIe siècle avant notre ère. Au cours des VIe et Ve siècles, l'ancien royaume féodal chinois se désagrège en une poussière de cités-Etats autonomes. Pendant cette même période, d'importants changements ont lieu: introduction de la technique de la fonte; forte augmentation des terres arables, soit par assainissement, drainage et défrichement, soit par la conquête de terres "barbares"; essor des techniques de fortification; développement du commerce; explosion démographique enfin. Parallèlement à tous ces bouleversements, un changement de mentalité se prépare. L'ordre nouveau que les experts techniques de tous bords tentent d'amener doit être pensé. Les premiers philosophes de la Chine ancienne ne peuvent concevoir un possible ordre nouveau en dehors du cadre d'un empire chinois (ré)unifié. La recherche de l'union politique et sociale est ainsi le présupposé inavoué de toutes les réflexions philosophiques conduites à l'époque des Royaumes Combattants.

Le philosophe des Royaumes Combattants est lui aussi, en quelque sorte, un expert. Mais il n'est pas, comme la plupart des autres experts, un spécialiste de techniques de fortification ou de travaux de drainage: c'est un expert en techniques de gouvernement. Le philosophe n'est cependant pas, ou du moins pas essentiellement, un politicien. En fait on peut dire que le philosophe chinois est au politicien ce que le compositeur est au chef d'orchestre.

Le point de départ de la réflexion est partout le même: la situation présente est intenable. Le pays est exposé à une crise à la fois politique, morale et économique. Le philosophe se voit appelé à repenser l'ensemble des valeurs, qu'elles soient morales, politiques, économiques ou sociales. Cette situation de départ explique l'éventail extrêmement large des doctrines philosophiques qu'on rencontre à l'époque des Royaumes Combattants. Il y a tout d'abord les philosophes qui cherchent à construire un ordre nouveau par une réforme des hommes eux-mêmes, donc par l'établissement ou le rétablissement de valeurs morales. A ce premier groupe appartiennent les confucianistes et les moïstes. Confucius prend pour guide dans cette crise des valeurs le code d'honneur de l'ancienne société féodale. MOZI repense entièrement le système des valeurs morales et propose un modèle de société basée sur la notion de l'utile. L'utile est défini par MOZI comme le bien du peuple. Le bien du peuple — et donc celui de l'Etat — est pour MOZI l'étalon de toutes les vertus morales. A ces deux premières écoles, qui aspirent à une réforme "par l'intérieur" font pendant les écoles légistes[1], qui voient dans les techniques administratives la méthode parfaite pour gouverner. Les taoïstes, quant à eux, s'opposent à l'une et à l'autre de ces méthodes. Toute action — qu'elle soit intérieure ou extérieure — sur le développement spontané de la nature humaine brise cette spontanéité et ne peut, en fin de compte, que rendre ingouvernable l'individu, qui ne sait plus comment s'orienter devant la pléthore des interdits sociaux et moraux. L'homme est capable de se prendre en charge lui-même. En suivant ses aspirations spontanées, pour autant que celles-ci n'aient pas encore été déformées par l'éducation moralisante ou par les techniques de gouvernement, l'homme doit nécessairement atteindre le bonheur. Les premiers taoïstes sont, bien entendu, des reclus qui refusent de se lancer dans les tourbillons de la politique; les taoïstes de la fin des Royaumes Combattants par contre prêtent volontiers main forte aux légistes, qui cherchent un fondement métaphysique à leurs doctrines politiques.

La tâche du philosophe des Royaumes Combattants diffère radicalement de celle de son homologue grec. Héraclite, Parménide, Socrate et Platon évoluent dans une société qui possède des cadres bien établis. Rien de tel à la période des Royaumes Combattants. Le philosophe chinois est un constructeur de société avant tout. Il est attendu comme un sauveur, comme celui qui va établir un modèle de société. C'est avec cette mission que le philosophe chinois est appelé à la cour de souverains locaux, qui cherchent à étendre, ou simplement à maintenir, leur position politique et économique. Ces princes locaux sont à la tête de véritables "Cités-Etats"[2] autonomes. Les plus forts aspirent à dominer l'empire chinois tout entier et à se proclamer empereur à la place de l'empereur ZHOU déchu. Les plus faibles, pour éviter de se faire broyer entre les grands, mènent une politique d'alliance et de louvoyements politiques.

La guerre, véritable fléau de la population, est à l'ordre du jour. Le problème de la guerre prend par conséquent une place importante dans la réflexion des philosophes: Confucius tolère la guerre "juste"; les moïstes n'acceptent que la guerre défensive; les légistes estiment que la guerre est nécessaire au maintien et au développement de l'Etat; seuls les taoïstes restent, par apolitisme, à l'écart de la guerre. Au souverain cependant se pose un autre problème: s'il mène trop de guerres, la population de son territoire risque non seulement d'être décimée, mais les chances de la voir décamper et se fixer ailleurs sont grandes. Les terres en friche ne manquent pas et les souverains en manque de colons non plus. Cette situation explique les tendances nomadisantes d'une grande partie de la population de la Chine ancienne[3]. Il n'est peut-être pas inutile de rappeler dans ce contexte que la Grande Muraille, dont la construction a été entreprise vers la fin des Royaumes Combattants, a servi autant à garder les Chinois à l'intérieur du territoire chinois qu'à en éloigner les "barbares". Le type de société ou d'Etat que recherche le souverain doit par conséquent obéir aux deux conditions suivantes: l'Etat doit "fonctionner" et le mode de vie qui en découle doit être agréable ou, du moins, supportable. Ces deux exigences délimitent donc également la tâche du philosophe.

De ce tableau succinct et nécessairement incomplet, il ressort que les souverains des différentes Cités-Etats avaient tous intérêt à écouter les discours de ceux qui se proposaient de mettre en place une société nouvelle, source possible de richesses et de puissance. Dans ce climat d'effervescence intellectuelle, l'intensité des débats philosophiques a dû être extrême. Malheureusement, du nombre incalculable de textes composés à cette époque, il nous reste peu de chose, et nombreuses sont les oeuvres dont nous ne connaissons plus que le titre[4].

Comparé à celui de la Grèce ancienne, le volume des écrits logiques produits en Chine ancienne est faible. Le document principal est formé par six chapitres (les chapitres 40 à 45) du livre *Mozi*. Les chapitres 40 à 43 sont plus communément rassemblés sous le titre collectif *Mojing* ("Canon moïste" ou "livre canonique moïste"); les chapitres 44 et 45, intitulés respectivement *daqu* ("grand choix" de fragments) et *xiaoqu* ("petit choix" de fragments) sont, comme leur titre l'indique, des collections de fragments.

L'architecture du *Mojing* est assez particulière: le *Mojing* est formé de 180 énoncés catégoriques (que nous appellerons "théorèmes") et d'autant de preuves ou d'arguments correspondants (que nous appellerons "explications"). Comme ces théorèmes ne sont pas numérotés, un des éditeurs des écrits logiques néo-moïstes a inventé une méthode très astucieuse pour retrouver l'explication correspondant à chaque théorème: il a fait précéder chaque explication par le mot initial (c'est-à-dire par l'idéogramme représentant ce mot) du théorème correspondant. Cette particularité s'explique par le fait que le *Mojing* ne fait pas suivre chaque théorème de son explication, mais énonce d'abord tous les théorèmes (groupés dans les chapitres 40 et 41), ensuite toutes les explications (chapitres 42 et 43). Voici un exemple. Le théorème initial statue:

> La *cause**[d'un événement] est ce qui doit être réalisé afin que cet événement puisse se produire. (théorème)
>
> [La cause]** La petite cause: si elle agit, l'événement ne se produit pas nécessairement; si la petite cause n'agit pas, l'événement de nécessité ne se produira pas. (...) La grande cause: si elle agit, l'événement se produira de nécessité; si elle n'agit pas, l'événement de nécessité ne se produira pas. (...) (explication)[5]

Le *Mojing* a ensuite passé par plusieurs réarrangements typographiques différents. L'artifice conçu par son premier éditeur est vite tombé dans l'oubli et ces idéogrammes, dont la seule fonction a été d'aider à retrouver les démonstrations des différents théorèmes ("head characters"; voir ci-dessus) ont été traités comme des mots faisant partie du texte. On s'imagine aisément l'embarras devant lequel devaient se trouver les premiers traducteurs, surtout si l'on songe au fait que la syntaxe du chinois classique était encore assez mal connue. Nous ne savons rien des auteurs du *Mojing*, et nous ne connaissons pas sa date de composition exacte. La forme de l'écrit nous fait présumer qu'Il s'agit d'un document destiné à être

---

* Premier mot du théorème.

** L'explication reprend d'abord le mot initial du théorème. Ce mot initial ne doit évidemment pas être traduit!

utilisé à l'intérieur de l'école moïste, d'une sorte de compendium fréquemment mis à jour. La version qui nous est parvenue date probablement du -IIIe siècle. Le texte le plus fameux est sans aucun doute le "Discours sur le cheval blanc" (*Baimalun*) de GONGSUN LONG[6]. La limpidité (apparente) de sa syntaxe et la transparence (illusoire) de son argumentation en ont tôt fait l'échantillon représentatif de toute la pensée logique en Chine ancienne. Le texte commence ainsi:

> "Est-il permis de dire que cheval blanc n'est pas cheval?"
> – C'est permis.
> – Comment donc?
> – Cheval, c'est ce qui désigne la forme; blanc, c'est ce qui désigne la couleur. Ce qui désigne la couleur n'est pas ce qui désigne la forme. Voilà pourquoi je dis que cheval blanc n'est pas cheval."

Ce texte est à l'origine de maints débats sur la valeur de la logique chinoise. Il y a tout d'abord ceux qui ne voient dans cet argument qu'un plat sophisme (confusion entre prédication et identité), qui démontre à satiété la vanité des recherches logiques entreprises en Chine ancienne[7]. D'autres chercheurs se sont évertués à montrer que GONGSUN LONG prouve seulement que la couleur blanche est différente du cheval, et non pas que, comme il le prétend, "cheval blanc n'est pas cheval"[8]. Il y a aussi ceux qui cherchent à sauver le raisonnement de GONGSUN LONG en le coulant dans différents moules de logique mathématique (logique des classes, algèbre de Boole etc.)[9]. Finalement, il vaut la peine d'attirer l'attention sur une interprétation récemment proposée par A.C. Graham: le comportement logique des termes "cheval" et "cheval blanc" doit être aligné sur celui de "tout" et de "partie"[10]. Nous proposerons toutefois, après avoir considéré les formes d'argumentation propres aux philosophes chinois, une interprétation toute nouvelle du raisonnement de GONGSUN LONG.

GONGSUN LONG, l'auteur de ce texte curieux, a vécu dans la première moitié du IIIe siècle avant notre ère. Il a été l'un des protégés du seigneur PING YUAN, riche mécène qui entretenait une cour entière de savants et de philosophes, tous experts en rhétorique et en diplomatie.

Le texte qui a suscité le plus d'intérêt auprès des commentateurs occidentaux sont les deux listes de paradoxes transmises par le chapitre XXXIII du livre *Zhuangzi*. La première liste est attribuée à HUI SHI, grand homme politique et infatigable débatteur de l'époque des Royaumes Combattants. Voici ses fameuses "dix thèses":

I L'absolument grand n'a pas d'extérieur: appelez-le le grand Un.
L'absolument petit n'a pas d'intérieur: appelez-le le petit Un.
II Le "sans épaisseur" ne peut être empilé, mais sa grandeur est de mille Li.
III Le ciel est aussi bas que la terre: montagnes et marais sont à un même niveau.
IV Le déclin du soleil commence au zénith, un être meurt dès qu'il naît.
V Une grande ressemblance qui diffère d'une petite ressemblance, voilà ce qui est appelé une petite "ressemblance-et-différence".

Les Dix mille êtres qui à la fois se ressemblent et diffèrent, voilà ce qui est appelé une grande "ressemblance-et-différence".

VI Le Sud n'a pas de limite, mais il est limité.

VII Je vais à YUE aujourd'hui, mais j'y suis arrivé hier.

VIII Des anneaux enchaînés peuvent être séparés.

IX Je connais le centre du monde: il est au Nord de YAN et au Sud de YUE.

X Il faut aimer sans distinction les Dix mille êtres, car le ciel et la terre forment un seul corps.

Selon l'interprétation courante, ces dix propositions énigmatiques formeraient un "système" prouvant que toutes les distinctions spatiales et temporelles sont illusoires. Ce principe de l'identité essentielle des choses serait expliqué par la thèse X: "Tous les neuf paradoxes sont destinés à montrer que l'univers est un et que nous devons aimer également toutes choses. En d'autres termes, les paradoxes forment une tentative faite pour asseoir la doctrine moïste de l'altruisme universel sur une base métaphysique", écrit H. Maspéro[11].

Le point faible de cette interprétation réside dans l'absence de toute considération doxographique. Même si l'on admet que c'est HUI SHI qui a été l'auteur de ces dix thèses, il reste le problème de la forme sous laquelle ces dix thèses se présentaient dans le texte primitif de HUI SHI. On peut montrer assez facilement[12] qu'un lecteur contemporain de HUI SHI ne pouvait en aucun cas lire ces dix thèses comme un texte circonscrivant un système métaphysique niant la réalité des différences temporelles et spatiales. Trois thèses au moins sont, en effet, des proverbes tirés soit du *Livre des Mutations* (*Yijing*) ou du *Livre des Rites* (*Liji*). Il en faut conclure que les compilateurs du chapitre XXXIII du *Zhuangzi* — si toutefois ceux-ci disposaient d'un texte rédigé par HUI SHI — ont dû opérer un choix. Or nous ne sommes pas du tout renseignés sur les principes qui ont pu guider ce choix. Tout au plus pouvons-nous présumer que l'attitude des compilateurs face à HUI SHI n'a pas été des plus favorables. L'objectivité de ce choix et partant son caractère représentatif est donc sujet à caution.

Une réinterprétation des dix thèses est par conséquent nécessaire. On pourrait voir ainsi, comme nous l'avons suggéré[13], en HUI SHI l'originateur d'une forme de rationalisme scientifique. Les dix thèses peuvent en effet très bien être lues comme un rudiment de système cosmologique.

Les paradoxes de la deuxième liste du chapitre XXXIII du *Zhuangzi* ("les vingt et un paradoxes") se présentent davantage comme des sophismes ou des jeux de mots que comme des antinomies sérieuses. Voici quelques exemples: "La tortue est/a une vie plus longue que le serpent" (l'expression chinoise *chang yu* pouvant signifier à la fois "est plus long que" et "vit plus longtemps que"); "Le poulain orphelin n'a jamais eu de mère" (sinon il ne serait pas orphelin) etc.

La plupart des paradoxes de cette deuxième liste ont été extraits du *Mojing*[14]. Arrachés à un contexte dans lequel, malgré leur apparence paradoxale, ils s'imbriquaient parfaitement, ces théorèmes néo-moïstes se transforment, dans leur nouvel entourage du chapitre XXXIII du *Zhuangzi*, en purs sophismes, et c'était bien là le résultat recherché par les compilateurs. De ces considérations, il ressort que ces deux fameuses listes de paradoxes ne sont nullement représentatives de la pensée logique et rationnelle en Chine ancienne.

L'accès aux textes logiques produits en Chine ancienne est par conséquent semé d'embûches philologiques et le lecteur aura vu que toute étude, dans ce domaine, doit être précédée d'un travail de critique textuelle et d'analyse doxographique.

En dépit de ces obstacles, plusieurs savants occidentaux s'étaient attaqués à la traduction de ces fragments logiques. Mais tant H. Maspéro [16] que A. Forke [8] avaient sous-estimé les problèmes philologiques et leurs traductions demeurent souvent approximatives. La même critique vaut pour les passages du *Mojing* traduits par J. Needham dans son ouvrage monumental *Science and civilization in China* [17].

Depuis cette époque de "pionniers", de notables progrès ont été accomplis dans l'édition et dans l'interprétation de ces textes logiques. C'est surtout grâce aux travaux d'A.C. Graham que l'élément rationnel propre à la pensée chinoise est de nos jours de mieux en mieux connu. La publication récente des écrits logiques moïstes [4] ne nous donne pas seulement une édition et une traduction fiables de ces textes, mais il nous invite également à découvrir les éléments logiques et rationnels que renferment l'oeuvre de penseurs considérés jusqu'ici comme de purs moralistes (MENCIUS, XUNZI) ou mystiques (ZHUANGZI), ou encore comme des politiciens pragmatistes (HANFEIZI). Ainsi, on aurait tort de croire que la pensée rationnelle en Chine ancienne est restée confinée à ces quelques textes. Il faut savoir, d'une part, que GONGSUN LONG et les néo-moïstes cherchent à codifier les règles d'un art déjà largement répandu, à savoir l'art de la discussion, de la "dialectique" (au sens grec du terme!); d'autre part, que les méthodes développées par les néo-moïstes ont à leur tour fortement influencé la pensée philosophique de la fin des Royaumes Combattants (notamment le chapitre *zhengming* du livre *Xunzi*).

3. Aspects de la rationalité chinoise

La rationalité, en Chine ancienne, n'est pas une découverte faite au terme d'un effort solitaire par quelque penseur génial. Elle est l'oeuvre collective des "philosophes" de l'époque des Royaumes Combattants. Certes, ce sont GONGSUN LONG et les néo-moïstes qui se sont engagés le plus loin sur la voie du rationalisme. Ce sont eux qui, les premiers, ont cherché à codifier les règles du débat rationnel. Mais il ne faut pas oublier que les néo-moïstes ont cherché ainsi à mettre de l'ordre dans une discipline qui existait bien avant eux et à laquelle ont eu recours la presque totalité des penseurs de leur époque. Ainsi, MENCIUS se lamente d'être obligé par ses adversaires d'avoir recours à la dialectique; ZHUANGZI, le fameux penseur taoïste, excelle dans l'art de retourner la logique contre elle-même.

La rationalité chinoise s'est manifestée sous deux formes principales. La première a trait à ce que je nommerai le principe de l'argumentation analogique; la seconde concerne les tentatives amorcées pour construire (ou reconstruire) de manière axiomatique (*more geometrico*) une éthique philosophique. Ce dernier projet est une entreprise purement néo-moïste; d'autres écoles (notamment les taoïstes, et plus tard XUNZI) ont tenté de l'imiter, sans toutefois pouvoir rivaliser avec le chef d'oeuvre des dialecticiens néo-moïstes. L'argumentation analogique par contre a été pratiquée par tous les philosophes de la Chine ancienne. Il ne saurait être question ici de faire une étude détaillée de ce phénomène. Seul les aspects logiques de cette forme d'argumentation nous intéresseront.

Quand on parle de logique chinoise, il importe de préciser d'emblée qu'il ne s'agit pas seulement de l'étude des raisonnements tels qu'on les trouve chez les premiers philosophes chinois. Ces raisonnements, disons-le une fois pour toutes, ne sont ni moins, ni plus illogiques que ceux de n'importe quelle autre civilisation. Il s'agira avant tout d'étudier les manières dont ces penseurs s'y sont pris pour vérifier la validité d'un raisonnement donné et pour comprendre les mécanismes gouvernant l'inférence correcte. Ce n'est qu'au moment où la réflexion se replie sur ses capacités et les examine en vue de s'assurer de leur fiabilité que naît, à proprement parler, la pensée logique.

Il me faudra commencer cependant par un aperçu sur les procédés narratifs qu'utilisent les premiers philosophes chinois. En effet, l'organisation quasiment géométrique du texte des néo-moïstes constitue l'exception plutôt que la règle. La plupart des textes sont conçus comme des dialogues entre le philosophe et un roi (ou quelque notable) disposé sinon à mettre en pratique les directives du philosophe, du moins à leur prêter oreille. Font partie de cette forme littéraire les *Entretiens* de CONFUCIUS, le *Mencius*, de larges portions du *Zhuangzi*, du *Hanfeizi*, du *Lüshichunqiu*. Rare dans le *Mozi* (on ne la trouve que dans les chapitres 46 à 50), la forme dialogique fait complètement défaut dans le *Xunzi*, oeuvre qui se compose en fait d'une série d'essais ou de traités, généralement assez courts, sur les problèmes fondamentaux qui ont agité les philosophes des Royaumes Combattants. On trouve aussi des dialogues mettant en scène non pas un roi et son conseiller, mais deux philosophes appartenant à des écoles de pensée ennemies. On peut citer à ce propos les fameux entretiens entre MENCIUS et GAOZI sur la nature humaine du chapitre VI, A du *Mencius*. Mais la particularité de cette forme d'expression ne réside pas uniquement dans le caractère dialogique des textes, mais surtout dans l'appel quasi constant qui est fait aux analogies et aux paraboles pour imposer ou défendre une opinion. Voici un exemple:

> BAI GUI dit au roi de WEI: "Prenez un chaudron fait pour cuire un boeuf et mettez-y un poulet. Si vous utilisez beaucoup d'eau, la viande deviendra fade et immangeable. Si vous ne mettez que peu d'eau, le poulet noircira à l'extérieur et restera cru à l'intérieur. Ce chaudron est beau à voir, mais il est inutile. En cela, il ressemble aux paroles de HUI SHI." Quand HUI SHI l'apprit, il dit: "Supposons qu'une armée affamée campe à côté de ce chaudron. Si elle peut l'utiliser pour y cuire du riz, rien ne lui sera plus utile qu'un tel chaudron".

(*Lüshichunqiu*, chap. 18.7; notre traduction)[15]

Comme il est possible de varier à l'infini le procédé d'argumentation analogique, beaucoup d'interprètes occidentaux ont désespéré de pouvoir y trouver une quelconque régularité, renonçant du coup à en chercher les règles constitutives[16]. Pourtant, l'argumentation analogique renferme, comme nous allons le montrer, une part de raisonnement logique.

3.1 *L'argumentation par analogie*

L'intention purement logique de l'argumentation analogique est surtout visible dans des cas comme celui-ci:

> GONGMENGZI dit: "Il n'existe ni phantômes, ni esprits." Mais il affirmait aussi: "L'homme de bien doit étudier les cérémonies des sacrifices." MOZI remarqua: "Etudier les cérémonies des sacrifices tout en affirmant qu'il n'y a pas d'esprits, c'est comme si on étudiait les cérémonies de l'hospitalité sans jamais recevoir d'hôte; c'est comme si on tressait des filets quand il n'y a pas de poissons."
>
> *Mozi*, chap. 48, p. 275/1; notre traduction

Une représentation schématique de ce type d'analogie conduit au résultat suivant: en somme nous avons affaire à deux discours, dont l'un, que j'appellerai la *présentation* de l'analogie est toujours pleinement exprimé, alors que l'autre discours, que j'appellerai *l'application* et qui représente le cas pratique qu'il s'agit de mettre en relief, ne l'est que rarement. Les éléments constituants de ce discours double sont les *termes* de l'analogie. A chaque terme de la présentation correspond un terme de l'application. Ainsi, dans notre exemple, au terme "tresser des filets" de la présentation correspond le terme "étudier les cérémonies des sacrifices" de l'application de l'analogie. L'existence de fantômes et d'esprits est mise en parallèle avec celle des poissons et des hôtes. La particularité de l'argumentation analogique réside dans le fait qu'il y a des relations logiques entre les termes à l'intérieur d'un discours. L'ensemble de ces relations logiques définissent ce que j'appellerai la structure logique ou les prémisses de l'analogie. Les inférences enfin que l'on peut tirer des prémisses de l'analogie déterminent les propriétés logiques de l'analogie.

Le but argumentatif de l'analogie est clair: il s'agit de faire reconnaître à son interlocuteur que le discours développé dans la présentation, et dont la validité est sinon évidente du moins présupposée, est de la même nature qu'un autre discours, à savoir l'application (de l'analogie), et que s'il accepte le premier, il n'a pas de bonnes raisons pour refuser le second. La contradiction qui apparaît dans la présentation de l'analogie se transmet, par isomorphisme, à l'application.

La structure logique de l'analogie par contradiction est très simple. On pourrait la schématiser comme suit:

Presentation	Termes	Application
"Tresser des filets"	P	"étudier les cérémonies des sacrifices"
"il n'y a pas de poissons"	Q	"il n'y a pas d'esprits"

Structure logique	
1. $P \equiv Q$	[On tresse des filets si et seulement si il y a des poissons]
2. $P \ \& \ \neg Q$	[Quelqu'un tresse des filets alors qu'il n'y a pas de poissons]

Proprietes logiques

Des deux prémisses 1. et 2., on peut dériver une contradiction:

(i) $(P \supset Q) \ \& \ (Q \supset P)$: 1., def.
(ii) $P \supset Q$:(i), simpl.
(iii) P :2., simpl.
(iv) Q :(ii), (iii), mod. pon.
(v) $\neg Q$:2., simpl.
(vi) $Q \ \& \ \neg Q$

Nous avons par conséquent affaire à une sorte de réduction à l'absurde. Or, *ex falso sequitur quodlibet*: la contradiction de la présentation appelle celle qui se manifeste dans l'application et vice versa. Le recours à l'analogie par contradiction est extrêmement fréquent dans les textes des premiers philosophes[17]. Voici quelques autres exemples que nous présentons sans en faire l'analyse.

> "La loi est la vie même de la nation, le pilier du gouvernement, le garde-fou des citoyens. Aussi espérer gouverner sans son secours est-il aussi vain que de vouloir calmer sa faim en jeûnant, se protéger du froid en ôtant ses vêtements, marcher vers le couchant pour aller au levant."
>
> [*Shangjunshu*, chap. 26, p. 43 ZZJC; trad. de J. Lévi, *Le livre du prince Shang*, Paris : Flammarion, 1981, pp. 183–184]

Une autre propriété logique souvent représentée à l'aide d'une analogie est celle d'une condition nécessaire. Elle est introduite par la "formule analogique" consacrée à cet effet: "Qui peut tenir quelque chose de chaud sans avoir auparavant mouillé ses mains?" (*Mencius*, IV, A, 7)

Les exemples abondent. En voilà un tiré du *Mozi*:

> Qui peut tenir quelque chose de chaud sans avoir auparavant mouillé ses mains? — Cela explique pourquoi ceux qui, dans l'antiquité, étaient à la tête

du pays ou qui étaient des seigneurs féodaux ne pouvaient pas ne pas s'attacher ceux qui excellaient à apporter leur soutien aux officiers et ministres et tous ceux qui étaient au service de l'Etat.

(*Mozi*, édition *Zhuzijicheng*, p. 30/9 et suiv.; notre trad.)

Ce type d'argumentation analogique, très fréquent en Chine ancienne, est quasiment absent des textes philosophiques grecs. Les quelques analogies fameuses que Platon a développées dans ses dialogues (le mythe de la caverne dans la *République* ou l'image du colombier dans le *Théétète*) sont de nature tout à fait différente et ne visent pas de but proprement logique.

Un pas important vers la rationalité a été franchi au moment où on a cherché à "régimenter" l'argumentation analogique et à déterminer les conditions dans lesquelles une analogie peut être considérée comme valide. Un texte du *Mencius* nous servira d'illustration.

MENCIUS dit au roi XUAN de QI:
- Supposons que quelqu'un vienne vous dire qu'il est assez fort pour soulever un poids de cent livres, mais prétend qu'il est incapable de porter une plume; ou qu'il vous dise qu'il a la vue perçante au point de voir un duvet d'automne, mais prétend être incapable de voir un chariot plein de fayard. Seriez-vous d'accord?
- Non, dit le roi.
- Or, reprit MENCIUS, votre compassion est assez forte pour s'étendre aux animaux, mais ses effets ne se font pas sentir auprès du peuple. En quoi votre cas est-il différent du précédent? Si l'on est incapable de soulever une plume, c'est parce qu'on n'y applique pas sa force; si on est incapable de voir un chariot plein de fayard, c'est qu'on refuse d'ouvrir les yeux. Si le peuple se trouve démuni de soutien, c'est que vous n'y appliquez pas votre compassion. Ainsi, si vous ne réussissez pas à être un vrai roi, ce n'est pas que vous n'en seriez pas capable, c'est que vous ne mettez pas en pratique votre compassion.

(*Mencius*, I,A,7; notre traduction)

Pour MENCIUS, bien entendu, les trois termes de cette analogie font partie d'une seule et même "catégorie": l'acuité visuelle, la force physique et la compassion sont des *facultés*; à ces trois termes on peut appliquer le principe "qui peut le plus, peut le moins". Ainsi, il est plus difficile (dans l'optique de MENCIUS) d'aimer des animaux que d'aimer les hommes. Or le roi XUAN s'était apitoyé sur le sort d'un animal (cf. *Mencius*, I,A,7, début). Le roi qui éprouve de la compassion pour une pauvre bête et qui reste insensible devant les maux qui frappent son peuple ressemble à l'athlète qui se déclare incapable de soulever une plume.

La présentation de l'analogie part des deux prémisses suivantes:

1. Si l'on est assez fort pour soulever un poids de cent livres (= P) tout en étant

incapable de porter une plume (= Q), c'est qu'on refuse d'appliquer sa force (= R).

2. La force est suffisante pour soulever un poids de cent livres (= P). Conclusion: si l'on ne refuse pas d'appliquer sa force (= $\neg$ R), on peut porter une plume (= $\neg$Q).

1. $(P \& Q) \supset R$
2. P

   Conclusion: $\neg R \supset \neg Q$

D'un point de vue strictement formel, ce raisonnement est valide[18]. Sur ce point, MENCIUS et le roi XUAN sont d'accord. Le désaccord éclate au moment où MENCIUS se propose de remplacer les termes de la présentation par ceux de l'application. Ainsi, la compassion vient prendre la place de la force physique; l'ncapacité de soulever une plume est remplacée par l'ncapacité de s'apitoyer sur le sort du peuple; la force suffisante pour soulever un poids de cent livres cède sa place à la capacité de s'apitoyer sur le sort d'une pauvre bête.

Ce texte du *Mencius* est, à notre connaissance, un des premiers en Chine-ancienne à soulever le problème de la validité d'une analogie. Le problème logique que pose MENCIUS est celui-ci: le terme "compassion" (*en* en chinois) est-il du même type ou fait-il partie de la même *catégorie* que "force physique" et "acuité visuelle"? Le roi XUAN doute de la validité de l'analogie. Selon lui, son incapacité à se montrer bienveillant envers le peuple résulte d'une impossibilité physique, et non pas, comme le prétendait MENCIUS, d'un simple refus d'agir. Sur ce, MENCIUS explique au roi XUAN en quoi réside la différence (logique) entre ces deux types d'impossibilité. Il y a d'abord l'impossibilité du type "prendre le mont TAI sous le bras et sauter par dessus la mer du Nord"; d'autre part, il y a celle du type "masser les articulations d'une personne âgée". Ne pas exercer sa bienveillance relève du dernier type d'impossibilité: le roi refuse tout simplement d'agir et de mettre en pratique la doctrine qui consiste à gouverner avec bienveillance. Mais il n'y a aucune impossibilité physique, comme "porter le mont TAI sous le bras" qui l'empêcherait de faire montre de bienveillance.

Le *Mencius* expose d'autres analogies, dont celle-ci, particulièrement complexe. Nous avons compté pas moins de six termes et de quatre prémisses.

CHUNYU KUN dit à MENCIUS:

- N'est-il pas prescrit par le rituel que l'homme et la femme ne doivent pas se toucher en se donnant quelque chose de main à main?
- En effet, répondit MENCIUS.
- Si votre belle-soeur est en train de se noyer, ne la saisirez-vous pas par la main pour la sauver?
- Ne pas sauver sa belle-soeur si celle-ci est en train de se noyer serait faire preuve d'une grande cruauté. Il est certes prescrit par le rituel que l'homme et la femme ne doivent pas se toucher en se donnant quelque chose de main

à main. Si l'on saisit la main de sa belle-soeur pour la sauver de la noyade, c'est qu'on évalue la situation [c'est-à-dire on choisit le moindre mal].

– A présent, reprit CHUNYU KUN, le monde entier est en train de sombrer. Pourquoi n'étendez-vous pas votre main pour le sauver?

– Si le monde est en train de sombrer, on le sauve par le Voie (*dao*); si ma belle-soeur est en train de se noyer, je l'aide en saisissant sa main. Voudriez-vous que je sauve le monde avec ma main?

(*Mencius*, IV,A,17; notre traduction)

Toute cette analogie repose sur l'opposition qu'il y a entre les termes "Voie" et "rituel". Pour CHUNYU KUN, ces deux termes font partie d'une seule et même catégorie. MENCIUS de son côté insiste sur la différence fondamentale qui les sépare. Si l'on n'observe pas les principes du rituel, c'est pour obéir aux impératifs d'un principe encore plus haut, à savoir celui de l'humanité. Mais la Voie n'a pas, au-dessus d'elle, d'autre principe. Ce qui veut dire que toute forme de compromis, "d'évaluation de la situation", est exclue dans ce dernier cas.

Il faut noter toutefois que la structure de l'analogie présentée par CHUNYU KUN est valide[19]. La conclusion que CHUNYU KUN veut faire accepter à MENCIUS (à savoir "si le monde est en train de sombrer et si vous n'acceptez pas de faire une entorse à vos principes rigides pour le sauver, alors vous faites preuve d'une grande cruauté") est la conséquence logique des prémisses de l'analogie. Le différend qui a éclaté entre les deux philosophes touchait uniquement au problème de la classification catégorielle des termes "Voie" et "rituel".

L'argumentation analogique comporte donc un axe syntagmatique et un axe paradigmatique. Au niveau syntagmatique, on ne considère que l'agencement logique des termes de l'analogie. L'analyse logique de ce type d'inférence ne présente aucune difficulté et s'effectue aisément à l'aide de la logique des prédicats inanalysés. Le système dit de la "déduction naturelle" suffit à rendre compte de manière tout à fait adéquate des inférences que l'on rencontre. Les raisonnements conduits à l'aide de ce support analogique peuvent atteindre des sommets de complexité vertigineux et qui vont largement au-delà des capacités d'intuition logique occidentales, comme le montre l'analogie proposée par CHUNYU KUN à MENCIUS. Il faut cependant remarquer que l'élément purement formel du raisonnement n'a jamais retenu l'attention des philosophes chinois. Nous reviendrons sur ce problème. Au niveau paradigmatique, par contre, les premiers logiciens chinois se sont assez vite heurtés au problème de la substitution des termes à l'intérieur d'un schéma d'inférence reconnu comme valide. La majeure partie du travail logique des philosophes chinois a été consacrée à ces problèmes de substitution de termes à l'intérieur d'une analogie et à la recherche de critères permettant de reconnaître la similitude ou la différence entre les termes d'une analogie. MENCIUS avait bien vu ce problème, mais au lieu de chercher une solution globale, il s'est contenté de noter les différences isolées qu'il repérait au fil de l'argumentation. Sa méthode était vouée à l'échec, car il est évident qu'une telle liste de différences entre termes est potentiellement infinie. De plus, il est plutôt rare qu'on puisse réutiliser une distinction une fois établie. Un autre désavantage de la méthode de MENCIUS est celui-ci: n'est-il pas fastidieux pour le philosophe de chercher à chaque fois une analogie appropriée à l'opinion qu'il veut défendre? La puissance créatrice a ses limites. D'autre part, il n'est même pas sûr s'il existe toujours une analogie plausible pour représenter une situation donnée. Il n'est donc pas étonnant de constater que d'autres penseurs ont tenté de détacher l'argumentation analogique de son support et de son enracinement textuel et ont cherché à comprendre la nature de la différence qui existe entre les termes de l'analogie. En effet, les quelques exemples d'argumentation analogique que nous venons de donner font voir que la valeur d'une analogie gravite généralement autour d'une seule paire de termes ("rites" et "Voie"; "compassion" et "force physique" etc.), d'où l'idée de faire une étude méthodique de toutes les différences possibles entre termes.

Une des manières — la plus simple peut-être — pour entreprendre une telle étude est de placer les deux termes en question dans des contextes syntaxiques ou sémantiques rigoureusement identiques et d'étudier ensuite les comportements logiques de ces termes. C'est là notamment la méthode choisie par les néo-moïstes. Il faut tout de suite préciser que cette technique se trouve, chez eux, incorporée à une théorie générale de la dénomination correcte, ou de la description correcte des choses. Cette discipline remplace, en quelque sorte, l'ontologie des philosophes occidentaux: au lieu de se demander *ce qu'est une chose* (le courage, par exemple, dans le *Lachès* de Platon), les premiers philosophes chinois cherchent plutôt à décrire de manière cohérente tous les actes courageux ainsi que les personnes appelées courageuses. Cette affirmation se fonde sur la constatation que beaucoup de problèmes qualifiés d'ontologiques par un philosophe occidental réapparaissent en Chine sous l'emblème de problèmes de *dénomination*. Nous verrons notamment que le problème de l'*a priori* se pose, en Chine ancienne, en termes exclusivement linguistiques et nullement, comme en Occident, en termes ontologiques ou gnoséologiques. La notion d'un *a priori* linguistique est d'ailleurs récente en Occident et a été introduite par les philosophes de l'école analytique (Z. Vendler; W. V. O. Quine). La discipline qui, chez les néo-moïstes, remplace l'ontologie de la philosophie occidentale est une sorte de logique de la dénomination correcte. Les questions que posent ces philosophes sont de l'ordre: "peut-on appeler noir un chien blanc qui a des yeux noirs?", "peut-on appeler grand un chien qui a de grands yeux?" ou "faut-il aimer une infinité d'hommes pour être dit aimer les hommes?" etc. Cette approche du problème, comme nous allons nous en rendre compte, mène tout droit à une syntaxe logique du chinois classique.

Le moyen auquel les néo-moïstes ont le plus souvent recours est celui de mettre en parallèle deux phrases à structure syntaxique identique afin d'en faire ressortir les différences d'ordre logique. En chinois classique, en effet, le comportement logique d'un mot est très souvent le seul critère dont on dispose pour établir des différences entre classes grammaticales. Un exemple très simple illustrera nos propos. La différence entre un verbe décrivant une activité et un adjectif numérique n'est pas marquée dans tous les contextes syntaxiques. Une phrase du type *ren dou* "l'homme se bat" présente fondamentalement la même structure que *shi sanbai* "les Odes sont au nombre de trois cents". Dans de tels cas, la grammaire peut induire en erreur le philosophe en masquant la différence logique qui existe entre un verbe et un adjectif numérique. Pour découvrir la différence en question, il suffit de mettre les deux termes sur un "banc d'essai" logique: ici, il suffit d'ajouter un quantificateur aux deux énoncés pour en établir les comportements logiques divergeants. En effet, on peut dire *ju dou* "tous combattent", mais on ne peut pas dire *ju er* "tous sont deux"[20]. Autre cas semblable: les termes *bai* "blanc" et *miao* "aveugle" peuvent prendre tous les deux la fonction d'un adjectif: *ma bai* "le cheval est blanc"; *ma miao* "le cheval est aveugle". Pourtant il existe une différence d'ordre logique entre ces deux termes. Celle-ci saute aux yeux si on met l'une à côté de l'autre les deux structures suivantes: *bai ma duo bai* "un cheval blanc est blanc en sa majeure partie" et *miao ma duo miao* "un cheval aveugle est aveugle en sa majeure partie". Le premier est un énoncé tout à fait normal, alors que le second ne l'est pas.

Il est remarquable de constater à quel point la phraséologie de la langue chinoise favorise une telle approche. La méthode des propositions parallèles est en effet basée sur la propriété qu'a la langue chinoise de présenter de nombreux cas où une anomalie d'ordre logique est en même temps une anomalie grammaticale. L'intuition grammaticale coïncide donc souvent avec l'intuition logique.

Le progrès réalisé par les néo-moïstes est double: d'une part, ils sont les premiers à tenter de ramener ces différences catégorielles isolées à un nombre restreint de catégories fondamentales (voir ci-dessous); de l'autre, les néo-moïstes avancent également une explication quant à la nature de ces catégories: ce sont des catégories de *noms*.

Dans le traité "*Noms et objets*"[21], le dernier qui nous soit parvenu de l'école néo-moïste, nous assistons à une tentative remarquable pour réduire les différences catégorielles à des différences entre types de noms:

> Le nom et l'objet ne vont pas nécessairement ensemble. Si cette pierre est blanche, chacune de ses parties, si vous la cassez, sera également blanche; mais si cette pierre est grande, on ne peut pas dire que chacune de ses parties sera grande".
>
> (*Noms et objets*, NO 1; notre traduction)

Chacune des parties d'une pierre blanche est blanche, mais les parties d'une grande pierre ne sont pas nécessairement grandes. L'explication que les néo-moïstes donnent à cette asymétrie est celle-ci: le terme "grand" fait partie de la catégorie des noms qui désignent une quantité ou une grandeur; le terme "blanc" appartient à la catégorie des noms qui désignent une forme ou un aspect[22]. Il peut paraître intéressant de noter que ces deux premières catégories sont identiques aux catégories aristotéliciennes de la quantité et de la qualité; la troisième catégorie néo-moïste, c'est-à-dire les noms donnés d'après la provenance ou la destination, rappelle la catégorie aristotélicienne de l'espace. Le traité néo-moïste sur les noms et les objets est malheureusement incomplet, ce qui nous prive de la possibilité de vérifier si la liste des catégories chinoises aurait été en tous points identique à celle d'Aristote. Pour une discussion des rapports qu'il y a entre les catégories aristotéliciennes et les catégories néo-moïstes, le lecteur se reportera à notre étude "Greek and Chinese categories" [24].

Cette théorie élémentaire des catégories de dénomination est en fait le complément naturel du raisonnement par analogie: le récit qui soustend la présentation et l'application de l'analogie garantit la dimension syntagmatique (agencement logique des termes de l'analogie); le problème de la substitution des termes est pris en charge par la théorie des catégories (axe paradigmatique).

Il est permis de supposer que l'énigmatique "Discours sur le cheval blanc" (*baimalun*) de GONGSUN LONG constitue lui aussi la *présentation* d'une analogie. En effet, comme nous l'avons montré ailleurs[23], on peut reconstruire ce traité comme un essai ingénieux pour montrer qu'un criminel n'est pas un homme. "Criminel" se dit en chinois *dao ren* (litt. "criminel homme"). "Cheval blanc n'est pas cheval" peut ainsi être mis en parallèle avec cette autre proposition, tout aussi paradoxale, à savoir, "homme criminel n'est pas homme".

L'attitude à adopter face aux criminels a posé de graves problèmes aux néo-moïstes, école dont GONGSUN LONG doit lui-même être issu, puisqu'il en utilise les termes techniques et la méthode dialectique. L'amour universel (*jian ai*), un des dix principes néo-moïstes (voir ci-dessous) enjoint aux moïstes d'aimer tous les hommes; d'autre part, les criminels doivent être exécutés afin de sauvegarder l'ordre public. Les néo-moïstes, de leur côté, affirment qu'un criminel est un homme, mais que tuer un criminel n'est pas tuer un homme. Ils justifient cet argument au moyen de l'analogie "tuer un chiot n'est pas tuer un chien". En effet, si l'on tuait tous les chiots, la race canine ne s'en trouverait pas éliminée pour autant. GONGSUN LONG, quant à lui, prend le taureau par les cornes et commence par déclarer qu'un homme criminel n'est pas un homme, probablement pour en tirer la même conclusion que ses confrères nèo-moïstes, à savoir qu'il faut à tout prix exécuter les criminels. Toute l'argumentation développée dans le "Discours sur le cheval blanc" peut être transposée, comme l'avait déjà suggéré HU SHI[24], dans les termes de *dao ren* "homme criminel" et *ren* "homme", en mettant partout "criminel" et "homme" là où GONGSUN LONG écrit "blanc" et "cheval". L'échange initial du "Discours sur le cheval blanc" (voir ci-dessus) pourrait alors se reformuler ainsi:

- Est-il permis de dire que homme criminel n'est pas homme?
- C'est permis.
- Comment donc?
- "Homme", c'est ce qui désigne la forme [= disposition innée]; "criminel", c'est ce qui désigne la couleur [= propriété acquise]. Ce qui désigne la forme n'est pas ce qui désigne la couleur. Voilà pourquoi je dis que homme criminel n'est pas homme.

L'analyse de toutes ces différentes formes d'argumentation analogiques révèle deux faits assez surprenants. Nous apprenons, d'une part, que la validité formelle d'une analogie ne fait jamais problème aux premiers penseurs chinois (exception faite du "Discours sur le cheval blanc"), même si l'analogie en question est fort complexe et qu'un lecteur occidental doit faire appel au calcul des prédicats pour suppléer à son intuition logique défaillante. D'autre part, les textes nous montrent que le véritable problème logique que soulève l'argumentation analogique se situe au niveau de la substitution d'un terme de la présentation par un terme correspondant de l'application. Les premiers philosophes chinois sont tout à fait conscients du fait que la simple validité formelle d'un raisonnement ne suffit pas. Ils savent qu'il faut en plus que les termes de part de d'autre soient en quelque sorte "commensurables", bref qu'ils soient de la même catégorie (*lei*). Il s'agit donc de savoir si on peut mettre sur le même plan l'existence d'esprits et l'existence de poissons, ou si l'acuité visuelle est une entité du même type qu'une qualité morale, ou encore si le terme "Voie" est semblable à celui de "rite". L'approche choisie par les premiers logiciens chinois les force donc à s'attaquer non pas au problème de la construction d'une logique formelle, mais à celui, infiniment plus complexe, de l'élaboration d'une syntaxe logique de la langue chinoise. Ce n'est pas la logique en tant que telle qui les intéresse, mais plutôt la manière dont la logique opère à l'intérieur même du langage.

Au terme de ce rapide coup d'oeil sur le développement de la pensée logique en Chine ancienne, le lecteur ne manquera pas d'être étonné par une apparente incohérence: il semble en effet qu'en Chine ancienne une intuition logique hors pair ait vécu côte à côte avec une totale absence de tentatives faites pour élaborer une logique formelle. Il importe de noter auparavant que cette façon de poser le problème est assez nouvelle, car l'intuition logique des chinois et le caractère logique de leurs analogies n'a, pour ainsi dire, encore jamais été mis en relief dans les travaux consacrés à la logique chinoise. On s'est plutôt contenté de constater l'absence d'une logique formelle en Chine ancienne. Afin d'esquisser une solution à cette apparente incohérence dans le développement intellectuel de l'esprit chinois, nous allons brièvement comparer le raisonnement analogique chinois au raisonnement syllogistique grec.

En premier lieu, il faut regarder de plus près la notion d'intuition logique. Notre analyse a montré, nous l'espérons, que la transparence logique d'un argument est un facteur très relatif. Ce qui est transparent pour un philosophe chinois du -IIIe siècle ne l'est pas nécessairement pour nous aussi, et vice versa. Il faut plutôt admettre que la transparence logique dépend de facteurs peu connus, subconscients, habitudes culturelles, linguistiques etc. que seule une analyse contrastive pourra mettre à jour.

Le syllogisme peut paraître, à maints égards, comme étant la forme la plus directe et la plus naturelle de présenter une inférence correcte. Mais la transparence logique, à elle seule, ne démontre rien. Il se peut très bien, en effet, que nous acquiescons au syllogisme de la forme BARBARA par simple habitude culturelle. Il serait par conséquent fallacieux de vouloir déclarer incapables de logique les chinois, parce qu'on ne trouve pas de syllogismes en Chine ancienne[25]. Le syllogisme, tout comme d'ailleurs les formules de la logique mathématique, ne sont que différentes façons *d'exprimer*, ou de *représenter* la logique, c'est-à-dire la pensée ou le discours conduit de manière cohérente.

Il y a en fait une explication autre et beaucoup plus plausible que celle de la présumée incapacité logique des chinois pour rendre compte de l'absence de syllogismes en Chine ancienne. Il a été observé que la forme d'argumentation syllogistique s'exprime assez mal en chinois[26]. En effet, il n'est pas naturel du tout pour un chinois de complètement déterminer, comme dans un syllogisme, tous les éléments de la phrase par des quantificateurs. On serait tenté de dire qu'en chinois classique la plus petite unité signifiante n'est pas le mot, ni même la phrase, mais le texte. Et puis absolument rien dans un texte chinois n'est sur-déterminé. N'y figure aucune information qui pourrait être devinée d'après le contexte. La langue chinoise a, comme l'a bien mis en relief Chr. Harbsmeier[27], comme fonction essentielle celle de *communiquer*, au contraire des langues indo-européennes, qui sont *descriptives* avant d'être communicatives. Cet aspect rend compréhensible le nombre élevé de présupposés qui sont nécessaires pour aborder un texte chinois. Il est bien connu qu'on peut faire (et même qu'on doit faire) la construction d'une phrase latine ou grecque avant d'en chercher le sens. Situation impossible en chinois, où il faut d'abord avoir compris le sens, ou du moins le contexte, du texte entier avant de pouvoir faire la construction syntaxique des phrases dont ce texte est composé.

La notion de grammaticalité prend de ce fait un sens très spécial. En effet, des phrases comme le fameux *Colourless green ideas sleep furiously* ("Des idées vertes incolores dorment furieusement") de N. Chomsky sont, contrairement à nos habitudes linguistiques, agrammaticales en chinois classique, parce que le lecteur n'arrive pas à les traduire faute de trouver un contexte où les insérer. Placé devant *wu se zhi lü xiang nu shui* (L'équivalent chinois de la phrase précitée), le lecteur bute d'abord sur la contradiction évidente qu'il y a entre *wu se* "incolore" et *lü* "vert". Cette contradiction l'empêche de lire la phrase comme un tout. La réaction normale du lecteur est même celle de prendre *wu se* "incolore" comme un nom propre! La syntaxe, en chinois classique, même si elle est très rigide, reste subordonnée à la sémantique. S'il n'y a pas de sens il ne peut pas y avoir de syntaxe non plus. Inversement, même une construction syntaxique très dépouillée rend un sens si elle est placée dans un contexte approprié. Voici un exemple frappant noté par Chr. Harbsmeier.

er ba shi liu san si shi er

Cette phrase n'est, à première vue, rien d'autre qu'une succession de huit chiffres. Transcrite en chiffres arabes, elle se lit:

2 8 10 6 3 4 10 2

La traduction est simplement: $2 \times 8 = 16$; $3 \times 4 = 12$. Il faut donc avoir compris le sens de la phrase avant de pouvoir traduire, car du point de vue purement syntaxique, des traductions comme $2 \times 80 = 6$; $3 \times 40 = 2$ ou $2 \times 86 = 3 \times 42$ seraient tout à fait possibles[28]. L'essentiel est de comprendre qu'il s'agit d'une opération arithmétique. En partant de là, la bonne interprétation de la phrase vient toute seule. Ce genre de phrase ne constitue pas du tout une exception en chinois classique. Simplement, la langue est ainsi faite qu'elle omet tout élément (comme par exemple × ou =) qui peut être deviné d'après le contexte. En chinois *bai ma ma ye* (littéralement "blanc cheval = cheval") ne précise pas si l'on parle de tous les chevaux ou seulement d'un. C'est le contexte qui suggère au lecteur l'une ou l'autre de ces interprétations. Mais cette économie dans l'expression n'empêche pas les philosophes à raisonner correctement, même lorsqu'ils doivent tenir compte d'éléments logiques (comme des quantificateurs par exemple) qui ne sont pas marqués grammaticalement. Ici encore, l'interlocuteur ou le lecteur supplée ces éléments sans peine d'après le contexte. L'énoncé même d'un syllogisme, où tous les éléments sont en quelque sorte "surdéterminés" paraît étrange à un lecteur chinois et peut éveiller sa suspicion. On trouve en fait, par ci par là, quelques syllogismes épars dans la littérature philosophique. Voici un bel exemple. Le *Lüshichunqiu* (chapitre 18.4) relate le cas d'un serviteur qui refusait d'accompagner dans la mort son souverain, comme la coutume l'exigeait. Un ami de longue date reprocha à ce serviteur sa conduite lâche: "Quoi, tu n'es donc pas mort?" Sur quoi le serviteur lui répondit:

> "Tout service rendu à d'autres a comme motif la recherche d'un profit.
> Or, la mort n'est pas [un service] profitable.
> Voilà pourquoi je ne suis pas allé mourir."

Ce texte n'est pas seulement intéressant pour le joli syllogisme qu'il renferme, mais aussi et avant tout pour le commentaire qu'y ont ajouté les auteurs du *Lüshichunqiu*: "Si une telle attitude commence à être connue, personne ne voudra plus mourir pour son prince et son supérieur, et un grand tort en résultera. *Pourtant, les paroles du serviteur sont irréfutables.* Cela montre clairement que les paroles ne sont pas suffisantes pour traiter de telles affaires."

Le syllogisme du serviteur est irréprochable, parce que valide. Le commentaire de celui qui est confronté à cette façon d'argumenter montre néanmoins que, pour un chinois, il y a hiatus entre le raisonnement et le contexte dans lequel ce raisonnement vient s'inscrire. En fait, pour un chinois, la forme syllogistique *détache* l'argument de son contexte naturel. Ce détachement est senti par l'interlocuteur qui ne peut pas ne pas y voir un acte volontaire et de mauvaise foi de la part du locuteur, qui voudrait ainsi détourner l'attention du contexte véritable de l'argument. Les textes de GONGSUN LONG, les seuls qui se rapprochent, quant à l'explicité de leur argumentation, de l'idéal occidental, ont toujours paru, pour la raison indiquée ci-dessus, suspects aux lecteurs chinois. Et si ceux-ci objectent à GONGSUN LONG qu'il sépare les mots des choses, on n'a pas voulu lui objecter autre chose que cette séparation du texte avec son contexte.

Cela ne veut bien évidemment pas dire que le chinois soit inapte à représenter correctement une suite logique d'arguments, mais plutôt que la forme syllogistique va à l'encontre des principes de base de la phraséologie de la langue chinoise. Il faudrait alors plutôt se demander si les philosophes de la Chine ancienne n'ont pas raisonné à travers des moyens d'expressions différents, mieux adaptés aux structures de leur langue. Il est bien connu que le chinois est une langue isolante, presque monosyllabique, dépourvue de toute morphologie et qui, de ce fait, utilise souvent le parallélisme entre deux structures syntaxiques comme moyen d'expression. Nous venons de voir comment cette particularité de la langue chinoise a été mise à profit par les logiciens néo-moïstes. La réduction de tous les raisonnements à une structure canonique unique, telle que la pratique la théorie du syllogisme, serait donc impossible en chinois, puisque cette langue emploie l'ordre des mots comme principe générateur de sens. Un syllogisme ne pourrait jamais capter la richesse et la diversité des raisonnements possibles en chinois classique.

Ces quelques observations nous permettront maintenant de proposer une hypothèse nouvelle quant à la nature des réflexions logiques conduites en Chine ancienne. La phraséologie des langues indo-européennes en général, et celle du latin et du grec, véhicules de la culture occidentale, en particulier, aurait tendance à inhiber le flux naturel des facultés logiques en enchaînant celles-ci aux contraintes d'une syntaxe dominée par la morphologie, alors que la phraséologie du chinois classique favoriserait la représentation linguistique de structures logiques, du fait qu'il y a souvent recoupement entre intuition logique et intuition grammaticale.

Cette hypothèse rendrait compte à la fois de la réaction des philosophes grecs, qui ont régimenté le discours philosophique en le mettant dans l'enclos de la logique, et de l'attitude des premiers philosophes chinois, qui n'ont pas senti le besoin de codifier les règles d'une activité qui pour eux s'exécutait spontanément. La transparence logique qui habite les structures de la langue chinoise dispense les philosophes chinois de la tâche fastidieuse d'élaborer, comme les Grecs, une logique *extérieure* au langage. Voilà, à notre avis, la raison profonde pour laquelle on ne trouve pas de logique formelle en Chine ancienne.

Un résultat semblable au nôtre a été obtenu, mais en partant d'un angle d'approche différent, par le chercheur polonais J. Chmielewski[29]. D'après J. Chmielewski, la phraséologie de la langue chinoise se rapprocherait, de par son uniformité, de la notation de la logique symbolique et fournirait de ce fait un support non négligeable aux capacités d'intuition logique. Mais il ne faut pas, à notre avis, confondre logique et notation symbolique. La notation symbolique qu'utilise J. Chmielewski n'en est qu'une parmi beaucoup d'autres. La transparence logique de la phraséologie du chinois ne doit rien à la notation de la logique symbolique. Le chinois classique n'est pas non plus une "anticipation" de la logique symbolique. Il faudrait plutôt dire que la notation symbolique et la phraséologie du chinois sont deux manières semblables de représenter les règles d'inférence logiques. Mais en fait, la logique, ou plutôt la syntaxe logique, du chinois englobe des facteurs beaucoup plus complexes que les seules règles d'inférence. Nous venons de voir, entre autres, que les logiciens chinois se sont également occupés du problème des catégories, un domaine entièrement négligé par J. Chmielewski.

Un raisonnement chinois a toujours besoin d'une base textuelle — d'un récit — pour s'exprimer. Une logique formelle et abstraite, en dehors du texte, ne pouvait pas naître dans une civilisation qui ne connaît pas de texte sans contexte. Cette particularité nous fait comprendre également l'importance que prend le raisonnement par analogie en Chine ancienne. Cette forme de raisonnement est en effet la seule qui ne détache pas le raisonnement de son contexte tout en laissant entrevoir l'armature logique du récit. L'argumentation analogique ne peut être efficace que dans une langue dans laquelle il est possible de construire des énoncés rigoureusement parallèles, d'un point de vue syntaxique autant que d'un point de vue logique. Or une telle situation ne peut se donner que dans une langue qui démarque la fonction grammaticale des expressions au moyen de leurs positionnements respectifs dans la phrase. Ceci explique la raison pour laquelle l'argumentation analogique est relativement fréquente en chinois et presque absente en grec. En grec, au contraire, le parallélisme grammatical est le plus souvent générateur des pires confusions logiques, ainsi que peut le montrer une lecture de l'*Euthydème* de Platon ou des *Réfutations Sophistiques* d'Aristote. Aussi les philosophes grecs se sont-ils détournés du langage et ont cherché en dehors de celui-ci un niveau de réalité plus "profond", effort qui allait les mener à la découverte de l'ontologie.

Si l'on compare, à une grande échelle, le développement de la logique occidentale à celui de la logique chinoise, on notera, mis à part le caractère éphémère des études logiques en Chine (menées indépendamment de l'influence du bouddhisme), que les Chinois ont abordé le problème de manière beaucoup plus directe que les Occidentaux. La syllogistique aristotélicienne, considérée encore par Kant comme le *nec plus ultra* de la science logique, est apparue aux pionniers de la logique mathématique moderne (Frege, Russell, Peano) comme un outil inadéquat et primitif. L'essor des systèmes formels a fait en sorte que la syllogistique aristotélicienne ne figure dans cette panoplie de systèmes plus que comme un sous-système de moindre importance. Les systèmes formels sont ensuite restés longtemps entre les mains des mathématiciens et méta-mathématiciens avant de rejoindre, à travers les grammaires transformationnelles et la syntaxe logique, la philosophie (analytique). Les études logiques occidentales ont par conséquent mis du temps pour se rapprocher à nouveau du milieu dont elles étaient issues, à savoir celui de l'analyse des structures logiques qui s'expriment dans le langage naturel.

En regard du cheminement occidental plutôt erratique, la voie suivie par les premiers logiciens chinois paraît plus directe, encore qu'elle fût moins efficace de par ses résultats. Si beaucoup de philosophes occidentaux souffrent, à l'heure actuelle, de la trop grande séparation de la logique d'avec la syntaxe et sont en mal de caractériser et d'expliquer les relations exactes entre les deux, on peut dire que les premiers logiciens chinois n'ont pas su découvrir l'indépendance de ces deux disciplines. Ainsi, les chinois ont évité de faire le "détour" par la syllogistique, mais ils se sont attaqués à une tâche infiniment plus complexe que celle que s'était proposée Aristote et devant laquelle ils ont finalement échoué, à savoir décrire la syntaxe logique de leur langue.

3.2 *Les philosophes-géomètres de la Chine ancienne*

La pensée rationnelle se manifeste encore sous une autre forme en Chine ancienne, non pas cette fois-ci dans le discours argumentatif, mais dans l'agencement logique d'un discours doctrinal. Il a été observé que la pensée chinoise manquait d'une des propriétés essentielles à toute pensée philosophique, celle de savoir organiser un ensemble de connaissances dans un *système*[30]. Ce jugement s'appuie, bien entendu, sur la forme littéraire sous laquelle se présentent la plupart des textes philosophiques chinois. Les *Entretiens* de CONFUCIUS adoptent celle d'une collection d'aphorismes, d'autres textes se présentent sous la forme d'une suite de dialogues plus ou moins brefs portant, pour un lecteur non averti, sur des problèmes de détails casuistiques. Les premiers moïstes exposent leur doctrine sous forme d'un véritable "manifeste" en dix points programmatiques: honorer les sages et les capables; instauration d'une moralité commune; amour universel; contre la guerre d'agression; économie dans les dépenses; économie dans les funérailles; croyance en la volonté du Ciel; croyance en l'existence d'esprits; contre la musique; contre le fatalisme.

La thèse de l'absence d'une pensée systématique se vérifie pour les deux premiers siècles de l'époque des Royaumes Combattants, mais pas pour la suite de cette même époque. En effet, au début du -IIIe siècle, les néo-moïstes élaborent la première éthique vraiment philosophique en Chine ancienne en reconstruisant *more geometrico* les dix recettes ou préceptes moraux prônés par MO DI (-Ve siècle), le fondateur du mouvement moïste. Cette innovation intervient à la suite d'une nouvelle orientation de la pensée philosophique. La pensée philosophique des Royaumes Combattants fut secouée, en effet, vers le milieu du -IVe siècle, par une véritable crise, provoquée par l'entrée du taoïsme sur la scène philosophique. La crise éclate au moment précis où les taoïstes affirment qu'ils ont eux aussi une solution à proposer au désordre social et économique: que chaque chose, que chaque être vivant suive le cours naturel de son existence, que l'homme ne s'ingère pas dans ces développements spontanés, et tout rentrera dans l'ordre, puisque la nature (*tian*, litt. "le Ciel") a tout disposé en un ensemble cosmique harmonieux où chaque être trouve sa place, pourvu qu'il accepte spontanément le cours des choses. Transformer l'homme par l'éducation morale ou par des techniques administratives détraque la merveilleuse mécanique des choses et tout finira dans le désordre, l'homme étant incapable d'imiter le mouvement parfait de la nature. Et YANG ZHU, le fondateur du taoïsme philosophique, de donner une formulation très pointée de cette pensée en déclarant: "si personne n'a besoin de donner, ne fût-ce qu'un cheveu pour le Royaume, alors seulement le Royaume est bien gouverné." Les adversaires de YANG ZHU, MENCIUS notamment, déforment bien entendu cette pensée et taxent le philosophe d'égoïste ne *voulant* pas donner un cheveu pour sauver le Royaume. Mais ce qui provoque la crise proprement dite est la manière dont YANG ZHU, et après lui ZHUANG ZHOU, fondent leur pensée. YANG ZHU, en effet, soutient que la nature humaine est ainsi faite qu'elle aspire à la santé, à la longévité et au plaisir des sens plutôt qu'aux honneurs et richesses que fait miroiter le service pour le bien commun. Par rapport aux premiers taoïstes, ce n'est donc pas l'attitude du refus de participer aux affaires de l'Etat qui est nouvelle, mais bien le fait de donner à cette attitude un fondement philosophique[31]. La crise vient de ce que MENCIUS et MO DI fondent des vues diamétralement opposées à celle de YANG ZHU sur le même principe que ce dernier, à savoir sur la nature (*tian*). Conscients de cette impasse, les néo-moïstes bannissent de leur vocabulaire cette notion de *tian*, qui ne peut plus remplir le rôle de pierre angulaire de l'édifice moïste, puisque d'autres écoles philosophiques l'avaient utilisée pour justifier une doctrine tout à fait opposée à la leur.

Les néo-moïstes avaient reconnu assez tôt que l'attitude de YANG ZHU et de ses nombreux sectateurs était irréfutable aussi longtemps qu'on faisait découler une quelconque action ou vertu morale de la nature. Ce qui était requis, plutôt, était une analyse plus poussée des données fondamentales qui constituaient cette nature, et en particulier la nature humaine. Ainsi, le point de départ de l'éthique néo-moïste ne sera plus, comme chez MO DI, la nature (*tian*), mais des tendances bien observables de la nature humaine: le désir et l'aversion. Le désir et l'aversion seront ainsi, pour les néo-moïstes, des données irréductibles, et nous verrons comment ces penseurs les posent comme axiome de leur système éthique, construit *more geometrico*.

Il est instructif de constater que c'est sur la méthode géométrique que se calquent, tant en Grèce qu'en Chine, les premiers essais de réflexion philosophique méthodiques. La géométrie ne manque pas d'exercer un pouvoir de fascination formidable sur l'esprit humain. La progression sûre du savoir ainsi que l'organisation systématique des connaissances confèrent à la méthode géométrique son attrait tout particulier.

L'origine de la géométrie en Chine ancienne est assez obscure. On peut tout au plus affirmer que la géométrie n'y apparaît jamais comme séparée de l'astronomie. Or l'astronomie atteint très vite un niveau élevé. Au -IVe siècle, en tout cas, le calcul des équinoxes par exemple ne semble plus poser de problèmes (voir *Mencius* IV,B,26). Beaucoup de cours locales entretiennent un astronome officiel, qui doit fixer chaque année le début des saisons et faire un rapport sur d'éventuels mouvements inhabituels des corps célestes. Les textes géométriques et astronomiques de la période ancienne sont tous perdus, exception faite d'un bref essai d'astronomie géométrisée dans le *Huainanzi*. Les recherches purement théoriques en géométrie commencent avec les moïstes, mouvement philosophique issu des corporations d'artisans et d'ingénieurs. Les moïstes ont l'habitude de manier compas et équerre, et le chapitre 35 du *Mozi* fait allusion au gnomon. Dans le *Mojing*, nous trouvons une quinzaine de théorèmes géométriques, fragment d'un ensemble plus grand selon toute probabilité, mais construit de manière rigoureuse: chaque théorème démontré sert à son tour de point de départ à une nouvelle démonstration. Le fragment qui nous a été préservé permet néanmoins de se faire une image suffisamment nette des contours de la géométrie néo-moïste pour que nous en risquions une présentation sommaire.

La géométrie, telle que la conçoivent les néo-moïstes, est très proche d'une théorie générale de la mesure (pour surfaces planes). Il faut imaginer des artisans et des ingénieurs maniant le compas et l'équerre, le fil à plomb et le niveau d'eau, ainsi que la règle servant d'unité de mesure. Le fragment de géométrie que l'on va exposer n'est manifestement pas destiné à un usage pratique. Preuve en est le caractère non seulement élémentaire, mais surtout *fondamental* des axiomes et théorèmes. Il s'agit donc bien d'un travail *épistémologique*, probablement consacré à la mise en forme d'un ensemble de connaissances géométriques déjà acquises.

En tête de la géométrie néo-moïste, nous trouvons une définition de la ligne droite:

"Etre droit, c'est être aligné (can)" (Théorème A 57)

Can "aligner" est un terme utilisé en astronomie pour désigner l'action d'aligner deux piquets et un corps céleste observé. Il y a ici, par conséquent, recours implicite à l'intuition. Cette définition de la ligne droite est ensuite utilisée pour fixer les conditions sous lesquelles deux trajets peuvent être considérés comme égaux:

> "[Deux trajets] sont de longueur égale quand ils se remplissent mutuellement (sans restes) lorsque posés droit" (théorème A 53)

Ce théorème, que les néo-moïstes ont illustré, en guise de démonstration, par un exemple — le cadre de la porte et le battant de la porte, qui sont de longueur égale — peut être considéré comme l'équivalent de l'axiome de congruence dans la géométrie euclidienne. Après avoir défini les conditions sous lesquelles deux droites sont à considérer comme égales (l'une remplissant l'autre sans restes), les géomètres néo-moïstes s'avancent vers les grandeurs qui sont le multiple d'autres grandeurs. Le théorème A 60 envisage le cas du double: doubler, c'est aligner deux règles de longueur égale l'une à la suite de l'autre. Mais ici intervient une difficulté, sans doute pratique à l'origine: comment, en effet, concevoir le point de raccord entre les deux règles? La mesure serait faussée s'il y avait un intervalle entre les deux raccords et aussi s'il y avait recoupement à l'endroit du raccord. Les néo-moïstes trouvent une solution en introduisant la notion d'un "point de départ" (*duan*): "Le point de départ est l'unité sans dimension qui précède toutes les autres" (théorème A 61)[32]. Ainsi est résolue la difficulté des deux règles alignées: le point d'arrivée de la première règle est en même temps le point de départ de la seconde. Mais comme ces deux points sont "sans dimension" (ou "sans étendue"), le raccord entre les deux règles se fait au point "idéal", le seul qui ne fausse pas les mesures.

Sur la notion de "longueur égale" est bâtie ensuite celle de "centre": le centre est le lieu qui sépare toutes les distances en longueurs égales (théorème A 54). Les notions de "centre" et de "longueur égale" entrent à leur tour dans la définition du cercle: le cercle est défini comme un ensemble de longueurs égales partant d'un seul centre (théorème A 58).

Ces quelques théorèmes montrent assez nettement le mode d'opération de la géométrie néo-moïste. Il ne s'agit pas, à vrai dire, de théorèmes au sens strict du terme, mais plutôt d'une série de définitions, au cours desquelles des concepts complexes (le cercle, p. ex.), sont construits à partir de concepts simples (centre, longueur égale). On voit néanmoins la parenté qu'a cette procédure avec la démarche géométrique. Les premières définitions sont posées comme des axiomes, et fonctionnent aussi en tant que tels. Dans notre cas, la ligne droite et l'axiome de congruence de la géométrie néo-moïste permettaient la définition du concept de "centre"; celle de "centre" permettait, à son tour, de s'avancer vers celle de "cercle".

L'intérêt épistémologique de la démarche est patent. La méthode des néo-moïstes permet de dériver d'un petit nombre de termes simples des notions de plus en plus complexes et abstraites. Cette méthode renferme, pour ainsi dire, sa propre justification. Si l'on parvient à rattacher un terme complexe par l'intermédiaire d'une chaîne de définitions à un ou à plusieurs termes simples, on aura en même temps *démontré* l'exactitude de contenu de cette définition. Une fois acquise l'idée d'un système de définitions interdépendantes, il a été naturel pour les néo-moïstes de s'avancer vers celle d'une connaissance indépendante de l'expérience, d'une connaissance *a priori* (*xian zhi*; litt. "connaître à l'avance"). Après avoir découvert que de la simple définition d'un objet géométrique peut découler tout un nombre de propriétés, qui sont alors connues sans qu'il faille les vérifier "empiriquement", c'est-à-dire "après coup", sur l'objet même, les néo-moïstes vont généraliser cette méthode et la réappliquer, comme nous le verrons, à la reconstruction de l'éthique moïste, deux millénaires avant Spinoza.

Les concepts moraux peuvent, tout comme les notions géométriques, être connus *a priori*, parce qu'on peut les décomposer eux aussi en notions simples. Mais avant de passer en revue le système éthique néo-moïste, il vaut la peine de s'arrêter sur leur conception de l'*a priori*. Le sens en est expliqué dans le théorème A 93: on peut connaître *a priori* ce qui s'exclut et ce qui s'implique. Il s'agit ici d'un *a priori* linguistique, lié aux contenus cognitifs implicites des mots qu'expriment les définitions. Afin de faire connaître en quoi consiste cette connaissance *a priori*, les néo-moïstes imaginent un objet placé de l'autre côté d'un mur et se demandent ce que nous pouvons savoir sur cet objet en n'en connaissant que le nom. Ainsi, il sera évident que la nature du cercle sera connue sans qu'il faille aller vérifier empiriquement ses propriétés "derrière le mur".

Le contenu cognitif du nom n'est, bien entendu, pas une image mentale, ou un faisceau de connotations attachées au nom. Comme ce contenu ne peut être trouvé que par l'intermédiaire de la décomposition du nom en ses éléments, nous avons affaire plutôt à une analyse de type logique. Les néo-moïstes sont tout à fait conscients du fait que la connaissance *a priori* est un type de connaissance à part, différent de la connaissance empirique. L'exemple qu'invoquent les néo-moïstes est suggestif: "Savoir que quelque chose de carré ne peut pas tourner, c'est là connaître par explication" (théorème A 80). La connaissance par explication (*shuo*) est donc celle qui nous livre les contenus cognitifs implicites des noms. La notion de carré exclut celle de rond; la notion de cercle implique celle d'un centre et d'un ensemble de points équidistants de ce centre. Du même coup se dégagent les fondements logiques sur lesquels repose la méthode géométrique des néo-moïstes. Il y a encore un autre fait qui vient à l'aide des philosophes-géomètres chinois. En Chine ancienne n'existe pas la même prolifération de termes techniques qu'en Grèce ancienne. Alors que les philosophes grecs créent tous leur propre vocabulaire, les penseurs de la Chine ancienne utilisent chacun les mêmes termes: *ren* "humanité", *yi* "justice", *xiao* "piété filiale etc., tout en les remplissant de contenus différents selon l'école à laquelle ils appartiennent. C'est cette uniformité terminologique qui contribue d'une part à délimiter l'éthique comme discipline philosophique autonome, et qui de l'autre délimite la tâche des néo-moïstes en leur imposant les notions à décomposer en éléments plus simples. Le but des néo-moïstes sera de montrer que leur façon de définir les concepts éthiques fondamentaux est la seule qui puisse garantir la cohérence de la pensée éthique.

L'analyse de la constitution fondamentale de la nature humaine est le centre autour duquel gravitent toutes les préoccupations philosophiques de l'époque des Royaumes Combattants. En simplifiant quelque peu les données, on pourrait dire, en gros, que MENCIUS professe que la nature humaine (*xing*) est bonne (voir *Mencius* VI,A,1), que pour GAOZI, le grand antagoniste de MENCIUS, elle est indifférente et que ce sont les circonstances qui la pousseront à faire le bien ou le mal; que pour XUNZI enfin, la nature humaine est mauvaise, et qu'il est absolument indispensable de l'éduquer si on veut la mener vers le bien. YANG ZHU, l'individualiste, de son côté, décrète que la nature humaine aspire à la longévité et aux plaisirs des sens. Ce qu'ont en commun toutes ces approches en apparence si diverses est la manière de fonder leur doctrine. Tous ces penseurs, en effet, déclarent que l'homme tient sa nature du Ciel (*tian*).

Les néo-moïstes sont les seuls à pleinement assumer la crise philosophique qui résulte de cette situation. En prenant pour guide la méthode géométrique, ils décomposent d'abord le concept de nature humaine en éléments plus simples, puis, forts de ce point de départ inébranlable, s'avancent vers la redéfinition des notions-clé de l'éthique.

Les tendances fondamentales de la nature humaine sont d'une part le désir (*yu*), de l'autre, l'aversion (*wu*). A partir d'ici, les néo-moïstes définissent les notions de *li* "bienfait, avantage" et de *hai* "préjudice". Cela signifie que toute action morale a comme moteur l'une ou l'autre de ces deux tendances fondamentales. En d'autres termes, chacune des vertus chinoises traditionnelles peut être décomposée, à l'aide de la méthode géométrique, en ses éléments simples, et parmi ces éléments simples figurera obligatoirement, selon les néo-moïstes, la tendance qu'a la nature humaine à rechercher l'avantage et à fuir tout ce qui lui est préjudiciable. Ainsi par exemple, *ai* "l'amour" se définira comme "désirer faire du bien à l'autre" et "éviter ce qui est préjudiciable à l'autre". *Ren* "l'humanité" est l'amour au niveau individuel. *Yi* "la justice" est l'intention de faire bénéficier l'autre. La fameuse piété filiale (*xiao*) enfin, n'est rien d'autre que les bienfaits prodigués à l'intention de la famille. Il est inutile ici de faire l'exposé complet de l'éthique néo-moïste, la méthode rationnelle ayant présidé à son élaboration est la seule à nous intéresser ici. L'éthique est, tout comme la géométrie, une discipline dont les contenus peuvent être dérivés au moyen d'une méthode logique infaillible à partir d'un petit nombre d'éléments primitifs agissant en guise d'axiomes. Il y a par conséquent ici incontestablement passage d'une morale à une véritable éthique philosophique. Cette observation nous permet en même temps de définitivement repousser la thèse selon laquelle la Chine ancienne n'aurait pas connu de pensée vraiment philosophique. La géométrie a appris aux chinois, tout comme aux Grecs d'ailleurs, à réfléchir méthodiquement. Autrement dit, la pensée géométrique est la première méthode philosophique rationnelle. C'est grâce à cette méthode que les philosophes chinois découvrent qu'il existe des vérités *a priori*.

Arrivé au terme de notre exposé sur la pensée géométrique en Chine ancienne, il nous faut résumer maintenant les points essentiels de cette démarche intellectuelle:

1) la pensée procède par étapes, et non par intuitions globales ou par "sauts";
2) l'étape précédente sert de point d'appui à l'étape suivante;

3) chaque étape de la démarche est accompagné d'un procédé de justification;
4) le point de départ, posé en axiome, ne peut pas être justifié, mais sert d'appui à toutes les déductions futures.

Il faut seulement observer que les philosophes-géomètres de la Chine ancienne ne semblent pas avoir possédé de conception rigoureuse de la notion de preuve, mais se contentent de rendre explicites les contenus cognitifs implicites des termes d'une pyramide conceptuelle. Il faut admettre que la croyance sous-jacente à cette manière de procéder est celle en un "ordre naturel" des termes définis. Il y aurait une voie naturelle qui mène du cercle à la ligne droite et au centre, et qui mène de la recherche de l'avantage à la piété filiale. Dans ce cas, les premiers principes posés par les néo-moïstes délimiteraient chacun un domaine, tel que la géométrie ou l'éthique par exemple, et on se rapprocherait des théories épistémologiques qu'a exposées Aristote dans les *Seconds Analytiques* (chapitre 1.28).

NOTES

1. Le terme "légiste" est formé d'après l'expression chinois *fajia* "école des lois". Mais le sens primitif du terme *fa* est "méthode", dont est dérivé celui de "loi". Les légistes sont donc les philosophes qui cherchent à développer une méthode pour gouverner.
2. Voir à ce sujet l'étude de V. Rubin, "Tzu-Ch'an and the city-state of ancient China", [3].
3. Cf. *Lunyu* (= *Entretiens*) 13.4 et 13.16; *Lüshichunqiu* 2.5.
4. Le premier catalogue bibliographique de la Chine ancienne, à savoir le chapitre *yiwenzhi* du *Hanshu* n'énumère pas moins de 189 philosophes ayant produit quelque 4.500 chapitres (*pian*, divisions du texte de longueur variable).
5. Toutes nos références aux écrits logiques néo-moïstes se rapportent à l'édition magistrale d'A.C. Graham, *Later Mohist logic, ethics and science* (London ; Hong Kong, 1978) [4]. Le théorème initial (A 1) se trouve à la p. 263. La "petite cause" désigne bien entendu une condition nécessaire; la "grande cause", une condition nécessaire et suffisante.
6. Le "Discours sur le cheval blanc" fait partie du livre *Gongsunlongzi*, le "livre de maître GONGSUN LONG", qui est en réalité un faux de l'époque des Six Dynasties (+IIIe au VIe siècles). Cette question a été brillamment élucidée par A.C. Graham (cf. bibliographie, [5]). Le *Gongsunlongzi* authentique que mentionne le *Hanshu* et qui comportait quartorze chapitres a été perdu assez tôt. Mais il s'est trouvé un faussaire pour "réécrire" ce livre, probablement pour pouvoir le présenter à l'empereur WU DI (502–549) de la dynastie des LIANG, grand amateur de logique et de sophistique. Le faussaire a pu mettre la main sur deux textes authentiques des GONSUN LONG, le "Discours sur le cheval blanc" et le "Discours sur la désignation des choses", ainsi que sur un certain nombre de fragments, dont un petit dialogue sur le problème du changement. C'est à partir de ces éléments que le faussaire a ensuite composé son texte. Pour combler les lacunes, il s'est copieusement servi du *Mojing*, texte qui discourait en maints endroits de problèmes semblables à ceux posés par GONGSUN LONG. Seulement — et c'est l'un des indices par lesquels s'est trahi le faussaire — l'édition du *Mojing* sur laquelle il se basait était déjà corrompue: les fameux "head characters" (voir ci-dessus) avaient glissé dans le texte principal et il est manifeste que le faussaire, comme il les copiait servilement, ne comprenait même pas le sens de ce qu'il composait. Voilà aussi l'explication du clivage assez net qui existe entre les deux discours authentiques mentionnés plus haut et les autres: les discours authentiques sont rédigés en une syntaxe très régulière, d'organisation stricte, alors que les textes restants sont plutôt lâches et tendent à présenter des caractéristiques stylistiques qui portent indubitablement la marque d'une composition tardive.
7. Voir p. ex. J. Lohmann [6]; ZHOU WENYING [7]; A. Forke [8].

8. Voir I. KOU [9]; A.C. Graham [10].
9. Cf. J. Chmielewski [11]; H. Greniewski [12].
10. A.C. Graham [13].
11. H. Maspéro, [14], p. 439.
12. Voir à ce sujet nos remarques dans *Les fondements philosophiques de la rhétorique chez les sophistes grecs et chez les sophistes chinois*, [15], pp. 350–378.
13. *Ibid.* pp. 378–382.
14. *Ibid.*, pp. 435 et suiv.
15. Pour une étude détaillée de la place que HUI SHI occupe dans le *Lüshichunqiu*, voir notre étude *Les fondements...* [15], pp. 301–331.
16. Comme le font p. ex. A. Waley, [18], pp. 193–5; H.O.H. Stange, [19]; D. Bodde, [20].
17. Voir à ce sujet l'étude de D. Leslie, [21].
18. Comme nous l'avons montré dans notre article "Analogical reasoning..."[22].
19. *Ibid.*
20. Les termes "verbe" et "adjectif" sont à prendre ici dans un sens purement fonctionnel, et non pas dans un sens morphologique. Une des particularités de la grammaire chinoise est justement la propriété qu'ont les mots de passer d'un type à l'autre. Sur ce problème, voir surtout Chr. Harbsmeier, *Wilhelm von Humboldts Brief...*, [23], pp. 155–217.
21. Reconstruit par A.C. Graham, [4], pp. 469 et suiv.
22. Cf. les propositions NO 1 et NO 2; A.C. Graham, [4], pp. 470–71.
23. Voir notre *Les fondements philosophiques...*, [15], pp. 407–413.
24. HU SHI, *The development...*, [1], p. 123.
25. Voir p. ex. M. Granet, *La pensée chinoise*, [25], pp. 277–279; HU SHI, *op. cit.*, [1], p. 97; I. KOU, [9], pp. 125–128.
26. A.C. Graham, *Two dialogues...*, [10].
27. Chr. Harbsmeier, *op. cit.*, [23], pp. 114–115.
28. *Ibid.*, pp. 109–110; le cas est déjà cité dans la grammaire de Von der Gabelentz, [26], p. 127.
29. J. Chmielewski, [11], IV, pp. 103 et suiv. et p. 111.
30. Cf. FENG YOULAN, *Précis d'histoire de la philosophie chinoise*, [27], pp. 32–35.
31. FENG YOULAN, [2], p. 135.
32. Le "point de départ" n'est donc pas tout à fait l'équivalent de notre point géométrique, comme beaucoup l'ont, un peu hâtivement, pensé.

BIBLIOGRAPHIE

* Les chiffres entre crochets carrés [] indiquent l'ordre dans lequel les ouvrages en question ont été cités dans le texte de l'exposé.

Billeter, Jean-François, "Pensée occidentale et pensée chinoise: le regard et l'acte". In *Différences, valeurs, hiérarchie: textes offerts à Louis Dumont*, pp. 25–51. Paris 1984.

Bodde, Derk, *China's First Unifier*. Hong Kong: University Press, 1967[2]. [20]

Chan, Wing-Tsit, *A Source Book in Chinese Philosophy*. Princeton 1973.

Chao, Yuanren, "How Chinese Logic Operates". *Anthropological Linguistics* 1(1959): 1–8.

Cheng, Chung-Ying, "Logic and Language in Chinese Thought". In *La philosophie contemporaine*, Vol. III, pp. 335–347. Firenze 1969.

Chmielewski, J., "Notes on Early Chinese Logic I-VIII". In *Rocznik Orientalistyczny*, 26–32 (1962–1969). [11]

Cikoski, J.S., "On Standards of Analogical Reasoning in the Late Chou". *Journal of Chinese Philosophy* 2(1975): 325–357.

Feng, Youlan, *A History of Chinese Philosophy* (trad. du chinois par D. Bodde). Princeton 1952. [2].

——, *Précis d'histoire de la philosophie chinoise* (trad. de l'anglais par G. Dunstheimer; préf. de P. Demiéville). Paris : Ed. Le Mail, 1985. [27]

Forke, Alfred, "The Chinese Sophists". *Journal of the China Branch of the Royal Asiatic Society* 34(1901/2): 1–100. [8]

Gabelentz, Georg von der, *Chinesische Grammatik*. Halle: Niemeyer, 1960 (4e éd.). [26]

Gernet, Jacques and Vernant, Jean-Pierre, "L'évolution des idées en Chine et en Grèce du VIe au IIe siècle avant notre ère". *Bulletin de l'Association Guillaume Budé* (1964): 308–325.

Graham, Angus Charles, "The Composition of the Gongsuen Long tzyy". In *Asia Maior* 5(1957): 147–183. [5]

____, "Kung-Sun Lung's Discourse Re-read as Argument about Whole and Part". *Philosophy East and West* 36(1986). [13]

____, *Later Mohist Logic, Ethics and Science*. London, Hong Kong 1978. [4]

____, "Relating Categories to Question Forms in Pre-Han Chinese Thought". *Tsing Hua Journal of Chinese Studies* (1986).

____, *Studies in Chinese Philosophy and Philosophical Literature*. Singapore: The Institute of East Asian Philosophies, 1986.

____, "Two Dialogues in Kung-Sun Lung Tzu: 'White Horse' and 'Left and Right'". *Asia Maior* 11(1965): 128–152. [10]

Granet, Marcel, *La pensée chinoise*. Paris, 1934, 1968. [25]

Greniewski, H. and Wojtasiewics, O., "From the History of Chinese Logic". *Studia Logica* 4(1956): 241–243. [12]

Hansen, Chad D., *Language and Logic in Ancient China*. Ann Arbor 1983.

Harbsmeier, Chr., *Wilhelm von Humboldts Brief an Abel Rémusat und die philosophische Grammatik des Altchinesischen*. Stuttgart: Frommann Holzboog, 1979. [23]

Hu Shi, *The Development of the Logical Method in Ancient China*. Shanghai : Commercial Press, 1922 (reprint New York 1968). [1]

Kou, Pao-Koh Ignace, *Deux sophistes chinois: Houei Che et Kong-souen Long*. Paris 1953. [9]

Lau, D.C., "On Mencius' Use of the Method of Analogy in Argument". *Asia Maior* 10 (1963): 173–194.

____, "Some Logical Problems in Ancient China". *Proceedings of the Aristotelian Society* 53(1952/53): 189–204.

Leslie, D., "Argument by Contradiction in Pre-Buddhist Chinese Reasoning". In *Occasional Paper No. 4. Center of Oriental Studies* (Canberra), Canberra 1964. [21]

Liou, Kia-Hway. "The Configuration of Chinese Reasoning". *Diogenes* 49(1965): 66–96.

____, *L'esprit synthétique de la Chine*. Paris 1961.

Lohmann, J., "Der Sophismus des Kung-Sun Lung". *Lexis* 2(1949): 3–11. [6]

Maspéro, H., *La Chine antique*. Paris, 1927, 1965. [14]

____, *Histoire et institutions de la Chine ancienne: des origines au XIIe siècle après J.-C.* Paris 1967.

____, "Notes sur la logique de Mo-Tseu et de son école". *T'oung Pao* 25(1927/28): pp. 1–64. [16]

Needham, Sir Joseph, *Science and Civilization in China*, Vol. 2: *History of Scientific Thought*. Cambridge : University Press, 1956. [17]

Reding, Jean-Paul, "Analogical Reasoning in Early Chinese Philosophy". *Etudes Asiatiques* 40(1986): 40–56. [22]

____, *Les fondements philosophiques de la rhétorique chez les sophistes grecs et chez les sophistes chinois*. Berne: P. Lang, 1985. [15]

____, "Greek and Chinese Categories : A Re-examination of the Problem of Linguistic Relativism". *Philosophy East and West* 36(1986): 349–372. [24]

Rosemont, H. (Jr.), "On Representing Abstractions in Archaic Chinese". *Philosophy East and West* 24(1974): 71– 88.

Rubin, V.A., "Tzu-Ch'an and the City-state of Ancient China". *T'oung Pao* 52(1965): 8–34. [3]

Stange, H.O.H., "Chinesische und abendländische Philosophie". *Saeculum* 1(1950): 380–396. [19]

Trauzettel, R., "Zum Problem der chinesischen Ontologie unter dem Aspekt der Sprache". *Zeitschrift der deutschen morgenländischen Gesellschaft* 119(1970): 270–277.

Waley, Arthur, *Three Ways of Thought in Ancient China*. London 1939. [18]

Wu, J.S., "Chinese Language and Chinese Thought". *Philosophy East and West* 19(1969): 423–434.

Zhou, Wenying, *Zhongguo luoji sixiangshi gao* [= Abrégé d'histoire de la pensée logique en Chine]. Beijing 1979. [7]

Central themes of Buddhist philosophy[1]

JENS BRAARVIG
University of Oslo

What is really the philosophy of Buddhism? One often tends to categorize and treat different traditions of philosophy as homogeneous entities quite easily assessible, overlooking the fact that the great philosophical and religious traditions — be they Greek, Indian, Christian, Chinese, Jewish, Japanese, European, or whatever — may change much more through history than they are different from each other.[2] For the present, however, we will make an attempt to treat at least some of the elements which make the Buddhist philosophical tradition a continuous historical entity, and as such defining its viewpoints as different from those of other traditions. It seems that a few basic assumptions constute the *crux* of a philosophical tradition, the agreed starting point from which thinking develops and is bound to relate to — if it is to keep itself within the tradition. For historians of philosophy it is essential to detect such central themes, as they generate the framework within which a tradition of thinking allows itself to move. Thus Judeo-Christian philosophy, and also the Muslim tradition, always had to relate to the idea of *God* as a basic truth. A tradition seldom criticizes it basic assumptions — even though they may be paradoxical or logically unfounded — this is usually hindered by authoritative doctrines and traditional creeds. So only quite rarely are the *praemissae propositiones* of a tradition seriously criticized — when it happens, the critics are usually persecuted and fall into oblivion, if they are not the founders of a new tradition of thinking, based upon a new set of basic assumptions. The last mentioned possibility is the case with Buddhist thinking. Buddhism originated as a criticism of earlier philosophical and religious traditions, and has thus been as acutely critical in its philosophical approach throughout its history, as it was polemical in religious matters.

So what are the basic assumptions which generate the Buddhist tradition of thought? The two main points on which the Buddha differed from the main-stream thinking in his own time, were connected to two incidents in his life, as depicted in his "biography" — or rather the legends of his ideal life piously handed down in the Buddhist canon to suit later doctrines.

During his life before becoming a Buddha, the prince Siddhārtha practiced ascetism and meditation with five other ascetics, and nearly fasting himself to death, he attained all states of consciousness possible to experience in the whole universe, but still he attained no final liberation. He then decided to give up strict ascetism, and received food from a woman who was to become his first lay disciple, thus realizing that neither the ordinary life of pleasure, nor the life of extreme

Contemporary philosophy. A new survey. Vol. 7, 195–206.

ascetism, would bring release from the world of suffering. Rather the middle way, the *madhyamā pratipad*, between worldliness and ascetism, would develop the one aspiring for freedom to extinction, *nirvāna*. At a later stage this insight was developed in the way that was also to be relevant for thinking; the one following the Buddha's path should cleave to no extreme in his thinking, neither the viewpoint of existence, *bhāva*, nor of non-existence, *abhāva*, of substance, *vastu*. Thus avoiding the extremes, and, eventually, being in accordance with non-duality, *advaya*, became a basic tenet of Buddhist tradition in its later development; this principle was even applied to the duality of *nirvāṇa* and *saṃsāra*, being described as a dualistic, conflicting thought-construction and essentially the same in being empty of substance. But also *integrating* the extremes was an important theme all the way from the beginning of the Buddhist tradition, without the correct view, *samyakdṛṣṭi* — first part of the eightfold noble path — the two last parts, awareness, *smṛti*, and mediatation, *samādhi*, would be of no avail. So insight, *prajñā*, under which the two first parts of the eightfold path are subsumed, and meditation, *samādhi*, the three last parts, when integrated and supported by an ethical life, *śīla*, the third, fourth and fifth parts, lead the practitioner of the Buddhist path to freedom.[3] Thus also expanded vision, *vipaśyanā*, should be integrated with peaceful meditation, *śamatha*, during the Buddhist path.

The most important element, however, which made the Buddha differ from earlier brahmanical tradition, was the negation of the Self, *ātman*. There is no permanent entity which constitutes the core or the continuity of the individual; the personality is only a succession of moments of existence, mainly of mental states, propelled by actions, *karman*, and their results, by desire, *rāga*, aversion, *dveṣa*, and bewilderment, *moha*, the three roots of our continued existence. In this endless succession of states of existence, or *dharmas*, which is their technical term, there is no constant. In the basic suffering, *duḥkha*, however, of all existence, the living beings try to find such a constant, believing it to be part of existence, trying to find "the self where there is no self", believing it to be release from the suffering of existence, *saṃsāra*. It is nothing but ignorance, *avidyā*, however, which binds living beings to *saṃsāra*, the ignorance which the Buddha dispelled at the moment of his *bodhi*, his awakening. Thus understanding that all moments of existence, all *dharmas*, are without self, *anātma*, and thus impermanent, *anitya*, and empty, *śūnya*, constitutes the basic insight, *prajñā*, by which the Buddha was able to free himself from and completely give up his attachment to all possible moments of existence, which he had realized through his meditation, his *samādhi*. Part of the first "noble truth" of the Buddha's teaching, then, the truth of suffering, *duḥkha*, is selflessness, impermanence, and emptiness. Though early Buddhism shared the view that all is suffering with most of the other philosophical systems in brahmanical speculation, it was strongly in opposition to the view that the individual is given continuity by a self which is the core of the personality as an eternal spirit or soul. This was the teaching of the Upaniṣads, which taught that the individual self, *ātman*, is really identical with the universal self, *brahman*, and that realizing or knowing the individual self as such, is final release from existence. This theme was elaborated throughout the history of Indian philosophy by Sāṃkhya, Vedānta and others, and the self/non-self or substance/non-substance controversy became one of the crucial problems throughout history, both within Buddhism itself and in the Buddhist discussion with other traditions of thought, though the form of the theme changed. In the following we will show how some of the important forms of these two themes, *viz.* the theme of non-self or non-substance, and the theme of the middle way, non-duality or integrating the opposites, have appeared in the Buddhist tradition of philosophy.

By the dialectical process of the Indian philosophical tradition it seems that several ontological possibilities were investigated. The Upaniṣads postulated the eternally existent principle of individual existence, the Self, *ātman*, and laid the foundation for later philosophical speculation. The philosophy of Yoga and *Sāṃkhya* both admitted the *puruṣa*, more or less equivalent with the *ātman*, as an eternal individuating principle, as they also admitted the reality of the states of counsciousness, of sense perception and the objects of sense perception. But, as we have seen, Buddhism denied this principle, but still it, at least in part, admitted the existence of the *dharmas*, the momentary mental states which in endless succession make up the stream of consciousness, which we mistake for a self. Vedānta, especially the school developing after Śaṅkara, maintained that the *ātman-brahman* is the sole existent reality, and that existence beside it is nothing but an illusion, *māyā*. The Vedānta was, however, partly inspired by the thinking of Mahāyāna Buddhism, the adherents of which maintained the ontological — or rather anti-ontological — philosophical viewpoint — which they stated really not to be a viewpoint — that all existence is empty, *śūnya*, of substance. There is no self, but also the moments of existence, the *dharmas*, are empty of substance, thought-constructed, and no more real than an illusion, *māyā*, or a *fata morgana*. Thus the early Mādhymika philosopher Nāgārjuna repudiated the view of the Sārvāstivāda sect that the *dharmas* were momentarily real, as he repudiated other substantialist views of the Brahmanical traditions. The philosophy of Buddhism, then, should be seen in the perspective and in the dialectical tension of these four ontological possibilities: 1. the self, or the eternal spirit, as well as the world of its experience exist, as in Sāṃkhya-Yoga; 2. the self is non-existent, but the world of experience and sense-perception is momentarily existent, as in several of the Buddhist sects; 3. the Self exists as the sole reality, while all other things are non-existent, as in Vedānta; 4. both the self and the world of experience are unreal, empty and non-existent, as in Mahāyāna Buddhism.

As mentioned, selflessness, *anātma*, impermanence, *anitya*, emptiness, *śūnyatā*, and suffering, *duḥkha*, are the four qualities of existence subsumed under the first of the four truths, which make up the *nucleus* of the Buddha's teachings.[4] Thus the presupposition of selflessness, understanding and, through meditation, realizing that there is no self in any of the parts of the personality, in the *skandhas*, are basic in Buddhist teachings, the realization of which is the basic prerequisite for liberation, *nirvāṇa*. The *anātma*-doctrine was originally a denial of the upaniṣadic *ātman*, but also the only way to real freedom. Among the momentary, impermanent and empty moments of existence, the *dharmas*, a finite number[5] of which existence is constituted, there is no lasting core, no permanent entity which constitutes the personality. One tends to postulate and cling to such entities for safety and pleasure, but the real truth is that any entity, anything existent, vanishes the next moment, which state of things is felt as suffering, *duḥkha*, really rooted in the misunderstanding and ignorance, *avidyā*, that a self exists.[6]

So the principle of selflessness is the basic supposition related to by all the schools of thought in Buddhism. The Mahāyāna, however, emphasized the principle of emptiness, also part of the first truth describing existence as suffering. Thus early Mahāyāna thinking asserted that the stream of consciousness which make up the personality is not only empty of self, of *ātman*, but empty also of any substance, empty of any positively self-existent nature, *svabhāva*. Existence is seen as an illusion, *māyā*, a *fata morgana*, unreal as an echo. The *dharmas*, atoms of existence, are only thought-constructions, *vikalpa*, superimposed by discursive thinking, *prapañca*, on existence which is nothing in itself. Everything is relative, *pratītyasamutpanna*, mutually, but not relative to any basic cause, since any such postulated cause must also be thought-constructed and empty, on the supposition of universal emptiness. This viewpoint of universal relativity — which is nothing, by consequence, nothing but a relative expression of language conventions — is basic in Mahāyāna thinking from the oldest Mahāyāna sūtras on. It is expressed as follows in Nāgārjuna's *Mahyamakakārikā*[7]: "What is relative origination (*pratītyasautpāda*), we call emptiness (*śūnyatā*), this is an expression of language (*prajñapti*) we use, and this is really the middle way." The theme of selflessness, emptiness and substancelessness, *niḥsvabhāvatva*, is the basis of Buddhist dialectics, a dialectic method which does not argue for any viewpoint, but has as its objective to free the opponent from any viewpoint taken to be absolutely consistent and true. No such viewpoint can exist, since all viewpoints are really thought-constructions, and to "deconstruct" the views of the opponent in discussion is to free him from the suffering of existence, since suffering is also a mental construction of something existent. Things such as they really are, their "suchness", *tathatā*, and emptiness, *śūnyatā*, are beyond conflicting mental constructions, *nirvikalpa*.[8] By having the opponent's presupposition as the starting-point of the dialectical process, the Mādhyamika-dialectic wishes to show that any viewpoint, if analyzed properly by arguments and *reductio ad absurdum*,[9] is really conflicting with itself and inconsistent, and thus to be given up — every thought-construction being by nature in conflict with any other thought-construction. The theme of selflessness, emptiness and substancelessness, *niḥsvabhāvatva*, is also the basis of Buddhist language philosophy. The Sarvāstivāda and other *abhidharma*-schools of the non-mahāyana tradition would assert that the word *ātman* refers only to an illusion, or to a conglomerate of conditions, of *dharmas*, building up what is believed to be an entity, which, however, it is not. Words, though, may refer to existent entities, namely to the *dharmas*, or to the groups they may constitute. According to Mahāyāna sūtras, and to the later philosophies of Mahāyāna, the Mādhyamika and the Yogācāra, or the *cittamātratā*, "thought only" philosophy, words are only empty conventions, constructed as *signs* working in a given context, for the best in the context of freeing man from the thought-constructed (*kalpita* etc.) illusion they themselves constitute. Accordingly, the *dharmas* are non-existent, and empty, and existing only as imagined (*parikalpita*). Language, then, is only a constructed convention, and the meaning, or the *referendum*, of a word is really nothing, the "thing"[10] is mere emptiness on which an illusory convention is superimposed. Buddhism, accordingly, was taught in may languages, and the Mahāyāna asserted that any language which conveys the meaning of emptiness, is really the true teaching of The Buddha, regardless its form. This view strongly opposes the Hindu philosopher's view on language, namely that the *artha* is organic with the word of the true and holy Sanskrit language. The question, however, was posed to the Buddhists: "How can language have meaning if it does not refer to anything but emptiness?" The Buddhist side would answer that language at best would discover its own illusory nature and free man to the understanding of emptiness, thus bringing about the understanding of the conventional and empty nature of the very expressions that communicate this understanding, much in the same way as Wittgenstein makes the main project of philosophy the abolition of philosophy itself. "But still, how can the single word refer to anything different from what other words refer to if there is no entity different from other entities refered to?" The Buddhist side, as represented by the famous Buddhist philosopher Dharmakīrti[11] would answer that the word does not refer to an existent entity, but excludes all other thought-constructed and illusory meanings, and thus establishes a distinct meaning. Deciding what something is not without saying anything of the thing itself is the possibility of language which Buddhism uses throughout its history, and thus, in accordance with the doctrine of selflessness, avoids saying something positively definite about such basic concepts as, e.g., *nirvāṇa*. The philosophy of Mahāyāna accordingly also asserts that truth — that is emptiness — is inexpressible, and beyond the four ontological possibilities contain in the *catuṣkoṭi*: Neither 1. being (*bhava*), nor 2. not being (*abhava*), nor 3. being and non-being (*bhavābhava*), nor 4. neither being nor non-being. Truth is possible to demonstrate only by negative dialectics, which itself, however, also is a game of language.[12] Thus real truth, *paramārthasatya*, is not found as apprehended through thought-constructions, discursive thinking, and conventional verbal expressions. These, however, are necessary to communicate real truth, necessary to free man from being entangled in just these conventional thought-constructions. As Nāgārjuna says[13]: "The teachings of the Buddha is based on two kinds of truth, *viz.* the conventional truth of the world (*lokasaṃvṛtisatya*), and the truth in accordance with the highest reality (*paramārtha*).... The highest truth cannot be taught if one is not based upon the conventional truth, but *nirvāṇa* is not realized if the highest truth is not understood." The doctrine of emptiness, *śūnyatā*, put the adherent of the Mahāyāna into a difficult position when forced to defend the truth of this doctrine: if he asserts everything to be empty, his arguments for the case of emptiness would also be empty and not valid; if the arguments are valid, they cannot be empty. Thus everything is not empty — at least not these very words expressing the arguments — and the position is inconsistent. The Mahāyāna adherent would answer that the statement negating the substantiality of all things would *also* be relative (*pratītyasamutpanna*) and empty, being merely an illusory and thought-constructed convention expressing emptiness — as empty of substantial meaning as all other expressions, but as a convention pointing out the emptiness of all other expressions. As a comparison[14] to the statement of emptiness is given the example of a man who is created by magical illusion: "Suppose ... that a magic man (*māyāpuruṣa*) should prevent another man created by his *own* magic [from doing something]. Of the same nature would be this negation.... the magic man who is prevented is void, and he (the magic man) who prevents is also void. In the like manner, a negation of the intrinsic nature of all things by my statement is possible, even though this statement is void." The argument, which is the core of Nāgārjuna's *Vigrahavyāvartanī*,[15] is of course very important in the Mādhyamika philosophy, which does not assert any basic premiss to be true, it merely uses the, according to them, thought-constructed premisses of the other party in the discussion to show by *reductio ad absurdum* (*prasaṅga*) that these premisses are without substance, thought-constructed and fundamentally inconsistent. "If I asserted a proposition of mine [to be *substantialiter* true] I would be wrong, but I do not...."[16] This kind of discussion, of course, also had a religious aspect, since freeing all beings from their thought-constructed reality, believed to be true, is the aim of the Mahāyāna path of religious development.

The traditional sects, like the *Sarvāstivāda*, would assert that the moments of existence, the *dharmas*, exist as entities, however momentary (*kṣaṇika*), disappearing the moment they originate, this view being in accordance with the basic doctrine of self-lessness, *anātmatā*, and impermanence, *anityatā*. The emptiness, *śūnyatā*, of the *dharmas*, was, however, more stressed by the Mahāyāna, and in their emphasis on the thought-constructed and illusory nature of reality, they would reinterpret traditional Buddhism in most of its aspects. One example of this is their reinterpretation of the twelve-fold chain of dependent origination, the *pratītyasamutpāda*. The traditional form was: from ignorance, *avidyā*, the factors forming consciousness, *saṃskāra*, arise; then consciousness, *vijñāna*, from which concepts and forms, *nāmarūpa*, arise; then, in sequence, the six senses, *ṣaḍāyatana*; contact, *sparśa*, between the sense and the objects; feelings, *vedanā*; thirst, *tṛṣṇā*, for continued existence; grasping, *upādāna*; existence, *bhava*; birth, *jāti*; ageing and death, *jarāmaraṇa*, etc. Thus understanding how these factors arise in relation to each other would bring man nearer to freedom, seeing that there is no self, nor any substantial and lasting reality, neither within nor beyond dependent origination. The Mahāyāna interpretation stresses that old age and death is created through our mental activity, and that the moments of existence as part of dependent origination are also *empty* and *thought-constructed*, as the *dharmas* are nothing but concepts superimposed upon nothingness: from apprehending (*upalambha*), things as something, be it the parts of personality (*skandha*), the fields of sense-perception, (*dhātvāyatana*), etc., the complexity of discursive thinking, "the net of inner talkativeness (*prapañcajāla*)" arises, from which the net of thought-constructions, (*kalpanājāla*), or conflicting thought-constructions (*vikalpa*), arise, which create adherence to the dual conflict between the constructed subject and object, misunderstood as *substantialiter* existent,[17] thus being the cause of the mass of [constructed] vices (*kleśa*) and actions (*karma*), keeping living beings bound to *saṃsāra*, which consists of birth, ageing and death, the two last members of the traditional *pratītyasamutpāda* mentioned above. Understanding the development, then, of mental complexity is also to understand its emptiness, things as they really are, *tathatā*, their "suchness" as beyond mental constructions, *nirvikalpa*[18]; while apprehending reality as non-empty and existent as substance is, through the development of the complex mental activity, the cause of suffering. *Avidyā*, ignorance, is not so much emphasized in the Mahāyāna interpretation as the cause of suffering as is the mental activity of construing an illusory reality based on the collective imputation of concepts on empty suchness. This interpretation of course has an element of polemic, as ever in Mahāyāna scriptures, towards what was defined as the inferior vehicle, the hīnayāna, and their highly abstract and theoretical speculation, but the interpretation is certainly also basic in Mahāyāna ontology and epistemology. Thus the vices, (*kleśa*) are not innate (*nija*), as one would tend to assert in traditional Buddhism, but the constructed product of a mental dialectical process between a constructed subject and a constructed object. Thus the old saying of Buddhist tradition: "Desire, I know your root, you are born from a construction (*saṃkalpāt*), if I do not construe you, you will not be", is the basis of Nāgārjuna[19]: "Desire, aversion and bewilderment are said to have their origin in thought-constructions, in relation to (*pratītya*) the erronous idea (*viparyāsa*) of pure and impure." The vices, then, are born from nothing but the dualistic thought-constructions which constitute the mental activity of man, or of "living beings", as Buddhism would express it; they do not really exist.

The Mādhyamika tradition was often accused of being nihilstic, *nāstika*, an accusation that was denied by asserting that the middle way does not take a stand on neither being, nor non-being, it only teaches the true middle way between the two in demonstrating the emptiness and non-final nature of all viewpoints. To make the doctrine of emptiness into a definite viewpoint might amount to nihilism, but such a standpoint is repudiated with strong emphasis, as compared with a medical remedy to be used with the greatest care: "In the same way, *Kāśyapa*, [seeing] the emptiness of all viewpoints is the way out [of the sickness of existence], while for the one, Kāśyapa, who holds emptiness as a viewpoint (*śūnyatādṛṣṭi*), I say there is no cure." And, as echoed by Nāgārjuna: "Emptiness wrongly conceived destroys those with lesser intelligence as a snake grasped wrongly, or an incantation wrongly performed".[20]

On the grounds that the Mahāyāna morality is not mentioned by what are considered the main works of Nāgarjuna, the *Madhymakakārikā* and the *Vigrahavyāvartanī*, and scarcely by Āryadeva and Nāgārjuna's later followers, it has been asserted that Nāgārjuna is not necessarily an exponent of the Mahāyāna.[21] According to the view of the present writer, Nāgārjuna, in the two works mentioned, is the systematizer of the dialectical, or even rhetorical,[22] method of winning adherents to the Mahāyāna faith by means of argumentation and debate, a method which is employed in the numerous dialogues depicted in the different Mahāyāna sūtras. If the *Ratnāvāli*, a work on the general teachings and the morality in the Mahāyāna, is also accepted as having been written by Nāgārjuna, as well as others of the works under his name,[23] there is no doubt that his negative dialectic is only a *part* of the general Mahāyāna teachings, as contained also in the Mahāyāna sūtras, which he propounded.

So far we have treated the Buddhist philosophical theme of selflessness, of impermanence, etc., as related to the central doctrine of *anātma*; we will now treat the second theme mentioned at the beginning of this paper, that of integrating extremes and non-duality. As we have seen, Buddhism, and especially the Mahāyāna, had a negative ontology, which has strong tendencies towards nihilism. Basic in the Mahāyāna faith, however, is also a strongly positive morality, positive in the way of being directed towards all living beings. The Mahāyāna made moral action into an end in itself as the realization of emptiness, polemicizing against earlier tradition and their ascetic ideal of getting rid of all action to bring about extinction of all life, *nirvāṇa*. Thus the perfection of generosity, *dānapāramitrā*,[24] is giving to all what they need; food to the poor, medicines to the sick, lodging to the homeless, etc. etc. one should even give one's kingdom, one's wife and only son when asked, even the hands, the feet, the head and the marrow of one's body. But, as the *Prajñāpāramitā* literature has it, this is still the kind of gift which is part of the world, only the gift which is *trimaṇḍalapariśuddha*, "pure in the three fields"—that is having no apprehension or thought-constructions connected to any part of the three parts or "fields" of generosity, namely 1. the subject of giving, 2. the person receiving and 3. the action of giving—is the expression of emptiness and non-duality, *advaya*, the nature of Buddha's actions. Thus the bodhisattva advancing on the spiritual way also gradually gets rid of his thought-constructions: in the practice of friendliness, *maitrī*, the bodhisattva on the first stages practices friendliness with all living beings as the object, *sattvālambanā*; the more advanced friendliness has all moments of existence as the object, *dharmālambanā maitrī*; while the most advanced bodhisattva is *one* with friendliness, *one* with emptiness, and beyond the duality of subject and object—his friendliness is without any object whatsoever, *anālambanā*. In this way action for the sake of all beings became the means of understanding emptiness. While earlier tradition emphasized the integration of expanded vision, *vipaśyanā*, and peaceful meditation, *śamatha*, as a team, *yuganaddha*, the Mahāyāna—in its project of ethical action—emphasized another pair to be integrated, namely that of knowledge of emptiness with compassionate action for the sake of all beings. Thus insight, *prajñā*, and *upāya*, expedient means in compassionate action, are compared to a team of two horses which draws the Great Vehicle, the Mahāyāna, the vehicle is *prajñopāyayuganaddha*.[25]

Thus, as the famous Mahāyāna sūtra *Vimalakīrtinirdeśa* states,[26] knowledge, or meditation, united with expedient means, *viz*. action for the good of all beings, is the real freedom of a bodhisattva; while practicing only one of them is bondage: "What is bondage (*bandha*), what is liberation (*mokṣa*)? Grasping for liberation from existence without expedient means (*upāya*) is the bodhisattva's bondage, while entering existence with expedient means is liberation; tasting meditation (*dhyāna*), concentration (*samādhi*) and states of meditation without expedient means is the bodhisattva's bondage, while tasting meditation and concentration with expedient means is liberation; insight (*prajñā*) without expedient means is bondage, while insight with expedient means is liberation; expedient means without insight is bondage, while expedient means with insight is liberation." The bodhisattva should also not stick to any viewpoints or realizations of freedom, of the end of suffering, but give them up through actions; actions which he should transform into *bodhi*, awkening for sake of helping and maturing all beings spiritually. Thus actions and understanding should balance as the realization of emptiness, which commitment for the sake of all beings should be imperishable, *akṣaya*, or endless. The dialectical process between knowledge — which may coagulate into dry and meaningless doctrines if not employed in action—and action — which may be without aim if not supported by understanding — should be the bodhisattva's obligation for limitless time. Time, however, is also nothing but an inexistent and relative category according to Mahāyāna philosophy: both the past and the future is relative to the present, and the present is non-existent and turning into past as soon it is constructed in thought. Thus the aim of the bodhisattva's practice, awakening or *bodhi*, is not something to be attained in the distant future, it is, with the full realization of emptiness, the ever-present possibility of being, which, however, can never be grasped, or constructed into something definite to be attained.[27]

NOTES

1. *The Central Philosophy of Buddhism* is the title of two books. T.R.V. Murti in his splendid book of 1955 has the sub-title A *study of the Mādhyamika System* and treats the dialectical philosophy of Nāgārjuna (first century, according to others the second), about which he says (p. vii): "Considering the rôle and the importance of the Mādhyamika, I have ventured to appraise it as the Central Philosophy of Buddhism". David J. Kalupahana's *Causality: The Central Philosophy of Buddhism* (Honolulu 1975), is full of criticism of Murti, who, according to Kalupahana (p. 177) "wrongly attributed certain theories, such as the theory of real elements, to the early Buddhists and even to the Buddha himself." This is no doubt right, and even his most prominent predecessor in interpreting Buddhist philosophy, who has influenced almost all modern studies of this topic, namely Fedor I. so Stcherbatsky, superimposed certain viewpoints on the opponents of Nāgārjuna. Stcherbatsky is also the author of the book *The Central Conception of Buddhism* (London, 1923), which title is the background for the two mentioned. But superimposing viewpoints on their opponents was also one of the weak points of the whole Mahāyāna movement, portraying a picture of the Hīnayāna with which the traditional form of Buddhism would not agree. In a later book, an interpretation of Nāgārjuna's philosophy as set forth in the Mādhyamakakārikāḥ (*Nāgārjuna. The Philosophy of the Middle Way* [New York 1986]), Kaluphana tries to show that Nāgārjuna was not necessarily an adherent of the Mahāyāna faith. By him, he says (p. 5) "an attempt will be made to present Nāgārjuna merely as a grand commentator on the Buddha-word and to show that he did not try to improve upon the teachings of the Buddha", implying he was not an exponent of the early Mahāyāna faith, but a non-sectarian commentator with his basis in traditional Buddhism as handed down to us in the Pāli-canon. In modern Buddhology there also seems to be a tendency to lessen the importance of the Mahāyāna faith, making it a somewhat marginal movement in the first two or three centuries A.D. After the enormous impact of Edward Conze's work, which made the Mahāyāna not only an object worthy of study, but cast doubt even on the accepted fact that the Pāli tradition would be "original Buddhism", many researchers again find the Mahāyāna movement only a marginal part of "Mainstream Buddhism" in the first centuries A.D. Though the present writer is working on the early history and formation of the Mahāyāna movement, he does not at this stage of his research find it sound to take a firm stand on the question of the historical development of the beginnings of Mahāyāna, and he does not aim at engaging in any kind of modern

Mahāyāna/Hīnayāna controversy. He is however, fairly well convinced that Nāgārjuna, really the most celebrated individual philosopher of Buddhism, was a systematizer of the philosophy found in the early Mahāyāna sūtras, and possibly also had a function in formalizing the Mahāyāna canonical scriptures, which, during the three or four centuries A.D., suffused most of the sects of Buddhism, excepting the most traditionalist and, in a sense, those most true to the earliest teachings of Buddhism. The present paper, then, will present some themes which the writer considers to be central to the shaping of the framework within which Buddhist thinking has moved through history, giving it continuity as a tradition, *not* presenting any kind of *the* central philosophy of Buddhism, which may easily become an arbitrary and intentional construction in conflict with historical information.

2. I will not here go into the complex, but interesting, discussion of how the views of different philosophical traditions may be "the same" or "identical". In relating to traditions of philosophy, however, which have originated in a different language context than the one we are going to explain it through, we are bound to reflect on the use of terms we use in representing, in our case, Buddhist philosophical terminology, which reflection necessesarily involves a comparative element. Let it for the moment suffice to say that comparison, as well as translation, is always intentional, as is the assumption of "identity" or "difference". From one point of view aspects of Buddhist philosophy may be said to be identical with (or at least similar to) elements of the philosophy of Kant, Bradley, Heidegger, Wittgenstein, Derrida etc. — all of which Buddhist philosophy has been compared to, and whose terminology has been employed by translators in the attempt to convey Buddhist philosophy to a western language context. But certainly the same elements in most cases may be said to be different, both in terms of language, culture, tradition and scope. For literature on the subject, see the bibliography.
3. The *credo* of Buddhism is seeking the protection of the so called *triratna*, "the three jewels", *viz.* the Buddha himself, his teaching, the *dharma*, and his followers, the *saṃgha*. The *dharma* consists in the four *āryasatya*, "noble truths", namely 1. the truth that all is suffering, *duḥkha*, 2. the truth on the origin, *samudaya*, of suffering, 3. the truth that there is an end, *nirodha*, to suffering, namely extinction, *nirvāṇa*, and 4. the eightfold path, *mārga*, to the end of suffering, consisting of 1. right insight, *samyakdṛṣṭi*, 2. right intention, *samyaksaṃkalpa*, 3. right speech, *samyagvāk*, 4. right action, *samyakkarmānta*, 5. right way of earning one's living, *samyagājīva*, 6. right mental effort, *samyagvayāyāma*, 7. right awareness, *samyaksmṛti*, and 8. right concentration, *samyaksamādhi*.
4. See note 3.
5. The different schools of Buddhism postulated different numbers of these "existential atoms" which make up the stream of existence, of suffering. The *Sarvāstivāda* had seventy-five as the number, as described in the Abhidharmakośa, while the Vijñānavāda, would have one hundred *dharmas* to make up existence. The *dharmas* are classified in several ways, as subtle and coarse, wholesome and unwholescome, etc. Examples of *dharmas* are sensation, memory, concentration, faith, ignorance, anger, attachment, etc. etc.; common to them all are impermanence, selflessness, emptiness and suffering, and momentariness. For such lists, see e.g. A. Hirakawa, *Index to the Abhidharmakosabhāṣya* (Tokyo 1973).
6. Recent literature on the subject of the absence of self in Buddhism is Steven Collins, *Selfless Persons: Imagery and Thought in Theravada Buddhism* (Cambridge 1982), and on the Hindu-Buddhist controversy on the self, Claus Oetke, *'Ich' und das Ich*, Alt- und Neuindische Studien (Stuttgart 1987). See also bibliography.
7. XXIV, 18.
8. The verbal root *kḷp* means "to construct" in a variety of meanings.
9. Sanskrit *prasaṅga*.
10. Sanskrit *artha* means "aim, thing wanted", then "the thing to which a word refers", "meaning".
11. *Floruit* ca. 650. This later development of the theme of non-substiantiality was aimed polemically against the (Hindu) philosophers who asserted the existence of the universals. See D. Sharma, *The Negative Dialects of India* (Leiden 1970) with references to other literature on the subject; p. 111: "... the concept 'cow' can only be determined by all other instances of reality from which it is excluded". Says Dharmakīti in his *Pramāṇavārttika*: "A universal is in its very essence exclusion of

the other (*sāmānyam anyavyāvṛttilakṣaṇam*)", see *ibid*, and also by the same author, *The Differentiation Doctrine in Indian Logic* (The Hague 1969). See also bibliography.

12. A negative proposition is of course much more difficult to argue against than a positive, a fact Buddhist dialectics and rhetorics exploited to attack other philosophical viewpoints.
13. *Mūlamadhyamakakārikā*, XXIV 8ff.
14. The comparison (*upamāna*) is regarded as the third of the five terms contained in the syllogism in early Indian logic — the fifth being the final conclusion — and thus being of much more importance than a merely poetical figure. This fact, however, is one of the many indications that early Indian "logic" has as strong rhetorical element.
15. Verse XXIII, ed. E.H. Johnston, A. Kunst, translation by K. Bhattacharya (Delhi 1978).
16. *Ibid*. XXIX
17. *ahaṃ mamety abhiniveśāt satkāyadṛṣṭiḥ*
18. For a fuller treatment see J. Braarvig "Development of Mental Complexity as Suffering in Sāṃkhya and Buddhism", in *Kalyāṇamitrārāgaṇam*. (Oslo 1986), p. 49ff.
19. In *Madhyamakakārika* XXIII, 1, *Prasannapadā* (ed. L. de la Vallé Poussin, Bibliotheca Buddhica 1903–13), p. 451.
20. *Kāśyapaparivarta* ed. A. v. Staël-Holstein (Beijing 1925), § 65; *Madhyamakakārikā* XXIV, 11.
21. See note 1.
22. On rhetorics in the early Mahāyāna see my paper, "*Dhāraṇī* and *Pratibhāna*: Memory and Eloquence of the Bodhisattvas". JIABS, Vol. 8, No. (1985) p: 17ff.
23. On the problem of authorship of the works under the name Nāgarjuna, v. C. Lindtner, *Nāgārjuniana* (Delhi 1987). *Nāgārjuna* possibly also had a role in formalizing the early Mahāyāna scriptures as a *corpus*.
24. The *pāramitā*, perfections or virtues of the bodhisattva are basically six: 1.generosity (*dāna*), 2. general morality (*śīla*), 3. tolerance (*kṣānti*) 4. spiritual courage (*vīrya*) 5. meditation (*dhyāna*) and 6. insight (*prajñā*).
25. As in the *Gaganagañjaparipṛcchāsūtra*.
26. In É. Lamotte's translation (Louvain 1962), p. 233.
27. The philosophy of imperishability, *akṣayatā*, and the Mahāyāna ethics as propounded on the background of the integration of knowledge and action, the *prajñopāyayuganaddhavāhī mārgaḥ* as called in Sanskrit, has been treated by the present author in *The Akṣayamatinirdeśasūtra and the Tradition of Imperishability in Buddhist Thought*, to be published 1992. On the one hand the commitment of the bodhisattva to the weal of all beings, and his virtues, are imperishable, *akṣaya*, as lasting forever. The sphere of living beings, *sattvadhātu*, will last forever, thus also the generosity, and the other *pāramitās* of the bodhisattva also must last forever. On the other hand, what is unoriginated and empty cannot perish, and is thus imperishable. These are the two basic meanings given to the concept *akṣaya* in the philosophical tradition making this a key concept.

BIBLIOGRAPHY

This bibliography is far from being a complete list of books on Buddhist philosophy; it is rather an attempt to present some recent writings in the field which treat themes related to those of the present paper.

On the concept "Central Philosophy of Buddhism" cf. Fedor I. Stcherbatsky, *The Central Conception of Buddhism and Meaning of the Word "Dharma"* (London 1923); T.R.V. Murti, *The Central Philosophy of Buddhism: A Study of the Mādhyamika System* (London 1955), and David J. Kalupahana, *Causality: The Central Philosophy of Buddhism* (Honolulu 1975) and note 1.

There is a *flora* of books interpreting Buddhism in the light of western philosophers, mentioned are Stcherbatsky's, Kalupahana's and Murti's works; others are K.N. Jayatilleke, pupil of Wittgenstein, with his book *Early Buddhist Theory of Knowledge* (London 1963); C. Gudmundsen, *Wittgenstein and*

Nagarjuna (London 1977); A. D. P. Kalanasuriya, *Philosophical Analysis of Buddhist Notions: The Buddha and Wittgenstein* (Delhi 1987); R. A. F. Thurman, "Philosophical Nonegocentricm in Wittgenstein and Candrakirti in Their Treatment of the Private Language problem", *Philosophy East and West*, No. 3 (1987); D. Loy treats Wittgenstein, Heidegger and Derrida in *Nonduality: A Study in Comparative Philosophy* (New Haven 1988); R. Magliola, *Derrida on the Mend* (Indiana 1984); and C. W. Huntington, *The Emptiness of Emptiness: An Introduction to Early Indian Mādhyamika*. G. R. Welbon treats the early interpretations of *Nirvāṇa* in the West in *The Buddhist Nirvāṇa and Its Western Interpreters* (Chicago and London 1968); and Per Haraldsen gives an overview of the interpreters of Nāgārjuna and their background in western philosophy in *Mūlamadhyamakakārikā: Theodore Stcherbatsky og Frederick Strengs tolkninger belyst i et idéhistorisk perspektiv* (Bergen 1991) (Thesis, still unpublished.)

On the problem of self/non-self Steven Collins, *Selfless Persons: Imagery and Thought in Theravada Buddhism* (Cambridge 1982); and Claus Oetke, *'Ich' und das Ich*, Alt-und Neuindische Studien (Stuttgart 1987); further R. A. F. Thurman, mentioned above; Serge Christophe Kolm, "The Buddhist Theory of 'No-Self'", in *The Multiple Self*, ed. Jon Elster (Cambridge 1985).

For an overview of Mahyamaka literature see D. S. Ruegg *The Literature of the Madhyamaka School of Philosophy in India*, Vol. vii in *A History of Indian Literature*, ed. Jan Gonda (Wiesbaden 1981); C. Lindtner, *Nagarjuniana Studies in the Writings and Philosophy of Nāgārjuna*, 2nd ed. (Delhi 1987); P. D. Santina, *Madhyamaka Schools in India* (Delhi 1986); cf. also his "The Madhyamaka and Modern Western Philosophy", *Philosophy East and West*, Vol. XXVI, No.1 (1986). F. J. Streng, *Emptiness: A Study in Religious Meanings* (New York 1967) is, besides the books already mentioned, an important interpretation of the Madhyamaka, as is M. Sprung, *Lucid Exposition of the Middle Way* (London 1979) and David J. Kalupahana, *Nāgārjuna: The Philosophy of the Middle Way* (New York 1986). Kenneth K. Inada, *Nāgarjuna, A Translation of His Mūlamadhyamakakārikā* (New York 1970), and K. Bhattacharya, *The Dialectical Method of Nāgārjuna: Vigrahavyāvartanī* (Delhi 1978) are standard translations of Nāgārjuna's main works. A short essay epitomizing the main concern of the Mādhyamika is J. May, "On Mādhyamika Philosophy", *Journal of Indian Philosophy*, Vol. VI (1978): 233ff. For present research on Buddhist language philosophy see *Buddhist Logic and Epistemology*, Studies in the Buddhist Analysis of Inference and Language, Eds. B.K. Matilal and R.D. Evans (Dordrecht 1986).

On *vikalpa*, and *nirvikalpa*, and the Buddhist variant of "deconstruction" see F. J. Streng, "Reflections on the Attention Given to Mental Construction in the Indian Buddhist Analysis of Causality", *Philosophy East and West* (1978); and Jens Braarvig, "Development of Mental Complexity as Suffering in Sāṃkhya and Buddhism", in *Kalyāṇamitrārāgaṇam*, ed. Eyvind Kahrs (Oslo 1986).

On the *yuganaddha* doctrine and its relevance for hermeneutical dynamics and for the morality of the Mahāyāna see D. S. Ruegg, *Buddha-Nature, Mind and the Problem of Gradualism in a Comparative Perspective: On the Transmission and Reception of Buddhism in India and Tibet* (London 1989); and Jens Braarvig, *The Akṣayamatinirdeśasūtra and the Tradition of Imperishability in Buddhist Thought* (Oslo 1992) (forthcoming); and *ibid.* for the *akṣaya* doctrine and the Mahāyāna morality of ethical commitment to all beings. Cf. also R. W. Perett, "Egoism, Altruism and Intentionalism in Buddhist Ethics," *Journal of Indian Philosophy* Vol. XV (1987): 71ff. For Mahāyāna Buddhism generally, see Paul Williams, *Mahāyāna Buddhism: The doctrinal foundations* (London and New York 1989), and its extensive bibliography.

Taiwan

Taoist Philosophy compared to European Philosophy

PETER KUN-YU WOO
National Taiwan University

Introduction

Although the Taoist Philosophy can hardly be divided into Epistemology, Metaphysics and Ethics, like the Aristotelian-Thomistic Philosophy, the content of the Taoist thought is really comprehensive enough to embrace all the philosophical disciplines in search of true knowledge, of cosmic principles and of life-wisdom.

In the western traditional division of philosophy, epistemology has been considered as the entrance, metaphysics as the essence, and ethics as the practical principle of all sciences. In the Taoist Philosophy, there is also a similar threefold division, though its main emphasis seems to be on Life-philosophy, especially on self-perfection. In the Philosophy of life, however, the Taoists make themselves understandable not directly through life-experience, but with and in the philosophical literature to show their epistemological method and their methodological contents. The methodology used by the Taoists seemed to be mystical as well as supra-ontological; its fundamental grounds were concerned more with personal perfection in the existential sense than with ontological meanings.

The Taoist morally-oriented way of thinking was fundamentally very different from that of the Confucianist philosophy, even though the latter also flourished in the same period, i.e., in the period of the Warring States of Ancient China. The Confucian ethical philosophy was introduced directly from the metaphysical foundation. Their normative principles of moral postulation were directly introduced from the innate consciousness of human conscience. This was an upward path arising from the epistemological subsumption to the moral postulate of doing good and avoiding evil. Taoist Philosophy, however, was not directly oriented towards the innate conscience of human nature; the Taoist tried to free himself from the mundane spirit and to transcend himself to the sphere of subjectivity of self-perfection through the aesthetical life-experience. The Confucianist is a moral subject, and his study is anthropological investigation. The Taoist, on the other hand, is an aesthetical person, and his study is a transcendental research transiting from the anthropological to the ontological sphere. Therefore, the Taoist "Chin-jen" (The authentic man) has an ontological meaning, in which a man should be united with "Tao" (the cosmic and ontological principle of the universe). Thus, the main concern of Taoist Philosophy is ontology via moral practice.

It has been thought that the most comparable theme between Chinese and

Contemporary philosophy. A new survey. Vol. 7, 207–222.

Western Philosophy was Confucian Ethics. Fung Yu-lan, the famous historian of Chinese Philosophy, has compared Confucius with Socrates. He maintains that these two great philosophers were comparable in the perspective of moral philosophy.[1] Of course, the Socratic verse "Knowledge is virtue" is episteme-oriented, whereas the Confucian text "virtue is knowledge" is more moral-oriented. This interchange between subject and predicate reveals the prevalence of virtue by the Confucian philosophy. In this sense the Taoist Philosophy is compatible with the Confucian thought. Chuang-tzu's verse, "There must be the authentic man, and then there is the authentic knowledge",[2] is an approval of this orientation.

In the theme of comparative philosophy there are innumerable topics to be discussed. Concerning the Taoist Philosophy as compared to European Philosophy, the topic of methodology should be raised first of all. Fundamentally speaking, methodology was properly a main concern pertaining to the western philosophy. From Descartes on, philosophers placed heavy stress on methodological research. The Chinese traditional scholars who uphold the Taoist Philosophy as their central work, on the other hand, maintain that the Taoist theory is oriented towards the sublimation of human nature. But this orientation would be valid only if the aspect of historical development of Taoism Philosophy is considered. In fact, Taoism has degenerated into superstitious belief and consequently into the faith in the fairy land with bodily life. I have published two articles, using modern methods to analyse the negative terms used in *Tao-te-ching*[3] on the one hand, and comparing the Husserlian and Taoist Philosophy[4] on the other hand. These two articles reveal the possibility of an Oriental-Occidental comparison on epistemology and of understanding the Taoist Philosophy.

Of course, in the earlier time of my dissertation entitled "Begriffsgeschichtlicher Vergleich zwischen Tao, Hodos und Logos bei Chuang-tzu, Parmenides und Heraklit"[5] I have shown the methodological identity and difference between Chinese and Western philosophy. Today I repeat the same theme in this article on Taoist Philosophy, hoping to attract more detailed discussions, with more profound points of view.

This article concentrates mainly on the system of Taoist Philosophy, attempting to lead to an eventual comparison with European Philosophy in the spheres of epistemology, metaphysics and ethics.

1.

Before we start to make the methodological comparison, let us look at the circumstances in which the Taoists developed their research method in order to build up their philosophical system. The rise and development of Taoism took place in Ancient China in the Spring and Autumn period (722–481 B. C.) and in the Period of the Warring States (480–222 B. C.). It was a time of degeneration in which the political and social order were declining. Lao-tzu (ca. 571–467 B. C.), the first Taoist Philosopher, and his followers Chuang-tzu (ca. 369–286 B. C.) and Lieh-tzu (ca. 500 B. C.), were alarmed by the gradual decline of the political authority of the

House of Chou (1122–249 B.C.). They were conscious of their mission of showing a salvific way in such a circumstance. In the philosophical contents of Lao-tzu's *Tao-te-ching* and of Chuang-tzu's and Lieh-tzu's writings, they took care to emphasise the socio-political order on the surface; however, they possessed a more profound existential understanding, assigning an important position to the individual person. On this point the Taoists are similar to the western existentialists who are conscious of individual existence in each circumstance, trying to make self-transcendence from the existential level to the ontological one. Both the Taoists and the existentialists can ask the same question of Being from the perspective of personal existential experience. Martin Heidegger is a typical example who seeks "Sein" through the analysis of "Da-sein". We will discuss this theme later.

Lao-tzu was born during the Spring and Autumn period. This period was one of degeneration, as Mencius (372–289 B.C.) said:

> "The world fell into decay, and principles faded away. Perverses speakings and oppressive deeds waxed rife again. There are instances of ministers who murdered their sovereigns, and of sons who murdered their fathers." [6]

The society fell into decadence. This disorder persecuted the life of the people, further provoking the sence of mission of the scholars. Philosophers in the East as well as in the West concerned themselves with the passions of the people and tried to build up the social order. Plato did so, Confucius did so.

The Confucian follower, Mencius, after describing the social decay, pushed forward the redemptive spirit of Confucius. He said:

"Confucius was afraid, and composed the 'Spring and Autumn'."[7] But, Confucius' attempt to revive traditional virtues, especially the "Li" (Ritus) through the "Rectification of Names",[8] had the side effects of encouraging hypocrisy and vanity. On this occassion Lao-tzu took his opposite stand. He said:

> "Not to value nor to employ men of superior ability is the way to keep the people from rivalry among themselves; not to prize articles which are difficult to procure is the way to keep them from becoming thieves; not to show them what is likely to excite their desires is the way to keep their minds from disorder."[9]

> "When the Great Tao ceased to be observed, benevolence and righteousness come into vogue. Then appeared wisdom and shrewdness, and there ensued great hypocrisy. When harmony no longer prevailed throughout the six kinships, filial sons found their manifestation; When the states and clans fell into disorder, loyal ministers appeared." [10]

In the logical order the Taoist thought has its origin after the Confucian. Thus from Lao-tzu on, through Chuang-tzu, until Leih-tzu, the "Via negativa" was founded to balance the "Via positiva" of the Confucian doctrine.

In Lao-tzu's *Tao-te-ching* we can clearly recognise a number of negative terms.

It is worthwhile to note that there are more than 500 negative terms in *Tao-te-ching*, a composition which is only a little more than 5000 words. I have explained this point entirely in my article, "The Negative Concepts Used in *Tao-te-ching*". [11]

Lao-tzu began from "Wu-wei" (Non-action) to oppose the Confucian "Yu-wei" (Action) and rested upon the ontological "wu" (Nothingness), which is supposed to be the substance of Tao.

Lao-tzu's epistemological "Wu-chi" (No-knowledge), ethical "Wu-Yu" (No-desire), socio-political "Wu-wei" (Non-action) and institutional "little state" and "small population" are the gradual negation, which introduces the ontological term "Wu" (Nothingness) as the ultimate aim of the philosophical process. In my article "A Comparative Study of Lao-tzu and Husserl: A Methodological Approach" I have used this point to compare Lao-tzu and Husserl, especially in the methodology. I have maintained that the process of Lao-tzu's term "Wu" should be analogical to Husserl's "Epoche". I said:

> "Lao-tzu used the metaphysical 'Wu' to support the physical 'Yu', and used the ethical 'Wu-wei' to sustain the moral 'Yu-wei'. If the ontological foundation be 'Wu', which is higher than all the phenomenal 'Yu', then, as a consequence, 'Wu-wei' is higher than 'Yu-wei' in moral praxis. .... Husserl's 'Epoche' in its negative form seems close to *Tao-te-ching*'s negative term. Husserl used his 'bracket' to avoid the epitemological doctrines of modern philosoply from Descrates to Kant, likewise, Lao-tzu put aside all the attempts to restore the decaying social order in the Chun-chiu (Spring and Autumn) period. The methodological 'bracket' here seems to have an identical meaning of seeking reality."[12]

We can hardly possess Lao-tzu's negative way of thinking clearly and distinctly, because of the very condensed content of *Tao-te-ching*. But there is at least one thing that is clear, namely, that Lao-tzu identifies the term "Wu" with the substance of Tao. He said:

> "Ten-thousand-things under heaven were born from *Yu*, and *Yu* was born from *Wu*." [13]

This cosmogenic expression evidently has a supra-ontological or meta-ontological function. "Ten-thousand-things under heaven" means the whole sensible world. The sensible world comes from "Yu", which is equivalent to Being in the western terminology. This is an ontological expression. But, for Lao-tzu "Yu" is not the ultimate principle. "Yu" was born from "wu", which literally means Nothingness. Of course the term "wu" does not exactly signify absolute nothingness, otherwise it would be excluded absolutely from existence. According to the Aristotelian-Thomistic philosophy this "Wu" should be a potency which can be actualised and can exist without any inner contradiction with existence; it has the potency to generate "Yu".

On the other hand, this meta-ontological understanding of Lao-tzu's philosophy did not create any epistemological function, but it leads man to recognise Tao. On this point Lao-tzu's follower, Chuang-tzu, made a clearer explanation.

The Warring State period in which Chuang-tzu lived was a more decadent and more disordered situation than the Spring and Autumn period in which Lao-tzu lived. Therefore, Chuang-tzu's critique of his time was radically stronger than Lao-tzu's. On the other hand, concerning the idea of philosophical development and progress, Chuang-tzu tried to be freed from mundane circumstances; he used the anthropological existential experience as a standpoint through which human nature would be elevated, step by step, toward unity with the Supreme Tao.

The anthropological method which Chuang-tzu used is different from Lao-tzu's epoche, which emphasises freedom from the external sensible world. Chuang-tzu's method relies mainly on the intrinsic self-negation and self-transcendence; his reflectional function was clarified in two methods, namely: "Heart fasting" and "Sitting and forgetting". His spiritual transcendence was revealed in the article "Hsiao-yao Yu" (Enjoyment in Untroubled Ease) in which the big bird Phang is depicted as flying in the highest heaven.

The modes of "Sitting and forgetting" was a follows:

> "Yen Hui said, 'I am making progress.'
> Confucius replied, 'What do you mean?'
> 'I have ceased to think of benevolence and righteousness,' was the reply.
> 'Very well; but that is not enough'
> Another day, Hui again saw Confucius, and said,
> 'I am making progress.'
> "What do you mean?'
> 'I have lost all thought of ceremonies and music.'
> 'Very well, but that is not enough.'
> A third day, Hui again saw (the Master), and said,
> 'I am making progress.'
> 'What do you mean?'
> 'I sit and forget about everything.'
> Confucius changed countenance, and said,
> 'What do you mean by saying that you sit and forget?'
> Yen Hui replied, 'My connexion with the body and its parts is dissolved; my perceptive organs are discarded. Thus leaving my material form, and bidding farewell to my knowledge, I become one with the Great Pervader. This I call sitting and forgetting all thing.'
>
> Confucius said, 'You are free from all likings; so transformed, you become impermanent. You have, indeed, become superior to me! I must ask leave to follow in your steps.'[14]

Chuang-tzu began with forgetting about surrounding things. Through forgetting about things of the word, he finally managed to forget about the thinking self. Through this act of forgetting Chuang-tzu reached the milieu of authentic knowledge and authentic self.

Chuang-tzu's method is really comparable to Husserl's. Their similarity lies in their using of the bracketing of *Um-welt-ding* and *Welt-ding* as a starting point for gaining true knowledge. Their difference should be in their attitude about the thinking subject: Husserl, following Descartes, emphasises the indubitable character of the subject; Chuang-tzu, however, maintains that in the authentic knowledge there must be a subjective unconsciousness. For Chuang-tzu, the real knowing subject is not the human mind, but the enduring Tao. The knowing subject has to forget about everything so as to let Tao think in him.

The Tao in itself and by itself is not knowable by human mind, as Lao-tzu declared:

> "The Tao that can be trodden is not the enduring Tao." [15]

So Chuang-tzu made his mind unconscious and let the Tao function fully.

This ascetic expression of knowing Tao was balanced with the active doing of the aphorismic transcendence. In the article of "Hsiao-yao Yu" Chuang-tzu used the big bird Phang as a symbol of the spirit of transcendence, always aiming at the zenith of spirituality, i.e., unity with the enduring Tao. This way of the transcendence of the spirit by Chuang-tzu is the same as the Parmenidean "hodos ano", through which Parmenides can face the mystery of Truth. Here the Parmenidean "visio veritatis" has the same epistemological meaning as Chuang-tzu's "unio mystica" with Tao.

The Parmenidean saying "Being is being, non-being is non-being"[16] identified the criterion of Truth with the principle of identity, and at the same time determined the way of falsity through the principle of non-contradiction. Chuang-tzu's philosophy would not like to remain on this epistemological level, but tried to transcend himself by flying upwards like the big Phang toward unity with Tao. This understanding of unity was as mysterious as the philosophy of Nikolaus Cusanus who emphasised the "coincidentia oppositorum" in his vision of "sub specie aeternitatis". It seems to me that Chuang-tzu's harmonious philosophy should be comparable to that of Cusanus.

Chuang-tzu's transcendental method becomes free from the epistemological differentiation of true and false, right and wrong. His thinking process belongs not so much to the epistemological investigation, but to the anthropological and even mystical and ontological. In Chuang-tzu's philosophy, the man who tries to know the way of transcending the mundane thing and himself is, of course, an epistemological man, like Husserl. But his transcending method is something more intrinsic, more ethical, more practical than a pure epistemological speculation. In other words, Chuang-tzu's method is not only epistemological but also anthropological.

Chuang-tzu tries to transcend himself in the existential experience. At the same time he transcends Lao-tzu's life-philosophy, and directly goes into the relation between Tao and himself, hoping to reach the milieu of the union with Tao. This is exactly the process from the antropological to the ontological sphere.

Only in this sense would Chuang-tzu's philosophical purpose of becoming an Authentic man, a Godlike-man, a Super-man and a Holy-man be fulfilled.

In the writings of Chuang-tzu, epistemology, anthropology, ontology, and other branches of human knowledge, like ethics or spirituality, are not distinctly separated. If we use the Cartesian postulate: clear and distinct ideation as the axiom of epistemology, Chuang-tzu's philosophy would certainly not be easy to classify. But if we let philosophical wisdom be freely functional, if we assent that the perfection of mankind can also be one of the main purposes of philosophy, then his positive contribution, especially his existential feeling, are highlighted:

> "Heaven, Earth, and I were generate together, and Ten-thousand-things and I are one."[17]

This feeling is cosmological and ontological. The man who possesses this feeling is a perfect man, an authentic man who has mastered the true knowledge. For Chuang-tzu, knowledge is the effect of the authentic man who is united with Tao.

In the study of the development of the Taoist epistemology there is still one thing to be usefully observed; that is, the rhetoric style. Lao-tzu used monologue, whereas Chuang-tzu frequently employed dialogue. In Lao-tzu's *Tao-te-ching*, monologue was expressed not only in its form but also in its contents. In *Tao-te-ching*, a composition containing approximately 5000 words, there is not a single personal pronoun "you" or "he" to be put forward as an object for conversation. Lao-tzu said 27 times "I", the first personal pronoun, or "the Saint" to express his heavy stress on subjectivity. He maintained that his subjective knowledge should be the only true knowledge for recognising the enduring Tao. Apparently this transcendental subjectivity, which has no relation to other persons nor to other thing, has the same meaning as the Husserlian "pure Ego" (*reines Ich*), which is free from all empirical affairs.

The rhetoric style which Chuang-tzu frequently used was dialogue through which true knowledge would be found, e.g. the dialogue between "great" and "small" in his aphorism to distinguish "authentic knowledge" from "pseudo-knowledge". Finally he concluded that the anthropological and ascetical "authentic man" should be the ultimate standard of "authentic knowledge". Of course, this "authentic knowledge" is not the intuition of an individually attained wisdom, but the creation of collectively shared wisdom which is produced from profound thinking and discussion. Chuang-tzu made more progress than Lao-tzu in epistemology.

The Taoist epistemological method has its development from Lao-tzu's monologue, via Chuang-tzu's allegoric dialogue, to Pao-pu-tzu's (Go Hung, 253–333 A. D., who seems to belong to the fourth generation of Taoism) dialectical discussion, especially the chapter Tzie-pao (Investigate Pao Chin-yen) which manifests the total operation of the Taoist method. The content of the chapter Tzie-pao is oriented to the controversy between monarchy and anarchy; it it a reflection on and a total investigation of the socio-political philosophy between Confucianism and Taoism.[18] Pao-pu-tzu's dialectic method seems to be something like Plato's dialogue. The author of this Taoist work finally lets the theory of monarchism triumph over anarchism. Following the Chinese traditional Confucian orientation, the process of discussion, however, contains a very profound logical construction comparable to that of western philosophy.

2.

As was mentioned above, in the discussion on epistemology, Chuang-tzu's dialogical method was in some ways profounder than the method of Lao-tzu. But Lao-tzu's self-perversion on the "Ego" may overcome Chuang-tzu's idea of self-forgetting so as to construct the Transcendental Ego. The construction and the conservation of the Transcendental Ego is a guarantee for Taoist philosophy to transit from anthropology to ontology. Lao-tzu's "Ego", which exists without any relation to "you", to "him", manifests the autonomy of his own personality. Lao-tzu put the self-consciousness in place of the unknowable Tao. As was mentioned, the relation between Ego and Tao was not clear by Lao-tzu's philosophy, but the Ego's intelligibility, consciousness and conceivability were strongly maintained by him. This obstinacy promotes a transcendental effort through which the knowing subject makes his way upwards from the epistemological to the ontological sphere through his consciousness. Lao-tzu said: "The Tao that can be trodden is not the enduring Tao."[19] This sentence clearly maintains the inconceivable character of Tao. But it does not conclude that the Tao might not exist. Contrarily, it shows the endurance of Tao. While the enduring Tao in itself was unknowable, Lao-tzu introduced his own Tao as follows:

> "My Tao is very easy to know,
> and very easy to practice."[20]

But the people do not find it easy to know or to practice. Thus,

> "The great Tao is very level and easy;
> but people love the byways."[21] And
> "There is no one in the world, who is able to know and to practice it."[22]

In other words, the objective Tao is unknowable and inconceivable. Lao-tzu's subjective Tao, however, is knowable and can be practiced. There should be a coincidence between epistemology and ontology in the sphere of subjectivity.

Lao-tzu's autonomy and authenticity of subjectivity allowed all the relationship to be sacrificed. Therefore, in Lao-tzu's socio-ethical philosophy, the social constitution would be changed into the "little state and small population",[23] and the interpersonal relationship would also be changed into the "people who live to old age, even to death, having no intercourse with each other".[24]

This suspension and bracketing of all relationships has the sole purpose of manifesting the authority of subjectivity. Such a philosophical orientation is also the central theme of western philosophy from the 16th century on. Descartes' "Cogito" was seen as the starting point and the arche of all genuine sciences. This mode of thinking by Rationalism has its proper development. For Husserl, the phenomenologist alone can change the thinking subject to subjectivity; he solemnly declared that the "pure Ego" would be the beginning of all knowledge, and at the same time would be its ontological foundation.

Of course, in the cognitive order there are some differences between Lao-tzu and Husserl: the former has his intuition towards the enduring Tao, whereas the latter's is towards the essence of things. Lao-tzu's follower Chuang-tzu could unify the cognitive subject with Tao, and in this union he arranged ontology to support anthropology for the fulfillment of Lao-tzu's philosophy. Husserl's followers, however, can neither unify the subject with the essence of things, nor arrange the theory of "All-einheit" to fulfill the phenomenological ultimate purpose. Moreover, Husserl's phenomenology shows a way from subjectivity to intersubjectivity, and finally in the theory of Merleau-Ponty the corporeal interrelation became interpersonal relationship to explain all the characters of "Mit-sein". [25]

Lao-tzu and his follower, however, from the beginning to the end could never accept the corporeal world. Lao-tau's words reveal this attitude. He said:

> "Great calamity is to be regarded as body. What makes me liable to great calamity is my having the body; if I had not the body, what great calamity could come to me?" [26]

This attitude forms the Taoist tradition up to the third generation. As its representative, Lieh-tzu tried to break all corporeal bondage by going into the fairy-land in which the human being transmigrates as an immortal fairy.

Although the Taoist substance is limited by the body, the ideal Tao would be purely spiritual and incorporeal. Its ideal Tao is indeed in the world, but it does not belong to the world. According to Chuang-tzu's thought, although the "Tao in ten-thousand-things" signifies "Immensitas", much more than "Ubiquitas", the Tao which operates in the material world is not material. Contrarily, Tao is a saviour who redeems the material world from its matter, and lets the material world participate in the spirituality of Tao. The Taoist idea of union between the Tao and the world is not substantial, i.e. Tao-and-world-in-one is not consubstantiative, but only "Tao in the world". The Taoist philosophy has never accepted the theory of "Tao becomes the world". In western philosophy, however, Logos becomes flesh. This idea of incarnation was accepted and developed by Christian faith, and by many contemporary philosophers like Whitehead, Marcel, Merleau-Ponty, etc.

The genuine fundamental character of Chinese philosophy has no such vexation of the mind-body dualism. In *I-ching* (the Book of Change), the prototype of Chinese philosophy, the Yin and Yang, are not opposite to each other, but complement each other. Taoist philosophy, though it had its origin in I-ching, yet declined the pneuma-centricism, and became absent-minded in the participation of body in spiritual existence. It is worthwhile to show that the superstitious faith in fairies in the Wei-chin Dynasties (233–420 A.D.) and the way of leaving one's home in the Hinayana Buddhism in the Sui-Tang Dynasties (589–90 A.D.) could flourish as folk religions in China.

Since the metaphysical question on the substance of Tao became epistemological, Lao-tzu's answer was simply negative. Tao is not the object of knowledge, and our epistemological potentiality can never reach the milieu of Tao. From this negation of the objective Tao there arises a subjectivity which, through the act of negation, can create in itself and by itself a relation with the enduring Tao which is above the subject-object opposition. It is only in the personal subjective consciousness that one can imagine the ultimate principle of heaven and earth and ten-thousand-things. Although the substance of Tao in itself is not knowable, one can describe it in some way. It is obscure, and its genuine face has never been revealed to our knowledge. Yet one can describe it as follow:

> "Who can of Tao the nature tell?
> our sight it flies, our touch as well.
> Eluding sight, eluding touch.
> The forms of things all in it crouch;
> Eluding touch, eluding sight.
> There are their semblances, all right.
> Profound it is, dark and obscure;
> Things' essences all there endure."[27]

For the philosophers who habituate the western thinking mode, the most difficult thing lies in the fact that the impossibility of understanding of Tao in epistemology becomes mysteriously understandable in ontology, or in meta-ontology. (Lao-tzu identified Tao and Nothingness.) This explication of Tao in *Tao-te-ching* has its foundation in the epistemological "Wu-chi" (no-knowledge), "Wu-chio" (no-sensation), coming from "Wu" (nothingness) of experience. But Lao-tzu can finally find a path from knowledge to Tao; in other words, Lao-tzu creates a road from "Wu-chi" and "Wu-chio" to "Wu". He said:

> "Ten-thousand-things under heaven were born from Yu (Being), and Yu was born from Wu (Nothingness)."[28]

"Wu" is therefore the archetype of all things, but it is identified with Tao. Lao-tzu said:

> "Tao generates one, one generates two, two generates three, three generates ten-thousand-things."[29]

Tao is the generator of all things. Wu is also the generator of all things. Lao-tzu identified Tao with Wu. Lao-tzu's cosmogony clarified the term "generation" to coincide the Tao and Wu. This fact reveals that what is inconceivable in epistemology becomes conceivable in ontology. In other words, the term Wu signifies real Being. For Lao-tzu, Yu and Wu (Being and Nothingness) become a paradox. It is the high-point of the taoist sophistication which cannot be verified in western logic.

3.

Now we reach the essential point. The term "Wu" has a profound meaning in the practical philosophy towards which the Taoist philosophy is oriented. The term

"Wu" would combine with "Wu-wei" (non-action) in the socio-political life. "Wu-wei" means literarily non-action. But, essentially it means "doing everything through doing nothing." "Wu" literarily means "nothingness", but essentially it means "being itself". Lao-tzu said:

> "(The saint) diminishes himself and again diminishes himself, until he arrives at doing nothing (Wu-wei).
> Having arrived at this point of non-action (Wu-wei), there is nothing which he does not do."[30]

"Wu-wei" is therefore not non-action. In social affairs, it is non-human or non-artificial action in which one lets Tao function freely. Thus the essence of Wu-wei lies in the Wu-ju (non-concupiscence) which means that one is not limited by any intentional purpose; one is not tempted by mundane honour and merit. The un-intentionality and the un-consciousness are the real essence of Wu-wei. From this point of view, the characteristics of Wu-wei can be both "non-action" and "acting upon everything through non-action".

Thus the activity of Wu reveals a practical orientation. Its ethical inclination is evident. Loa-tzu's Wu has a essential difference in its significance in Ethics from Sartre's "le Néant" which has a more psychological meaning than an ontological one. Sartre said:

> "The being by which nothingness comes into the world must be its own nothingness."[31]

> "Man is the being through whom nothingness comes into the world."[32]

Of course, Sartre has hoped to construct, through the psychological feeling of nothingness, an ontological nothingness in spite of traditional term "Being".

The Taoist theory of "wu-wei", realised in the socio-political affairs, was just opposite to the Confucian Yu-wei (action). This opposition has endured for several centuries. Because of Tung Chung-shu's (179–140 B.C.) extreme adoration of Confucius, the Chinese socio-political praxis was oriented towards Confucian "Yu-wei" from the Han Dynasties (206 B.C.–220 A.D.) on, without any interruption. For personal cultivation, the emperors as well as the ministers or even the simple people take the Taoist "Wu-wei" for granted. "Wu-wei" here means being in accordance with natural harmonies and avoiding evils. To be in harmony with nature and to avoid evil, one must cultivate oneself as rigorously as the Confucianists who practise "Yu-wei". The essential difference between these two theories lies mainly in this: the Confucianists concern themselves with interpersonal relationships, whereas, the Taoists concentrate on personal cultivation.

Therefore the practical usage of Wu signifies the negation of mundane affairs, but it does not mean the avoid once of the spiritual milieu. This fact manifests itself by the development of the Taoist philosophy from Lao-tzu to Chuang-tzu. It begins with the epistemological intuition of Tao in Lao-tzu's theory; it culminates in

Chuang-tzu's "Hsiao-yao Yu" (Enjoyment in Untroubled Ease) in which the big bird Phang is described as being freed from all material bondage, transcending to the "height of the celestial", finally achieving the "Enjoyment with the Creator". Just like Thomé Fang said:

> "The true meaning, I think, should be made out in the light of the philosophy of Infinity under discussion by following up the clues indicated by Chuang-tzu himself in the context of their close relevance.... The perfected and perfect man is now what he is in virtue of his identifiability with the 'Creator' imparting his potency to all the world without becoming the centre of fame, the contriver of plans, the director of works, and the claimer to knowledge. He embraces infinity with levity and freedom. He fulfills all that is natural in him without a sense of gain. In the spirit of vacuity, he employs his mind and heart like a mirror, impartially reflecting all that there is in the world without showing a trace of lure, dejection, or injury."[33]

Evidently this is a profound description of Chuang-tzu's transcendental way of thinking. His longing for the Creator was similar to the Augustinian seeking God. St. Augustine explains his longing with the following words: "and our heart is restless, until it repose in Thee."[34] There is of course a difference between Augustine and Chuang-tzu. The former was seeking God with his heart of penitence; the latter, however, longs for the Creator through admiring nature and enjoying the sense of beauty. From the perspective of the way of transcendence, we may say that the spiritual contents of the Parmenidean way of truth, the Augustinian way of confessions, and of Chuang-tzu's way of longing, have the same meaning, analogically.

The status of an authentic man, which Chuang-tzu has attained through his way of longing, is thus freed from all mundane bondage. The way of longing becomes possible only through the fasting of the heart and the process of sitting-for-getting in which the existence of self would be bracketed. If the Husserlian method of epoche is fulfilled by intentionality, then Chuang-tzu's fasting of heart would be attained through ecstasis, which is identified with un-intentionality, unconsciousness. Husserl's way, starting from consciousness, through intentionality, finally rests on the awakening of subjectivity. Chuang-tzu's way, however, starting from the fasting of the heart, through unconsciousness, finally rests on the attaining of the status of an authentic man. Chuang-tzu's authentic man creates an authentic ego. Both egos are the grounds of ontology as well as of ethics.

Chuang-tzu's ontology was based on his anthropology, whose theory of man has been clarified both in the "Hsiao-yao Yu" (Enjoyment in Untroubled Ease) and in the "Ta Zung Shih" (The Great and Most Honoured Master). The former offered three kinds of man, namely the perfect man, the spirit-like man, and the sagely-minded man;[35] the latter presented the "authentic man" as the ultimate and highest grade of humanity. The "Hsiao-yao Yu" highlighted the big bird Phang as a symbol of a mystic who transcends himself through transformation, taking the way upwards to engender a vital force aiming at freedom, breaking up all the restrictions and limitations until absolute freedom is achieved and union with Tao is completed.

> "The perfect man has no self, The spirit-like man, none of merit, The sagely-minded man, none of fame."[36]

"Ta Zung Shih" made use of the authentic man to explain the process of the manifestation of Tao and to show how the Tao is involved in human affairs. While Lao-tzu manifests Tao in the natural operation, Chuang-tzu lets Tao operate in the authentic man; and, through the observation of the authentic man, Tao would be manifested. His opinion is like that of Heidegger, who tries to uncover the countenance of Sein through Da-sein.[37] Heidegger's method promotes the human being from the anthropological level to the ontological one.

Therefore Chuang-tzu's authentic man is a synthesis of nature and Tao. This synthesis has the characteristic of transcendence. Chuang-tzu said:

> "The authentic man of old could pass through water without being made wet by it; he could go into fire without being burnt; so it was by his knowledge that he ascended to and reached the Tao."[38]

Here the authentic man seems to be conscious of ascension by his knowledge. Nevertheless Chung-tzu build up a concept of "Ego" and transmigrated it into the "authentic ego". In the Taoist philosophy the concept "Ego" is founded first by Lao-tzu's "solus ipse", without any personal relationship with "you" or "he", as was previously mentioned. This "solus ipse" which is realised in the concrete society would be practised as an isolated function. That is rather different from Heidegger's understanding of his Da-sein, which seems to be "Mit-sein" or "das man" being fully involved in interpersonal relations.[39]

But, according to its height and depth, the Taoist authentic ego, especially Chuang-tzu's authentic man, was really comparable to Nietzsche's Übermensch (Superman).

The authentic man is a transcendental ego. Chuang-tzu said:

> "The authentic man of old knew nothing of the love of life or of the hatred of death; the exit from it provoked no resistance. Composedly he went and came."[40]

The spirit of "considering life and death as having neither beginning nor end"[41] made Chuang-tzu's thought far from Heidegger's, who has adopted the tradition of Christianity and maintains that death should happen only once, and this unique opportunity makes man go into psychological anxiety and fear. [42]

The Taoist philosophy had no psychological anxiety nor fear like Heidegger's. Its mode of thinking was not something like the Hegelian dialectique. But the Taoists made use of the consciousness and experience of self-existence to develop the authentic ego in which the enduring Tao appears in its excellence in this sensible world. Thus, when the Taoist described the authentic ego, his description is similar to the way which Husserl understands his pure ego in his descriptive phenomenology.

The concept of "ego" which appears 27 times in *Tao-te-ching* undoubtedly has the meaning of subjectivity and autonomy. But the structure of Lao-tzu's ego is under the corporeal bound, besides its having the difficulty of intuitting the substance of Tao. Hence Lao-tzu tried to bracket the "Um-welt-ding" first, then interpersonal relations, and finally the bodily existence of his own ego.

The content of the concept of ego by Chuang-tzu should be much broader, and much easier to classify. Professor Thomé Fang has his division. He said:

> "The self means either the essential nature of the spontaneous spirit, characterized as the great function of reason, or the eternal Presence of Mind, identifiable with infinite Tao itself. The former is called by Chuang-tzu the centre of mind (Lin Tai) or the seat of the spirit (Lin Fu) .... The latter is called either the Absolute Mind as the supreme disposer of all things in knowledge (Chin-Tzai, Chin-Chun) or the eternal Spirit ultimately discerned by the Mind among all minds (Chang-Sin)."[43]

It is true, according to Chuang-tzu's philosophy, that there are five different egos, namely the corporeal ego, the psychic ego, the psycho-mechanic ego, the spiritual ego and the authentic ego. In fact, the former three are pseudo-egos, the latter two, however, are indeed real egos.

Thus all of Chuang-tzu's effort lies in the removal of the pseudo egos and the construction of the real egos. Lieh-tzu's effort was a further step towards getting rid of all the limits of the mundane world and sensible affairs. He then tried to go into the sphere of the fairy — the pure spiritual sphere.

From the above discussion we find clearly that the Taoist philosophy is indeed oriented towards praxis-philosophy, just like the Confucian philosophy. But its stress lies only in personal perfection of the authentic man without any positive socio-political concern. Furthermore, the taoists tried to avoid all interpersonal relationship, and consequently denied the constitutional monarchy of that time and yearned for the age of anarchism.

Concerning the path leading to the acceptance and further development of Buddhism in China, Taoist philosophy should be a great, meritorious medium.

0.

In conclusion, the Taoists have selected the enduring Tao as their philosophical object. In epistemology, however, they have denied the possibility of knowing the objective Tao. As methodological research, both Lao-tzu and Chuang-tzu, the primordial Taoists, were taking the *via negativa* to explain the substance of the enduring Tao; they held that Tao is unknowable, unintelligible and inconductible. Finally they maintained that Tao should be called "Wu" (nothingness). From the result of the epistemological *via negativa* Taoism built up its ontology. Tao is "Wu". This is more an ontological than epistemological statement.

Although Tao is "Wu", and its function is "Wu-wei" (non-action), the operation of Tao goes on without interruption. Since "Ten-thousand-things under heaven are born from Yu (Being), and Yu is born from Wu (Nothingness)," [44] Tao generates all things without any exception. In the metaphysical sense, therefore, Tao is "Wu". But it is dynamic, with infinite potentialities to create all things. This dynamic function of Tao pushes human beings in the way upwards to the acme of perfection — to be united with Tao. The authentic man practises Wu-wei in his mundane life, following the example of Tao.

"Tao" is the central concept of Taoism. Tao's substance — "Wu" is an effort of Taoist epistemological *via negativa*, and at the same time is the name of the ontological Being. Tao's operation — "Wu-wei" is the example and the normative principle for human life in the sensible world.

"Cheng-jen" (the authentic man), who is unified with Tao, is the final purpose of Taoist philosophy oriented towards life-praxis.

Tao, which appears in nature and in the authentic man, would indeed be unknowable directly in the human self and by the human self. However, it can be described through the *via negativa* and can be mainfested through human cultivative operations, like fasting of the heart, sitting forgetting etc.

The question of how far and how deeply one can compare the Taoist type of philosophy to European philosophy remains open and needs further discussion.

NOTES

1. Cf. Fung Yu-lan, *A History of Chinese Philosophy* (Hong Kong: Pacific Book Co., July 1959), pp. 77–78.
2. *Chuang-tzu*, Book VI Tai Zung Shih, trans. by James Legge, *The Writings of Chuang-tzu*, VI, I.
3. Kun-yu Woo, *The Negative Concept Used in Tao-te-ching*. Proceedings of the International Conferences on Sinology, Academia Sinica, Taipei, 15–17 August 1980, pp. 791–800.
4. Kun-yu Woo, *A Comparative Study of Lao-tzu and Husserl: A Methodological Approach*. In Analecta Husserliana, ed. by A.T. Tymieniecka (Dordrecht: D. Reidel Publishing Company, 1984), Vol. XVII, pp. 65–73.
5. Peter Woo K.Y., *Begriffsgeschichtlicher Vergleich zwischen Tao, Hodos und Logos bei Chuang-tzu, Parmenides und Heraklit*, Universitas, Series IV (Taipei, April 1969).
6. James Legge, *The Works of Mencius*, Book III, Pt. II, Ch. 9, Vers. 7 (Oxford 1892), p. 281.
7. *Ibid.*
8. *Confucian Analects*, Book XIII Tzu-lu, Ch. III, Vers. 2.
9. *Tao-te-ching*, Ch. III, Vers. 1.
10. *Ibid.* Ch. XVIII, Vers. 1,2.
11. See note 3.
12. See note 4, p. 67.
13. *Tao-te-ching*, Ch. 40, Vers. 20.
14. *The Writings of Chuang-tzu*, Book VI, Ta Zung Shih (The Great and Most Honoured Master), 14, trans. by James Legge, pp. 304–305.
15. *Tao-te-ching*, Ch. I Vers. 1.
16. Diels-Kranz; *Die Fragmente der Vorsokratiker*, Bd. I (Dublin and Zürich: Weidmann, 1966; 12. Aufl.), 28 B. 6. Bd.I.

17. *The Writings of Chuang-tzu*, Book II, "Chi Wu Lun" (The Adjustment of Controversies, 6).
18. Peter Woo, *Reflections on Taoist Thought During the Wei-chin Period*. Paper presented at the Fifth International Congress of Chinese Philosophy, 12–17 July 1987, San Diego, Calif., USA.
19. See note 15.
20. *Tao-te-ching*, Ch. 70, Vers.1.
21. *Ibid.* Ch. 53, Vers. 2.
22. *Ibid.* Ch. 70, Vers. 1.
23. *Ibid.* Ch. 80, Vers. 1.
24. *Ibid.* Vers. 5.
25. Nien-feng Chiang, Whitehead and Merleau-Ponty: Two Critiques of Cartesianism. *Tunghai Journal* 28 (Taichung, June 1987): 187–200.
26. *Tao-te-ching*, Ch. 13, Vers. 2.
27. *Ibid.* Ch. 21.
28. *Ibid.* Ch. 40, Vers. 3.
29. *Ibid.* Ch. 42, Vers. 1.
30. *Ibid.* Ch. 48, Vers. 2.
31. Jean-Paul Sartre, *L'Etre et le Néant*. *Essai d'ontologie phénomènologique* (Paris, 1943). (Translated by H. Barnes as *Being and Nothingness* (New York, 1956; London, 1957), p. 59; Engl. p. 23.
32. *Ibid.* p. 60, Engl. p. 24.
33. Thomè H. Fang, *Chinese Philosophy: Its Spirit and Its Development* (Taipei: Linking Publishing Co. Ltd., 1981), pp. 131–132.
34. St. Augustinus, *Confessiones*, I, 1.
35. James Legge, *The Writings of Chuang-tzu*, p. 217.
36. *Ibid.*
37. Martin Heidegger, *Sein und Zeit*, p. 63.
38. James Legge, *The Writings of Chuang-tzu*, pp. 285–286.
39. Martin Heidegger, *Sein und Zeit*, p. 149 ff.
40. James Legge, *The Writings of Chuang-tzu*, p. 286.
41. *Ibid.* p. 668.
42. Martin Heidegger, *Sein und Zeit*, p. 260 ff.
43. See note 33, p. 140.
44. *Tao-te-ching*, Ch. 42, Vers. 1.

Korean Confucianism reconsidered

YOUN SA-SOON
Korea University

1. Introduction

The purpose of this paper is to examine the philosophical trend of Korean Confucianism in the modern period. As the history of Korean Confucianism is as long as that of Chinese Confucianism, the content and the characteristics of Korean Confucianism are very rich and variegated. But an examination of the whole history of Korean Confucianism is beyond the scope of this paper.

As our concern here is limited to the "modern period", it seems necessary to specify, first of all, the span of this period. Generally, in the history of philosophy, we call the 20th century the modern age. But in the history of Korean Confucianism, it is reasonable to denote by 'modern period' the span from 15 August, 1945 to the present, for the reason given below.

From 1910 to 1945, Korea was ruled by the Japanese Imperialist, who tried to distort the history of Korean thought. During this period Korean thought came to lose its uniqueness, and could not be considered to be the object of profound study. Moreover, during this period some Koreans regarded Confucianism as the cause of national ruin, for the reason that it had been the governing ideology of the Yi Dynasty, which was responsible for the fall of the nation. This was really a dark age in the history of Korean Confucianism. Therefore, the present writer regards the year of 1945, i.e. the year of Korean liberation from the Japanese rule, as the beginning of the modern period.

There is another point to be mentioned in our methodological consideration. This is the problem concerning the interpretation of "the philosophical aspect" of Korean Confucianism in the modern period. First of all, we mean by the philosophical aspect of Confucianism the influence of Occidental philosophy on Confucianism,[1] or the approach to Confucian knowledge and methodology from the standpoint of Occidental philosophy. Some scholars try to study Confucianism in order to complement the shortcomings of Occidental philosophy; others try to introduce its merits to Confucianism. Such attempts can be said to be common practice among Korean Confucian scholars in the modern period, for most of them have already gained a considerable understanding of Occidental philosophy through their university education.

But we do not mean by the philosophical aspect of Korean Confucianism simply the characteristic resulting from the influence of Occidental philosophy. Rather, we

Contemporary philosophy. A new survey. Vol. 7, 223–238.

mean by it the very characteristic of Confucianism that is distinguished from its religious characteristic. As Confucianism is a branch of philosophy, it naturally follows that it should have some inherent philosophical aspect. This aspect is the object of study that attracts many Confucian scholars of Korea these days.

2. The beginning of the modern interpretation of Korean Confucianism

The first scholar in the modern period who attempted to systematise the history of Korean Confucianism was Professor Hyŏn Sang-yun 玄相允 (1892–?), whose study was crystallized in his book, *The History of Korean Confucianism*[2] published in 1949.

His predecessors' approach to the history of Korean Confucianism had only been to confirm the relationship between teacher and disciple, or between friends.[3] But his was a fairly objective and critical approach: unlike his predecessors, he tried to understand the history of Korean Confucianism on the basis of a philosophical point of view. His characteristic approach is presented in his book, *The History of Korean Confucianism.*

According to him, the influence of Korean Confucianism on Korean culture can be examined in two ways, i.e. with respect to its merits and its demerits. With respect to its demerits, he thought that Confucianism is responsible for the following trends or events: Koreans' general condescension to China, the clan or family first policy, partisan spirit, the effeminacy of men of letters, contempt for the laboring class, industrial decline thereby, and the final surrender to the Japanese Imperialist. On the other hand, with respect to its merits, he thought that Confucianism inspired Koreans to practice ethical and virtuous behaviour, and to pursue profound philosophical thought, such as Neo-Confucianism.

So he thought, like some of his contemporaries, that Confucianism was the cause of national ruin. However, as is suggested by his question, "Is it [national ruin] due to Confucianism itself, or contemporary Koreans?" he attributed the fundamental cause of the ruin not to Confucianism itself but to its abuse by some Koreans.

Professor Hyŏn, by warning his contemporaries against misunderstanding of and disbelief in Confucianism, intended to re-establish Confucian value—though his intention cannot be said to be so positive. It is in this context that he laid emphasis on the philosophical aspect of Korean Confucianism. According to him, Korean Confucianism is characterised first of all by its philosophical aspect, although it also retains some characteristics of politics and literature. He meant by the philosophical aspect the theory of Neo-Confucianism, which originated from Ch'eng-Chu's

doctrine. He said that a lot of Confucian scholars in the Yi Dynasty (from the beginning of the 15th century to the early period of the 20th century) left many academic achievements on the basis of the Neo-Confucian concept of *li-ch'i* 理氣 (principle-material force). He saw that representative Neo-Confucian theories in the Yi Dynasty are implied in the following debates: the debate on "the interpretation of Four Beginnings and Seven Feelings 四端七情 "between Yi Hwang 李滉 (pen name, Toegye 退溪; 1501–1570) and Yi Yi 李珥 (pen name, Yulgok 栗谷 ;1536

–1584); the debate on "the identity or the difference between man's nature and thing's nature" between Han Wŏn-chin 韓元震 (pen name, Namdang 南塘; 1682–1751) and Yi Kan 李柬 (pen name, Woiam 巍巖; 1677–1727); the debate on "the interpretation, on the basis of *li* or *ch'i*, of the characteristics of the universe." Im Sŏng-chu 任聖周 (pen name, Nokmun 鹿門; 1711–1788) is a representative scholar who interpreted the characteristics on the basis of *ch'i*, while Ki Chŏng-chin 奇正鎭 (pen name, Nosa 蘆沙; 1798–1879) is a representative scholar who interpreted them on the basis of *li*.

In fact, the above debates, which began respectively from the 16th, 18th and 19th century, and continued until the 20th century, attracted numerous scholars of the Yi Dynasty. Therefore, Professor Hyŏn might well say that these debates imply the representative Neo-Confucian theories in the Yi Dynasty. However, it cannot be denied that he did not carry out any in-depth research into these debates.[4] Nevertheless, his identification and brief introduction of important theories may be said to have been a pioneering contribution to the modern interpretation of Korean Confucianism.

3. Revaluation of Korean Confucianism

The present writer (尹絲淳; 1936–) is one of the scholars who take aggressive attitudes toward the revaluation of Korean Confucianism. First of all, he denies that Confucianism was the cause of national ruin, for it was also the governing ideology of Japan, the invading country. The decisive cause for the fall of the Yi Dynasty, he maintains, is not Confucianism itself, but the abuses of the governing class. He ascribes it to the inferiority complex of the subjugated people to think that Confucianism was the cause of national ruin, thus useless, and worth nothing.[5] Moreover, he emphasises that such a self-abusing prejudice was instigated by some Japanese scholars, for they faked, as a means of justifying Japanese colonialism, the myth that Korea has not had her own thought, including Confucianism.[6] Therefore, according to him, the revaluation of Korean Confucianism corresponds to the correction of such a prejudice or myth as well as the objective understanding of Confucian theories.

In his attempt to evaluate Korean Confucianism, the present writer emphasises, most of all, the independence or the uniqueness of Korean Confucianism.[7] According to him, this uniqueness refers not only to philosophical theories developed or originated by Korean scholars, but also to the creative application of Chinese theories. For instance, though loyalty and filial piety have been the general Confucian concepts since Primitive Confucianism, the concept of loyalty was propagated in Korea for the safeguard and prosperity of national life[8]; and the concept of filial piety was used in Korea as a means of establishing social order and, thereby, of enhancing the self-respect of the nation.[9] In short, these concepts were the principles by which Koreans sustained their independent and self-regulating lives. In this connection, the present writer points out, Confucian theories, regardless of their origin, may be classified as Korean thought or philosophy as occasion demands.[10]

In order to evaluate Korean Confucianism objectively, the present writer also suggests his unique view on the history of Confucianism, including that of Korean Confucianism. In Korea there were few, if any, systematic views on the history of Confucianism. The most prevalent one, at best, was the view of the history of Korean Confucianism as a static development, which was fabricated by some Japanese scholars for the purpose of justifying Japanese colonialism. This view is similar to that of Western Orientalists in that they deny the concept of development or improvement in Asian history. Against such a view, the present writer suggests his own for the objective understanding of the history of Korean Confucianism.[11]

He thinks that Confucianism retains many characteristics of agricultural life, for it originated from an agricultural community. As agricultural life is characterised first of all by its adaptation to and harmonisation with nature, the concept of time or history in an agricultural society comes to be founded on the seasonal order of nature. In other words, the concept of time or history is based on seasonal change. As the change of the four seasons is a kind of circular movement, an agricultural society comes to have a concept of time or history based on circular movement. For this reason, many concepts of Confucianism are also based on the circular movement of nature, as is evidenced by many examples in *I-ching* 易經. However, unlike linear movement, circular movement is not subject to ceaseless advance and development. So long as emphasis is laid on the analogy between nature and history based on circular movement, many historical concepts of an agricultural society, including the concept of Confucian history, should be interpreted with respect to the view of history as static movement.

However, the present writer thinks that Confucianism, as well as agricultural life, pursue not only the harmonisation with nature and circular seasonal change, but also the increase of production and the prosperity of the nation. As can be seen by the teaching of Confucius, Confucianism, more than any other branch of thought or philosophy, has strong characteristics of this-worldliness and practicality. So the present writer affirms that such characteristics can and must be interpreted with regard to the view of history as improvement and development. However, as Confucian concepts based on circular movement cannot be discarded completely, he says, it may be reasonable to consider that the Confucian view of history is based not on ever-advancing linear movement, but on gradual and spiral development.

He emphasises that the history of Korean Confucianism agrees with the view of history discussed above. According to him, Korean Confucianism, as time has flowed from ancient through middle to modern period, came to overcome its religious and dogmatic characteristics, and develop into a full-fledged philosophy, which is focused not on the lives of the aristocratic and governing class but of the middle and governed class.

4. Revaluation of *Silhak*

The academic achievements of the Confucian circle since the 1960s have enabled us to evaluate some trends of Confucianism from quite a different angle from that of the Confucian circle before the 1950s. One such trend is *Silhak* 實學 (practical learning). Though we cannot, at this point, make any final evaluation of new approaches to *Silhak*, which are still actively going on, the achievements up to now show clearly that the attitudes are sharply different from those before the 1950s.

Professor Hyŏn Sang-yun represents the past attitudes. Professor Hyŏn saw that some scholars in the late period of the Yi Dynasty tried, with their outstanding practical sense, to realise their ideas such as "contribution of learning to society 經世致用," "provision for the welfare of the people 利用厚生," and "pursuit of truth on the basis of objective fact 實事求是." And he thought that *Silhak* is concerned with ideas of political, economic, and social reform. He even called the ideas of those scholars adhering to *Sihak* a "new politics."

But the new attitude since the 1960s regards *Silhak* as a kind of philosophy with which to overcome and complement the shortcomings of Neo-Confucian theories. Such a view is implied in various works such as "Chŏng Yak-yong's 丁若鏞 Study of Chinese Classics" by Professor Yi Ŭl-ho 李乙浩[12], "Ch'oi Han-ki's 崔漢綺 Empiricism" by Professor Pak Chong-hong 朴鍾鴻[13], and Professor Youn Sa-soon's 尹絲淳 "A Study on Pak Se-dang's 朴世堂 *Silhak*,[14]" "*Silhak*'s Characteristic Interpretation of Chinese Classics,"[15] "Silhak's Philosophical Aspects"[16] and "The Philosophical Basis of Silhak."[17]

The above works show that the most important factor for the understanding of *Silhak*'s characteristics is not the term *Silhak* itself, for it was not the term used for the first time in the late period of the Yi Dynasty. Neo-Confucian scholars also attempted to explain, in terms of *Silhak*, concepts of Neo-Confucianism, or even of Primitive Confucianism. Thus the term *Silhak* may be called "Practical Confucianism". Therefore, what matters is to determine which system of thought or philosophy is opposed to *Silhak*. In other words, when a branch of Confucianism of a certain period is called *Silhak*, what matters is to determine which branches of learning or philosophy can be considered to be "vacant", or "impractical". In fact, Neo-Confucian scholars called their learning *Silhak*, assuming the philosophy of Lao-tzu and Chung-tze and Buddhism to be impractical. But some scholars even thought that Neo-Confucianism (Ch'eng-Chu's Doctrine) is impractical, and, in an attempt to pursue more practical learning, called their own *Silhak*.

In the late period of the Yi Dynasty (from the early period of the 17th century to the late period of the 19th century), *Silhak* was advocated by such scholars as Yi Su-kwang 李睟光 (pen name, Chipong 芝峯; 1563–1628), Yu Hyŏng-wŏn 柳馨遠(Pankye 磻溪; 1622–1673), Pak Se-dang 朴世堂 (Sŏkye 西溪; 1629–1703), Yi ik 李瀷 (Sŏngho 星湖; 1681–1763), An Chŏng-pok 安鼎福 (Sunam 順庵; 1712–1791), Hong Tae-yong 洪大容 (Tamhŏn 湛軒; 1731–1783), Pak Chi-wŏn. 朴趾源 (Yŏnam 燕巖; 1737–1805), Pak Chae-ka 朴齊家 (Ch'ochŏng 楚亭; 1750-?), Chŏng Yak-yong 丁若鏞 (Tasan 茶山; 1762–1836), Kim Chŏng-hi 金正喜 (Wandang 阮堂; 1786–1856), Ch'oi Han-ki 崔漢綺 (Hyekang 惠岡; 1803–1879). All these scholars assumed common attitude, which may be summarised below:

a) They pursue extensive and practical learning so that their field of interest includes not only Neo-Confucianism but also astronomy, geography, mathematics, agriculture, and tactics.

b) As is shown by ideas such as "From below to upward," and "From the close to the distant," they greatly value the practical characteristics of Primitive Confucianism.

c) Their learning is based on a positivistic methodology of seeking after classical evidence.

The above attitude may be interpreted as a new methodology of those scholars who, in their pursuit of practical learning, tried to revitalise the depressed and stereotyped atmosphere of the Neo-Confucian circle of their own time. *Silhak* was nothing other than the crystallisation of such a pursuit.

Of course, there were slight differences of opinions among scholars adhering to *Silhak*. Though most of them laid primary emphasis on *ch'i*,[18] several of them did not do so.[19] But the main tendency of their philosophy, by contrast with that of Neo-Confucianism, may be summarised as follows.[20]

4.1 *The unique nominalism*

While Neo-Confucian scholars attached more importance to appearance, title, and form than utility, efficiency, and content, scholars advocating *Silhak*, such as Pak Se-dang and Hong Tae-yong, did the opposite. Ch'oi Han-ki is an outstanding example of this attitude. He said, "As a name denotes an actuality or a thing, an actuality is the cause of a name, but not vice versa." The attitude may be said to represent the unique nominalism of *Silhak* which is opposed to Neo-Confucian realism. On the basis of such an attitude, scholars advocating *Silhak* criticised the characteristic realism of Neo-Confucianism.

4.2 *The empirical epistemology*

Neo-Confucian epistemology is generally based on the nature of *li-ch'i*, which constitutes man's mind; not on feeling arising from the five sensory organs. But the epistemology of those adhering to *Silhak* stands at the opposite pole. For instance, Chŏng Yak-yong thought that man's perception is an automatic process starting from the sensory organs. Ch'oi Han-ki maintained that man's perception corresponds to man's reasoning which is induced by feeling. Therefore, it may be said that the epistemology of *Silhak* contains a kind of empirical theory based on the nominal way of thinking.

4.3 *The dualistic view on man and thing*

Scholars adhering to *Silhak* deny the Neo-Confucian view of identifying man with thing. Pak Se-dang denied this view by pointing out that the universe or nature is not influenced by man's mind but by *ch'i*. Chŏng Yak-yong affirmed that man and nature are distinguished from each other not only by "kind" but also by "the mode of existence". That is, he saw that nature exists by necessity, while man exists by autonomy. To be brief, by nature he meant physical nature. Scholars such as Pak Che-ka and Pak Chi-wŏn, on the basis of such a dualistic view, regarded nature as the object of utilisation.

4.4 *The view of man as the subject of desire*

As scholars adhering to *Silhak* had greater confidence in man's autonomous ability than Neo-Confucian scholars, the former showed a stronger tendency to view man as the subject of free will than the latter. Moreover, they generally believed in the purity of feeling, and did not hesitate to view man as the subject of desire. They even considered sexual desire to be natural, and rejected the ascetic attitude of emphasising the cultivation of mind.

4.5 *The view of ethics and morality as changeable*

Neo-Confucianism regarded morality as the inevitable manifestation of human nature, and attached absolute value to it. On the other hand, *Silhak*, as is shown by the thought of Chŏng Yak-yong, regarded human nature as a kind of propensity, and denied the view of ethics and morality as the inevitable manifestation of human nature. According to Chŏng, benevolence and righteousness do not refer to essential human nature; they are only the names for corresponding patterns of behaviour. So he thought that ethics and morality are subject to change in accordance with man's necessity.

4.6 *The view on politics and society*

In the Yi Dynasty, the "Three Bonds and Five Moral Relations" 三綱五倫 were not merely the items of ethical and moral obligations but also the ideas which were popularised for the maintenance of the governing system. So it would not be far-fetched to say that *Silhak*'s view of ethics and morality as changeable is in line with its view of the governing system as variable. In fact, scholars adhering to *Silhak* emphasised Mencius' motto, "Down with despotism", and advocated reformation by the people and for the people. In this connection, they insisted on land reform and the abolition of slavery. Of course, such an insistence was intended for the extention of civil or human rights.

As has been examined above, *Silhak* is similar to modern empirical philosophy. If *Silhak* had been applied positively to every field of contemporary society, the modernisation of Korea would have been accelerated and would have proceeded in a uniquely Korean way. Unfortunately, however, the process of Korean modernisation bears many faults and shortcomings. In order to guide Korean modernisation into a desirable direction, it is high time that Korean *Silhak* was caused to be born again.

5. The View of the Universe and Human Nature

Korean Neo-Confucianism has enjoyed a long history of some 700 years, from the late period of the 12th century to the late period of the 19th century. It was not only the governing ideology of the Yi Dynasty but also the main stream of Korean Confucianism. The metaphysical theories of Neo-Confucianism in the Yi Dynasty, especially, are the main object of study of modern Confucian scholars in Korea. As a matter of fact, the enthusiasm for the study of Neo-Confucian metaphysics exceeds that of *Silhak*.

Sometimes, a personal view on the universe provides modern Neo-Confucian scholars with an attractive object of study. A paper, entitled, "Yi Hwang's View on the Formation of the Universe" by the present writer is a good example. Yi Hwang regarded *li* as the basis of the rise and fall of *ch'i* by which every phenomenon occurs, and *T'ai-chi* 太極 (the Great Ultimate), which is a kind of *li*, as the ultimate source of the universe. Basically, he agreed with the idea that the mobility of *T'ai-chi* is the cause of the creation of the primordial *ch'i* of yin-yang 陰陽, the collection and dispersion of which is in turn the cause of the formation of *wu-hsing* 五行 (the five elements) and all things, including man. In order to justify his idea, he said, "When *li* moves, *ch'i* originates accordingly". And then, in order to justify the mobility of *li*, he defined *li* in terms of *t'i-yung* 體用 (substance and function), "*Li*, with respect to *t'i*, does not move, but, with respect to *yung*, moves".[21]

Yi Hwang's view above implies problems as to whether *li* exists independently, and whether *li* is mobile. The present writer pointed out that a problem is also implied in the idea that "*Li* originates *ch'i*."[22] What matters here especially is the interpretation of the verb "to originate." That is, if "originate" is interpreted as a transitive verb, a problem arises as to how a heterogeneous being (*li*) bears another heterogeneous being (*ch'i*), for, in general, the phenomenon of generation refers to the successive reproduction of homogeneous species. On the other hand, if "originate" is interpreted as an intransitive verb, it means that *ch'i* comes into being of itself, and implies an antinomy of approving an effect without a cause.

So the present writer explains that the idea "*Li* originates *ch'i*" implies a self-contradiction.[23] But he does not consider this idea useless. Rather, he thinks that there may be a hidden meaning — concerning value judgment — in this idea. Concretely speaking, he sees that *li* refers not merely to the natural law of necessity but also to the principle by which to judge good and evil. Likewise, *T'ai-chi*, according to him, is not only the principle of the formation of the universe but also the ultimate standard of the good towards which the universe moves.[24] A detailed study of value judgment is presented in the present writer's article, "Yi Hwang's Identification of Be and Ought" to be introduced later.

Theories on the formation of the universe were studied by some Neo-Confucian scholars in the Yi Dynasty. But some philosophical topics attracted the whole Neo-Confucian circle in the Yi Dynasty. As Professor Hyŏn introduces in his book, *The History of Korean Confucianism*, "the interpretation of Four Beginnings and Seven Feelings on the basis of *li-ch'i*", "the identity or the difference between man's nature and thing's nature" are representative examples of such topics. The former was studied and debated for about 350 years from the middle period of the 16th century to the late period of the 19th century; the argument on the latter continued for some 200 years from the early period of the 18th century to the late period of the 19th century. These topics continued to be an overwhelming concern of the Neo-Confucian circle for such a long period that they not only brought about the formation of schools—*Chulipa* 主理派 and *Chukipa* 主氣派; *Hopa* 湖派 and Nakpa 洛派—but also deepened the study of Neo-Confucianism. In fact, it may be said that Korean Neo-Confucianism is characterised by these topics. Even these days, they are the main research subjects of Neo-Confucian scholars.

A series of arguments on "the interpretation of Four Beginnings and Seven Feelings on the basis of *li-ch'i*" began from the debate between Yi Hwang and Ki Tae-sŭng 奇大升 (pen name, Kobong 高峯; 1527–1572). An outline of this debate may enable us to grasp the main point and characteristics of the arguments. Once, one of Yi Hwang's acquaintances (Chŏng Chi-un 鄭之雲) said, "Four Beginnings issue from *li*, and Seven Feelings issue from *ch'i*". Yi Hwang amended this, and said, "Four Beginnings are the issuances of *li*, and Seven Feelings are the issuances of *ch'i*" This is Yi Hwang's first interpretation of Four-Seven. Ki Tae-sung opposed Yi Hwang's interpretation for the following reasons: as Seven Feelings, which represent man's whole gamut of feelings, include Four Beginnings, which represent some of man's good feelings, Four and Seven cannot be contrasted with each other; as all feelings originate from the combination of *li* and *ch'i*, the interpretation of Four-Seven with respect to either *li* or *ch'i* is wrong; it is reasonable to apply the term "issuance", which denote a kind of movement, to *ch'i*, while it is not to apply the term to *li*. Against this refutation, Yi Hwang advocated his position as follows: as Four Beginnings are different from Seven Feelings in that the former refers to good feelings, and the latter refers to those feelings which are not subject to the judgment of good and evil, it is no unreasonable to contrast the former with the latter; as the characteristics of *li* and *ch'i* correspond respectively to those of Four Beginnings and Seven Feelings, it is reasonable to match those which have the same characteristic, i.e., *li* with Four and *ch'i* with Seven; the term "issuance" can be applied to both Four Beginnings and Seven Feelings, for the former refers to the natural issuances of the *li* of mind, while the latter refers to the issuances caused by the touch of an outer object. Nevertheless, Ki Tae-sung tenaciously reputed Yi Hwang's interpretation. So Yi Hwang modified his original interpretation of Four-Seven, and said, "Four Beginnings are the issuances of *li* followed by *ch'i*, while Seven Feelings are the issuances of *ch'i* upon which *li* rides". His modified interpretation suggests that the main causes of arousing Four and Seven are respectively *li* and *ch'i*, but that Four and Seven can be explained not in terms of either *li* or *ch'i* but of both *li* and *ch'i*. Besides, in order to justify the mobility of *li*, he not only applied, as he did when he theorized his view on the formation of the universe, *t'i*-

yung to the explanation of *li* but also insisted on "the simultaneous issuances of *li* and *ch'i*."

The present writer analyses the above argument as follows.[25] Originally, Four Beginnings refer to good feelings, as was classified by Mencius, while Seven Feelings refer to general human feelings. Therefore, it may be said that the former refers to abstract and ideal feelings, while the latter refers to concrete and actual feelings. So it may be said that the former is concerned with the idea of value, while the latter with fact. The different interpretations of Four-Seven result from whether one lays emphasis on the aspect of good and evil of Four-Seven or on the origin of Four-Seven. Therefore, the great problem implied in the interpretation of Four-Seven arises from the attitude of regarding the aspect of ideal value and of concrete fact on the same level. In other words, the identification of ideal value and concrete fact is the reason why the argument around the interpretation of Four-Seven could not easily be settled.

The biggest problem implied in the interpretation of Four-Seven on the basis of *li-ch'i* was how to justify the "issuance of *li*." It may be said that Yi Hwang had some intention when he tried, by all means, to justify the mobility of *li*. Considering that Yi Hwang accepted Ch'eng-Chu's 程朱 idea, "*hsing-chi-li* 性即理 (human nature is no other than the principle)", we may say that he believed in the spontaneous manifestations of Four Beginnings — i.e. benevolence, righteousness, propriety, and wisdom — and that he intended to emphasise the spontaneity of manifestations. We may say so the more because, traditionally, Confucian morality is believed to be realised by the spontaneous manifestations of human nature. Once, Yi Hwang said, "If human feeling were related only with *ch'i*, having nothing to do with *li*, man would be blinded by avarice, and no better than a beast". Therefore, it may be said that his interpretation of Four Beginnings with respect to the issuance of *li* implies his view of human nature as fundamentally good. To put the matter the other way round, he intended to protect man from depravity by awakening him to the manifestation of essential human nature.

His view of human nature as fundamentally good is also implicated in his interpretation of Seven Feelings. If four Beginnings refer to good, Seven Feelings, which are contrasted with Four Beginnings, should refer to evil. But he assumed that Seven Feelings are not subject to good and evil. At first Yi Hwang said, "The good and evil of Seven Feelings cannot be determined". Later on he said, "Seven Feelings are essentially good, but easily liable to evil". Considering that he regarded Seven Feelings, which are contrasted with Four Beginnings, as essentially good, we can conclude that he did not approve of the existence of evil at all. Therefore, it may safely be affirmed that his interpretation of Seven Feelings as "essentially good" was intended for the consolidation of his view of human nature as inherently good.

Traditionally, Confucianism regards man's inborn nature itself as good. This tendency is evidenced by a saying of *I-ching* 易經, "The formative process of physical nature corresponds to the good", and by Ch'eng-Chu's definition of the universe as "the system of one *li* called *T'ai-chi*". Thus it is true that Yi Hwang's view of human nature as fundamentally good agrees with the traditional thought of Confucianism.

This view has an idealistic bias which has to do with the contemporary social situation. It was such a turbulent period that political antagonists killed one another. Yi Hwang himself, as a government official, tried, even until his later years, to realise a political ideal, as is evidenced by some of his writings. So it may be said that his view of human natures was intended for the cure of contemporary social ills. Similarly, Mencius' teaching and the thought implied in *I-ching* mentioned above may be examined with regard to the social situations at the time when Mencius lived, and *I-ching* began to be formed. Supported by such classical evidence, the present writer thinks that Confucian views of human nature as inherently good are all based on an idealisation, i.e. the ideal attitude of reforming social ills and abuses. He regards Yi Hwang's view of human nature as a typical example of Confucian idealism.[26]

Debates on "the identity or the difference between man's nature and thing's nature" began in the early period of the 18th century among the disciples of Kwŏn Sang-ha 權尚夏 (pen name, Suam 遂菴 ;1641–1721), especially between Yi Kan 李柬(pen name, Oiam 巍巖; 1677–1727) and Han Wŏn-chin 韓元震 (pen name, Namdang 南塘; 1682–1751).[27] Yi Kan represents those who insisted on the identity, while Han Wŏn-chin represents those who advocated the difference.

The present writer has pointed out in a study that they themselves knew that the difference of their thoughts resulted from the difference of their viewpoints. That is to say, they knew that a transcendental and general viewpoint would result in their identity, while a concrete and specific viewpoint would result in their difference.[28]

Both Yi and Han agreed with each other on points as follows: man, thing, and mind originate from "the combination of *li* and *ch'i*"; both *T'ai-chi* and *T'ien-ming* 天命 (the Heavenly Order) refer to *li*, and "Human nature is no other than *li*"; human nature is divided into *pen-jan chih-shing* 本然之性 (original nature) and *ch'i-chih chih-hsing* 氣質之性 (material nature); the concrete manifestations of *pen-jan chih-shing* are benevolence, righteousness, propriety, and wisdom. But they differed in the interpretation of *hsing* 性 (nature), especially *pen-jan chih-hsing*. Yi Kan thought that *pen-jan chih-hsing* is the *li* of *T'ai-chi* and *T'ien-ming*, while Han Wŏn-chin asserted that *pen-jan chih-hsing* refers to the *li* of mind, i.e., benevolence, righteousness, propriety, and wisdom. Thus the former, interpreting human nature with a transcendental point of view, maintained the identity between man's nature and thing's nature, while the latter adhered to the opposite position.

Their different positions were respectively supported by classical evidence. According to *Chung-yung* 中庸, all things, including man, are born with *hsing* which originates from the *li* of *T'ien-ming*. This view supports the identity of man's nature and thing's nature. On the other hand, Mencius thought that *hsing* is manifested by benevolence, righteousness, propriety, and wisdom, which distinguish man from the other things. This view supports the difference between man's nature and thing's nature.[29]

What is worthy of our special attention here is that the present writer ascribes one of the motives of the debate to the incompatible two meanings implied in the term, "*pen-jan chih-shing*". On the one hand, the term implies "essence". With respect to this meaning, it may be concluded that man's nature is different from thing's nature. On the other hand, it implies transcendental and ultimate source. With respect to this meaning it may be said that, ultimately, man's nature is identical with thing's nature. The classical evidence of *Chung-yung* and Mencius, introduced above, proves that *pen-jan chih-hsing* implies such incompatible meanings. The present writer thinks that this debate would continue endlessly, unless an eclectic interpretation of the imcompatible meanings were to be adopted.

He also thinks that, though this debate has not come to a conclusion for so long a period, it cannot be said to be good for nothing. He draws special attention to remarks such as "Not only *wu-ch'ang* 五常 (five constants) but also *T'ai-chi* and *T'ien-ming* are precious and good" by Yi Kan, and "Man's nature is more precious than thing's nature" by Han Wŏn-chin. He explains that these remarks are intended for the enhancement of original nature and, thereby, the dignity of the human being.[30] It may be concluded that this debate is of great significance, like the argument on the interpretation of Four-Seven on the basis of the issuance of *li*.

6. Revaluation of Confucian ethics and its ground

After the Korean War (1950–1953), waged by the ideological confrontation between north and south, Korean philosophers came to be greatly interested in humanism. Influenced by this trend, the Neo-Confucian circle of Korea tried to illuminate Confucianism with an humanitarian point of view. Professor Yi Sang-ŭn's 李相殷 (1905–1976) paper, "Confucian Thought with regard to Humanism", published in 1961, is a representative example of such an attempt.[31]

He thinks that, fundamentally, Confucianism is based on Confucius' thought, and that the core of his thought is the concept of "benevolence". He emphasises that the recovery of the spirit of benevolence is required in this age more than anything else. This spirit, according to him, is the ground of humanitarianism, for the Chinese character "benevolence", 仁, symbolises two persons, and means, as Confucius himself defined, "love of others". He concludes that the recovery of the spirit of benevolence as the source of love would enable us not only to correct the loveless human relation based on egoistic individualism but also to surmount the problem of alienation stemming form modern civilisation marked by mechanisation and mass production.

The present writer tries to interpret the concept of benevolence from an ethical point of view. As is well known, the ethical virtues of Confucianism consist of the five constants 五常 of benevolence, righteousness, propriety, wisdom, and trustworthiness 仁義禮智信, which are the indispensable elements of the ethical and moral system of five human relations 五倫. Accordingly, Mencius emphasised "benevolence, righteousness, propriety, and wisdom" as the human nature of Four Beginnings. But Confucius' terms of ethical virtue include only wisdom and benevolence, not propriety. This seems unreasonable, when we consider that Confucius attached importance to propriety more than anyone else, and that he is said to have pigeonholed the customs and practices of propriety. The present writer attributes this seemingly incongruous attitude of Confucius to the implication of the term "benevolence". That is to say, he sees that Confucius' term "benevolence" implies propriety as well.[32] Thus, the present writer tries to interpret the concept of propriety implied in the term "benevolence".

He thinks that Confucius had already suggested the way of realising benevolence in terms of "magnanimity", by saying "Don't make others do what I do not want to", or "Treat others as if they were my own self". These remarks imply Confucius' lesson that one should respect others and practice altruistic deeds before one can love others. So, it may be said that the practice of benevolence is to take care of others with generosity and modesty. And this practice corresponds to the realisation of propriety which is based on love. This is the very reason why Confucius defined only benevolence as "love of others" and did not think it necessary to enumerate propriety as one of the ethical virtues. Similarly, Mencius said that benevolence is realised through compassion, and propriety through modesty or concession. The present writer interprets Mencius' remark as the hint of accepting the synonymous meaning of propriety and benevolence.[33]

The above attitude of laying emphasis on benevolence reflects an attempt to internalise the emotionally biased ethics of Primitive Confucianism. Basically, however, in order that Confucian ethics should be tenable even today, the validity of its theoretical basis must be presented. The present writer's paper, "Toegye's Identification of Be and Ought" was written for this purpose.[34] The outline of this paper is introduced as follows.

Since ancient times, Confucianism has considered that ethics and morality are nothing other than the manifestations of human nature. For instance, a sentence of *Chung-yung* 中庸 reads, "*Tao* 道 is to follow human nature". Neo-Confucianism also maintained such a view of ethics. Concretely speaking, Neo-Confucian scholars thought that the manifestations of human nature, called five constants (benevolence, righteousness, propriety, wisdom, and trustworthiness), correspond to the five cardinal principles (intimacy, loyalty, distinction, priority, and faith) which should be observed in the five relations of father and son, sovereign and subject, man and wife, the young and the old, and friends. To take another example of such a view of ethics, there is an old Confucian belief that if one lives according to *tao-li* 道理 (the principle of the way), one comes to correspond with *T'ien-tao* 天道 (the way of Heaven), to contribute to the formation of physical nature. Such a view of ethics is expressed, in terms of Western philosophy, as "the identification of be and ought", or "the identification of fact and value" ; and in terms of Neo-Confucianism, as "the identification of *so-i-jan* 所以然 (cause, or be) and *so-tang-jan* 所當然 (normative law, or ought)". As Neo-Confucianism is the most profound philosophy of Confucianism, only when the validity of the identification of *so-i-jan* and *so-tang-jan* can be presented could Confucian ethics be admitted, even in this age.

There was an argument in Yi Hwang's time on whether *so-i-jan* and *so-tang-jan* refer to the same *li* or not. Scholars such as Ki Tae-sŭng 奇大升 (pen name, Kobong 高峯; 1527–1572) asserted that they are different, for *so-i-jan* refers to *li*, while *so-tang-jan* refers to *shih* 事 (affair). But Yi Hwang thought that ultimately, they are so identical that they may be considered to refer to one and the same *li*. His thought represents Neo-Confucian theories on the identification of *so-i-jan* and *so-tang-jan*.

First of all, supported by Mencius' view that Four Beginnings as the human nature of benevolence, righteousness, propriety, and wisdom are not manifested by force, or that the manifestations of Four Beginnings such as pity and shame arise of itself, Yi Hwang took five relations 五倫 as the evidence of the spontaneous manifestations of human nature. For example, if one sees a child going toward a well, one would naturally protect the child from falling into the well. This, according to him, is due to the spontaneous manifestation of the human nature, benevolence.

However, beside the human nature of benevolence, righteousness, propriety, and wisdom, there is the human nature of instinctive and selfish desire, such as appetite and lust. Such a desire, if left alone, would most likely instigate evil deeds. In this respect, Neo-Confucianism distinguishes *pen-jan-chih-hsing*, which is not subject to *ch'i-chih* 氣質 (physical basis), from *ch'i-chih-chih-hsing* which is subject to *ch'i-chih*, and thinks that the evil originates from *ch'i-chih* or *ch'i-chih-chih hsing*. Then it may be said that the existence of *ch'i-chih-chih-hsing* is itself the evidence of the difference of *so-i-jan* and *so-tang-jan*. In other words, it may be said that in the case of *pen-jan-chih-hsing*, *so-i-jan* and *so-tang-jan* agree with each other, but that in the case of *ch'i-chih-chih-hsing*, they don't.

Thus, Yi Hwang had to prove that man has control-especially, the spontaneous control-of *ch'i-chih-chih-hsing* or selfish desire. He said that the evil is occasioned by the function of *ch'i* which is not in accordance with the *li* of *so-tang-jan*. And he admitted that frequently occurring selfish desire is the cause of the evil. Nevertheless, he believed that such a desire can be controlled by the concentrated consciousness of *ching* 敬 (seriousness or reverence). But here arises a problem in that one should make and "artificial" attempt in order to maintain the state of *ching*. But we may suppose here that Yi Hwang seems to have kept in mind not the artificiality but the autonomous will of the attempt. In other words, he seems to have considered that man's rational and autonomous will is itself the natural state, and that, regardless of *ch'i-chih-chih-hsing*, the identification of *so-i-jan* and *so-tang-jan* can hold true. In fact, he believed that, ultimately, *li* is far superior to *ch'i* in that only the former conforms to the rational and autonomous principle.

In retrospect, however, Neo-Confucian ethics does not remain within the scope of normative theory, for, basically, it seeks for the practical realization of human nature as is implied in the view that *tao-li* accords with *T'ien-tao*, and that this accordance contributes to the formation of physical nature. It may be affirmed that such a view presupposes a cosmological or ontological metaphysics in its background. Therefore, Neo-Confucian ethics should be examined and justified, even on cosmological and ontological bases.

Actually, Yi Hwang applied both *so-i-jan* and *so-tang-jan* to things such as ship or wheels, as well as man. For example, a ship "is made to sail on the water (*so-i-jan*)", and so, it "should sail on the water (*so-tang-jan*)". So, if a ship sails on the water, it means that the identity of the *so-tang-jan* and *so-i-jan* of the ship is realised. On the other hand, if a ship is driven ashore because of, say, a storm, it means that the ship is against its *so-tang-jan*, though it still retains its *so-i-jan*. This is the case when an evil phenomenon occurs because of the discordance of *so-i-jan* and *so-tang-jan*. In view of the example of a ship, Yi Hwang says, "The evil occurs when *li* does not realise its *so-tang-jan* owing to the hindrance of *ch'i*'s function".

He ascribes the state of "anti-*so-tang-jan*", in which a ship is driven ashore, not to the inferiority of *li* to *ch'i*, but to the outer condition called *shih* 勢. Therefore, the identity of the *so-i-jan* and *so-tang-jan* of a certain thing should be understood with regard to its relation with other things, and further, with the whole universe. Then what matters now is to discover the ground on which the identification of *so-i-jan* and *so-tang-jan* can be extended to the scope of the universe.

It can be said that the law of *so-i-jan* is applicable when a thing is formed necessarily (naturally), while the law of *so-tang-jan* is applicable when a thing is formed teleologically. Therefore, in order to meet the two laws at the same time, all the things in the universe should be formed necessarily (naturally) and, at the same time, teleologically. This is the essential condition on which the identification of *so-i-jan* and *so-tang-jan* can be maintained.

In *I-ching*, in which Yi Hwang has a firm belief, the characteristic of the universe is defined as life, and all the changes of things as "endless formation". And the book explains the cause of these changes in terms of *yüan heng li chen* 元亨利貞 (origination, flourish, advantage, and firmness). The notion implied in such a view of change suggests that fundamentally, as is evidenced by *I-ching*, Confucianism regards the universe as an organism (living entity). Although an organism seems to change in accordance with necessity, it also changes in accordance with some purpose. For instance, the root, branches, leaves, and blossoms of a tree are closely interrelated. There are many instances that Yi Hwang, on the basis of *I-ching*, regarded the universe as an organism. Furthermore, he agreed with the notion that the universe is "a system of *T'ai-chi* (*li*) 理一分殊", in which *so-i-jan* is identified with *so-tang-jan*.

Seen in this respect, it may be said that Yi Hwang's view of morality as the natural manifestation of life and his identification of *so-i-jan* and *so-tang-jan* are based on his belief in man's rational and autonomous will and on his theory of the universe as a harmonious organism. However, it cannot be said that Yi Hwang's thought discussed above is impeccable. The weakest point is his notion that evil is a kind of indispensable "necessary evil" which cannot easily be corrected by man's positive will. Moreover, considering that the formation of the universe and its purpose are beyond man's knowledge, we cannot easily understand the basis on which Yi Hwang extended his ethical view to the scope of the universe. Nevertheless, his firm belief in man's ability to reason and the necessity of its manifestation, together with his intention of maintaining harmonious relation with nature, should be held in high esteem in the context of modern life.

NOTES

1. Youn Sa-soon, "The Conceptual Basis of Korean Confucianism," *A Study on Modern History of Korea*, ed. the Korean Society of History (Ulyumunhwasa, 1986), pp. 41–52.

2. Hyŏn Sang-yun, *The History of Korean Confucianism* (Hyonmunsa, 1982).
3. Chang Chi-yŏn, *The Source of Korean Confucian Religion* (1922).
4. Hyŏn Sang-yun, *op. cit.*
5. Youn Sa-soon, "Some Problems of Korean Confucianism," *A Study on Korean Confucianism* (hyonamsa, 1980), pp. 8–25.
6. *Ibid.*
7. *Ibid.*
8. *Ibid.*
9. Youn Sa-soon, "The Conceptual Basis of Korean Confucianism," *op. cit.*
10. See note 1, and Youn Sa-soon, "Traditional Ethical Views of Koreans," *Studies on Spiritual Tradition*, Vol. 22 (The Academy of Korean Studies, 1984), pp. 3–11.
11. Youn Sa-soon, "Some Problems of Korean Confucianism," *op. cit.*
12. Yi Ŭi-ho, "Chong Yak-yong's Study of Chinese Classics," *The Journal of Korean Studies*, Vol. 6 (Yonsei University Institute of Korean Studies, 1963).
13. Pak Chong-hong, "Ch'oi Han-ki's Empiricism," *The Journal of Asiatic Studies*, Vol. 20 (Korea University Asiatic Research Center, 1965).
14. Youn Sa-soon, *A Study on Korean Confucianism*, pp. 188-284.
15. Youn Sa-soon, *ibid.*, pp. 168–187.
16. Youn Sa-soon, *A study on Korean Confucianism*, pp. 311–339.
17. Youn Sa-soon, "The Philosophical Basis of Silhak," *Study of Korean History of Thought*, Vol. 1, ed. The Association of the Study on Korean History of Thought (Sasayon, 1987), pp. 115–140.
18. *Ibid.*
19. *Ibid.*
20. *Ibid.*
21. Youn Sa-soon, "Toegye's View on the Formation of the Universe," *A study on Toegye's Philosophy* (Korea University Press, 1980), pp. 40–72.
22. *Ibid.*
23. *Ibid.*
24. Youn Sa-soon, "Toegye's Theory of Value," *ibid.*, pp. 134–219.
25. Youn Sa-soon, "Toegye's Theory of Mind and Heart," *ibid.*, pp. 73–133.
26. *Ibid.*
27. Youn Sa-soon, "The Korean Controversy over Chu Hsi's View on the Nature of Man and Things," *Chu Hsi and Neo-Confucianism*, ed. Wing-tsit Chan (University of Hawaii Press, 1986), pp. 570–583.
28. *Ibid.*
29. *Ibid.*
30. *Ibid.*
31. Yi Sang-ŭn, "Confucian Thought with Regard to Humanism," *The Journal of Asiatic Studies*, Vol. IV–2 (Korea University Asiatic Research Center, 1961).
32. Youn Sa-soon, "Confucian Ethics Reconsidered: The Concept of Humanity in Confucian Ethics," *International Symposium on Confucianism and the Modern World Proceedings* (Taipei 1988), pp. 707–723.
33. *Ibid.*
34. Youn Sa-soon, "Toegy's Identification of Be and Ought: Toegye's Theory of Value," *The Rise of Neo-Confucianism in Korea*, ed. Wm. Theodore de Bary and JaHyun Kim Haboush (Columbia University Press, 1985), pp. 223–242.

Taoismus in Korea

SANG-RAK NAM
Sung Kyun Kwan Universität, Seoul

1. Einführung: Entstehung und Entwicklung des Taoismus in China

Die europäische Philosophie begann mit den griechischen Naturphilosophen, die versuchten, durch die Frage nach der *Arche*, dem Ursprung und Prinzip aller Dinge, den Kosmos zu begreifen. In China hingegen begann die Philosophie mit Konfuzius,[1] dem Begründer der konfuzianischen Schule und Laotse,[2] dem Urvater der Taoisten. Bei Konfuzius und Laotse bildete das Tao, als Ursprung des Kosmos oder Wesen aller Dinge in der Metaphysik und als Wahrheitskriterium oder Richtschnur des menschlichen Denkens und Handelns in der Ethik den zentralen Begriff ihrer Philosophie. Beide Philosophen, wie auch die nach ihnen, in der Zeit der 'Frühling- und Herbst-Periode' und der 'Kämpfenden Reiche[3] vor der Ch'in Dynastie (221–206 v. Chr.) aufgetretenen sogenannten 'Vielen Meister und Hundert Schulen'[4] versuchten, das Wesen des Tao zu erkennen und dann nach den Prinzipien bzw. im Rahmen des Tao zu leben. Nur unterschied sich der Taobegriff, der die chinesische Philosophie erst möglich machte, je nach Meister und Schule.

Die Philosophie Laotse's und der Taoismus sind, wie der Name besagt, aus dem Taobegriff entstanden. Der erste Satz aus Laotse's *Tao Te King*, das als berühmtes, aber schwer verständliches Werk zur Weltliteratur gehört, beginnt mit folgendem Zweizeiler:

> "Könnten wir weisen den Weg, es wäre kein ewiger Weg, könnten wir nennen den Namen, es wäre kein ewiger Name ..."[5]

Wir begegnen in diesem kurzen chinesischen Satz des ersten Verses dem Wort 'Tao' dreimal. Dabei ist jedes 'Tao' in einem anderen Sinn benutzt.

Man kann an der Beziehung zwischen dem ersten und dem zweiten 'Tao' leicht erkennen, dass die Bedeutung des 'Tao' sich ständig ändert. Dieses ständig sich verändernde Tao soll oder muss sich auf den ewigen Weg beziehen, oder dieser sogar sein, sonst würde das Tao seine Bedeutung verlieren. Was ist somit ein ewiger Weg? Das chinesische Wort Ch'ang (   ) des Ch'ang Tao bedeutet eigentlich 'immer' oder 'unveränderlich', d.h. es meint den ewigen Weg in dem obengenannten Satz, den 'unveränderlich bleibenden'.

Der Taobegriff Laotse's diente dazu, die Naturordnung oder den Ursprung des

Contemporary philosophy. A new survey. Vol. 7, 239–251.

Kosmos zu bestimmen. Die Welt der Erscheinung ist ständig in Veränderung, wobei die Gesetze dieser Veränderung immer gleichbleibend sind. Nach Einblick in die vielschichtige Natur des Tao soll der Mensch das Kriterium oder die Richtschnur seines Handelns in der Ethik und Politik auf das Tao ausrichten.

Der Weg als Grund des Seins aller erscheinenden Welt, einschliesslich der Menschen, läuft nach eigenen Gesetzen. Das nennt Laotse 'Tzu-jan' ('selbst so, selbstbildend, das was aus sich selbst ist',[6] umgangssprachlich bedeutet es: Natur).

In Kapitel 25 des *Tao Te King* heisst es:

"Der Mensch nimmt zum Gesetz die Erde;
Die Erde zum Gesetz den Himmel
Der Himmel zum Gesetz den Weg;
Der Weg nimmt zum Gesetz den Weg."[7]

Der Weg, der der Ursprung und das Wesen aller Dinge ist und nach dessen Prinzip sie sich verändern, wurzelt in der Natur. Der 'Weg', der den Charakter der Natur in sich trägt, hat aber weder konkrete Form noch Inhalt und verbindet sich bei Laotse mit dem Terminus 'Wu-wei', wörtlich mit 'Nicht-Tun'[8] übersetzt. Weil der Weg Laotse's naturgemäss ist, kann man ihn nicht bewusst gehen. Von der Seite des Wegs selbst betrachtet, ist der Weg der Natur nicht 'Nicht-Tun', sondern naturgemäss 'Tun'. Aus diesem Weg, der naturgemäss und 'nicht-tuend' ist, entstand ein neuer Weg des Menschen in der Ethik und Politik. Laotse rät dem Menschen genügsam und in seinen Vorstellungen nicht vermessen zu sein und meint dazu:

"Ein kleines Land! Volk gering an Zahl!
Friedlich ihr Wohnen,

Fröhlich die Lebensweise!
Man sieht von weitem wohl das Nachbarland,
Die Hähne sind, die Hunde noch zu hören;
Das Volk wird alt, und wenn sie sterben;
War dennoch keiner, der zum Nachbarn fand.[9]

Chuangtse, der allgemein als Nachfolger des Laotse gilt, hat über den Ursprung des Kosmos und die Entstehung der Menschen, ferner über den Weg und die Tugend ganz ähnliche[10] Gedanken geäussert. Nur liegt sein philosophisches Interesse etwas anders als das des Laotse.

Chuangtse versuchte in seinem philosophischen System, das menschenzentrierte Denken zu vermeiden und eine Aufhebung zwischen Leben und Tod, 'Ich und Du' anzustreben.[11] Dadurch erkenne man den Weg aller Dinge, d.h. die wirkliche Wahrheit und geniesse Glückseligkeit.

In der späten Han Dynastie, unter Kaiser Ming (58–74 n. Chr.) gelangte der Buddhismus von Indien nach China. Die rasche Ausbreitung dieser Fremdreligion und die Stärkung des Konfuzianismus durch seine Erklärung zur Staatsdoktrin unter Kaiser Wu (142–87 n.) mögen zur Gründung einer 'Taoistischen Religion', auf chinesisch Tao Chiao genannt, durch Chang Tao-Ling 張道陵 geführt haben.[12]

Chang Tao-Ling liess seine Anhänger Amulette tragen und bei Krankheit den Göttern geweihtes Wasser trinken. Trat Heilung ein, musste man zum Dank fünf Schüsseln Reis bezahlen, was der Bewegung den Namen "Fünf-Schüssel-Reis- Religion" einbrachte. Nach dem Tod Chang's wurde sie von seinen Nachfolgern weiterentwickelt und später von Ko Hung (葛洪283–343 n.Chr.) systematisiert.

Ausser den Anhängern und Nachfolgern des Chang Tao-Ling beschäftige sich ein anderer Chang, Chang Chiao 張角 mit der Gründung einer taoistischen Religion, um ein taoistisches Reich höchsten Friedens zu errichten. Die Zahl seiner Anhänger wuchs rasch, so dass die Bewegung von der regierenden Dynastie als staatsgefährdend angesehen wurde. Der darauf folgende Aufstand ging der gelben Kopfbedeckung der Rebellen wegen, als 'Aufstand der gelben Turbane' in die Geschichte ein.

Obwohl sich die beiden Changs bei ihrer Religionsgründung der Lehren und Namen Laotse's und Chuangtse's bedienten, unterschied sich ihre Lehre doch grundsätzlich. Während die Taoisten vor der Ch'in Dynastie versuchten, den Kosmos und das natürliche Wesen aller Dinge zu begreifen und die falsch gebildete Kultur umzuformulieren, ging es den Taoisten der späteren Han Dynastie um Techniken zur Lebensverlängerung und Herstellung des Friedens in der Umwelt. Beide Richtungen werden unter dem Begriff 'Taoismus' zusammengefasst.

2. Überlieferung des Taoismus nach Korea

Korea liegt geographisch an der nordöstlichen Grenze Chinas, wobei die Halbinsel Liaotung des heutigen China in der alten Zeit zum koreanischen Staat Koguryo (Kao-Kou-Li 高句麗 37 v. -668 n.) gehörte. Durch die zahlreichen Kriege, diplomatischen und wirtschaftlichen Beziehungen etc., kam es zu einem engen kulturellen Autausch.

Wann der Taoismus nach Korea kam, ist nicht genau belegt. Mit Hilfe von ethnologischem Material, Fragmenten in koreanischen oder chinesischen Geschichtsbüchern und Vergleichen anderer Religionen, kann man die Überlieferung des Taoismus nach Korea zur Zeit des Reiches Koguryo annehmen.

Zur Überlieferung des Buddhismus findet sich ein eindeutiger Hinweis im Buch *Geschichtsschreibung der Drei Reiche*, auf koreanisch: 'Samguk Sagi' (三國史記).[13] Hier lautet ein berühmter Ausspruch des Königs Sosurim (小獸林王371–384 n.) im 2. Jahr seiner Herrschaft über das Reich Koguryo:

> "Im Sommer (Juni) des zweiten Jahres schickte der Herrscher der Ch'in Dynastie, König Fu-Chien einen Boten mit einer Figur Buddhas und buddhistischen Kanons. Der koreanische König schickte als Dank ebenfalls einen Gesandten mit einem Geschenk, das in Korea gefertigt war. Ferner errichtete er eine konfuzianische Schule und liess dort junge Leute erziehen."[14]

Das Jahr 372 n. Chr. ist somit sehr bedeutsam zum Verständnis der koreanichen Kultur und Geschichte, da man annimmt, dass der Taoismus auch um diese Zeit nach Koguryo kam und sich, wie der Buddhismus, von dort in die südlicher liegenden Königreiche Peakche (百濟18 v. -660 n.Chr.) und Silla (新羅 57 v. -936 n.Chr.) ausbreitete. Es gibt jedoch keine Dokumentation in der Geschichtsschreibung von Koguryo, die diese Annahme stützt.

Aus der Zeit des Königs Kun Kusu von Paekche (375–384 n.Chr.) stammt ein Ausspruch seines General Mak Gohae auf dem Kriegsfeld:

> "Ich habe einmal einen Satz aus dem Taoismus gehört:
> Wer Genügsamkeit kennt, ist ohne Schande,
> wer Einhalt kennt, ist ohne Gefahr.
> Gerade gewannen wir sehr hoch. Warum versuchen wir nochmals zu gewinnen?"[15]

Dieser Ausspruch ist der letzte Vers im 42. Kapitel vom Laotse's *Tao Te King*. Aufgrund dieses Ausspruchs, mit dem der General den Krieg zu beenden suchte, wissen wir, dass die taoistische Philosophie in Peakche verbreitet war.

Aus Silla ist mehr Material über den Taoismus überliefert als aus Peakche, darunter eine Inschrift (Die Geschichte tapferer junger Menschen) des Philosophen Ch'oe Ch'i-Won[16] auf einem Gedenkstein Nan-Rang (鸞郎碑):

> "Es gibt einen mystischen Weg in unserem Land. Dieser Weg heisst Pung-Liu. Die Art der Unterweisung ist in alten Geschichtsbüchern eingehend beschrieben. In der Tat umfasst er die Lehre der drei Religionen. Im Kontakt mit vielen Menschen finden sie ihre Anwendung. Das heisst, die Lehre des Konfuzius von Lu, dass junge Menschen zu Hause ihre Eltern verehren und in der Gesellschaft dem Staat dienen, des Laotse von Chou, dass sie im "ohne Tun" und nach der Lehre des Schweigens leben und Verbesserung erfahren durch die Lehre des Buddha aus Indien, indem das Böse nicht mehr aufkommt und sie das Gute praktizieren."[17]

Die Philosophie Laotses und Chuangtses war in allen drei Reichen Koreas verbreitet und in vielen Lebensbereichen von den Koreanern angewandt. Während die obere Schicht, die die taoistischen Kanons zu lesen vermochte, die taoistische Philosophie verstand, berührte der später einfliessende religiöse Taoismus eher die unteren Schichten.

In China erfuhr der Taoismus Unterstützung durch die T'ang Dynastie (nicht zuletzt durch die Übereinstimmung des Familiennamens von Laotse mit der kaiserlichen Familie und dem damit verbunden und damals in Ostasien verbreiteten Brauch gegenseitiger Hilfe). Die T'ang Dynastie förderte den Taoismus nicht nur im eigenen Reich sondern aussenpolitisch auch im Nachbarland. Im Jahr 624 n. entsandte der Kaiser Kao-Tsu der T'ang Dynastie seinen Justizminister nach Korea, um den koreanischen König Yong nyu (榮留 618–641 n.Chr.) anerkennen zu lassen. Dabei schickte der chinesische Kaiser taoistische Meister mit einer Figur des Laotse und methodischen Büchern des Taoismus. Sie kamen nach Korea und lehrten die Lehre Laotse's.[18]

Ein ähnlicher Hinweis findet sich in dem berühmten Werk *Überlieferungen der Drei Reiche* (三國遺事), koreanisch 'Samguk yusa[19]:

> "In den letzten Jahren des Staates Koguryo, zwischen Wu-Te[20] und Chen-Chuan[21] glaubten die Leute fanatisch an die Fünf-Schüssel-Reis-Religion. Als der Kaiser Kao-tsu der T'ang Dynastie diese Nachricht hörte, schickte er taoistische Meister mit einer Figur Laotse's. Sie kamen und lehrten das koreanische Volk das Tao Te King. Auch der König Koreas und seine Gefolgsleute nahmen an den Vorlesungen teil. Es war das siebte Jahr des Königs Yong nyu (榮留 618–642 n. Chr).
>
> Es war das siebte Jahr des Kaisers Wu Te. Der König schickte Gesandte ins Reich der T'ang, um Buddhismus und Taoismus zu studieren. Der Kaiser (Kao Tsu) erlaubte das."[22]

Die oben zitierten Textstellen lassen den Schluss zu, dass der religiöse Taoismus etwa zwischen 620 u. 650 n.Chr nach Koguryo gelangte und sich von dort ausbreitete. Durch seine Gleichberechtigung neben Buddhismus und Konfuzianismus erreichte er breite Schichten der Bevölkerung und konnte somit einen reichen Beitrag zur Ethnologie, Literatur . . . leisten.

3. Geschichtliche Übersicht über den Taoismus in Korea

Im Jahr 668 n.Chr. hat Silla mit Hilfe der T'ang Dynastie das Reich Koguryo zerstört und dann Korea geeinigt. Von dem in der Periode der vereinigten Silla Dynastie berühmten Gelehrten Kangsu (强首) wird berichtet, sein Vater habe ihn als Jüngling folgendes gefragt:

> "Willst Du Buddhismus studieren oder Konfuzianismus? Dieser antwortete seinem Vater: "Ich habe gehört, der Buddhismus sei eine Lehre für das Jenseits. Ich bin ein Mensch, warum soll ich den Buddhismus studieren? Ich möchte den Konfuzanismus studieren." Die Erwiderung seines Vaters war "Folge Deiner Neigung."[23]

Diesem fast märchenhaft anmutenden Dialog zufolge besass man in der damaligen Zeit wissenschaftliche Freiheit und konnte sich unter den in der Gesellschaft gleichberechtigten Richtungen für eine oder gar eine Kombination aller entscheiden.

In der neu begründeten Koryo Dynastie (918–1392 n.Chr.) wurde vom königlichen Hof dem Buddhismus zur geistigen Untermauerung der Staatsideologie der Vorzug gegeben.

Der Taoismus erfuhr eine Veränderung vom früher mehr philosophischen zum religiösen Taoismus. Der Begründer der Koryo Dynastie, Wang Kon (王建 posthum T'aejo) hinterliess seinen Nachfolgern die "Sip hunyo (十訓要10 Lehren), die dem Reich und der Bevölkerung zu Glück und Wohlstand verhelfen sollten. Darin heisst es im zweiten Kapitel:

> "Alle Tempel waren nach der Wahrsagung des Mönches Tosǒn über eine geeignete Lage in der Landschaft gebaut. Baute man einen Tempel entgegen seiner Wahrsagung, würden die Tugend der Erde und das Glück des Landes verloren."[24]

Dieses Zitat aus den "Zehn Lehren" über den taoistischen Mönch Tosǒn (道詵) zeigen auf, dass die Koryo Dynastie vom religiösen Taoismus stark beeinflusst war. Eine mystische Lehre dieser Zeit war die Geomantie (風水地理說) auf koreanisch: 'P' ung su chi li', nach der sich auch der König richtete. Dieser Lehre zufolge bringt es Glück, sein Haus an einem günstigen Ort zu bauen oder die Toten in 'guter Erde' zu bestatten.

Wegen der schon aufgezeigten starken Verbreitung des religiösen Taoismus, dessen Zeremonien unter König Yejong (毅宗1146–1170) ihren Höhepunkt erreichten,[25] unterteilte man den damaligen Taoismus in drei Stufen: die höhere philosophische, die mittlere wahrsagerische und die untere schamanistische.

Da Koryo nach der Mystik taoistischer und buddhistischer Lehren regiert wurde, geriet es allmählich in Wirrungen.

Die letzte königliche Dynastie in Korea (1392–1910) wurde von General Yi begründet und nach ihm benannt. Gegen Ende der Yuan Dynastie kam der Konfuzianismus von China nach Koryo und wurde zur Staatsdoktrin erhoben. Taoismus sowie Buddhismus wurden verboten.

Die konfuzianischen Beamten versuchten mit Hilfe des praxisbezogenen, realistischen Gedankenguts Ordnung im Staat zu schaffen, was dazu führte, dass der Taoismus eher als eine Art ethnologische Religion von den unteren Schichten des Volkes gepflegt wurde. Trotzdem war es dem Einzelnen möglich, in kleineren Tempeln Buddhisnus und Taoismus zu studieren bzw. Zeremonien abzuhalten. Auch unter den konfuzianischen Gelehrten studierten einige neben dem Hauptstudium noch Taoismus, jedoch nicht die ethnologisch-schamanistischen Aspekte, sondern als hochgeistige Philosophie.

Die Vertreter des Taoismus in dieser Zeit waren Kim Si-sup, der Neo-Konfuzianer Yulgok, der buddhistische Meister Sǒsan u.a.m. Kim Si-sup (金時習 1435–93) ist in der Stabilisierungszeit des neu gegründeten Reiches geboren und in einer konfuzianischen Familie aufgewachsen. Mit 20 Jahren erlebte er die Absetzung und Ermordung des 6. Königs der Yi Dynastie. Wie Platon, der sich wegen der politischen Situation im damaligen Griechenland mit der Philosophie beschäftige, zog sich Kim Si-sup nach dem Tod seines Königs aus der Gesellschaft zurück und wurde Mönch in einem buddhistischen Bergtempel. Dort studierte er nicht nur Buddhismus sondern auch Taoismus, weshalb er, wie auch Ch'oe Ch'i-won[26] grosses Wissen über alle drei Richtungen d.h. Konfuzianismus, Buddhismus und Taoismus besass.

Der Neo-Konfuzianer Yulgok beschrieb ihn einmal mit den folgenden Worten: "Sein Herz gehört dem Konfuzianismus, aber er widmet sich dem Buddhismus und Taoismus."[27] Kim Si-sup studierte sowohl den religiösen als auch den philosophischen Taoismus, verwehrte sich jedoch gegen die Anwendung der Praktiken zur Lebensverlängerung. Er wollte nach den Naturgesetzen leben und äussert dazu:

> "Glückliche oder schlechte Gesetzmässigkeiten zu erlernen, bedeutet lediglich, eine andere Form des Lebens und Sterbens zu kennen. Wie aber kann man ausser seinem Leben in dieser Welt ein anderes Leben stehlen?"[28]

Kim Si-sup, dessen Einfluss späteren Philosophen der Yi Dynastie nicht nachstand, hat seine konfuzianischen und buddhistischen Lehren aus seinem Verständnis des Taoismus heraus entwickelt.

Als weiterer Erforscher des Taoismus gilt Yulgok (栗谷1536–1584), ein Zeitgenosse von T'oegye (退溪1501–73), die beide berühmte Theoretiker des Neo-Konfuzianismus waren. Yulgok war politisch sehr aktiv und einige Male Staatsminister. Mit eigenen philosophischen Methoden verfeinerte er die kosmische Lehre im 'Li-Chi' und die Lehre der 'Vier Kanten und Sieben Emotionen'.[29]

1974 wurde in der Nationalbibliothek ein Werk von ihm über Taoismus gefunden, das 'Sun ŏn' (醇言). Darin hat er das 81 Kapitel umfassende *Tao Te King* des Laotse nach seiner Vorstellung des Taoismus neu geordnet, auf 40 Kapitel zusammengefasst und jedes Kapitel mit einem Nachwort versehen. Die Wahl des Titels *Reine Worte* wird im Nachwort begründet: Demnach wurden die unnötigen Worte des *Tao Te King* weggelassen und nur die bedeutenden Worte neu zusammengefasst. Im ersten Kapitel heisst es:

> "Der Weg schuf die Einheit, Einheit schuf Zweiheit, Zweiheit schuf Dreiheit. Dreiheit schuf die zehntausend Wesen. Himmel und Erde, wie gleicht ihr Zwischenraum einem Blasebalg! Er fällt nicht ein ob noch so leer: je mehr bewegt, gibt aus er um so mehr. Die zehntausend Wesen tragen das dunke Yin auf dem Rücken, das lichte Yang in den Armen. Der Atem des Leeren macht ihren Einklang." [30]

Dieses erste Kapitel des Buches *Reine Worte* entstand durch Auswahl bzw. Vermischung des 5. und 42. Kapitels von Laotse's *Tao Te King*. Hierin versteht er den Weg des Taoismus als *Arche* und und Yin-Yang als Zentrum der werdenden Dinge, einschliesslich der Menschen, in der phänomenologischen Welt.

Im Verhältnis zwischen Tao und Yin-Yang ist das eine das Wesen des anderen. Beide sind aber nicht verschieden sondern gleich. Daher darf man sagen, dass Eins gleichzeitig Zwei ist oder umgekehrt, vergleichbar der griechischen Theorie 'hen kai pan', 'Einheit und Vielheit'.

Die Kapitel 2 bis 40 sind ähnlich aufgebaut. Man bemerkt, dass auch Yulgok den Taoismus aus der Theorie des Neo-Konfuzianismus heraus verstand, ohne dabei dessen Bedeutung zu verfälschen.

Ein weiterer taoistischer Gelehrter war der buddhistische Mönch Meister Sŏsan

(西山大師 1520–1604 n.Chr.). Meister Sŏsan studierte Konfuzianismus, bestand die staatliche Prüfung nicht, wandte sich später dem Buddhismus zu und wurde Mönch. Als 1592 der Krieg zwischen Korea und Japan ausbrach und die Japaner das Land besetzten, organisierte er landesweit unter den buddhistischen Mönchen Freiwilligentruppen und kämpfte, damals schon 70 Jahre alt, an deren Spitze. Er hinterliess einige Bücher über Buddhismus, jedoch auch ein Buch über Konfuzianismus und Taoismus: 'Gu-Kam (龜鑑). Gu-Kam bedeutet: Spiegel, mit dem man sich selbst erkennen kann. In den '3 Spiegeln': Zen-Buddhismus, Konfuzianismus und Taoismus ist der letztere ähnlich aufgebaut wie Yulgok's *Reine Worte*[31], d.h. Meister Sŏsan hat aus der Weltanschauung des Buddhismus heraus das taoistische Gedankengut neu geformt und ausgedrückt. Im ersten Abschnitt heisst es:

> "Ein Wesen gibt es chaotischer Art.
> Das noch vor Himmel und Erde ward,
> so gross, so mystisch, leer und seelisch.
> Weil es so gross, breit und noch deutlich klar,
> ist das nach der Richtung den Ort nicht bestimmt,
> und darf man nach der Zeit und zahl die Lebenszahl nicht zählen.
> Ich kenne seinen Namen nicht.
> Wenn ich mit Mühe benennen soll, sage ich: Das Herz."[32]

Der erste Satz ist identisch mit Laotse's *Tao Te King* Kapitel 25, auch die Betonung ist ähnlich. Jedoch verwendet Meister Sŏsan in der letzten Zeile des obigen Zitats das Wort 'Herz' statt 'Gross' in Laotse's *Tao Te King*. Im Buddhismus gilt das Herz als unentbehrliches Instrument zum Erfassen der Wahrheit.

Im vierten Abschnitt spricht Meister Sŏsan über Chuangtse.

> "Das Leben der Menschen ist ein grosser Traum. In diesem Traum gibt es immer einen König, der Wahrheit im Herzen empfindet. Deshalb erkennt man beim Erwachen die Bedeutung des grossen Traumes. So waren Chuangtse und der Schmetterling ein Traum"[33]

Im letzten Teil des zweiten Kapitels 'Ausgleich der Weltanschauungen' im Buch Chuangtses' geht es in Chuangtses' Traum von einem Schmetterling um Wahrnehmung und Erkenntnis der Realität. Nach Meister Sŏsans Interpretation hat der Mensch die Möglichkeit, nach dem Erwachen sein eigentliches und wirkliches Ich zu finden. Kurz gesagt, Meister Sŏsan hat im 'Spiegel des Taoismus' die Philosophie von Laotse und Chuangtse zusammengefasst und nach buddhistischem Verständnis neu gedeutet.

Neben den schon erwähnten Philosophen der Yi Dynastie: Yulgok, Kim Si-sup und Meister Sŏsan, sind unter den zahlreichen konfuzianischen oder buddhistischen Gelehrten, die sich mit dem Taoismus auseinandersetzten, noch Hwadam (花潭 1489–1546) und Sŏge zu nennen. Daneben gab es Gelehrte, die versuchten die drei Richtungen zu vereinigen, unter ihnen z.B. Hamhŏ (涵虛 1376–1433). Was den religiösen Taoismus betrifft, wurde zur Zeit des ersten Königs der Yi

Dynastie der Tempel (So Kyok Sŏ 昭格署) gegründet, der auch von der Königin und ihren Dienerinnen aufgesucht wurde. Der Taoismus blieb jedoch offiziell verboten, der Tempel wurde im Jahr 1518, auf Protest und Forderung des radikalen Reformers Cho Kwangjo (趙光祖 1482–1519) und seinen Anhängern der konfuzianischen Schule hin, abgeschafft,[34] wieder aufgebaut und nochmals abgeschafft. Obwohl taoistische Tempel in Korea verschwunden sind, ist der Taoismus selbst bis heute überliefert.

4. Die heutige Forschungssituation des Taoismus in Korea

Vom Anfang dieses Jahrhunderts bis in die siebziger Jahre begegnete Korea innen- und aussenpolitischen Schwierigkeiten. Infolge der 35-jährigen Besetzung durch die Japaner wurde die koreanische Kultur fast ganz vernichtet. In den letzten Jahren der Besetzung zwangen die Japaner die Koreaner z.B., deren Sprache und Familiennamen zu benutzen.

Mit Ende des Zweiten Weltkriegs verschwanden die Japaner, und es begann die Wiederherstellung des eigenen Lebensstils, der eigenen Kultur. Fünf Jahre später brach der dreijährige Bruderkrieg zwischen dem Norden und Süden aus, in den sechziger Jahren gab es Studenten- und Militärunruhen, so dass sich unter solchen Umständen die wirtschaftliche oder politische Lage Koreas nicht verbessern konnte. Trotzdem wurden in der Zeit zwischen 1945 und 1959 zahlreiche Universitäten gegründet, an denen nicht nur westliche Wissenschaft, sondern auch traditionelles koreanisches Gedankengut, darunter Taoismus, gelehrt wurden. Das anfänglich geringere Interesse der Studenten an der Lektüre des *Chuangtse* oder *Tao Te King* ist heute, eine Generation später, nicht mehr vergleichbar.

Zur Erforschung des Taoismus wurden unlängst zwei Institute gegründet, darunter 1986 das 'Institut für taoistische Gedanken Koreas' (韓國道教思想研), die beide in der Zukunft sicher eine bedeutende Rolle in der Taoismusforschung spielen werden. Die Gründungsmitglieder sind etwa 30 Professoren sowie zahlreiche graduierte Studenten aus dem ganzen Land.

Die Tätigkeit des Instituts umfasst, neben Publikationen über den Taoismus, die Organisation von Sitzungen und Seminaren. Zweimal jährlich findet eine Hauptversammlung statt, einmal monatlich ein Seminar. Darüber hinaus treffen sich die Mitglieder wöchentlich, um gemeinsam im Original den taoistischen Philosophen 'P'o Pao-tse" (抱朴子)zu lesen.

Die Themen der seit September 86 abgehaltenen 11 Sitzungen sind:

1. Beschaffenheit des taoistischen Gedankenguts von Korea
2. Frühere und gegenwärtige Taoismusforschung in Frankreich
3. Situation und Aufgabe des Taoismusstudiums in der Koryo Dynastie (918–1392 n.Chr.)
4. Taoistische Gedanken in Korea
5. Taoismus im frühen koreanischen Roman
6. Religiöse Anschauung in der Cho-le Cheng-sa[35] Zeremonie im Buch, *Auswahl*

der ostasiatischen Literatur
7. Das Tao-Verständnis bei Chuangtse
8. Unsterbliche Wesen in den Werken von Lew Mong-in[36]
9. Wu (無)Verborgene Gedanken und Schönheit bei den Taoisten
10. Das Thema der Unsterblichkeit in Gedichten der Koryo Dynastie
11. Wiederaufleben der Lehren der Lebensverlängerung in der Yi Dynastie (1392–1910)

Die halbjährliche Sitzung im Februar behandelte die Themen:

1. Chin Tan Tao (金丹道)
2. Die Rolle der Charaktererziehung und des Schicksals in der Lehre des Lebens
3. Die Zerstörung des taoistischen Tempels 'So Kyok Sǒ'[37] und die Abschaffung des Taoismus
4. Mythos und Unsterblichkeit im *Buch der Landschaft*[38]

Die halbjährliche Sitzung im August behandelte die Themen:

1. Taoismus und die Technik zur Lebensverlängerung
2. Taoistische Gedanken in der Literatur.

Insgesamt sprachen bei diesen beiden Sitzungen, die z.T. der weiten Anreisen wegen ausserhalb Seouls abgehalten wurden, 11 Professoren.

Die Ergebnisse aller bisherigen Sitzungen sind in dem im December 1987 erschienenen Buch *Taoismus und koreanisches Gedankengut* veröffentlicht. Darüberhinaus erschienen weitere Bücher über den Taoismus, die hauptsächlich von den Mitgliedern des Instituts verfasst wurden.

Ein erwähnenswertes neuzeitliches Buch über den Taoismus stammt von Lee Nung-Hwa (李能和) *Die Geschichte des koreanischen Taoismus.* Das im Februar 1986 in koreanischer Übersetzung erschienene Buch wurde während der japanischen Besetzung auf chinesisch niedergeschrieben. Dieses Buch liefert uns nicht nur Material über den Taoismus, sondern spielt als Brücke zwischen dem alten und neuen Taoismus eine grosse Rolle.

Heute studiert man Taoismus 'ganzheitlich', d.h. ohne Trennung von Philosophie und Religion, was daher rühren mag, dass sich der Taoismus geschichtlich neben Buddhismus und Konfuzianismus entwickelte und wie diese seinen Beitrag zur Kulturentwicklung, besonders im Bereich der Literatur, Kunst und Ethnologie, lieferte. Niemals stand er im Vordergrund der koreanischen Geschichte, war aber seit der Zeit der drei Reiche bis heute ständig gegenwärtig und spielt auch heute noch in weiten Lebensbereichen unterschwellig eine nicht zu übersehende Rolle.

Seit dem 18. Jahrhundert hat sich Korea mit der hereinbrechenden westlichen Religion und Wissenschaft auseinanderzusetzen, wobei das westliche Gedankengut die traditionellen Wertvorstellungen vorübergehend erschüttert hatte.

Die Koreaner werden einerseits das westliche Kulturgut aufnehmen, andererseits das eigene Gedankengut weiterentwickeln, darunter auch den Taoismus, der neben den modernen Wissenschaften sicher verstärkt studiert werden wird.

ANMERKUNGEN

1. Konfuzius ist der latinisierte Name von Kung fu-tse (551–479 v.Chr) wörtlich übersetzt: Meister Kung. Vgl. G. Debon und W. Speiser, *Chinesische Geisteswelt*, S. 27 ff. und P.Do-Ding, *Konfuzius in Selbstzeugnissen und Bilddokumenten* (Reinbek bei Hamburg 1981), S. 20ff.
2. Wann Laotse lebte, bleibt Legende. Sein Eigenname war Li-Erh. Er wird allgemein mit Lao-tan identifiziert, einem Zeitgenossen von Konfuzius.
   Chuangtse, ebenfalls ein berühmter Taoist, hat in seinem gleichnamigen Buch (Kap. XIII,7) geschildert, Konfuzius habe Laotse in Tschou besucht und nach 'Li' gefragt. Vgl. G. Debon u. W. Speiser, *Chinesische Geisteswelt* (Baden-Baden 1957), S. 73, 86 und *Lao-tse*, hg. v. Lin Yutang (Frankfurt M. und Hamburg 1955), S.202ff. A. Waley, *The Way and Its Power: A Study of the Tao Te Ching and Its Place in Chinese Thought* (New York 1958), S. 104.
3. 春秋戰國。
4. 諸子百家。
5. "道可道 非常道 名可名 非常名" 。
6. Lin Yutang, *Laotse* (Frankfurt/M. und Hamburg 1955), S. 213, Anm. 87.
7. "人法地 地法天 天法道 道法自然"
8. Vgl. *Tao Te King*, Kap. 37.
9. Vgl. *Tao Te King*, Kap. 80.
   "小國寡民……，安其居，樂其俗，鄰國相望，鷄犬之聲相聞，民至老死，不相往來。"
10. Vgl. Fung Yu-Lan, *A History of Chinese Philosophy*, Vol.1. *The Period of the Philosophers*, transl. by D. Bodde (Princeton 1952), S. 223.
11. Vgl. 勞思光。⌊中國哲學史⌉S. 202.
12. Vgl. J.Needham, *Science and Civilisation*, Vol. 2 (Cambridge 1977), S.155ff. und J.J.M. de Groot, *Universismus* (Berlin 1918), S.133ff.
13. Das Buch 三國史記 wurde zur Zeit des Reiches Koryo (918–1392 n.Chr) von dem konfuzianischen Beamten Kim Pusik 金富軾 im Jahr 1145 verfasst. Dieses Buch ist nicht nur das älteste Buch für koreanische Geschichte sondern auch eines der vertrauenswürdigsten.
14. 金富軾，⌊三國史記⌉卷第１８，高句麗本紀第６，小獸林王２年條，"二年夏六月，秦王符堅遣使及浮屠順道，送佛像經文，王遣使迴謝，以貢方物，立大學教育子弟" 。
15. 卷第２４，百濟未紀第２，近仇首王條。
    "將軍莫古解諫曰，⌊嘗聞道家之言，知足不辱，知止不殆，今所得多矣，何必求多⌉，太子善之止焉" 。
16. Ch'oe Ch'i-Won (857 n.Chr.-?), ein berühmter Vertreter aller drei Richtungen, ging im Alter von 12 Jahren nach China, bestand dort nach 6-jährigem Studium die staatliche Prüfung und wurde Beamter. Bei seiner Rückkehr nach Korea befand sich Silla gerade in einer Zeit der Wirrungen.
17. Ebd. 卷第４，新羅本紀第４，眞興王條。
    "崔致遠鸞郎碑序曰，國有玄妙之道，曰風流，設教之源，備詳仙史，實乃包含三教，接化羣生，且如人則孝於家，出則忠於國，魯司寇之旨也，處無爲之事，行不言之教，周柱吏之宗也，諸惡莫作，諸善奉行，竺乾太子化也。"
18. Ebd. 卷第２０，高句麗本紀第８，建武王條。
19. Das "Samguk yusa' (三國遺事) wurde etwa ein Jahrhundert später als das 'Samguk sagi' (三國史記) von dem buddhistischen Mönch Il-Yeon (一然 1206–1289) n.Chr.) verfasst. Beide Werke, das eine von einem Buddhisten, das andere von einem Konfuzianer geschrieben, sind nicht

nur unentbehrlich für das Studium früher koreanischer Geschichte sondern auch für die Forschung über Konfuzianismus und Buddhismus.

20. 武德 bezeichnet die Regierungszeit des T'ang Kaisers Kao-Tsu (618–626 n. Chr.)
21. 貞觀 bezeichnet die Regierungszeit des T'ang Kaisers T'ai-Tsung (627–649 n.Chr.).
22. 一然，「三國遺事」卷３，興法第３，寶藏奉老，普德移庵條。
"麗季武德·貞觀間。國人爭奉五斗米教，唐高祖聞之，遣道上，送天尊像，來講道德經，王與國人聽之，即第二十七代榮留王即位七年 武德七年甲申也，明年遣使往唐，求學佛老。唐帝（謂高祖也）許之"。
23. 金富軾，「三國史記」，卷第４６，列傳第６，强首
"父欲觀其志，問曰爾，學佛乎學儒乎，對曰愚聞之，佛世外教也，愚人間人安用學佛爲願學儒者之道，父曰從爾所好"。
24. 高麗丈。２　太祖２６年
"其二曰，諸寺院，皆道詵推占山水順逆而開創"。
道詵云，"吾所占定外，妄加創造，則損薄地德，祚業不永"。
25. Vgl. 李能和，「朝鮮道教史」卷１０章。 Dong-Kuk Uni. (Seoul 1969), S.98ff.
26. Siehe oben S. 10, besonders note 17.
27. Das heisst 心儒跡佛 vgl. 「栗谷全書」１金時習傳 Sung Kyun Wan Uni. Verlag, 1971), S. 292
28. 「梅月堂集」卷１７，雜著，龍虎第７ (Seoul 1973), S. 298
29. 理氣說，四端七情論
30. "道生一，一生二，二生三，三生萬物，天地之間，其猶橐籥乎，虛而不屈，動而愈出，萬物負陰而抱陽，冲氣以爲和"
31. Siehe oben note 30.
32. 西山大師，「道家龜鑑」。
"萬物混成先天地生 至大至妙 至虛至靈 浩浩蕩蕩 歷歷明明。
方隅不可定其君 劫數不能窮其數 吾不知其名 强命曰心"
33. Ebd. "人世大夢也 大夢之中 必有大覺之王 故大覺然後 知此大夢也，然則莊周與胡蝶，俱爲夢也"
34. Vgl. 車柱環 Die taoistischen Gedanken von Korea (Seoul 1984), S. 241ff.
35. 醮禮青河
36. 柳夢寅
37. Siehe ben, S. 19.
38. 「山海經」。

LITERATURVERZEICHNIS

A. Texte

「老子道德經」	Hong Kong 1973.
「莊子」	Taipei 1982.
金富軾。「三國史記」	Seoul 1973.
一然。「三國遺事」	Seoul 1973.
金時習。「梅月堂全集」	Seoul 1973.
李珥。「栗谷全書」	Seoul 1971.
李珥。「醇言」	Seoul.
西山大師，「道家龜鑑」	Seoul.

B. Literatur

Cha, J.H.,	*Taoistische Gedanken in Korea*. Seoul 1976.
Cha, J.H.,	*Taoismus und koreanische Gedanken*. Seoul 1987.
Debon, G.,	*Laotse. Tao Te King; Das heilige Buch vom Weg und der Tugend*. Stuttgart 1961.
Debon, G. und Speiser, W.,	*Chinesische Geisteswelt*: Von Konfuzius bis Mao Tse dong. Baden-Baden 1957.
Fung, Yulan,	*A Short History of Chinese Philosophy*. New York 1966.
Lao ssu Kuang,	*Geschichte der chinesischen Philosophie*. 4 Bde. Taipei 1980.
Lee Nung-hwa,	*Geschichte des koreanischen Taoismus*. Seoul 1986.
Lew S.K.,	*Konfuzianismus in Korea*. Seoul 1976.
Lin Yutang,	*Laotse*. Frankfurt/M. und Hamburg 1955.
Mathews,	*Chinese-English Dictionary*. Harvard University Press 1975.
Nedham, J.,	*Science and Civilisation in China*. Vol. 2. Cambridge 1977.
Schleichert, H.,	*Klassische chinesische Philosophie*. Frankfurt/M. 1980.
Waley, A.,	*The Way and Its Power*. London 1934.

The phenomenology and psychology of Korean shamanism

BOU-YOUNG RHI
Department of Psychiatry, College of Medicine
Seoul National University

1. Introduction — An historical perspective

Whether the Korean folk religion in which the mudang plays a central role can be designated as shamanism in the strict sense (Eliade, 1951) is certainly a question which requires continuous study. However, we would not hesitate to classify the mudang's magico-religious system as shamanism and to cautiously call it "Korean shamanism", a specific type of shamanism which we classify as such because the mudang's belief system displays striking similarities with Central Asian and Siberian shamanism (Liu, D.S. 1975; Rhi, B.Y. 1976).

The evidence for the connection between Korean shamanism and Central Asian shamanism has been suggested not only through the comparison of certain ritual symbols (Son, J.T. 1948) but also through etymological comparisons of the Korean terms for female shaman (mudang) and male shaman (baksu) and the Altai terms for the female and male shaman (Udagan, Ubakan and Pahshi, Paksa or Baksi) (Torii, R. 1950; Clark, C.A. 1903/1961).

The Chinese word wu (mu in Korean) which is frequently identified with the shaman in Korea, China and Japan appears in the history books of the Three Kingdoms Period. (B.C. 57–A.D. 668). The first detailed descriptions of shamans and shamanic rituals, however, are found in the collected papers of Lee Kyu Bo (1168–1241) of the Koryo dynasty (918–1392), a Confucian scholar who bitterly criticized the mudang's belief system (Lee, K.B. 1958). The meticulousness of Lee's scholarship has been very helpful in that it revealed that Korean shamanism at that time showed great similarities to the present features of Korean shamanism. Due to the lack of detailed information on Korean shamanism before the time of Lee Kyu Bo, one may not claim with absolute certainly, as did Kim, I. H. (1987) that ancient Korean religion was identical with present-day Korean shamanism. However, it is not unreasonable to assume that the main trends of Korean shamanism can be traced to even paleolithic times (Hwang, S.Y. and Moon, M.D. 1984). The evidence for the long history of Korean shamanism can be found in cave drawings and more definitively in artifacts from the Bronze period in which archaeologists have found bronze mirrors depicting the sun, moon, bells and swords, evidence which is subjective when considering the three divine instruments of present-day Korean shamanism, the *myongdo* (mirror), divine bells, and the divine sword (Chang, J.G. 1983).

Contemporary philosophy. A new survey. Vol. 7, 253–268.

The myth of Tangun, in which the son of the heavenly god descended to a sacred tree, Shin Tan Su on the top of the mountain Taipaek and through marriage with a bear woman bore Tangun, the first mythical ruler of Korea, has been frequently mentioned in connection with shamanic ideology (Liu, D.S. 975). This myth is also regarded as one of the prototypes for a particular shamanic ritual performance in Korean shamanism (Kim, Y.K. 1976).

A piece of earthenware decorated with clay dolls and rich fertility symbols from the period of the Three Kingdoms and the fresco of a newly excavated tomb in Jang-Chon (formerly Man-choo) which supposedly originated from the early state of Koguryo both suggest the features of shamanic rituals of their respective periods (Kim,Y.K. 1986; Lee, N.Y. 1976).

The historical literature on the Three Kingdoms Period only sporadically reveals the nature of the magic-religious life of the time. In Puyo, Koguryo and Ye, the Kingdoms located in the northern areas of Korea, there were nationwide seasonal festivals which were dedicated to the heavenly deities where the people played music, danced, drank and told fortunes for days and nights at a time during certain parts of the year (Liu, D.S. 1975; Kim, Y.K. 1976). In all kingdoms there existed a Byol Up, a sacred city which was also called Sodo, in which there existed a long staff with bells and drums which was worshipped as God. It recalls the sacred staff of the village god, So nang dae, which represents the body of the protecting god of the village and which is carefully preserved in the sacred shrine, shindang. Thus, the proceedings of the festival in which the gods are invocated through the sacred staff seems to have been transmitted throughout the centuries until today.

It seems that the mu played a number of socially significant roles during the Three Kingdoms Period, such as priest, political leader and perhaps at the same time, exorcist. Thus, he or she sometimes enjoyed high social status during the Three Kingdoms period from the Shilla dynasty to the Koryo dynasty (918–1392). As well, shamanism or muism flourished in the Buddhistic festival named Palkwan Hoe during the Koryo dynasty.

During the Chosun dynasty in which Confucianism was adopted as the ruling ideology and patriarchal ancestor rituals (*Je sa*) prevailed, the mu cult was preserved mainly among women. Through suppression by the ruling aristocratic class and intellectuals of the Chosun dynasty, Korean shamanism gradually lost its influence upon men and the male-centered society and thus the result was the disqualification of shamanism as a religion, which was mentioned by Chang J.G. (1983).

However, in the late 19th century, foreign missionaries and scholars who came to Korea were surprised to find how embedded the mu-cult was throughout the country. Thus, among foreign visitors to Korea at that time, Korean shamanism became an object of study for scientific purposes and, for missionaries and colonialists, for the purpose of gaining greater ideological and spiritual control over the people. While it is true that the Japanese government and certain scholars were very interested in shamanism as an object of study, under the Japanese regime the mu-cult was generally rejected as "superstition" and was looked upon disdainfully. However, shamanism has served as a basic force in the initiation of new religious movements in many different periods from the late Han Kingdom to the mu-ism which thrived after the liberation of Korea from Japanese occupation. Shamanic village festivals were displaced during the 1960s by the "Saemaul" (new community) movement, a social reform movement initiated by the government in rural areas.

With academic interest in shamanic performances increasing since the late 1960s, the mudang's performances have gained a new appreciation among the Korean people as a national cultural treasure. Thus, the shamanic performances are not overtly prohibited and the institution of shamanism has been slowly revived.

The power of Korean shamanism has waxed and waned in accordance with the decree of the dominating ideology of the times, but the core experience of ecstasy that has been unique to shamanism seems now to have crossed over the boundaries of the mudang's rituals and penetrated various social movements, such as new Christian sects and political demonstrations by college students.

2. Investigation and research of Korean shamanism

Motivated by the practical problems of Christian missions in Korea, a number of protestant missionaries wrote about the phenomenon of Korean shamanism (Hulbert, 1906/1969; Clark, 1903/1961). In the 1920s and 1940s, the efforts of some Korean scholars in searching for the primordial origins of the national culture of Korea led them to shamanism (Choe, N.S. 1927; Lee, N.H. 1927; Son, J.T. 1948).

Japanese investigations of shamanism during this same period seem to have been motivated either by the necessity to understand the mentality of the Koreans in order to govern them or for pure academic interest. Among the latter, the study of Korean shamanism written by Akamatzu C.J. and Akiba T. (1937/1938), and the field research on Korean shamanism by Akiba from a social anthropological viewpoint were the most comprehensive studies of Korean shamanism ever published (Akiba, T. 1950).

After a long latency period, which lasted until the late 1950s, students of Korean literature and ethnology began to take an interest in Korean shamanism. Thus, inspired by pioneers in Korean ethnology, the cultural bureau of the Korean Government began to collect shamanic songs and performances during the late 1960s.

In the late 1960s the investigation of Korean shamanism became more active. Experts in the field of mudangs and the mu-cult who published ethnographic and ethnological studies on shamanic songs and cults were able to compile comprehensive analyses of Korean shamanism (Chang, J.G. 1975; Choe, K.S. 1978, 1981; Kim, T.G. 1981). Among those investigators, Choe, K.S. has focused mainly on the family relations of the hereditary mudang and has also done comparative analyses of ecstasy and the purity-impurity dichotomy in Korean shamanism. Kim T. G. has focused more on the initiation process and the presentation of his huge collection of mudang's songs. Hyon, Y.J. (1986) has compiled a comprehensive literature on shamanism on Cheju Island which is in every respect different from shamanism on the continent. Recently, Kendall (1985, 1987) through her observations of manshin, threw new light on the religious life of Korean women in their conducting of household rituals.

There is great diversity in the ways in which Korean shamanism has been viewed. Besides the cultural anthropological and ethnological view, there was a comprehensive review and analysis of Korean shamanism from a religious historical perspective which dealt with the relationship of shamanism to other religious cultures introduced to Korea from the outside and with the ideology of Korean shamanism as a whole (Liu, D. S. 1975). There have also been comparative mythological approaches to the shamanic epics, structuralist interpretations of shamanic rituals (Kim, Y.K. 1971, 1976, 1982) and dramatological studies of shamanism (Lee, S.I. 1981).

The study of the psychiatric aspects of Korean shamanism, which was begun in 1961 (Rhi, B.Y. 1965) has continued through the 1970s and 1980s. Jungian interpretations of the various aspects of Korean shamanism have been conducted and steadily published (Rhi. B.Y. 1982), and Freudian psychoanalytic formulations of the initiation disease and shamanic performances have also been diligently published (Kim, K.I. 1984).

A recent remarkable work on Korean shamanism is the work of Kim, I.H. (1987) who attempted to uncover the philosophy of education in Korean shamanism.

However, it is quite surprising to recognize that there have been no academic psychologists or philosophers who have examined shamanic culture in Korea utilizing the methods of experimental psychology or philosophy. The psychological aspects of Korean shamanism have been discussed mostly by medical psychological investigators as mentioned above.

Below, some characteristics of Korean shamanism are reviewed in connection with their psychological implications.

2.1 *Becoming a Mudang—Initiation disease, Mubyong*

In certain areas of Korea, there are learned mudangs who earn their status through hereditary transmission and do not suffer through an initiation disease. They are a sort of priest-artist and they practice mainly in the southern provinces and along districts by the east coast of the Korean peninsula including Cheju Island.

The spiritistic initiation process in which mental disturbances and ecstatic outbursts play a central role is found mainly among shamanistic societies in the middle and northern parts of Korea and these societies have been the main concern of research among the ethnologists and the psychiatrists (Chang, J. G. and Choe, K. S. 1967; Rhi, B. Y. 1965, 1969, Kim, K. I. and Kim, T. G. 1970; Kim, K. I. 1972b;).

According to the report by Akiba, in the 1930s the initiation disease, *Mu-byong*, was characterized initially by extreme withdrawal and emaciation due to the inability to eat, sudden ecstatic outbursts leading to bizarre behavior ending with the victim falling unconsciousness in the fields or mountains. Occasionally shamanic instruments, such as divine bells, mirrors and swords, are said to be found in the very place where the future *mudang* falls down. These *mudangs*-to-be were also said to suffer from auditory and visual hallucinations and nightmares, the content of which are related to shamanic deities and performances.

From the report of Chang J.G. and Choe K.S. (1967) and the cases of shin-byong (synonym of Mu-byong) observed by Kim T.G. (1981) in the 1960s, the main features of initiation disease in the 1960s were similar to the cases in the 1930s, though the ecstatic experiences are not as prominent as in the 1930s and the obtaining of the divine instruments in the ecstatic state was not as popular.

An old question, whether mu-byong is a specific disease was first raised in 1961. (Rhi, B.Y. 1965) The conclusion was that mu-byong seems to be a symptom complex with different physical and psychogenic origins and its religious meaning is more significant than its nosological status.

Mu-byong is essentially different from the animal spirit possession syndromes found among the Ainu, the Manchoo or in Japan and these differences can be elicited upon observation of invocation ceremonies (Rhi, B.Y. 1965). Among the Ainu, the Manchoo and in Japan, animal spirit possession was an invoked syndrome in which the victim became the animal that possessed him. Once the spirit was exorcised, the person was thought to be cured and there was no ultimate meaning to be derived from the possession itself. *mu-byong*, on the other hand, was not regarded by the shamanic society as necessarily a possession state in the strict sense, but rather as a sign of a client's contact with the world beyond, mediated by mythical images such as voices from heaven, old men or fairies and thus as an omen of divine selection for future *mudangs*.

The initiation process in Korean shamanism exhibits the universal formula of initiation: suffering, death, and rebirth. The stage of death is suggested by the sufferer's extreme loss of physical power and at times by a temporary loss of consciousness. The images of being cut up into pieces and the initiation operation seen in Siberian and Southeast Asian shamanic cults are rarely found in Korean shamanism. Instead, the experience, which is regarded as an inevitable divine call for the shaman, is characterized by extreme suffering, both physical and emotional, in which the sufferer meets the shamanic spirits in visions and dreams.

Originally, however, it seems that in Korean shamanism, Mu-byong alone, was not sufficient for the sufferer's transformation into a full-fledged mudang. In Korean shamanism, it is thought that the candidate is still possessed by many demonic powers, *hu-ju* (a false master), other than the shamanic deities and they should be driven away through the initiation ceremony, *naerim-kut*. Through this ritual, the shaman to be would receive his/her good protecting spirit otherwise known as the "the master of the body", *momju*. A long hard training process for the mudang candidates as god daughter, shinttal, under the supervision of the god mother, shinmo, follows thereafter. Therefore, the Korean *mudang* may not be regarded as possessed and mentally ill but should rather be seen as a person who was cured from mental and physical disorders. This supports the position of M. Eliade on the personality of the shaman (Eliade, 1951). It is important to note, however, that there are many self-made mudangs who have not received the appropriate initiation rites and training.

Kim K.I. (1972b) has identified among several mudangs psychiatric illnesses due to emotional conflicts which were temporarily healed through the performing of shamanic seances and he further made an attempt at a psychoanalytic explanation of the mechanism of the initiation disease. However it may be interpreted, it is noteworthy that the initiation process in Korean shamanism has great religious significance in that the initiation illness was regarded as a call by the transcendental powers to take on the role of shaman. In addition, the meaning of suffering is not as important, with respect to the power of the shaman to heal, as it is among the people of Central Asia, especially the Jakut tribes in previous times (Rhi, B.Y. 1969). That illness possesses transcendental meaning is exactly the point which has long been ignored by medicine and has only been awakened in the recent past by psychiatrists such as C.G. Jung and Viktor E. Frankl.

The meaning of suffering as necessary condition as well as opportunity for obtaining insight into one's psychological and existential problems and for becoming a mature person is not well understood by the shamanic society. And yet, the following fact is noteworthy. A profane woman becomes a sacred one through initiation disease and she then becomes able to communicate with the spirits, that is, through her contact with the world beyond. Shamanic deities in the world beyond may be designated as archetypal autonomous complexes in the unconscious in terms of Jung's analytical psychology. The deities or spirits represent highly emotionally charged unconscious contents. Thus, her psychic totality is established by making contact with her *momju*, her helping spirit, as long as she maintains the rituals. In this case, momju represents the internal personality in the unconscious and by communicating with *momju*, ego consciousness is renewed and widened to establish psychic totality. Naturally for the mudang, autonomous emotional complexes can be realized and experienced only as concrete images such as spirits or souls in the outer world the mudang's themselves do not recognize them as intrapsychic existences.

However, along with the depth psychological understandings of the mudang, we should not overlook the role of Korean women as the carriers of household religious life (Kendall, 1985) and the fact that Korean shamanism has been a way of socialization for certain women through taking on the role of mudang (Kim, Y.S. 1979).

2.2 *Kut: The shamanic ritual*

The shamanic ritual, *kut*, can be classified into two categories: home rituals and village rituals, both of which have been held either regularly or transiently on occasions such as child births, marriages, illnesses, deaths or other important rites of passage. As well, there are home rituals which are performed only by spiritistic mudangs for the sake of her protecting spirits (momju) and the initiation ceremony, *naerim kut*. These two ceremonies are called *shin kut* or spirit ceremonies (Kim, I.H. 1987).

The shamanic ceremony, *kut*, usually consists of 12 sessions in which 12 shamanic deities appear and give the oracle *k'ongsu* through the *mudang*. Each session can be divided into three steps: the summoning of the shamanic deity, possession of the mudang by the gods who then give oracle to the participants, and finally, the sending away of the deities. Shamanic songs, music and dance act as intermediaries to entice the gods to descend to the *mudang*. The songs contain the origin myth of each shamanic deity and the singing of the shamanic songs which is called bon puri, unravels these origin myths. The mudang wears the divine costume of each deity as each new session begins.

In the Korean shamanism, the term *man-shin* which literally means ten-thousand gods has come into use as a more polite term in referring to the mudang.

The cosmos is filled with numerous deities high and low. All 12 deities who are summoned to the *kut* by the *mudang* have their own individual characteristics. For example, *tae-kam*, the god of prosperity and fertility is ill tempered, greedy, obscene, indecent, and magnanimous, while the god of Buddhist origin, *po-sal*, is neat, fine and graceful. The military god *kun-wung*, characterized by his masculine courage, and *pyol song*, the god of smallpox by his solemn appearance. The ancestor god *jo sang* gives the most moving oracle. Thus, once the *kut* begins, one goes through all kinds of emotional experiences, not only the personal frustrations and grudges suppressed hitherto in the unconscious, but also the primordial archetypal condition of our psyche has the opportunity to express itself as it is elicited and mediated by the rich mythological implications of the shamanic songs and performances within a highly emotional atmosphere.

The *kut* has the very important role of reviving the unlived emotion in man's mind and it thus provides for a total experience, conscious as well as unconscious by relieving the participant of the social "face keeping" attitudes derived largely from the traditional Confucian way of life. Therefore, the effectiveness of the kut lies not only in the psychologic effects of amusement, pleasure or catharsis but also in its synthetic function of providing an emotional encounter comparable to a certain type of modern psychotherapy, a sort of psychodrama.

In *chosang kori*, the session in which the ancestor provides the most moving performance, the soul of a dead family member is supposed to possess the *mudang* and then speak through her to the family members giving them *nokturi*, the word of the soul. Sometimes a family member is told to hold a divine staff or *shin-dae* to invocate the chosang. Then, highly affect laden dialogues between the living and the dead are exchanged discharging thereby the guilt, hostility frustrations and regret of the clients as well as of the *mudang*. As seen in the somewhat stereotypic patterns of the oracle *kongsu* and from observations of actual *kut*, the relationships established between the gods and man seem to be quite flexible (Rhi, B.Y. 1976). Gods don't enjoy absolute authority or, in the *kut*, man can play and negotiate with them. On the one hand, this relativism in terms of good and evil characteristic to Korean shamanism enables man to relax and free himself from daily moral restrictions; but, on the other hand, it enhances the dependency needs of the client toward the pleasure of playing with supernatural beings and fosters a longing for self pity and benevolent reassurance by the gods without making an effort to live as a responsible independent individual with an ethical consciousness.

In the session of *mugam* the *mudang* tells particular participants to wear the divine cloth of a certain god and dance. The participants often fall into trance states and even give the oracle, *kongsu* to another participant. This may have psychotherapeutic effects for the client in terms of ventilation and abreaction (Kim, K. I. 1972a). Kendall, (1977) through her keen observations appropriately emphasized the religious meaning of *mugam* while also stressing the fact that a common person becomes sacred as a *mudang* and so *mudangs* train themselves in order to learn the ways of making contact with transcendental existence and the culturally sanctioned religious expressions.

Concerning seasonal village festivals or *dang kut*, Kim Y.K. (1971, 1985) has clearly described its releasing function (puri) of pent-up rancor or the *maetchim* of village people through receiving divine power, *shin-param* during *dang kut*. In this seasonal rite of passage of the community, the sacred world is encountered by the profane represented in unusually enhanced eroticism. "Absolute others" in the sacred world and in the profane world are united in the process of *dang-kut*. Thus *dang-kut* serves to enhance community cohesion by opening the path to a shared religious experience and thus providing a forum for the renewal of interactions between community members.

2.3 *Concepts of illness and the healing cult*

From the investigations of the folk concepts of illness which are closely related to the views of illness in shamanistic society (Rhi, B.Y. 1970a, 1972; Kim, K.I. 1972d; Chung K.C. and Rhi, B.Y. 1975) one finds similar concepts reported by F. Clement (1932) except for the idea of "evil eye" as a cause of illness. "Soul-loss" or *nok-nam* as a cause of illness seems to be popular mainly in Cheju island where the shamanic healing ceremony of *nok durim* (receiving soul) is preserved (Hyon, Y.J. 1986). The major cause of illness in shamanic society is spirit possession in the form of attack, *bomjob*, by the spirits due to a breach of taboo or neglect of the worship of shamanic spirits and deities. *Han*, or feeling of grudge of the dead plays a central role. Illness as a call, as mentioned in previous sections is a unique idea in the initiation disease of shamanic society.

Feelings of grudge of the dead result from premature death before marriage, any death outside the home or due to unnatural causes and the neglect of the worship of ancestors by the living. Such emotional complexes incorporated in the shamanic deities such as general Choe–Hyong from Koryo dynasty and Kwan-woo, a Chinese hero, who were killed unjustly, are regarded as the important protecting gods of a village along with *momju*, the protecting spirit of the *mudang*. This tradition is now somewhat fading away. However, the soul of a dead child, *myong-do*, who can cause disaster plays the crucial role in divination as a helping spirit of the fortune teller *chombachi* in the southern provinces. Although the concept of illness of the shaman in Korea does not make any sharp distinction between physical and mental disorders, mental disorder was once regarded as a case of possession by the demon of mental illness *chong-gi* who is the soul of the drama player (Akiba, T.

1950). Because the *mudang* of today well knows the effectiveness of modern medicine, illness is attributed to supernatural beings only when it is believed to be incurable by means of modern medical treatment.

The causes of the illnesses mentioned above are not only related to the illness itself but are also related to the various disasters and unhappy events affecting a family or a village. This indicates an archaic interrelatedness and overinclusiveness in viewing illness, disaster, personal matters and community affairs as cosmic events (Rhi, B.Y. 1981).

As seen from the previous section, every *kut* serves for the prevention and healing of illness as well as for prosperity and fortune. The disease rituals, *byong-kut* differ generally from other shamanic rituals in that it contains the additional element of exorcism. The psychotherapeutic effects of *kut* are recognized even though the explanation of healing factors is different (Rhi, B.Y. 1970; Kim, K.I. 1972c) but there are hazardous effects as well in that *kuts* may induce psychotic breakdown or dissociative disorders among those participants with weakened egos or some predisposition to psychic dissociation (Rhi, B.Y. et al. 1970b; Yoo, C.H. and Rhi, B.Y. 1983).

The *kut* serves as a forum for the reorganization of distorted family interactions by allowing family members to express their frustrations in the name of the gods and then become reconciliated with each other through the mediation of powerful and benevolent ancestors. Today, the healing ritual, *byong-kut* is rarely seen and carefully hidden from the researcher's eyes.

2.4 *Chinoki: A ritual for the dead*

Among shamanic rituals dedicated to the soul of the deceased, the ritual for the dead called *chinoki*, which has been held in the Seoul and Kyonggi areas of Korea, has unique features with respect to ways of dealing with the dead and the living. The wandering souls of the deceased which are believed to cause illness or disasters can be seen in terms of analytical psychology as symbolic manifestations of unconscious autonomous complexes (Jung, C.G. 1948; Rhi, B.Y. 1978). Therefore, the ritual treatment of the soul, as long as we avoid the question of the real existence of the soul, is comparable with the process of analytical psychotherapy in which unconscious complexes are recognized and assimilated to the consciousness. With this assumption, in order to investigate the characteristics of the healing process of the human mind as well as the process of psychotherapy, the phenomena revealed in *chinoki* was extensively studied and comparisons were made with parallel phenomena in folk beliefs from different cultures (Rhi, B.Y. 1968, 1970b).

chinoki is conducted when the misfortune of a family is attributed to the wandering soul of a dead family member. During the session of the dead, the *mudang*, in the cloth of the deceased and thus bearing the identity of the deceased, expresses the dead spirit's feelings of regret in a pathetic tone, and there arises a highly dramatic emotional encounter between the dead and the living. If *chinoki* comprises only this kind of "soul dialogue", called *nocturi*, the ritual has essentially no differences from the usual *kut*. However, *chinoki* surpasses the common cut through the session of Pari-Konju "the abandoned princess", who is the mythical ancestor of all Korean shamans. During the session, the mudang, in the costume of the princess recites the famous song of Pari-Kongju, while beating the drum slowly.

The song of Pari-Kongju is the story of the ancestor of all shamans told from her birth through her sufferings and hardship in training to her final achievement of the supernatural power of healing. This story reveals obvious initiation motifs but as well, it is a bitter critique of male centered culture. The "abandoned princess" was the seventh daughter of a King who wanted so badly to have a male heir that he could not stand having only daughters and so he ordered his seventh daughter to be banished from the court. Though left to die, the baby survived with the help of animals and sacred beings and in the end achieved supernatural healing powers after long and painful suffering (Akamatsu, C. and Akiba, T. 1937/38; Akiba, T. 1950; Kim, T.G. 1966). Later in her life she cured the King and Queen of a deadly illness and, at the same time, she brought seven sons from her marriage with a mythical monster fairy. The King wanted to give her honor in whatever way she desired but she refused the offer of her father and declared that she would be the King of ten thousand gods or manshin. After telling this story, the mudang recites that this Pari-Konju shall now guide the lost soul to paradise.

The song of Pari-Kongju contains rich buddhistic and taoistic terms. The story contains meaningful suggestions that only those who have suffered the initiation disease are able to guide the dead in completing their lives. That the living may not dwell in personal sorrow is shown in the session of *nocturi*. Personal mourning should be overcome by accepting the sacred meanings of suffering which have been experienced by the shamanic ancestor. *pari-kongju* incorporated the initiator and the psychopomp for the dead as well as for the living. The wandering frustrated dead thus become purified and placed in the shamanic pantheon as one of the deities, namely, an ancestor god. In other words, the archetypal contamination with personal complexes is clarified in order to get more clearly to the meaning of self. Symbolically interpreted, these steps are exact manifestations of the therapeutic process of analytical psychotherapy with its process of realization of the unconscious meanings of the self. This insight into the transpersonal meaning of suffering suggested in *chinoki* makes Korean shamanism religious and distances it from the magical tinges which are usually included in shamanism.

3. World view and the ideology of ecstasy

The world view of Korean shamanism does not seem to be constructed in three layers, as in Siberian and Central Asian shamanism. The highest deities dwell in heaven, thus the upper world does exist, whether or not it plays an important role. However, the world of the dead is not located in the underworld but far beyond this world, which is bordered by a river. Thus the world seems to consist of two layers, heaven and earth, and the idea of axis mundi, which in Siberian and Central Asian shamanism is the center of the world which connects the three worlds and by which the shaman descends to the underworld and ascends to the heavens, is scarcely observed in Korean shamanism. The world is physically rather plain and the worship of the tree as divine place is active, as is the notion of the divine staff through which the gods descend to the person who holds it. However, there is no evidence that these elements signify the center of the world. The holy trees almost always stand in pairs, as evidenced in the tree of grandmother *halmang dangsan* and grandfather *halbae-dangsan*. These two trees symbolically emphasize harmony and balance between the masculine and feminine principles, Yang and Yin, rather than suggest the idea of a center. On the other hand, the border of the village is a sacred place where a pair of wooden totems, the male heavenly general and the female earth general, also frequently a wooden bird is set on the top of the long staff, called *susaldae*, and it is believed to possess the power to drive evil away.

There are numerous deities in Korean shamanism that exist in heaven, in the house, the backyard, the trees, the ground, rocks, the spring, the river, and the sea. Emotionally however, there are few shamanic deities to whom *mudangs* and clients feel closely attached. The most important god in this respect seems to be the ancestor, *jo-sang* and the highest god in the heavenly palace is forgotten and becomes deus otiosus. The god of heaven and mountain and the master of the ancestors *song-jo*, actually the god of the house, have some connection to each other (Kim, T.G. 1969). *song-jo*, therefore, seems to be the central god of the house and thus represents the ancestor as well as the god of the high place. He is in fact worshipped on the beam of the great hall of the house and receives regular offerings.

From these descriptions one may speculate about the psychologic characteristics reflected in the world view of Korean shamanism. Korean shamanism seems to have been losing its archaicity and the world seems to be regarded not as in its plasticity in the vertical dimension but extended into the horizon. Consequently, man is more concerned with the boundary of the village beyond which the unknown world begins. On the other hand the harmonious relations of Yin and Yang instead of axis mundi has developed. And yet, while confining the place of the gods from the cosmos to the house, a separate microcosm is established under the beam of the house where the god *song-jo* dwells encountered by *dae gam*, the god of wealth, fortune and earth as a counterpart of *song-jo*. This difference of world view of Korean shamanism from Siberian types is also related to the following characteristics of the ideology of ecstasy in Korean shamanism.

Korean shamanism does not fit M. Eliade's definition of typical shamanism in which the magic flight of the shaman is a necessary condition of shamanic ecstasy, for the mudang never flies to heaven during the trance state. The shamanic deities always descend, upon the call of the mudang to the mudang and take possession of her. The journey beyond this world is performed through mythological psychopomp represented as the soul of a young boy Kang-im or the ancestor of all mudangs, Pari-Kongju journeying beyond through shamanic songs, but the mudang herself does not directly make this journey. Although the mudang demonstrates in dramatic ways the sending of the soul of the dead, she never says that her soul leaves her body and accompanies the dead to the world beyond.

Such ego-centered attitudes of the mudang in touching the world beyond is presumably connected with the fact that the concept of illness due to soul-loss and the initiatory dismembering to the bone reported in Central Asian and Siberian shamanism is rarely found in Korean shamanism. On the contrary, in Korean shamanism, possession by a false master (*hou-ju*) during initiation disease is dominantly believed.

This ideology of ecstasy shows us two ways of initiation which should be taken into consideration in the light of the nomadic and agrarian types of initiation described by M. Eliade (1951) and actually no differences as a psychological experience of shaman, namely, ego's contact with numinous supernatural beings. Perhaps we could find two characteristic ways of ego's attitude in reaching the single ultimate goal of communication between opposites; this side of world, *isung* and the beyond, *chosung*. Consciousness with the unconsciousness and it seems to me there exists some culturally determined differences of ego organization and ego's attitude toward the unknown transpersonal world.

3.1 *Ethical values*

Liu D.S. (1975) sees the characteristics of the value systems of present day Korean shamanism in its secularistic trends and finds in it an extreme egocentrism lacking ethical values in terms of relationships with fellow community members other than the shamanic village festival and the optimism that man's conciliation with god will establish a paradiesic state in this world.

Kim, I.H. (1987) argues that the original philosophical base of Korean shamanism emphasized harmony and balance in life within the cosmic order. He sees the ethical system as emphasizing protection and sacrifice for the sake of the family's security and the need for constant effort in adapting oneself to nature. Kim also emphasizes the shamanic appreciation of the present life and he pointed out the absolutism of human prestige and the relativism of the authority of the gods. He regards as positive aspects of Korean shamanism, its reconciliating and reconstructing functions in terms of the way shamanic plays facilitate a community's cohesion through emphasizing ethical duties and hope in the present life. In terms of negative aspects, he points out the lack of individual responsibility and self-reflection, the lack of historical and national consciousness, the lack of abstract metaphysical values and an extreme utilitarianistic attitude (Kim, I.H. 1987).

It is true that Korean shamanism shows more concern with the life of the living in this world and thus attempts to utilize the power of the spirits for the sake of the prestige of living men and women whose final goals are the achievement of long life, wealth, fame and fortune in this world. The meaning of life in shamanic society is concrete and simple. Though the meaning of suffering is stressed in initiation disease and in the tales of shamanic ancestors such as Pari-Kongju, it was not adequately incorporated into the value system of the life of the profane people in the shamanistic belief system. Since it is thought that evil always comes from the outside and thus should be driven away, it is difficult in this way of thinking to realize the fact that evil also exists in our minds and that evil can also be a subject of moral reflection and integration.

However, we may not overlook the fact that the world beyond plays an essential role in Korean shamanism in dealing with present life problems. Misfortune is regarded as the consequence of disorder in the relationship of man to the cosmic deities. This means that Korean shamanism deals mainly with present problems through reflection and assimilation with the past that is beyond consciousness through the incorporation of images of the dead and ancestors. Therefore, the emphasis of scholars on the secularistic elements of Korean shamanism seems to be one-sided and perhaps influenced by the fact that many of the scholars of shamanism to date have taken as their point of reference, the Christian value system.

In regard to the moral codes and value system of Confucian culture which has prevailed since the Chosun dynasty, Korean shamanism seems to represent a diametrically opposite system in which direct communications with the supernatural beings in concrete form are maintained, wild ecstatic performances are common, the authority of supernatural beings is relativized, and aggressive, sexual outbursts of obscenity which express instinctive drives are allowed. And, while it is important to note that the social etiquette and morality of Confucianism is not denied by Korean shamanism (Choe, K.S. 1971, 1978, 1981), it is also clear that Korean shamanism counteracted, at least since Chosun-dynasty, the one-sided patriarchal face-keeping value system of Confucianism. This was done through the activation of the other aspects of man's mind the element which is characterized by a more earthbound emotional life and is rooted in the primordial longing of mankind to keep in touch with the power of numinosity. The ethical values of Korean shamanism lie probably not in preserving the daily moral etiquette but rather in compensatory activity of the primordial ethos, an irrational and transcendental one which is thriving for union in a situation of psychic dissociation by disclosing the hypocrisy of the time (Rhi, B.Y. 1986). In this respect, it is noteworthy that Korean shamanism, consciously or unconsciously, preserved abundant symbols which represent the primordial archetypal images of man's mind in terms of Jungian analytical psychology. Thus, this magico-religious system became a vessel of mythos which preserved the symbols that have been a creative driving force in Korean society. Examples of these symbols are, the symbols of wholeness manifested in the images of a couple of male and female deities depicted in shamanic songs and hung as divine images in front of the altar, motifs of heiros gamos in the mudang's relations to her helping spirits and in her dreams, the importance of the soul of the child *myong-do*, the effective power of divination, the helping spirits, the labyrinthine symbols of the shamanic talisman and the structure of rituals (Kerényi, K. 1950; Rhi, B.Y. 1984).

4. Concluding remarks

Korean shamanism shows us complex ethnographic and historical features. It is hardly possible to render in this brief review all the subtle differences of opinion of investigators on this phenomenon. Therefore I have dealt here mainly with the "healing" aspect of shamanism through examining the healer's unique ways of interacting with the world beyond and have confined myself mainly to the spiristic *mudang* and her psychic manifestations, where the ecstatic experience or trance state is presupposed in either a real or simulated form.

The mythical ecstatic experience has been practiced in Korean history not only by *mudangs* but also by priests from other high religions in the late Dai Han empire. As the mindset of the people turns more toward rationalism, due partly to the rapid industrialization of Korean society, the mudang's *kut*, including its technique of ecstasy, is increasingly being displaced by Holy Ghost movements and the faith healing of certain protestant churches in Korea.

Korean shamanism as a concrete historical and social system is changing, with a gradual diminution of the intensity of religious ecstatic implications. However, shamanic ecstatic encounters with the world beyond bloom in various social, religious systems other than the *mudang's* magico-religious system. Furthermore, the shamanic complex seems to have been organized into the personality traits of the Korean (Rhi, B.Y. 1983). Further investigation of these issues may bring fruitful results.

BIBLIOGRAPHY

Akamatsu, C.J. and Akiba T. (1937/1938). *The Study of the Korean Mu-customs*, Vol. 1/2. Tokyo: Osakayago-shoten.

Akiba, T. (1950). *Field Studies of Korean Mu-customs* (Japanese). Tenri: Yotoku-Shya.

Chang, J.G. (1975). *The Korean Folk Belief* (Korean). Seoul: Ulyu-Munhwa-Sa.

—— (1983). "Folk Religion". In T.H. Lee, J.G. Chang and K.K. Lee. *Introduction to Korean Folklore* (Korean). Seoul: Hakyon-Sa, p. 157.

Chang, J.G. and Choe, K.S. (1967). *Shamanism in Kyong-Ki Province*. Seoul: Cultural Bureau.

Choe, K.S. (1927). *Essays on Shamanism*. Kyemong 19, pp. 2–51. Seoul: Shin-Mun-Sa.

—— (1971). *The Korean Mudang*. Seoul: Yolhwa-Dang.

—— (1978). *A Study of Korean Shamanism*. Seoul: Asia Munhwa-Sa.

—— (1981). *Studies on Korean Shamanism* (Korean). Seoul: Hyong-Sol Pub. Co.

Chung, K.C. and Rhi, B.Y. (1975). "The Attitudes of Rural People toward Folk Treatments — A Socio-psychiatric Study of Rural Society in Korea (V)". *Neuropsychiatry* 14:405–416.

Clark, C.A. (1903/1961). *Religions of Old Korea*. Seoul.

Clement, F. (1932). "Primitive Concepts of Disease". *American Archaeology and Ethnology*, Vol. 32, pp. 185–252. Berkeley: University of California Press.

Eliade, M. (1951). *Le Chamanisme*. Paris: Payot.

Hulbert, H.B. (1906/1969). *The Passing of Korea*. Seoul: Yonsei University Press.

Hwang, S.Y. and Moon, M.D. (1984). *Rock Pictures at Ban-gu-Dae*. Seoul: Tongguk University Press.

Hyon, Y.J. (1986). *A Study of Shamanism on Cheju Island*. Seoul: Jib-Mun-Dang.

Jung, C.G. (1948). "Die psychischen Grundlagen des Geisterglaubens". In C.G. Jung, *Psychische Energetik und das Wesen der Träume*. Zürich: Rascher 2. Aufl.

Kendall, L. (1977). "Mugam: The Dance in Shamanism Clothing". *Korea Journal* 7 (12):38–44.

—— (1985). *Shamans, Housewives and Other Restless Spirits*. Honolulu: University of Hawaii Press.

—— (1987). "Let the God's Eat Rice Cake: Women's Rites in a Korean Village". In L. Kendall and G. Dix (Ed.), *Religion and Ritual in Korean Society*. Institute of East Asian Studies, Center for Korean Studies, University of California, Berkeley.

Kerényi, K. (1950). *Labyrinth-Studien: Labyrinthos als Linienreflex einer mythologischen Idee*. Zürich: Rhein Verlag.

Kim, I. H. et al. (1982). *The Comprehensive Study of Korean Shamanism* (Korean). Korea University. Seoul: Minjok-Munhwa Yonku-So Press.

—— (1987). *A Study of Korean Shamanic Thought*. Seoul: Jib-Moon-Dang.

Kim, K. I. (1972a). "Psychoanalytic Consideration of Korean Shamanism". *Neuropsychiatry* 11:21–129.

—— (1972b). "Psychoanalytic Study of Initiation Disease". *Neuropsychiatry* 11:223–234.

—— (1972c). "Folk Psychiatry in Korea: Part II, Shaman's Healing Ceremonies". *Korean Cultural Anthropology*. 5, pp.79–106.

—— (1972d). "Traditional Concept of Disease in Korean Culture." *The New Medical Journal* 15:49–52.

—— (1984). *A Psychoanalytic Study of Korean Traditional Culture*. Seoul: Shi-In-Sa.

Kim, K. I. and Kim T. K. (1970). "Analysis of Spiritistic Dreams". *Neuropsychiatry* 9(1): 47–56.

Kim, T. G. (1966). *A Study of Shamanic Songs about the World Beyond*. Seoul: Shang-Woo-Sa.

—— (1969). A Study of Songju Belief. *Collected Papers of Problems in Underdeveloped Society*, Vol. II, pp. 279–301. Kyonghee University, Seoul.

—— (1981). *The Study of Korean Shamanism* (Korean). Seoul: Jib-Mun-Dang.

Kim, Y. H. (1979). *Six Korean Women: The Socialization of Shamans*. San Francisco: West Publishing Company.

Kim, Y. K. (1971). *Studies on Korean Folklore and Literature* (Korean). Seoul: Il–Cho-Kak.

—— (1976). *The Korean Myth* (Korean). Seoul: Il-Cho Kak.

—— (1982). "Korean Shamanism and Folklore". In *The Extensive Study of Korean Shamanism*, pp. 57–90. Korean University. Seoul: Minjok-MunhwaYonku-So Press.

—— (1985)."Kut and Shinmyong" (Korean). In *Kut of Korea* 5. Seoul: Yol-Hwa-Dang.

—— (1986). *The Koreans — Who Are We?* pp. 186–196. Seoul: Jayu-Munhak-Sa.

Lee, D. H. (1987). "Brassware Contribution to the Shamanism Instruments in the Initiation Rite". *Sadae Nonchong*, Vol. 34, pp. 75–104. Seoul University.

Lee, K. B. (1958). *Tongguk Lee Sang Guk Jib*. Seoul: Tongguk Munhwa-Sa.

Lee, N. Y. (1927). *A Study of Korean Shamanism* (Chinese). Seoul: Kyemong Club.

—— (1976). *The Clay Figures from Shilla* (Korean). Seoul: Sejong-Daewang-Kinyom-Saop-Hoe.

Lee, S.I. (1981). *Kut and Play* (Korean). Seoul: Mun-Um-Sa.

Liu, D. S. (1975). *The History and Structure of Korean Mu-religion* (Korean). Seoul: Yonsei University Press.

Rhi, B.Y. (1965). The Psychiatric Study of the Process of Shamanisitc Initiation in Korea (Korean). Collected Papers in Commemoration of the 60th Birthday of Prof. Myung Joowan. Department of Neuropsychiatry, College of Medicine, Seoul National University, Seoul.

—— (1968). "On the Phenomenology of the Dead in Korean Shamanism". *Neuropsychiatry* 7:5–14.

—— (1969). "Psychological Considerations of Some Characteristics of the Shamanist Initiation Process". *Korean Cultural Anthropology* 2: 111–122.

—— (1970a). "The Folk Psychiatry of Korea (I) Concepts of Mental Illness in Shamanistic Society in Korea". *Neuropsychiatry* 9:35–45.

—— (1970b). "The Psychological Study of Shamanistic Treatment of Spirits with Special Reference to Analytical Psychotherapy". *The New Medical Journal* 13:65 – 80.

—— (1972). "Folk Treatment of Mentall Illness in Shamanistic Society in Korea". *The New Medical Journal* 15:191–213.

—— (1976). "Shamanism and Mu-culture". In S.I. Lee et al., *The Origin of Korean Thought* (Korean). Seoul: Pak-Yong-Sa.

—— (1978). "Understanding Demonic and Spiritual Phenomena from the Viewpoint of Jung's Analytical Psychology". In S. I. Lee et al., *The Origin of Korean Thought* (Korean), pp. 287–331. Seoul: Pak-Yong-Sa.

—— (1981). "Illness and Healing in the Three Kingdoms Period". *Korea Journal* (English), 21:4–12.

—— (1982). "The Psychological Considerations of Korean Shamanism". In I.H. Kim et al., *The Comprehensive Studies of Korean Shamanism*. Korea University. Seoul: Minjok-Munhwa-Yonku -So Press.

—— (1983). "Psychological Considerations of the Personality Characteristics of the Korean". In *The Ethical Concepts of the Korean*, pp. 227–276. The National Academy of Korean Studies, Seoul.

—— (1984). "Korean Folk Religion and the Collective Unconsciousness". In B.J. Chung (Ed.), *Life* and *Folk Art* of the Corcans, pp. 97–118. Daidong-Munhwa-Yonkuwon, Sungkyun Kwan University, Seoul.

—— (1986). "Korean Popular Belief and Ethic Consciousness". In The Korean Cultural Research Center, Korea University (Ed.), *Modern Society and Traditional Morality: A Search for an Order for Future Society*. Seoul.

Rhi, B.Y. Lee, C.K. and H.I. Chang (1970). "Three Case Analyses of Mental Disorders Associated with Folk Beliefs". *Korean Cultural Anthropology* 3:5–32.

Son, J.T. (1948). *Studies of Korean National Culture* (Korean) Seoul: Ul-Yu-Mun-Hwa-Sa.

Torii, R. (1950). "The Primitive Religions of Neighbor Folks (Japanese)". Cited in Akiba, T. (1950) above.

Yoo, J.H. and Rhi, B.Y. (1983). "Psychiatric Illness Associated with Folk Religion in Korea". *Neuropsychiatry* 22:233–252.

Japan

The character of Japanese thought

TOMONOBU IMAMICHI

Prologue

If we want to think about thought we must observe at least three points of view.

First, we must make clear the positive way to think, because the thought is the effect of thinking. So we must discuss the dynamic structure of thinking.

Second, we must make clear what is produced through the dynamic structure in question, namely, the thought produced by thinking. And third, we may not forget to observe what is produced by the thought, namely, the result caused by thought, e.g. social institution or cultural appearance in general. Perhaps there is an interaction between the thought and the cultural appearance. This interactive phenomen must be included in this third part. So we begin with the first problem, the dynamic structure of thinking.

1. Reflection on thinking

1.1 *On analogia entis*

The thought is formed, as we have mentioned above, through thinking activity. The thinking activity is accomplished by the line of logics. The logic is inherent in the language, which is ruled by the grammar. Therefore we must regard the characteristic points of Japanese grammatical structure in order to discuss the characteristic colour of Japanese thought. The following reflection does not require any linguistic knowledge of the Japanese language, although it refers to the grammatical structure. In order to make what I will discuss about Japanese language accessible to the reader, I would like to use the comparative method with Western language.

Being is said in many senses (τὺ ὄν λέγεται πολλάχως), so wrote Aristotle in his metaphysics, and this is summarized by Thomas Aquinas in the following form "being is said in two senses (Ens dicitur dupliciter)".
It means the grammatical *isoform* with semantic *difference*. For example:

Here *is* a box.
(a)
In the box there *is* an animal.
(b)

Contemporary philosophy. A new survey. Vol. 7, 269–281.

This animal *is* a cat.
(c)

As we see these three sentences we can immediately understand the meaning of the word grammatical isoform of semantic difference in regard to the word "is". The "is (a)" means the existence of an inorganic thing (box). The "is (b)" means the existence of an organic, living object (animal). The "is (c)" does not mean the existence of something, but it functions as copula, which combines subject and predicate in a sentence. So we can say without hesitation that there are at least two different senses under the same form, namely, designation of existence and copulative function of same being. "Is (a)" and "is (b)" designate existence, and "is (c)" designates copula. These three sentences are said in Japanese in the following form:

Koko-ni hako-ga *arimasu* (here) (box) (is) (a)	(Here is a box.)
Hako-no-naka-ni dobutsu-ga *imasu* (in the box) (animal) (is) (b)	(In the box there is an animal.)
Kono dobutsu-wa neko *desu* (this animal) (cat) (is) (c)	(This animal is a cat.)

It is not necessary to think about the details. The important thing is the fact that in the Japanese language there are three different grammatical forms for the three different semantic phenomena. Namely, the verb of existence of a lifeless inorganic object like a box is said *arimasu*, the existence of living thing like an animal is said *imasu*, and finally the copula is said in the form of *desu*.
Here we must state at least two very important things about Japanese thinking.

(1) There is no possibility of analogia 'entis' like in Western language.
(2) In the sign of existence there is a clear formal difference between the lifeless thing and a living existence.

1.2 *Analogia nullius rei (I mean strictly the genitive form of nihil)*

As I mentioned above, it is very difficult to discover the grammatical possibility for analogia entis in the Japanese language. Instead of this fact, or in other words in contrast to this fact, there is analogia nullius rei, analogy of nothingness in the Japanese language.

Here is *no* box. (koko-ni hako-*ga arimasen*)
In the box there is *no* animal. (hako-no-naka-ni dobutsu-*ga imasen*)
This animal is *not* a cat. (kono dobutsu-wa neko *dewa arimasen*)

For the negative form of existence in the English sentence 'no' is used, and for the negative form of copula *not* is used. "No" and "not" come, naturally, from the same etymological origin, but at least they do not have completely the same form. By contrast, in the Japanese language, the three negative forms "masen" are the same. If we use the formal sentence (not the oral form used in everyday conversation), the negative form for three different situations are expressed as "nai", or in the older form "arazu"; they are the same.

We can easily say that there is a possibility of analogy of nothingness, because the philosophical origin of negation is negation itself, negation as such, nothingness. And, as is well known, not only in Japanese thinking but also in Chinese or all over the Asian world, the nothingness which is thought beyond the being is the key word for philosophy. I will not now enter into this problem but I would like to make clear the existence of an antipodic contrast between the occidental and the oriental world to which Japan belongs, through denoting the contrast of analogy of nothingness and analogy of being. By the way, I must also remark on the linguistic difference between organic existence and inorganic existence in the Japanese language because it also gives a very interesting reflection about the technological problem afterwards.

1.3 *Fundamental meaning of the truth*

As I mentioned above, I wish to attempt a comparative study in order to make clear the character of the Japanese thinking. I know the danger of comparative study which sometimes exaggerates some outstanding phenomena. So I avoid using the superficial comparison and the impressive observation, but I wish to go into the fundamental basis. Also here we want to discuss the concept of truth through the philosophical reflection of linguistic fact. The Greek tradition is one of the most important origins of Western culture in regard to philosophy. So I will take an example from the Greek philosophy.

The word 'truth' in Greek is aletheia (ἀλήθεια), which means either "what comes out from the forgetfulness" or "what is from the forgetfulness": Therefore we can safely say that this word means 'clear from being present in the consciousness". How can we come to see this form (idea)? It is through the logos, because Socrates says that logos is the most important organ of the philosopher. (ὁ λόγος μάλιστα ὄργανον τοῦ φιλοσόφου) As everyone knows the logos has many important senses which are connected to each other in logical structure, namely, word — concept — syllogism — rule — calculation and theory. These senses make one system of thinking in the identity of the word logos, namely, in order to make clear the *truth* we must express the object in the *word*, and through the *logical calculation* in the form of *syllogism* we must find out the *rule* of phenomen and finally we must formulate what we think in the proposition as *theory*.

Therefore the truth is stated in the objective form of concept. The truth is logical form. Naturally there are some exceptional opinions in the occidental world on another form of truth, but principally what I have depicted is the authentic type of truth. Now, in the Japanese language what words are connected with logos and aletheia? There is an interesting word connection of the group of *koto* (situation). In the Japanese language the truth is *makoto*. *Ma* signifies perfect or beautiful and *koto* denotes the situation. *Kotoba* denotes the word but etymologically means a little part of a situation, and *kotowari* denotes the theory but etymologically it means analysis of a situation.

1.4 *Concrete images of the two types of truth*

What I have said about the truth as aletheia and truth as *makoto* will be here exposed through the concrete method of imagery. Let's take as an example a child falling into the river. What is the truth in this situation? For the objective descriptive truth of aletheia in the Greek tradition we must make clear the proposition with strictly measured words. So, we must say a child about six years old fell into the river. His clothes were red and he is driven away by the current of the river. The speed of the current is 5 m per second, the temperature of the water is 5°C, the child will be drowned within three minutes, etc. Every proposition is correct. We must say that they are to some extent true. But if we make efforts to stick to the facts as closely as possible, in order to assess the complete truth of the situation in this way, the more we delay saving the child. By the time we have written down all the facts the child will be dead. Even then we can continue writing down the facts, like how much water the child has swallowed, what is the direct cause of his death, etc. The idea, the objective correctness, can in this way destroy the given situation.

What is the truth according to the Japanese *makoto*? I have already said that *makoto* signifies the perfect situation. If we are in the process of arithmetical calculation, the perfect situation as *makoto* must be accomplished after finishing the theoretical calculation. But the perfect situation is not always accomplished through theoretical activity. The situation in which the child faces the danger of death is a given imperfect situation. It is remote from the truth as perfect situation. We must realize the truth as a perfect situation in the middle of the injured given situation. To jump into the river and to make efforts to save the child is the way to truth as perfect situation.

This comparison is naturally too simple. Nevertheless I dare to say it is not meaningless. In the Western tradition there are many long, discursive books which make efforts to demonstrate the objective truth through logical arguments. By contrast, in the Eastern tradition there are many symbolic abbreviations in which the writer as a philosopher realized the perfect situation. That is the reason behind the Zen-buddhistic meditation, in which the argument in discussion is strictly forbidden, and in which the suggestive expression after a long silence is welcomed.

Perhaps we may say without serious danger that the typical form of the Western philosophical truth is objective science as knowledge and the typical form of the Eastern philosophical truth is subjective wisdom as salvation. So, there is inclination to wards natural science in the Western philosophy and the inclination towards religion in the Eastern philosophy.

In this word series of *koto* (situation) we must say the word (a little part of situation, *kotoba*) cannot grasp a given situation (*koto*), not to mention the perfect situation (*makoto*) as truth.

2. Thought

2.1 *Japanese mythology*

Every nation has its own mythology and mythology is the natural origin of the characteristic thought of every nation because it is the reflection of destiny and unified imagination of pre-historical legend, sustained by the primitive world view of the nation.

Japanese mythology has a remote origin and it had been conveyed through oral recitation. But the oldest known written record was edited at the beginning of the 8th century. The book is called *Kojiki* (the record of the old things) and contains many old oral traditions.

In this book we can discover the botanic view of the world or vegetal assimilation of all the processes of the world.

All myths have an imaginative exposition of the origin of the cosmos. In the Japanese myth according to *Kojiki*, there is no god as creator. In the beginning of the *cosmos* (heaven and earth) there arose at first the god of the center (the apotheosis of seed) and then secondly two gods of binding (apotheosis of ripening of the fruit).

In the beginning of the world there was a watery chaos, the unstable vagueness of a jellyfish, and from this chaos something came out like a sprout of reed. And this sprout has a name which denotes the name of a beautiful male god of the reed sprout.

This god is the incarnation of the three unseen powers as symbol of seed and ripening. And this god the reed sprout is the origin of all the natural phenomena. Therefore we can confirm that the unconscious ground of the Japanese traditional thought has been formulated with the botanic language. It means that the Japanese world view has the tendency towards the reduction of human history to the vegetable dimension.

The influence of this vegetable assimilation and humanity has, throughout Japanese spiritual history, been so strong that many ethical and aesthetic terms used in modern times were originally adjectives for the vegetable phenomenon and that all family emblems have always been designed using the leaves of trees.

2.2 *Philosophy of wind*

The vegetable existence does not usually move from one place to another. Its dynamic movement is normally divided into two kinds, namely, the first growing up and down because during growth they develop trunk, branches, leaves and roots; the second is the horizontal movement of leaves of branches through the wind. Sometimes the leaves are prophets of the seasonal wind through their trembling, sometimes the branches are the organ of the wind for its roaring, sometimes the groves are the defense against the wind for the village. In such a way the wind is a correlative phenomen of vegetable. So the moral philosophy or aesthetic of the vegetable viewpoint of the world requests the philosophy of the wind as their essential part. Actually, the wind as natural phenomen itself is very interesting: it is essentially invisible, but it functions very effectively, namely, when it blows strongly like a storm it destroys the human dwellings and sinks the ships; and when it blows gently, it gives consolation to the tired eye or the sad and sorrowful ear. So the wind can be used as a symbol of the transcendental principle.

At the end of the Japanese mediaeval time the wind was used as the most important key word for poetics, because in the poem or art in general, the unseen and untouchable atmosphere must be created through the visible form and audible voice. And because the wind is the most suitable word for this unseen, untouchable and influencing atmosphere.

In the course of the year the spiritual importance of the symbol of the wind has been used more and more intensively, so that the wind became the terminology in the domain of moral philosophy and buddhistic religion.

In my opinion, in the natural world the form of the things shows their essence and function so clearly, that the form in general can become one of the most important key words for describing the world. So, the "idea" of Plato as principle denotes the form, as is well known. But technology has changed such a natural world so extremely that there are now many isomorphic and heterofunctional machines, in which the decisive priority of form in regard with function must be rejected. Therefore we must seek out some other principle for new dimensions in this technological world. And for this purpose, the wind as principle of the world could be helpful. At this moment I hope at least the suggestive force of Japanese classical philosophy for the revival of thought is possible or recommendable.

I would like to show some examples of the vegetable influence on the aesthetic terminology in our tradition.

The oldest word for "beautiful" is "kuashi", which means dense and compact natural growth of the leaves. That can be demonstrated through the philological facts but I refrain from going into linguistic details on it in this text.

In the medieval poetics there are many symbolic terms for the aesthetic fulfilment of poetic creation, e.g. "hana" (flower) as rhetorical beauty of the word, "tane" (seed) as creative moment of poetic activity, "taketakashi" (tall tree) as sublime, etc.

This tendency of reduction to the vegetability of the artistic or poetical activity is continuously inherited in the later periods of our history.

At the end of the mediaeval time in the dramaturgic theory of the 'no' play an outstanding aesthetician *Seami* used also the word hana (flower) as the most sublime and deepest value of dramatic expression. And, moreover, as he taught the principal attitude of actors in drama, he wrote that, in the drama of mystic divinity, people should be like a gigantic cypress, in the drama of grace like a blooming cherry tree, in the drama of tragic love like a marble tree in autumn with red burning leaves.

And in modern times, *Basho*, poet of the "haiku" (short poem consisting of 17 syllables) and at the same time an able aesthetician, said, as he taught his disciples how to create a poem, "learn the spirit of a pine tree from the pine tree itself". The tree is for him the teacher. Not only in the domain of aesthetics but also, and I think this is more important, in the domain of ethics there is a strong tendency of reduction to the vegetability of human morality.

The pine tree is a symbol of steadiness of unchangeable loyalty, the bamboo a symbol of straightforwardness and strong will, the plum flower is a symbol of dignity and a forerunner, and cherry blossom stands for sacrificing death at the peak of one's life.

Such similar symbolicism of vegetable phenomena can be found all over the world but, I must say, in Japan it has been far more used than in any other place.

2.3 *World construction*

In the *Kojiki* there are dramatic developments of the world of gods. These mythological stories are very interesting due to the interaction of many gods, who are mostly the personification of natural phenomena. So, not only because of the polytheistic structure but also because of the visual description of dramatic construction, there is a resemblance between Japanese myth in *Kojiki* and Greek myth in Hesiod and Homer etc.; so said a French orientalist, René Grousset. Although *Kojiki* resembles Greek myth in such a way, as many scholars say, there are two remarkable differences between both traditional myths. What are they?

The first is that, in the Japanese myth there is no war on the sea, but by contrast the Homeric epos contains the history of the war and wandering on the sea. The lack of the consciousness of this war on the sea in Japanese myth suggests that the Japanese, even in prehistoric time, had no perspective of foreign culture and no consciousness of the existence of other countries. The Japanese myth is accomplished with in the limited little islands united in Japan proper. There was a world theory, but it was constructed only vertically within the autistic narrowness of Japan itself, namely, there are first the land of the heaven (*Takama-nohara*, that is to say grassland of high heaven); and second the middle country of reed land (*Ashihara-no-nakatsukuni*), namely Japan itself. This country, Japan, consists of several islands, which were children produced by the gods. So, the gods, islands as nature and human beings were constructed as a relationship of family: namely, the gods are like father and mother, the nature and human beings are brothers and sisters, or at least relatives.

There is the movement of cosmic eros through the universe. Therefore nature is not objectified enough to be observed or to be used or abused as tool of the egoism of humanity, at least in the traditional thought in Japan. That is one positive side of Japanese myth, but at the same time there is no possibility of the strong consciousness of human dignity, entirely different from cosmic animality. The individuality of the person as the essence of human being, was not evident in the mind of the people at that time. It was hidden under the anthropomorphic description of the natural world as the dimension of human relatives. And third, under this world of the middle land there was thought to be the land of Hades (*Yomi-no-kuni*) wherein the dead people were hidden.

2.4 *Colours*

Although Japanese myth, as I have stated above, is full of dramatic scenes, which are visually described, there are only four colours nominated in the old text of *Kojiki*: namely, white, red, black and blue. It is really curious that we find only these four colours in the *Kojiki* and that in *Manyoshu* (10,000 leaves — the oldest anthology of traditional poems), edited almost at the same time, we find about 20 different colour names. What is the reason for this remarkable differences in both texts in this point?

We can guess without danger that the *Kojiki* maintained the older and primitive oral tradition, which is more original and older than the many poems created by poets who can be identified individually.

So, according to my opinion, these four colours in *Kojiki*, namely, white, red, black and blue, belong to the original, primitive arrangement of Japanese colour consciousness. We may not say, however, that these four colours are fundamental only in Japan.

Within the Chinese influence there are four protecting guardian animals which are posed at the four gates (East, West, North, South) of the capital city. These animals have their own colour: white tiger, red pheasant, black turtle, blue dragon. These guard animals must protect the yellow emperor (in Chinese tradition gold is symbolised by the yellow colour and also the Chinese natural world is the perspective of yellow immensity.)

Therefore in China the colour yellow is the supreme colour under which the other four colours are dominated.

But in the *Kojiki* there is no name for yellow colour. And at least in *Kojiki* there is no influence of such four animals guarding the gates of the capital. The four colours are mentioned already in the process of the wandering history of gods, long before the establishment of a câpital city in Japan. Therefore we can confirm that these four colours are really the original colour arrangement of the Japanese. According to my research (details of which cannot be presented here and so I recommend my book *Aesthetic in the Orient*), the four colours mentioned above have the most interesting anthropological symbolic relations among them. White is used only for the glory of god, the cleanness of the heart and the vividness of life. Therefore it is the symbol of luck.

Red is used only for the blood, which means the danger of life or the testimony of life; therefore sometimes it is the symbol of unluckiness.

Black is used for badness, dirtiness and also for death and for the world of Hades, where there is no light. Therefore it is the symbol of misfortune.

These three colours construct the human subjective destiny. By contrast, blue is used for leaves of the trees or grass, for sky, mountains and water. Therefore it is the symbol of the natural environment wherein the human existence, consisting of three colours, is situated. (Blue denotes the environment, this fact tells us also that the vegetable things are the most important elements of Japanese land, where the vegetal world view has been collectively created.)

In *Kojiki* the colour is anthropologically or even philosophically so important that, after the acceptance of Buddhism, the Chinese character of colour has been used in Japan in place of the word *form* in European tradition. In European philosophy the form (Platonic idea, Aristotelean eidos) is the essence of the thing, whereas in Japanese tradition, the colour represents the existence of the thing.

3. Reflection on the produced thing

Until now we have studied the two moments of thought: the first, its subjective activity, thinking, namely, logical structure; and the second, the characteristic points of the original objective thought in *Kojiki*, what was thought through the logics mentioned above.

Now we enter into the third part of this article, namely, the reflection on the produced thing. By the phrase 'the produced thing' I understand the material crystallisation of the thought and the phenomenal emergence of the thought supported by thinking.

By the words 'material crystallisation' I understand here only the work of art and by 'phenomenal emergence' I understand here only religious liturgy and moral deed.

3.1 *On art*

Very clearly everyone can remark that there is an antipodal difference between the traditional occidental painting and the traditional oriental painting. The former is the representation of the objective world in regard to form and colour. In this type of paintings there is no unpainted nothingness, it is always finished, finito. We can find here the influence of the objective, descriptive method of thinking and observational spirit of scientific thought: that is the result of the idea of mimesis which is highly valued by Plato and Aristotle as a fundamental idea for art creation. In the latter, namely, in the oriental painting, there is a concentrated expression of the one phase of one thing behind which there is the waste white space of nothingness.

There is no perfect form and there is even no real imitation of colours: i.e. the paintings are sometimes monochronic, that is to say there is no objective description in regard to the form and colour of the phenomenal world. Such paintings almost all the Western scholars call non-finito, because externally there is no fulfilment of objective description of the external world. But it is, according to the oriental idea of art, really accomplished, finitissimo, because the painter expresses the most essential in his perspective by ignoring the unnecessary things: that is the result of the idea of expression (in Chinese *shai*). It is unnecessary to mention that the Japanese art belongs to oriental art. So it is very difficult to distinguish how far

Chinese and how far genuine Japanese influences appear in such paintings. But there is a great difference between Chinese and Japanese architecture in the classical tradition. In Chinese architecture, e.g. on the Chinese temple, there are many artificial curves in the design of the roof, plenty of ornaments and a lot of colours, but the Japanese temple is very simple and ascetic, without curves, without ornaments, and without colour. Also there is a dissimilarity in the Chinese and Japanese paintings. There is a basic difference in regard to the intention behind artistic creation. In the Chinese painting the painter expresses his feeling or the spirit of the world intuited internally. By contrast, in the Japanese painting the painter chooses only the most impressive theme and suppresses other things, although they are sometimes necessary. So the Chinese painting is the result of the expression of eloquence and Japanese art is the result of suppression through silence. We must remember that in Japanese language the word is *kotoba*, which means the small part of *koto* (thing), so that the word cannot reach the splendid fertility of reality.

We must meditate about it in deep silence. The art is the entrance to this silent secret of the universe.

3.2 *Religious liturgy*

As is well known, in Japan there are now four major religions, namely Buddhism, Confucianism, Christianity (both Catholics and Protestants and a few other groups, like Russian Orthodox) and Shintoism. Among them the first three religions come from foreign countries, and Confucianism, although it is the first religion that had been accepted in Japan, is usually not enumerated as a religion but as a humanistic cultural system which penetrates the fundamental consciousness of the Japanese. And Christianity is, generally speaking, the religion for the intellectual persons in the cities, which were opened to Western culture after the Meiji restoration, the so-called Japanese modernization. (Naturally we may not forget old Christian tradition which has survived through martyrdom in the face of long persecution by the government. But the families who belong to this tradition are very few.)

Therefore the religions in Japan that are actually widespread through the population are only Buddhism and Shintoism. Anyhow, as a result of its doctrine Buddhism has been accepted in Japanese thought without changing its principal character. But in regard to the liturgy it has been, to some extent, Japanized. The liturgical change in religion is usually achieved through an attempt at accommodation stimulated by other religions, especially by the native religion. Therefore we must say that it is enough for us to argue on the character of the liturgy of Shintoism, which is the only native religion in Japan.

Shintoism derives from the Japanese word *Shintō*, which means the way of god, but it is polytheism. So there are many gods and godesses; so that in the oldest text, *Kojiki*, 8 million gods were supposed to exist. What sorts of gods are they? Principally they are the natural gods: sometimes every natural phenomenon is regarded as a god and sometimes the hidden spirit behind the natural phenomenon was regarded as a god.

3.3 *Philosophical meaning of Shintoistic symbols in its liturgy*

We have affirmed in the previous section that the proper Japanese religion, Shintoism, is a polytheistic natural religion.

Like many other natural religions the principal god is the god of the sun, the sun-godess in Japan. Being in heaven, she dominates over the whole world. But curiously there is a shrine to this godess on earth, in Ise. What, then is the function of this shrine? We may pose such a question because the godess is in heaven and not in this shrine and, moreover, there is no clear doctrine of *ubiquitas Dei* in Shintoism. Therefore the shrine must be empty since the sun-godess is in heaven.

In order to solve this problem, the symbolism of the mirror is effectively used. The mirror which shines by reflecting the sun is on the one hand the symbol of the sun herself and on the other hand it is the symbol of the judgement of god because it shows the face of the person who stands in front of it, the symbol of god. (Perhaps because of this second reason almost all the shrines, not only the ones devoted to the sun-godess but also the ones devoted to other gods, have mirrors as a substitute for god.) Therefore the function of the shrine is the place for liturgical movement in front of the symbol of god. The shrine must in principle be constructed only of unpainted wood. The roof is made of straw. In the construction no nails are used for fixing the wood.

What is the meaning of this architecture? It is clear that this architecture is a concrete form of the vegetable point of view of the world and that this simple naturalness of the shrine takes away the luxurious secularity from the mind of the visitors. So the final end of the liturgy is the pure, silent encounter with god in a remote place, away from the common world. But in order to reach this place there are presupposed steps to the liturgy. One must enter through the gate, called *torii* (seat of the birds as messenger of god). This signifies that the visitor exists after this moment, not in the material world, but in the garden of the spiritual. In order to show it visually, small white sand stones cover the ground, and acoustically the trees bring the voice of the wind as a sign of the vividness of nature. Through this atmosphere the visitor already acquires an elevated mood. The visitor must wash his hands and clean his mouth with fountain water in the little hall. This is the symbol of purification of the body, but it is the reminiscence of the old style of purification when visitors had to enter a river to clean themselves.

In order to purify the mind from sin and crime, after the purification by water, the visitor comes to the front of the shrine and claps his hands twice in order to draw divine attention to himself. He bows and then the priest comes with a symbol of trees, made of white paper and wood, and swings it above the bowing head of the visitor in order to drive away sin and crime.

The origin of this liturgy of purification was perhaps the oldest ascetic deed in the river and in the mountains because, as I have mentioned, in Japanese natural religion the natural entities and phenomena like river, mountain and winds are gods as such. Therefore the liturgical form mentioned above is the practically compacted style of purification without ascetic essence, but it retains the psychological protection of the divinity. The central thought of this liturgy is clearly written in the text of "Norito" (prayer) of "Ōtsugomori" (the last day of the month) in "Engishiki" (one of the liturgical canons edited in the 9th century): the principal thought behind this canon is that when the sins and crimes of every sort become replete in the country of Japan, then the priests of all the shrines have to read this canon in order to evoke the divine winds which sweep out all the sins and crimes like dirt into the water of the sea where a certain ocean god swallows them and brings them to the bottom of the ocean in order to split them out of the mouth into the split of this bottom, so that all the world gets purified.

This optimistic idea of purification is like a legal institution, automatically operated without subjective regret, moral reflection, or religious sense of guilt. We can confirm that here the excessive longing for cleanliness is so strong that it is not permitted to reserve any space for subjective remorse which contains the shadow of dirty sin. So here we have a religion without inner repentance. The sins and crimes must be completely forgotten in order to live brightly in the difficult world. Although this superficial optimism for the clear dimension without repentance has been to seem extent remedied and deepened by the spiritual influence of Buddhism, *received* in the 7th century but especially *accepted* after the 12th century (by received I understand taken in hand objectively, and by accept I understand to take into the heart), in my opinion, even today we can perceive its influence in the excessive use of game centre and TV in the daily life of Japan, because such technological apparatus makes people forget every thought of internal conflict brought by personal imperfection and makes them recover their vitality in the real world.

So technology takes the place of religion in regard to the so-called "purification". The religious festival is also transformed into popular performance wherein the substitution of religious ecstasy is experienced in order to forget the spiritual stress.

In the realm of philosophy of religion in Japan we can also discover such an influence of so-called purification in two delicate ways. The first, in Shinran Buddhism there is an avoidance of relying on the institution and the accentuation of reliance of the Buddha only through prayer. Even good people receive salvation, not to mention the bad ones. This means that the good persons can enter paradise through their own good deeds without the grace of Buddha, which is very strong. The bad person, to whom Buddha gives his grace through sympathy, must be more definitely salvated than the good person without grace.

The second is the opposite of this thought, namely, the independent attitude of Zen-Buddhism which teaches salvation through ascetic meditation which enables us to be unified with the truth. One half of this doctrine is the complete reaction against the superficial attitude without asceticism, but the other half is the inheritance of the forgetfulness of phenomenal dimension found in the old Shintoism.

3.4 *On moral deed*

The simple optimistic purification of Shintoism without any guilty sense has been amended or moderated through the influence of Confucianism, which the Japanese spiritual world accepted from China in the 4th century. The moral philosophy of

Confucius teaches five principal virtues which are charity (love), responsibility, sublimity of deed, of which the summit is liturgy, cognition, and confidence. The very important fact is the inventive conciousness of the second virtue, responsibility. As I have already shown both philologically and philosophically in my many other articles, that virtue, responsibility, is of relatively recent origin in occidental history. One cannot discover any equivalent Greek word for it and there is no word *responsabilitas* in the classical or medieval Latin language. The word 'responsibility' which we find in John Steward Mill is almost punishability and the German word *Verantwortung* or *Verantwortlichkeit*, which was created in the nineteenth century, meant only *Zurechnung*. The real sense of responsibility as moral response of intersubjectivity has been realized in western society in the twentieth century. But in the oriental world, especially in China, the moral inter-subjectivity was realized already in classical times through the philosophical reflexion of Confucius, because in his opinion the fundamental fact of human existence is intersubjectivity or interindividuality. At any rate in Confucius the vertical responsibility to Heaven and the horizontal responsibility to fellowship were strongly known. The individuum who failed to be responsible in both senses means the existence outside of the intersubjectivity as human fundamental condition, that is to say, the individuum in question lives no more as human being and here one animal respires, because in the oriental world there had been no idea of "persona" until Wang Yang Ming, who proposed the concept "Ryang Tsi" as *Gewissen* in later times. The human being who failed to be responsible lost face as a human being. He lives no more as a human but as an animal. Where is the meaning of his life as a human being? This reflexion guides the Japanese to the ethical idea of suicide. Especially on the chivalric morality the suicide known as "Harakiri" is a moral idea for the chivalric person who lost face. But suicide as a moral deed of responsibility has been found very often, even in the class of normal merchantman, namely in the citizenship. The work distributed to the individuum seems to him sometimes to be the task to which he must consecrate his life in order to be responsible enough. This idea is on the one hand very good for the effective functioning of the collective project, but on the other hand very inhuman, because human like seems to be thought of as a functional part of a machine. It is the defect of a tradition without the concept of "persona". But even here we can discover the tragic beauty of the humanity which is consecrated to the virtue responsibility.

Bibliographical notes

For Section 1: Reflection on thinking, see my book *Betrachtungen über das Eine* (1986). For Section 2: Thought, see my books *Toyo no Bigaku (Aesthetics in the East)* (1980) and *Studia comparata de aesthetica* (1976). For Section 3: Reflection on the produced thing, see my book *Tozai no Tetsugaku (Philosophy in the East and in the West)* (1981) and my a rticle "Contrariete et compatibilite" in *Actes du Congres de Santa Margherita*, pp 55–71. Genova, 1990.

Le concept de *ma* et ses transformations sémantiques comme voie d'accès à l'esthétique japonaise

NORIKO HASHIMOTO

Homo est animal loquens: l'homme est un animal qui parle. Le langage est un des moyens de cristalliser la culture; il est propice à la créativité. La transformation sémantique du langage, c'est-à-dire sa modification fonctionnelle, conduit toujours au changement modal de l'être et de la conscience, et inversement.

Je dois préciser tout de suite le sens que je donne au terme langage. Ce terme peut désigner non seulement le mot, la phrase, le discours, la signification, la structure grammaticale, la configuration rhétorique, mais aussi l'aspect visuel du verbe, le signe, le caractère qui le représentent. Ce que j'entends ici par langage, c'est uniquement le mot et son expression. Il est donc inhérent au mouvement de la conscience. Le langage est pour moi l'explosion d'énergies potentielles, ces dernières se transformant elles-mêmes en expression. L'homme est le sujet même de cette transformation et sa conscience en est le lieu.

On a appelé *ma* ce lieu qui stimule ces énergies potentielles et transforme cette potentialité en réalité. Définir le concept de *ma* est une des voies d'accès à l'esthétique japonaise. La base du matériel de recherche que j'utilise est constituée de textes classiques japonais parus au VIIIe siècle, le *Kojiki* et le *Nihon-Shoki*. Ces textes relatent l'origine du monde et du Japon; ils contiennent en germe de nombreux éléments philosophiques.

1. Origine du mot en tant que phénomène

Dans la mythologie japonaise, après la séparation du Ciel et de la Terre, les sept divinités les plus anciennes firent leur apparition l'une après l'autre. Elles étaient unisexuées (*hitori-gami*) et invisibles. Il n'y avait pas d'environnement linguistique, donc pas d'histoire pour ces divinités. Il n'existait qu'une productivité cosmique. Après l'apparition du dieu de la Terre et de son épouse, arrivèrent les couples de divinités, c'est-à-dire les divinités-couples des racines végétales, celles de la reproduction animale et, enfin, la plus importante des divinités-couples, Izanagi et Izanami qui créèrent les îles du Japon. Ce sont la racine végétale et la reproduction végétale qui figurent la possibilité pour l'être humain de se développer dans les mêmes conditions que les plantes. La référence à l'animal est très rare: il n'est que le symbole de la puissance reproductive. On retrouve même une conception végétale du monde au commencement du *Kojiki*, comme le suggère le professeur Imamichi.

Contemporary philosophy. A new survey. Vol. 7, 283–289.

L'environnement linguistique est né avec l'apparition des couples de divinités. La mythologie, donc l'histoire, nous dit que les anciennes divinités ordonnèrent soudain au couple Izanagi et Izanami de quitter le Ciel pour créer un pays: «Réunissez les terres flottantes pour en faire une île.»[2]

La première forme de langage dans l'histoire japonaise fut ainsi le mode impératif. Cette expression impérative avait deux objectifs; le premier était d'exiger l'obéissance à l'autorité supérieure; le second était d'ordre esthétique: créer une belle île.

Il est intéressant de voir que dans la Bible aussi la première forme du phénomène linguistique est impérative: «Que la lumière soit!» Mais dans notre histoire, le mode impératif n'exprime pas la création à partir du néant, *creatio ex nihilo*, mais l'autorité céleste attachée aux divinités-couples personnifiées. Ce que firent Izanagi et Izanami n'était pas de la *création* mais de l'*instauration* qui est, si l'on se réfère à la définition de l'art d'Étienne Souriau, une forme d'activité artistique.

L'origine de l'éthique et de l'esthétique se trouve par conséquent dans la forme impérative de l'expression spontanée. Éthique et esthétique s'exprimèrent d'abord par le commandement, en vue de l'avènement de quelque chose d'important. Car l'expression impérative possède une force intentionnelle propice à la création. C'est la forme linguistique de la relation de l'autorité divine à la personne humaine.

Izanagi et Izanami se tenaient sur le Pont Flottant Céleste unissant la Terre au Ciel. Ils créèrent l'île à l'aide d'une lance céleste couverte de pierre précieuses en remuant le sel dans la mer en provoquant le son «*koworo, koworo*»[3]. Cela signifie que la première création des dieux incarnés fut accompagnée de sons.

Dans le *Kojiki*, «*koworo, koworo*» est la première imitation sonore, l'onomatopée. Ce son vient de la nature environnant les activités des dieux personnifiés; elles transforment la nature en instrument de musique. On en déduit que le premier son vient de la nature sous forme d'un instrument de musique.

Après avoir créé les îles, Izanagi et Izanami commencèrent à se parler; ils échangèrent des mots d'amour. Cette conversation amoureuse ne fait nullement recours au mode impératif. L'existence de cette relation transforma la mort d'Izanami en véritable tragédie pour son compagnon. Et cette tragédie fut à l'origine du dialogue, lequel dépasse la dimension de la simple conversation. La plus grande particularité du *Kojiki* réside en effet dans le dialogue vivant entre Izanagi et Izanami, lorsque Izanagi se rendit au pays de la mort pour y rencontrer son épouse. Là, ils parlèrent de la naissance et de la mort. Les dialogues créatifs qui portent sur ces thèmes sont à l'origine de la pensée esthétique.

Izanagi (divinité masculine) et Izanami (divinité féminine) prirent la forme d'un homme et d'une femme. La déesse donna naissance aux nombreuses divinités liées aux phénomènes naturels. À la naissance du dieu-feu, son utérus s'embrasa et elle mourut. Elle partit alors pour le pays des morts. Izanagi, divinité mâle, en fut accablé: «Ah, mon épouse bien-aimée! Le dieu-feu a provoqué sa mort et la voici redevenue enfant!»[4] Il voulut lui rendre la vie; en vain. Pleurant sa détresse, il se jeta à terre et se roula sur le cadavre de son épouse. Se tordre, s'agiter pour évacuer la souffrance est l'aspect corporel du désir: son expression linguistique est un combat acharné. Quand ce désir n'est pas satisfait, quand son objet est au-delà des capacités humaines, il est en toute logique lié au désespoir, lequel est la conscience de l'éternel éloignement de la transcendance.

Au contraire, le plaisir revêt une autre forme corporelle dans son expression linguistique. Dans la mythologie japonaise, le soleil est la principale déesse, fille de la divinité mâle Izanagi. Elle avait un frère extrêmement violent dont elle ne supportait plus la conduite immorale et la cruauté. Alors elle se cacha dans une grotte céleste; et ce fut soudain l'obscurité sans fin. Tous les dieux restèrent perplexes, se recroquevillèrent et se mirent en boule, angoissés par ce monde noir et froid, où plus rien ne se manifestait excepté les démons. Chaque divinité exprima physiquement un intense désir de lumière solaire.

L'une des divinités, appellée Omoikane-no-kami (dieu de l'intelligence humaine), donc plus maligne que les autres, trouva une ruse: elle fit danser une jeune déesse à moitié nue. Toutes les divinités en furent émues et elles se mirent à rire et à pleurer. Le Ciel fut bientôt rempli de bruits divers, les dieux se mirent à rire bruyamment, ce qui excita la curiosité de la déesse cachée. Elle se demanda ce qui pouvait bien se passer dans le monde. Lorsqu'elle apparut à l'entrée de la grotte, le visage de toutes les divinités fut éclairé par les rayons du soleil.

«*Omo Shiro*»[5] (le visage est blanc) fut le premier mot prononcé par les dieux lorsqu'ils se découvrirent un visage illuminé. Ce mot est à l'origine d'*omoshiroi* qui veut dire: amusant ou intéressant, particulièrement agréable. Les dieux se sentirent alors détendus et commencèrent à s'étirer. Ils dirent: «*Ana ta noshi*»[6] (Ah! les bras sont allongés) (*tanoshi* veut dire: agréable). Leur esprit pu se délivrer du désir pénible que provoquait la nuit profonde et froide. Soulagés, les dieux montrèrent un visage riant et étendirent leurs membres librement. Le plaisir s'exprime donc physiquement par un étirement agréable de tout le corps.

Il peut donc s'exprimer et se manifester selon les aptitudes de l'homme. Cette possibilité ne se trouve plus à une distance infinie de l'être humain, mais à une distance mesurable.

Je ne parle pas ici du problème du langage du corps, mais de l'aspect corporel du langage exprimé. Il ne s'agit pas du geste formel ou du signe, mais de la réaction corporelle. La réaction corporelle propre au désir est la fermeture sur soi, nécessaire à l'élan qui permet d'atteindre le but désiré. La réaction corporelle propre au plaisir est l'étirement, donc la tranquillité dans une situation d'accomplissement.

Pour ce qui est de l'aspect corporel du langage, plaisir et désir diffèrent donc totalement. La distance qui les sépare dans le monde naturel est immense. Ainsi le poète Issa Kobayashi, désirant acheter en plein hiver une poire pour son père mourant, la chercha en vain toute la journée durant. Ce désir endolori invita Issa à le cristalliser dans un journal poétique (*Chichi no Shuen-nikki, Journal de la mort du père*).[7] Mais dans le monde de haute technologie où nous vivons, lorsque nous cherchons quelque chose «hors saison», nous pouvons presque toujours l'obtenir, si certaines limites économiques et financières ne nous l'interdisent pas. Du même coup, la distinction essentielle entre désir et plaisir tend à disparaître et le fossé qui les séparait diminue de jour en jour.

Quand le professeur Imamichi nous proposa une réforme de la structure logique du syllogisme pratiqué dans un contexte technologique,[8] il montra que l'objet du désir humain ne peut plus être situé au-delà du monde séculier. Cet objet doit être logiquement limité au matériel, c'est-à-dire à l'immanent, autant que le permet la réforme de la structure logique. Mes recherches sur l'aspect corporel du langage concernant le désir et le plaisir pourraient répondre à l'étude logique du professeur Imamichi.

Les progrès technologiques permettent de transformer d'emblée nos désirs en plaisir satisfait. Le mot désir dans notre vie quotidienne a perdu toute la dignité qu'il contenait à l'origine. Dans la conversation, le mot désir tend presque toujours à se confondre avec vénalité et cupidité. Il évoque pourtant *desiderium sensibile*; et *desiderium* sous-entend besoin de transcendance. L'essence du désir demeure le besoin spirituel de transcender le monde. Mais si nous prononçons aujourd'hui le mot désir, l'objet de ce désir est dans presque tous les cas sensuel ou matériel. Cela n'est d'ailleurs pas péjoratif dans tous les cas. Mais, de mon point de vue, la confusion ou l'association du désir avec le plaisir ne doit pas dominer. C'est rabaisser le désir au matériel. Il devient alors difficile d'éveiller la conscience éthique et esthétique, car éthique et esthétique doivent reposer sur un désir spirituel pour quelque chose d'autre qu'un objet matériel ou une attirance sensuelle.

Je voudrais en venir maintenant aux trois points suivants: 1° l'aspect corporel du langage peut être méthodologiquement utile dans toute étude éthique ou esthétique; 2° *a*) le désir est l'attirance pour quelque chose de mieux, même dans des conditions pathétiques; *b*) le désir doit être gardé à distance du plaisir qui, lui, s'apprécie par l'auto-évaluation rétrospective d'un long effort; 3° cette distance doit être subjectivée par l'idée du *ma*. So concrétisation permet de définir la *perspective* d'abord, et l'intervalle ensuite; enfin, cet *intervalle* doit être transfiguré en *ma*.

2. Vocatif et *ma*

Le langage est l'expression d'une victoire remportée sur le silence. Le silence est l'attente intense d'une concentration parfaite. Il est l'expression du résultat d'une telle concentration. Il est donc la cristallisation d'un pouvoir d'intériorisation en tant qu'être, en tant qu'existence.

J'ai montré que la première description du phénomène linguistique est, dans la Bible aussi bien que dans les textes philosophiques japonais, le mode impératif. L'origine de l'éthique et de l'esthétique se trouve dans la forme impérative de l'expression spontanée.

Le professeur Imamichi a souligné l'importance du cas vocatif et le professeur Olivetti, celui de l'accusatif. Ce dernier a confronté son point de vue au concept «me voici» de Levinas.[9] Il a développé cette idée d'intersubjectivité à partir de la relation entre l'*ego* et l'*alter*.

En étudiant le *Kojiki*, j'ai cité le cas particulier du dialogue entre les deux divinités Izanagi et Izanami. J'ai signalé le phénomène éthique et esthétique de l'origine des mots désir et plaisir, et souligné l'importance de la coïncidence entre aspect corporel et expression linguistique. Je cite *kotodo* (le divorce) et *seiyaku* (le serment) pour définir la distance créative servant au dialogue et le rôle du vocatif dans le dialogue.

Dans le cas de la *creatio ex nihilo* dans la Bible, et aussi dans la création du pays dans le *Kojiki*, j'ai fait remarquer que l'expression impérative a une force intentionnelle propice à la création. Mais dans une dimension plus éthique et plus esthétique, je voudrais parler d'évocation entre deux personnes. Le vocatif, dans ce cas, est le silence avant l'expression. Finalement, j'ai recherché la structure du *ma*, non seulement en termes d'intersubjectivité entre *ego* et *alter-ego* mais aussi dans l'*ego* et l'expression de l'*ego*.

Izanagi et Izanami sont les divinités-couples; elles créent les terres et de nombreuses divinités de la Nature. Après la mort d'Izanami, son époux veut la revoir et descend au pays des morts (Yomi, Hadès). Lorsqu'elle vient l'accueillir à l'entrée, Izanagi dit: «Ô mon épouse bien-aimée, les terres que nous avions commencé à créer ne sont pas encore achevées; tu dois revenir!»[8] Izanami répond alors: «Comme je regrette que tu ne sois pas venu plus tôt! J'ai mangé au foyer de Yomi (pays de la mort et de l'après-vie). Mais, ô mon époux bien-aimé, quel malheur que tu sois entré ici! Je vais donc discuter de mon désir de revenir avec les dieux de Yomi. Mais surtout ne me regarde pas!»[11] Disant cela, elle retourna dans l'entrée, mais son absence fut si longue qu'Izanagi ne put attendre plus longtemps.

Je définis le désir par l'attente vaine de la transcendance. Dans le passage mentionné ci-dessus, le désir est celui de quitter la mort pour retourner au monde des vivants. C'est possible, dit l'histoire, à condition de ne pas regarder Izanami qui vit maintenant à Yomi (Hadès).

La phrase: elle a mangé la nourriture de Hadès, est claire. À Yomi, le désir n'est satisfait que grâce à l'interdiction de regarder Izanami (regarder en arrière). Le désir est donc l'envie de transcendance sans intentionnalité rétrospective.

Parce que Izanagi céda à la curiosité et à son amour, il regarda Izanami à la dérobée et la vit avec un peigne aux dents enflammées. La réaction d'Izanami fut: «Il me fait honte!»[12] et elle laissa les déesses courir après lui.

Finalement Izanami elle-même se mit à le pour-suivre. Alors il tira un énorme bloc de pierre et ferma l'accès de Hadès. Il se retrouvèrent face à face, de chaque côté du rocher, et rompirent leur serment. Izanami dit alors: «Ô mon époux bien-aimé, si tu agis ainsi, j'étranglerai à mort chaque jour mille habitants de ton pays.»[13] Izanagi répondit: «Ô mon épouse bien-aimée, si tu agis ainsi, je construirai chaque jour mille cinq cents huttes pour l'enfantement.»[14] C'est pourquoi mille personnes meurent chaque jour et mille cinq cents autres naissent.

Je voudrais revenir sur la rupture du serment liant Izanagi et Izanami. La poursuite de l'un par l'autre est la course de deux personnes dans la même direction. La situation d'une personne courant après l'autre se trouvant à une certaine distance n'est pas propice à la conversation. Pour renverser cette situation, Izanagi déplace un bloc de pierre et la relation des deux personnages devient un face à face. Le rocher définit la distance qui sépare Izanagi et Izanami. Dans le premier cas, poursuivant et poursuivi sont à distance mais dans la même direction; dans le deuxième cas, l'objet marque la distance quand il y a deux directions pour permettre une rencontre. Cela crée un espace pour le dialogue, mais c'est à ce moment-là que le serment est rompu.

Le divorce d'Izanagi et d'Izanami fut à l'origine d'une nouvelle guerre. Cette guerre fut le commencement de la lutte entre la mort et la naissance.

En résumé, l'objet placé entre deux personnes engendre la confrontation, il crée l'atmosphère propice au dialogue, mais cela finit par une déclaration de guerre.

Finalement, Izanagi eut trois nobles enfants, à chacun desquels il confia une mission pour gouverner le monde; à Amaterasu Omikami (déesse du Soleil), il confia le Takamagahara, le royaume du Ciel; à Tsukuyomino-mikoto, le royaume de la Nuit; et à Susano-no-mikoto, l'océan. Mais c'est là qu'eut lieu la première manifestation de désobéissance. Susano-no-mikoto ne gouverna pas l'océan, car il voulait aller au pays de sa mère, c'est-à-dire celui de la mort. À cause de cette insoumission, Susano-no-mikoto fut banni par Izanagi.[15]

Avant de devenir victime de cette expulsion, il monta au ciel pour rencontrer Amaterasu Omikami. Il dit: «Je suis venu pour vous faire part de mon intention de partir. Et il n'y en a pas d'autre.»[16] Amaterasu Omikami répondit alor: «S'il en est ainsi, comment pourrais-je être sûre que tes intentions sont pures et claires?»[17] (Dans le *Kojiki*, l'expression *kiyoku-akashi*: pur et clair, transparence et clarté du cœur — expression de loyauté —, est absolument essentielle.) Puis Susano-no-mikito dit: «Prêtons serment et donnons naissance à des enfants.»[18] Sur quoi, ils se tinrent de chaque côté de la rivière et prêtèrent serment.

Revenons sur cette dernière phrase: «Ils se tinrent de chaque côté de la rivière.»[19] Ici, la distance entre les deux personnages (les divinités) est matérialisée par la rivière, la rivière céleste. Existe-t-il une différence entre le bloc de pierre et la rivière? Quelle est la signification de la rivière? La rivière coule et change constamment. Elle est donc le symbole de la nature vivante. Dans le *Kojiki*, la rivière est un lieu de purification; ainsi, Susano-no-mikoto se purifie avant de rencontrer Amaterasu Omikami. La purification mène au cœur pur, clair, limpide. La rivière représentant la distance entre deux personnes prêtant serment est un symbole de responsabilité. Parce que la responsabilité est la condition préalable au serment.

La notion de séparation créée par deux personnes comme condition de dialogue réel doit être clarifiée. Dans le dialogue réel, il n'y a ni rocher ni rivière; rien ne sépare les deux interlocuteurs. Mais je prétends que ce rien n'est pas le vide, mais quelque chose de virtuel et de potentiel. Cette virtualité et cette potentialité font que cette distance devient un véritable espace de dialogue. Et cet espace ménage la possibilité de réalisation d'un néant riche.

Dans le système technologique, il existe un lien entre deux interlocuteurs. Deux personnes qui conversent, par exemple, au téléphone sont reliées par un réseau électronique, mais ce réseau disparaît de notre conscience. On peut obtenir quantité d'informations au moyen d'un ordinateur tout comme on peut les y introduire en manipulant simplement des touches. Dans le système technologique, la confirmation reste le seul acte personnel, quand même il se fait mécaniquement: confirmation veut dire: ordre de départ («Commencez!» propose la machine; «D'accord», répond-on en appuyant sur la touche de confirmation).

Tous ces systèmes mécaniques, électroniques, informatiques sont comme le rocher entre Izanagi et Izanami. Toutes les machines proviennent de minéraux extraits de différentes sortes de pierres. Pour utiliser une métaphore, l'esprit de la pierre gagne le corps en tant que système organique. Dans le milieu technologique, la distance séparant deux êtres est définie par un système mécanique comparable à l'esprit de la matière.

En conclusion, j'aimerais clarifier les cinq situations décrites plus haut du point de vue de la fonction vocative.

1° La poursuite: la personne qui court après l'autre (poursuivant) exprime un vocatif. Mais ce vocatif est unidirectionnel. J'affirme que la distance est un intervalle créé dans la même direction, mais un intervalle mouvant.

2° Le divorce: la distance représentée par le rocher est un intervalle immuable, mais l'expression du vocatif est échangée avec un autre. L'issue est la lutte.

3° Les serments: la distance de séparation est définie par la pureté et la clarté du cœur. Cette distance est l'intervalle de responsabilité.

4° La cohésion technologique: les réseaux mécaniques définissent la distance (intervalle décidé), il n'y a pas d'espace pour le vocatif, il y a seulement des confirmations.

5° Le dialogue réel: le *ma* créé librement. Cette distance suscite le vocatif vivant séparant deux êtres. C'est l'espace du vocatif latent, virtuel.

NOTES

1. Ma est un mot japonais qui veut dire "entre" au regard de la spacialité et de la temporalité. Spécifiquement au domain d'esthétique le mot japonais Ma est très souvent utilisé comme terminologie sans traduction. J'espère que l'on puisse entendre ce terme "ma" graduelement par mon explication de cet article.
2. Kojiki (Vieux Document), p. 53, Nihon-Koten-Bungaku-Taikei (Tokyo: Iwanami, 1983).
3. *Ibid.*, p. 53.
4. *Ibid.*, p. 61.
5. Kogo-shui.
6. *Ibid.*
7. Issa Kobayashi, Chichi no Shuen-nikki (Journal de la mort du père) p. 414–415 (Issa-shû) Nihon-Koten-Bungaku-Taikei 58 (Tokyo: Iwanami, 1982). L'idée d'eco-ethica pour l'innovation de la pensée morale, pp. 1–24 Journal of the Faculty of Letters, The University of Tokyo, Aesthetics, Vol. 6, Problema ethica et Eco-ethica.
8. Tomonobu Imamichi, *Betrachtungen über das Eine* (Tokyo 1968), p. 12.
9. Emmanuel Levinas. *Autrement qu'être ou au-delà de l'essence* (La Haye: Martinus Nijhoff, 1978), pp. 183–194.
10. Kojiki, pp. 63–65
11. *Ibid.*, p. 65
12. *Ibid.*, p. 65
13. *Ibid.*, p. 67
14. *Ibid.*, p. 67
15. *Ibid.*, p. 75
16. *Ibid.*, p. 75
17. *Ibid.*, p. 75
18. *Ibid.*, p. 77
19. *Ibid.*, p. 77

The concept of creation in the philosophy of Kitaro Nishida

B. RYOSUKE INAGAKI
Kyushu University Fukuoka

1.

In this essay, I shall present an intelligible account of one aspect of the traditional metaphysical thought of Japan. At the outset, I would like to make it clear that I am not writing as a specialist on the history of Japanese thought. Rather I am writing this essay as a Japanese philosopher who is concerned about his identity as a Japanese. I have reflected on the problem of Japanese spirituality in my search for the national identity, and this paper is a part of the fruits of that reflection.

It is often said that there is a radical opposition between the western metaphysical thought, which is said to be founded upon some basic affirmation of being and substance, and the Japanese metaphysical thought, which is characterized by the fundamental negation expressed by the notion of nothingness and void. Faced with this opposition, and with the aim of opening the way towards a dialogue, some Japanese philosophers have attempted to show that there are, at least, some common elements in these two traditions. They attempted to show, for example, that within the metaphysical tradition of the west, there is a distinct element of *philosophia negativa* developed by such mystics as Psd. Dionysius, Eckhart, Tauler, and Ruysbroeck, which is strikingly similar, according to them, to the Japanese metaphysical thought of nothingness.[1] Such an attempt, if carried forward successfully, could open a way towards a fruitful dialogue between two metaphysical traditions.

In this paper I have taken a similar approach, but in a different direction. I shall attempt to discover the idea of transcendence in the traditional metaphysical thought of Japan, which has been, in most cases, characterized as a form of immanentism. If I can show that the alleged immanentism of Japanese metaphysical tradition does not really exclude the element of transcendence, that is, the fundamental affirmation of the absolute which transcends the phenomenal world, it would mean that there is another possible way towards the mutual understanding between these two metaphysical traditions.

2.

First, I shall make some preliminary remarks for the clarification of the state of the question. Dr. Hajime Nakamura, a distinguished philosopher of Buddhism and renowned specialist in comparative philosophy, makes the following observation at

Contemporary philosophy. A new survey. Vol. 7, 291–301.

the beginning of "The Japanese Way of Thinking," which forms the fourth section of his opus magnum *The Oriental Way of Thinking*:

> One of the basic features of the Japanese way of thinking which attracts our attention is the direct affirmation of the surrounding world or the objective conditions in which he finds himself to live. He tends to regard the phenomenal world in which he experiences things and events directly as the absolute, and to reject that position which tries to recognize the absolute in the realm separate from the phenomenal world.[2]

Dr. Nakamura characterizes such a way of thinking as "the way of thinking which tries to recognize the absolute in the phenomenal world" (p. 13), "the way of thinking which tries to acknowledge the absolute significance in the phenomenal world" (p. 20) "the way of thinking which grasps the absolute according to the given phenomenal world." (p. 25) He observes that, in such way of thinking, "the ever-changing state of the phenomenal world has, in itself, the absolute significance" (p. 13) leading one to the way of living which "finds satisfaction in the given phenomenal world" (p. 17). He maintains that we can recognize the element of thorough-going affirmation and acceptance of the given, actual condition in such way of thinking (p. 19). He also remarks, in another place, that "honkaku", which originally meant the ultimate enlightenment to be attained beyond and above the level of the phenomenal world, "has been dragged down into the phenomenal world in Japanese thought" (p. 12).

These remarks by Dr. Nakamura undoubtedly serve as a clear characterization of the Japanese way of thinking. I must point out, however, that they raise a number of problems concerning possible interpretations of the Japanese way of thinking. What does it mean, precisely, to say "to regard the phenomenal world directly as the absolute?" It could mean either: (1) that the point of departure in inquiring into the absolute, or the place in which the inquiry is to be carried, is nothing else but this phenomenal world; or (2) a simple assertion that this phenomenal world is the same as the absolute. If I am not mistaken, Dr. Nakamura seems to approve both of these two interpretations.

Of these two interpretations, the second seems to be highly problematic. What does the assertion of identity of the phenomenal world with the absolute mean, precisely? Whether we assert simply that "the phenomenal world *is* the absolute", or express our thought in the form that "the absolute is nothing else than the phenomenal world", different senses must be assigned to the words "the phenomenal world" and "the absolute", if these two propositions are to be meaningful. Supposing that the phenomenal world and the absolute are the same *in subjecto*, the assertion of their identity would be meaningless unless different senses are given to them. Otherwise, the formula of identity "the phenomenal world *is* the absolute" could not mean anything.

If the preceding interpretation is correct, the formula of identity can be understood in the two following meanings. (1) The assertion of identity in the sense of complacency or satisfaction with the given, actual state, dragging down the absolute to the level of the actual, in giving up all sincere effort to seek the absolute. (2) Starting from the realm of phenomena, one attains the insight of identity of the phenomenal with the absolute in the very pursuit of the absolute.

Now, several passages in Dr. Nakamura's work seem to suggest that he accepts the first meaning. "Satisfaction with the given phenomenal world" and "dragging down the absolute into the phenomenal world" can be cited as examples. These words should not be taken, however, in the sense of the rejection of all effort to pursue the reality and the absolute. Because it is quite unreasonable to maintain that such efforts were absent in the Japanese spiritual tradition. The great zeal with which the Japanese nation, from the ancient period on, tried to adopt and assimilate Buddhism testifies to the presence of such efforts. Admittedly various secular and utilitarian motives were involved in that zeal. The development of uniquely Japanese Buddhist thought, however, shows that the Japanese, as a nation, originally had the spiritual capacity and hunger for the absolute.

The identification of the phenomenal world with the absolute in the Japanese traditional thought, therefore, should be understood as pointing to an insight attained in the pursuit of the absolute. To regard the phenomenal world "directly" as the absolute does not mean the complacent attitude of the satisfaction with the given, actual condition, giving up the unyielding effort to pursue the absolute, nor does it mean some magic whereby the phenomenal world is transformed into the absolute. Speaking more positively, the assertion of the identity, insofar as it is a meaningful assertion, does not exclude the transcendence of the absolute. How can the pursuit of the absolute be earnest, unless we recognize the transcendence of the absolute in earnest? Hence, we must say that the identity of the phenomenal world and the absolute was asserted in the traditional Japanese thought with the full recognition of, or in spite of the recognition of the transcendence of the absolute.

From the preceding considerations, we can conclude that to interpret the identification of the phenomenal world with the absolute as "satisfaction with the given phenomenal world" or "acceptance of the actual, given condition" is misleading as the characterization of the traditional Japanese thought. Again, the recourse to the formula "the phenomenon «soku» («is immediately») reality" does not, by itself clarify the matter. We must avoid, above all, the hasty judgement that the identification means that immanentism which excludes the transcendence of the absolute. We cannot pass the judgment *a priori* that the Japanese traditional thought which asserts the identity of the phenomenal world with the absolute on the one hand, and the Judeo-Christian thought which teaches that the world is the work of divine creation, with its implication that the absolute must be transcendent on the other hand, are utterly different, and that no dialogue is possible between these two spiritual traditions.

In the following, I shall reflect upon the idea of creation in the metaphysical thought of Kitaro Nishida (1870–1945). According to Professor T. Shimomura, who published an excellent study on the life and thought of Nishida, Nishida's philosophical effort all through his life was devoted to the development and articulation of the logic of oriental experience.[3] That Nishida has succeeded in that work of articulation is the justification for my taking up his philosophy in my study of traditional Japanese metaphical thought. Although many studies have been undertaken on the philosophical thought of Nishida, as far as I know, his idea of creation has not been made the object of serious study. It is the aim of this paper to cast some light upon one hitherto neglected aspect of traditional Japanese metaphysical thought, through a reflection upon Nishida's idea of creation.

3.

In his first book, *A Study of Good*, Nishida takes a negative position concerning the idea of creation. After defining religion, in a general way, as relationship between God and man, Nishida rejects the position which considers God as a transcendent being, essentially distinct from finite, human beings, and approves the position which asserts that God is the foundation and the intrinsic unifying force of the universe.[4] The assertion that God is our foundation means that we do not lose ourselves by being united with God, but rather we gain ourselves thereby, Nishida emphasizes. He maintains that the core and essence of the relationship between God and man is expressed by the words that "man truly is, lives and finds himself in God." On the other hand, "the idea that some transcendent God governs the world from out-side is not only in conflict with our reason, but any religion with such an idea cannot be said to have the depth of true religion" (Vol. I, p. 175), Nishida asserts.

To Nishida's mind, the position which considers God as the foundation of the universe, or the immediate ground of this real world is in opposition to the idea of creation. The idea that God is a creator who is transcendent and outside of the universe, that is, the conception of the relationship between God and universe in the analogy of an artist in relation to his product, shows an intellectual and moral immaturity, according to Nishida. He rejects such an idea as a kind of anthropomorphism. He maintains, against the view which identifies God as a transcendent creator, that the relationship between God and the universe should be understood as that between the substance and phenomenon, and the concept that best serves as the key for the understanding of this relationship is not that of creation, but manifestation. "From the movement of the sun, the moon and stars outside, and to the most delicate secrets of human heart inside, all of them are nothing but the manifestation" of God, (Vol. I, p. 178).

Here Nishida emphasizes that the statements that God is the unifier of the universe, and that the universe is the manifestation of God, are not merely metaphorical statements, but they are meant to express facts strictly. He says: "God is the greatest and final unifier of our consciousness. Not only that, but our consciousness is a part of God's consciousness, and its unity comes from the unity of God." (Vol. I, p. 182). In other words, Nishida simply asserts that we can have the intuition of God by reflecting upon and grasp our own unity. That is to say, he is saying that to grasp one's own unity is to see — not in a metaphorical sense, but literally — God, the Unifier, who is the ground of our unity. We can conclude from this that, in Nishida, the identity of the unified universe or the self and the Unifier God, or the identity of the universe as the manifestation of God and God Himself, is affirmed.

Nishida admits that this unifying force itself is something spiritual, and is fit to be called "person", insofar as spiritual phenomena are governed by this unifying force. Thought and will are the attributes ascribed to such "personal" God. He does not admit, however, that God, who is said to be "person", transcends beyond the universe and its processes. It is because the universe, which is God's manifestation, must be identified with His thought and will. Hence, God's love, too, cannot be a voluntary and selective love, but must be necessary, as His nature itself is. The self-development of God is the infinite Love of God towards us, and there is no Love of God except the Love of God expressed in the necessary development of all things (Vol. I, p. 185).

The preceding reflection makes, I believe, the reason why Nishida has rejected the position which explains the relationship between universe and God by creation, attempting to explain it in terms of manifestation, quite clear. The creation he rejects is the deistic view of creation which places God outside of the world created by Him. Similarly, the transcendence he rejects is the transcendence of such deistic-creator. According to Nishida, it belongs to the essence of God that the world is His manifestation, and the manifestation cannot be some accidental operation which God performs now and ceases to perform the next moment. Nishida rejects creation, therefore, insofar as creation is taken in the sense of an accidental operation, and not the creation in the sense of eternal creation. The latter is equivalent to what Nishida calls manifestation, and is never rejected by him.

In other words, the real reason why Nishida does not accept the position of explaining the relationship between God and the universe by creation, is that it is incompatible with his insight that God and the world, or God and the self, are identical in a profound sense (this insight was gained, as observed before, through reflection and inquiry into the ground of the unity of the self), and with his religious demand that God and the self must be one. The real reason, in a word, is that the concept of creation is incompatible with God's immanence in the world. The concept of manifestation, on the other hand, emphasizes precisely this immanence of God in the world. Because, this concept tells that the universe is God's self-manifestation or revelation, and that the manifesting and the universe manifested are one.

Nishida adopts the bold expression that "just as there is no world without God, so there is no God without the world" (Vol. I, p. 190), in order to say that the infinite, absolute God and this world are one and inseparable. Certainly, it will be difficult to find a more radical word to express the immanence of God in the world. It is clear that the transcendence of God in the sense of His separate existence from the world, is rejected by this expression. We must be careful, however, in interpreting this expression, precisely because its form is extremely radical, if not paradoxical. In other words, we must carefully determine the direction which this radical expression really intends. It is true that Nishida affirms the identity of God and the world in a profound sense, but he does not simply equate these two. Even though God is said to be immanent in, and identical with the world, a certain transcendence must be attributed to God insofar as God is to be the object of sincere pursuit and the union with Him is to be sought earnestly. This transcendence is not the transcendence in the shallow sense of separation from the world, which Nishida has rejected, but must be the transcendence in a more radical sense. When Nishida rejected the concept of creation in order to reject the transcendence of God in a shallow sense, therefore, he did not reject the transcendence of God in a more radical sense — indeed, if it were rejected, the pursuit of God would lose its meaning completely, thus making it impossible to talk about religion —, and the room is left for the affirmation of the conceot of creation in a more profound sense. As a matter of fact, Nishida's metaphysical thinking after *A Study of Good* took the direction, so it seems, of the affirmation of creation in a deepened sense.

4.

I cannot discuss here the development of Nishida's metaphysical thought after *A Study of Good* in a comprehensive way. Instead I shall quote a few passages in which Nishida himself tells about the development of his thought. In the Preface to the new edition of *A Study of Good*, published a quarter of a century after the first edition, he tells:

> The position of pure experience (which was central in *A Study of Good*) proceeded into that of absolute will in *The Intuition and Reflection in the Consciousness of the Self*, via Fichte's position of Tathandlung, and then in the latter half of *From the Acting to the Seeing* took a radical turn and advanced into the thought of 'place' via Greek philosophy. It was there, I think, that I obtained the initium of articulating the logic of my thought. The thought of 'place' was concretized as the 'dialectical universal,' and the position of the 'dialectical universal' was rendered immediate as the position of 'acting intuition.' What I called in this book (*A Study of Good*) the world of immediate experience or the world of pure experience, I now conceive as the world of historical reality. The world of acting intuition and the world of poiesis is truly the world of pure experience (Vol. I, pp. 6–7)

He makes a similar and somewhat more detailed observation in the Preface to *Philosophical Papers* (III). Towards the end, he says:

> I said that my aim has been, since *A Study of Good*, to discover the most fundamental way of seeing things and thinking about things. Now, I think that it must be the logic of the consciousness of the self of the historical life, the logic of the consciousness of the self of the poietic self. (Vol. IX, p. 7)

A clear feature of the development of Nishida's metaphysical thought is the gradual shift in his basic outlook from the world of pure experience as the concrete, actual world to the world of historical, social reality. Now, it is important to notice that the concept of creation is closely connected with such grasp of the historical world in his metaphysical thought. That is to say, as Nishida gradually deepens his grasp of this historico-social reality and the logic of its formation, he comes to adopt a more and more positive view concerning creation. Indeed, we can say that in the metaphysical insight attained by Nishida in the final stage of his philosophical career, the concept of creation occupies a central position.

In his article titled "Human Being", Nishida characterizes the human being as the "creative element in the creating world" (Vol. IX, p. 9) "creans creatum" (Vol. XI, p. 22). He considers this developing-evolving historical world as creative. On the one hand, man is formed and created in the creative process of this historical world. On the other hand and at the same time, however, man, insofar as he is a conscious and reflective being, is the creative element in the same historical world—that is, man creates. Man is creative in his act of cognition, namely in his 'seeing'. More precisely, man is said to be creative in 'seeing through making' or 'knowing through poiesis'. (Vol. IX, p. 1).

The same point is further elaborated in "On the Consciousness of the Self". Nishida emphasizes in this article that he is not trying to grasp the world from the position of the self, that is, from the egocentric or self-assertive viewpoint, but rather trying to grasp the self and his act from the position of the world. (Vol. X, p. 510). The 'seeing through making' 'knowing through poiesis' as the creative act of man, namely the 'acting intuition' should be interpreted from this position, he observes. In other words, the acting intuition is not an egocentric act, but a creative act mediated by the absolute negation. It is the fundamental insight and conviction of Nishida that in order for an act to be an act in the true sense, it must be mediated by the absolute self-negation. That is to say, creation as the highest act must be the creation through the absolute self-negation. Nishida interprets the Christian teaching that God creates from love in this sense. According to Nishida, the absolute self-negation and the creation as the highest act are inseparable. It is noteworthy that he observes that the idea of creation did not enter into the range of Buddhist thought because it over-emphasized the aspect of absolute negation. (Vol. X, pp. 498, 509).

Now, it is in his article "The Logic of Place and the Religious World-view", published three months before his death, that the metaphysical concept of creation is discussed in the most clear manner. Here Nishida attempts an approach to the concept of creation through reflection upon three stages of act. The first stage is the kind of act observed in the material world. It is pointed out that since the action-reaction observed in the realm of matter is reciprocal and reversible, it is impossible to find an act in the strict sense in this realm. Next, the act in the realm of life is purposive and irreversible. It can be called, therefore, an act in the truer and fuller sense. The final stage is a self-reflexive act, which is found in the historical world constituted by those subjects who are conscious of the purpose of their own acts. Since only the subject who is conscious of the purpose of his own act can work from the inside of his own being, it is only in such subject that the act in the most perfect sense is found. Furthermore, in this case, the act does not mean acting upon some other particular being, but rather means manifesting the whole of the world. And it is this kind of most perfect act that Nishida calls creation. (Vol. XI, p. 374).

Thus, according to Nishida, the true act, the act in the most perfect sense is worthy of the name 'creation', and the most perfect act in the world is the act of manifestation. In order to manifest something other than oneself, however, it is necessary to negate oneself, that is, to make void of oneself. When the self-negation is complete, precisely then the manifestation can be perfect. The act of creation, therefore, presupposes the absolute self-negation. In this way, starting with the act of lower stages, Nishida reaches to the most perfect act which is worthy of the name of creation, and discovers that this most perfect act is, in fact, mediated by the absolute self-negation.

The act of manifestation-creation performed by the human subject, however, is not the pure act of creation. Man comes to realize, as he deepens and purifies his self-reflection, that his act of manifestation-creation is really a participation in the self-manifestation and self-creation of the world. Man performs his act of creation insofar as he participates in the self-creation of the world. Even though Nishida speaks of the self-creation of the 'world', and not God's creation, it is certain that he considers the world's self-creation to be the divine creation. He observes, in another place, that the consciousness of the self is a partaking of God's consciousness of the self. (Vol. XI, p. 6). He also says that it is there that man touches the fact of divine creation. (Vol. XI, p. 432). In other words, when, in reflecting upon his own act of manifestation-creation, he reaches to the origin and principle of his act, precisely then he realizes that he is a created being. Man does not create simply, but he is creans creatum.

Thus, through the deepening of reflection upon one's own act, man comes to realize that the historico-human world, namely the reality in the most concrete sense, is founded upon the absolute self-negation, self-manifestation and self-formation of the absolute. Man can truly act and create, only through identifying himself with the self-negation and self-formation of the absolute, by completely negating and making void of himself.

The preceding reflection has made it clear, I believe, that in the later writings of Nishida, while preserving fundamentally the position of God's immanence in the world (the position expressed in *A Study of Good*), the transcendence of God in a deeper sense came to be recognized in an increasingly conspicuous way. What I mean is that he came to maintain an increasingly clearer way that the unification of man with God — namely, man's self-transcendence towards God — is to be mediated by man's complete self-negation. As a matter of fact, Nishida now speaks about the immanent-transcendence of God, instead of simply speaking about the immanence of God in the world (Vol. XI, p. 462). That he came to realize more and more the importance of the element of transcendence can be seen in his criticism of what he calls the 'shallow immanentism of the modern culture'. According to him, it is not surprising that the modern culture is showing the sign of decay and disintegration. It is the inevitable consequence from the fact that it has neglected the exigency of transcendence, and pursued the various cultural activities apart from God, he judges (Vol. IX, p. 460). In any case, it is beyond doubt that his metaphysical thinking took the direction in which the transcendence of God — in a deeper sense than the one he had rejected in *A Study of Good* — came to occupy an increasingly more central place. And it is this affirmation of the transcendence of God in a deeper sense that is reflected in his more positive attitude towards the concept of creation in his later writings.

5.

The concept of creation in Nishida's metaphysical thought raises a number of problems, the importance of which has not been sufficiently recognized so far. One of these problems would be the relationship between the concept of creation mediated by the absolute self-negation and the traditional Christian concept of creation, namely creatio *ex nihilo*. Nishida maintains that creation cannot be explained in terms of causality or through the concept of potentiality-actuality. Creation is, as it were, something circular, and must be an eternal creation, he observes (Vol. XI, pp. 162–163, Vol. X, pp. 556–557). In accord with this position, he rejects the creation which is 'from nothingness to being'. What is created by the absolute through the absolute self-negation — which is called the world or God by Nishida — is the absolute itself, and thus the eternal creation is nothing else than the eternal self-creation of the absolute. Now, if it is the absolute itself that is created by the absolute self-negation, what is the meaning of 'self-negation' here? Is it really appropriate to call this self-negation 'love' as Nishida himself does? The fact that, while speaking of creation through the absolute self-negation, Nishida was not able to give a positive account about creation *ex nihilo*, seems to suggest some difficulty involved in his concept of creation.

Concerning this point, the article by Professor M. Matsumoto, an outstanding metaphysician and a specialist in medieval philosophy, entitled "*Creatio ex nihilo* — A Study",[5] in which he attempts to interpret the Christian concept of creation, namely *creatio ex nihilo*, by taking into account the metaphysical thought of Mahayana Buddhism, is highly interesting, in that, while showing a certain affinity with the concept of creation in Nishida's metaphysics, it seems to go one step further than the latter. The aim of Professor Matsumoto is to deepen the understanding of '*nihilum*' in the formula '*creatio ex nihilo*' through a reflection upon the meaning of 'nothingness' and 'void' in the metaphysical thought of Mahayana Buddhism. According to Professor Matsumoto, there is a close connection between the idea of 'nothingness' in Mahayana Buddhism and the notion of *analogia entis* in the Scholastic ontology, and once we realize this connection, we will no longer be able to be satisfied with the stereotyped explanation that '*nihilum*' in '*creatio ex nihilo*' means the total exclusion of preexisting matter in the creation of the world.

Then, how does Professor Matsumoto understand '*nihilum*' in '*creation ex nihilo*' in the light of the metaphysical thought of Mahayana Buddhism? *Nihilum* means, according to his interpretation, that 'place' in which the absolute God faces with the relative, finite world, without thereby being Himself reduced into something relative. The 'place' as s medium which enables the absolute to relate Himself with the relative, must be open in some way both to the absolute and the relative, without itself having any essence, that is, without being 'something'. Because if the 'place' as the medium were itself something, it is bound to reduce the absolute into the relative. This 'place,' therefore, must be conceived as 'nothingness' in the very close sense with the 'nothingness' spoken in Mahayana Buddhism. Moreover, according to Professor Matsumoto, this 'place' of 'nothingness' corresponds to the '*ens commune*' or '*ens analogicum*' in the Scholastic ontology, which is open both to the infinite and finite beings. Hence, Professor Matsumoto concludes, we must understand the '*nihilum*' in '*creatio ex nihilo*' in the sense of *ens commune* or *ens analogicum*, which signifies the actuality of mere existence, devoid of determination or limitation by any essence.

After having pointed out that the '*nihilum*' as the 'place' which mediates the absolute and the relative is not a mere non-being, Professor Matsumoto goes on to show that this '*nihilum*' is the '*intentio ad nihilum*' which only the absolute, infinite being, God, is capable of performing. Whereas a finite being can merely intend and face with some finite being other than itself or non-being in a relative sense, only the divine absolute, precisely because there is only nothingness outside of himself, can intend the absolute nothingness. And, when the divine absolute intends nothingness, he posits the 'place' of nothingness within himself, in which it becomes possible that the absolute faces with the relative — that is, creation becomes possible — without the absolute being reduced into the relative, or the relative being engulfed into the absolute. This '*intentio ad nihilum*' is nothing else than that love in which God 'gives Himself', and thus the way is open towards the understanding of '*creatio ex nihilo*' in the sense of that creation which takes place through God's 'making Himself void in love'. In other words, the way is open to understanding '*creatio ex nihilo*', not merely in the sense of the act of production without any antecedent matter, but in connection with the central teaching of Christianity, namely 'God is love'. And it is precisely the metaphysical thought of Mahyana Buddhism that has opened our eyes to such deep meaning, hidden in the traditional concept of '*creatio ex nihilo*' in the Christian theology. In any event, we can say that Professor Matsumoto's interpretation of '*creatio ex nihilo*', while accepting Nishida's concept of creation through the self-negating love of the absolute, has given it a metaphysically more articulate expression.[6]

6.

I do not claim that I have given an adequate account of the concept of creation, which occupies a central place in the later stage of Nishida's metaphysical thought. I believe, however, that the preceding consideration could serve as a key to the understanding of the traditional Japanese metaphysical thought, or the Japanese way of thinking. The key is the notion of the absolute self-negation, which, according to Nishida, makes the act of creation in the true sense possible. We can participate in the divine creation whereby we ourselves are created, and can create in the true sense, only when we negate and make void of ourselves completely. To adopt a more general manner of speaking, we will not be able to reach to the absolute, so long as we pursue the absolute, outside of ourselves, in an egocentric attitude. We can meet the absolute only when we negate ourselves completely. It is only then that we experience that we are really the manifestation of the absolute, being one with the absolute, while remaining ourselves in the true sense.

In concluding this paper, I shall make a few comments upon Dr. Nakamura's observations concerning the characteristics of the Japanese way of thinking mentioned at the beginning of this paper. I think that one of the lessons we could learn from Nishida's metaphysical insight would be to suspect behind what appears to be, at first sight, a simple, unreflective affirmation of the identity of the phenomenal world with the absolute, a thorough self-negation. It is possible that this affirmation of identity is, in fact, mediated by the thorough self-negation on the part of the one who sincerely seeks the absolute. The more complete and thorough is the self-negation, the more fully the phenomenal world manifests the absolute, finally leading to the affirmation of the identity with the absolute. Obviously, many problems will be raised as to the interpretation of the nature of this identity. It is clear, however, that the transcendence of the absolute in the fundamental sense is not thereby excluded, insofar as the identification is mediated by the true self-negation.

What is, then, the true self-negation, and how can it be practised and achieved? I cannot offer an answer to this crucial question here. I can only say that the self-negation in the true sense is closely connected with the self-reflection and self-purification. How is it possible to reflect upon oneself, and purify one-self in the true sense? The traditional Japanese metaphysical thought seems to contain these questions, and answers to them as well, and to rediscover them will be, undoubtedly, one of the most important problems for the study of the history of Japanese thought.

NOTES

1. In this regard, Keiji Nishitani has made a significant contribution, cf. *The Works of Keiji Nishitani* (Sobunsha, 1985–87). 13 vols.
2. *Selected Works of Hajime Nakamura* (Tokyo: Shunjyusha), Vol. III, p. 11.
3. *Nishida Kitaro — Man and the Thought* (Tokai University Press), p. 196.
4. *The Complete Works of Kitaro Nishida* (Tokyo: Iwanami Shoten), Vol. I, p.173.
5. In *Problems of Ontology* (Tokyo: Iwanami Shoten).
6. Cf. "The Metaphysical Significance of 'analogia entis'". In *Problems of Ontology*.

Japan

Dogen's thought on the Way as meta-ethical principle

KAKICHI KADOWAKI S.J.
Sophia University, Tokyo

1. Introduction

Dogen (1200–1253) is the founder of the Soto Zen sect in Japan.[1] In modern times his principal work, Shobogenzo (The Eye and Treasury of the True Law),[2] has been regarded as one of the most outstanding in Japan by the Japanese philosophers, Tanabe[3] and Watsuji.[4] Recently philosophical thought, based on his unsurpassed religious experience, has attracted the attention of many westerners. So I would like to make his subtle, but difficult thought on the Way (Tao) understandable to western readers who might get stimulation and new insight from it.

We have to learn first of all in what sense Dogen uses the character for "*Tao* (the Way)" though he does not write in detail on the Way itself. We must deduce, therefore, the meaning of the Way as meant by Dogen from his brief statements on the Way.

Let us begin to study the general meaning of the Chinese character for "Tao (do)" standing for the Way.

The character for "tao (the way)" is rich in content, and indeed it has a very wide range of meaning. To mention just a few important examples, we have to refer to as many as those mentioned below.[5]

As a noun, the character for *tao* (the way) means (1) road, (2) reason, *logic* (*logical* sequence), (3) the right coherence of one's thinkings and doings, (4) the source of myriads of things in this universe, as taught in Taoism, or the entities of various phenomena, (5) act, mysterious working, (6) means, method, policy, (7) principle, thought, (8) theory, and so on. As a verb the character for *tao* (the way) means (1) pass, go, (2) conduct and (3) obey. The above are a group of various meanings of the character for *tao* (the way). Besides, the same character standing for the Way is pronounced "iu" in Japanese as a verb and has a great variety of meaning, though not known very well in general. It means (1) (1) say, tell, (2) reign, (3) rely, rest on, copy from, (4) from, through, (5) guide, (6) teach, (7) help, (8) open, and (9) be familiar with.

Judging from this, the character for *tao* (the way) is known to have come to assume a great variety of meanings in the course of the long cultural history of China. When those contributing to the culture of Chinese Characters use this character with its variety of meaning, they do so in terms of the most appropriate of all these meanings. In many such cases, the chosen specific meaning is impregnated with the subtle nuances of other meanings. For instance, when Lao-tzŭ (Roshi) and Chuahg-tzŭ (Soshi) use this character standing for the way to express the "source

Contemporary philosophy. A new survey. Vol. 7, 303–313.

of myriads of things in the universe," the same character comes to tacitly assume other meanings as well. The character for *tao* (the Way), then, simultaneously means the "source of myriads of things in the universe" and the reason penetrating all things between Heaven and Earth, the right logic to be observed by man, and the power operating on all things. As will be seen later, Dogen expounds his fundamental thought by skillfully using this Chinese character for *tao* (the way), which is so rich in meaning.[6]

2. The way is the fundamental power penetrated by the Great Compassionate Wish for Universal Salvation

Now, Dogen uses the Chinese character for *tao* (the way) in expounding his fundamental thought. Dogen returned to Japan from China in the autumn of 1127 and wrote Fukanzazengi (General Promotion of the Principles of Zazen) in Chinese in the same year. To begin with, he expresses his fundamental principle with the words "*Tazunureba sore do moto enzu* (Searching for the Origin, I find out that the Way originally permeates everything roundly)". The expression "*do moto enzu*" pertains to the Way as the fundamental force setting the universe and everything in it in motion. From this as the "first principle", Dogen derives the various rules and the concrete method of training in Zazen. *Fukanzazengi* is the book in which Dogen summarizes the fundamental essence of the concrete ways of Zen Buddhism which he had learned in China. There is something new in it that cannot be found in any other book on Zazengi previously published. Just try to compare this book with Tsung-tzê's *Zazengi* (*A Book on Zazen*) included in *Ch'an-Yuan Ch'Ing-Kuei*. The latter *Zazengi* does not deal with the fundamental principle of "*do moto enzu* (the Way originally permeates everything roundly)", as expressed by Dogen. Instead, it advises the reader to sit in Zazen, awakened to Great Compassion and pledged to the salvation of all mankind. It is true that a pledge to Universal Salvation stemming from Great Compassion should be the fundamental spirit of a disciplinant, but in so far as it is his subjective fundamental intentionality, he fails to conform to the viewpoint transcending the duality between subject and object, I should say. Dogen, on the contrary, begins with the fundamental principle called the Way, transcending all dualities, such as subject and object, enlightenment and delusion, and so on.

Well, what is Dogen's Way, then? First of all, let us try to see in what sense he uses the Chinese character for *Tao* (the Way) in his writings.

In his *Hotsubodaishin* (*The Arising of Bodhi Mind*), Dogen explains the word "*Bodhi* (*Bodai*)" with the remarkable words that the originally Sanskrit word "*Bodhi*" is translated by him into the Chinese word "*Tao* (the Way)" in spite of the fact that "*Bodhi*", also called "Supreme *Bodhi* (*Mujo Bodai*)", means the ultimate stage of Enlightenment, being the True Awakening or Right Enlightenment attained by Sakyamuni under the bodhi-tree in Buddha-gaya. While it is said to have been translated into other Chinese words, such as "*Chih* (*Wisdom*)" and "*Chiao* (Awakening)", why did Dogen choose the word *Tao* for *Bodhi*. (Bodai) instead?

There must be some good reason for this. In the first place, it is the "*Bodhi* Mind" that Dogen deals with in his *Hatsubodaishin*, when he expounds the fundamental intentionality for universal salvation with the use of the Chinese character for *tao* (the Way), for in *Bodhi* Mind Compassion is stressed rather than Wisdom. This should mean that Dogen impregnates the Way with the subtle sense of the Great Compassion of Buddha. Judging from this, Dogen's fundamental principle of the Way set forth at the beginning of his *Fukanzazengi*, which is far from Tsung-tsê's *TSO-CH'AN I*, alludes to the Great Compassion of Buddha embraced by the Way, as perceived by him. It is not too much to say that the Way is the fundamental force penetrated by the Great Compassion of Buddha for universal salvation.

The second reason for Dogen's consistent use of the character for *tao* (the Way) is that "Practice" is what concerns Dogen most in the various volumes of his *Shobogenzo*, including that of *Hatsubodaishin*. As the Chinese character for *tao* (the way) implies Practice, it is evidently more suitable for the expression of the fundamental idea of Practice than any other characters, such as those standing for *Chih* (Wisdom) and *Chiao* (Awakening). I would like to elaborate on this point again later, for it pertains to the fundamental thought of Dogen's concept of the Way.

In Dogen's *Gakudōyōjinshū* (a collection of precautions to be taken in the learning of the way), Dogen defines the Way as follows from the viewpoint of the preparatory students of the Way of Buddha:

> "The disciplinant should wish to practice the Way of Buddha not for his own sake, not for distinction and profit, not for good luck, not for Buddha's beneficial rewards, but only for the the sake of *Dharma*. This is what we call the Way."

The above passage is quoted from the conclusive part of Clause 4 of *Gakudōyōjinshū*. It is entitled "Do Not Practice *Dharma* with Any Wish for Acquisition." "Wish for acquisition" means the mind anticipating something good to get. Dogen especially admonishes the anticipation of enlightenment as a result of Practice. Practice for one's own benefits, for honor and profit, for good results, and for mysterious rewards, such as the recovery of illness and miracles, all belong to the "wish for acquisition". All of them are nothing but limited values of this world. Unless one is freed from them, one will never be able to attain the absolute values. "To practice *Dharma* only for the sake of *Dharma*" means the Practice of the absolute truth of *Dharma* only for the sake of its absolute truth.

The above act is neither in pursuit of cause and effect nor subject to the "calculated impulse of gain and loss", which corresponds to Heidegger's "das bezahlenden Denken"[7] as its affective aspect. Rather, it is effected by the vertical and super-causal Act of the Way.

The Way is the ultimate aim of all things on earth and keeps on leading them to itself. By the way, the ultimate aim, far from various other aims of relativity, is never subject to other purposes as their means. They all exist and work for the sake of this ultimate aim, while the latter exists for its own sake alone and works for itself as its own aim. *Dharma* is practically identical with the Way, as will be discussed later. Therefore, to "practice *Dharma* for the sake of *Dharma*" is the same as to "practice the Way for the sake of the Way". That is why to "practice *Dharma* for the sake of *Dharma*" expresses the free and absolute "Act" of the Way as the ultimate aim, working for itself as its own aim.

From the above study, we learn that the Way is the fundamental force setting everything in this universe in motion toward itself in the free and non-abiding way and that it is also the intrinsic motive power guiding the disciplinant to "practice the Way for the sake of the Way". It is further clear that the Way is the fundamental force penetrated by Great Compassion for universal salvation by means of such a selfless conduct as "practicing the Way for the sake of the Way".

By the way, *Dharma* pertains to the Right awakening experienced by Śākyamuni with all his body and all his mind under the bodhi-tree. This Right Awakening "Acts" more greatly than anything else. It is the dynamic "truth" that served to enlighten Śākyamuni into Buddha (the one awakened to the truth of being). It has also served to make all sentient beings Buddhas. This Right Awakening in such in such great "Act" is Supreme *Bodhi* (Enlightenment) and is called *Dharma* (the Buddhist Law) in so far as it is the universal "truth" to make all setient beings Buddhas.

In addition, the Way is *Bodhi*, as mentioned above, and it is identical with the abovementioned Supreme *Bodhi* (the Right Awakening). So it is the same as the Buddhist Law in content as well. The Way and the Buddhist Law thus represent different aspects of the one and the same *Bodhi*. This Buddhist Law, which is the "truth" realized by Buddha and the "truth" making sentient beings Buddhas, is the aspect of *Bodhi* seen from the angle of the "truth". Moreover, the Buddhist Law throws into relief not only the content of Buddha's personal and subjective Enlightenment but also its universal principle of turning all sentient beings into Buddhas. In contrast to it, the Way which pertains to the Practice of this "truth" with all one's body and the attainment to Buddhahood in person, is the aspect of the same *Bodhi* seen from the angle of Practice. Furthermore, the Way throws the following dynamic aspect of *Bodhi* into relief. The Way indicates to us the fact that the "Truth" realized by Śākyamuni is not the object to be known through tranquil contemplation, but the dynamic principle never to be realized by any one until he becomes a Buddha himself through Practice. To speak more exactly, the Way clearly indicates the point that it is the dynamic motive power to make all sentient beings Buddhas. As the Way is the dynamic motive power to make all sentient beings Buddhas, it is also the intrinsic motive power leading all sentient beings to Practice and actually making them Buddhas through their Practice.

What we must note here is that the "Truth" identical with the Buddhist Law is altogether different from the Western Philosophical view of the truth. In the first place, the truth conceived in Western Philosophy is the object of cognition through human reason, being the "*Adaequatio intellectus cum re*". The "Truth" of the Buddhist Law pertains to the truth of being realized through Practice, by becoming a Buddha. This means becomings one with the Way with one's whole body and mind. The former is the rational and metaphysical truth to be cognized by means of reason, whereas the latter is the wholly human and "vivifying" truth which concretely comes into being through Practice with one's whole body and whole mind.

There is another difference between the "Truth" of the Buddhist Law and the truth in Western Philosophy. In Western Philosophy truth is abstracted from sensory things to be cognized by means of reason. It is the abstract and universal truth detached from the senses and the world perceived with them. The "Truth" of the Buddhist Law, on the contrary, means becoming a Buddha with one's whole body and mind. It is the universal Truth concretely revealing itself in the very living human body bound to the transmigration between birth and death. These two different kinds of truth, however, are mutually identical in universality. The former is abstract universality, whereas the latter is concrete universality. Dogen often says, "Coming and going between birth and death is the truth human body". These words well express the true human body (meaning Buddha) into which the Way and the living body as it truly is are integrated.

3. "The Way originally permeates everything roundly" — the free and non-abiding act of the Way Setting Everything in This Universe in Motion

Now, let us go back to the starting point to inquire into Dogen's fundamental idea of "*do moto enzu* (the Way originally permeates everything roundly)" as set forth at the beginning of his *Fukanzazengi*. We have so far studied Dogen's various writings to see in what sense he uses the word "Way". Judging from the results of our study, his idea of "*do moto enzu*" is summarized as follows:

Sakyamuni was awakened to the "principle of becoming a Buddha (the Buddhist Law)" under the *bodhi* tree in Buddha-gaya, and became Buddha himself. He further became the Ultimately Enlightened One brimming with such dynamic "Motive Power (the Way)" as to awaken all sentient beings to this "Principle" to become Buddhas. This motive power permeating Sakyamuni is called the Way by Dogen. This is nothing other than the source of universal salvation, the mysterious operation to make all sentient beings Buddhas, that is, the Way to be walked and the Law to be kept by men. Therefore, it is the reality to be well represented by the multivocal Chinese character for *tao*, which variously means (1) the way, (2) principle, (3) the logic to be kept by men, the source of the universe, and all things in it, (5) mysterious operation, and so on.

Though the Way is what Sakyamuni became aware of for the first time in history, it has been originally at work, giving life to everything in this universe and keeping it as it originally should be. From the viewpoint of historical transition, the said Way kept on being transmitted, first from Śākyamuni to Kasyapa, then from Kasyapa to his disciples, and finally to Dogen through generations of his predecessors as perfectly as the "water shifted from vessel to vessel". However, speaking from the angle of the absolutely free Act of the Way, the Way, which is intrinsic in history operating in everything, transcends the historical sequence of events, for the Way is such a perfectly faultless, free, and non-abiding "Act (Actualization)" that it needs neither discipline nor Enlightenment, nor any various devices to attain it on one's part.

In order to indicate such absolute freedom of the Way, Dogen writes in addition, "Why should discipline and Enlightenment be necessary?" immediately after the passage on "*do moto enzu*" at the beginning of *Fukanzazengi*. Further, he explains the same situation in different words, "It is freely capable of converting setient beings. So why should they waste their time making devices for Enlightenment?" He goes on, "The Way as a whole far transcends the world of delusion (dust). So it is not necessary to resort to any means to brush off the dust of desires and agonies. As the Way never goes away from where one stands now, one does not have to roam from one place to another for the sake of discipline for it".

In the terminology of Western Philosophy, this reality of the Way may be called "*a priori*" in the sense that its Act is antecedent to Practice and experience. Strictly speaking, however, we must say it is different from the *apriority* of Western Philosophy in the following points. In the first place, experience in Western Philosophy pertains to such experiences of cognition and desire, as indicated by Kant's *apriority*, whereas experience in the case of the Way pertains to the Practice of the Way with one's whole body and mind. In accordance with this difference in meaning of the word experience, the term *a priori* or *apriority* has different meanings to us.

There is another greater difference between Kant's *apriority* and that of the Way. In contrast to the former, lying within the subject, the latter is capable of bringing subject and object into one by transcending their mutual opposition, as will be discussed later. Kant's *apriority* is ultimately in close contact with the freedom of the subject, whereas the *apriority* of the Way operates in complete freedom and independence from all beings in their world, including the subject. The *apriority* of the Way, moreover, is penetrated by a compassionate wish for universal salvation even in its absolute freedom. Kant's *apriority*, on the contrary, has nothing to do with such a compassionate wish. Another important difference between them is that as long as the *apriority* of Western Philosophy is discussed in philosophy, religious faith in the true sense of the word is out of the question. On the contrary, in the case of Dogen's idea of "*do moto enzu*", faith forms the "original basis" of Practice.

Then what is the absolute transcendentality of the Way? What I must say first of all is that Dogen never turns away from the actual scenes of his Practice. The actual scene of Practice pertains to the concrete historical world where the Way is operating on living human beings. Therefore, Dogen does not think that the Way exists in another idealistic world beyond this historical one. From the very standpoint of Practice, Dogen perceives the Way as a whole, revealing itself to this historical world through Śākyamuni and Buddhist Patriarchs. It is in this sense that the Way is immanent in the historical world. And yet the Way is transcendental in that it is free and independent of concrete historical individuals, including various historical

Buddhas. The transcendality of the Way is, then, the transcendality immanent in history. However, since the Way is beyond all kinds of duality, it should transcend even the duality between the immanent and the transcendental.

What is closest to this immanent transcendentality of the Way is probably that of Heidegger's existence (*das Sein*)[8] in Modern Thought. This, however, does not mean that it is exactly the same as that of the Way. Due to space limitation I must restrict my discussion to their greatest difference, without going into details. Unlike Heidegger's "existence (*das Sein*)", the Way is the dynamic source of everything beyond all dualities, such as being and non-being, unifying and integrating them. It is because the Way is the source of all inconsistencies and oppositions, and it is something to bring them into one, working freely and independently of them, basing them upon itself as their source, and letting them live in their mutual oppositions all the same.

Last of all, let us note the particular character of the immanent transcendentality of the Way, whose counterpart is altogether absent in Heidegger's "existence (*das Sein*)". In fact, I have so far never dealt with the unique historical character of the Way in my efforts to stress the transcendentality and the vertical "Actualization" of the Way, but the particular character of the Way is part of the essence of the immanent transcendentality of the Way. The reason why I neglected to deal with this fact is because, in my observation of the Way which simultaneously transcends and operates vertically on the historical sequences of events in which the authentic Buddhist Law has been transmitted from Sakyamuni to Kasyapa, and so on, I neglected to note that even the historical sequences of events in the "correctly transmitted Buddhist Law" are due to the vertical "Actualization" of the Way.

From the angle of the fundamental principle of "*do moto enzu*", we learn that this fundamental principle of the Way acts vertically as well as horizontally in the direction of historical events. It develops on one hand the Buddhist Law authentically handed down from Śākyamuni to Dogen along the historical sequences of events into what Ju-ching (Nyojo) calls the "Great Way of the Buddhist Patriarchs". The Way, on the other hand, not only vertically operates each moment on these historical sequences of events to give life to and base on itself such historical events as "face-to-face transmission" and "inheritance of the Buddhist Law", but also infuses the Practice of each disciplinant with the "life of Buddha" to sublimate it into the Buddhist Practice. These two directions of development of the Way stand for the two aspects of one and the same "Act (Actualization)" of the Way. They are two but one, and one but two. The direction in the historical sequences of events forms the immanent side, and the vertical direction forms the transcendental side of the Way. These two sides of the Way interpenetrate and interact with each other; they are one and inseparable. For the Way, which is the fundamental source of all things, gives birth to mutually opposing things and integrates them into its fundamental self in such a way as to keep them alive as such. Because of this integrating "Act (Actualization)" of the Way such dualities as the Buddhist Law and Buddha, Enlightenment and Delusion, subject and object, and so on are brought into one.

4. Dogen's Fundamental Standpoint — on the Scene of the manifestation of the Original Enlightenment through Practice on the part of a disciplinant beyond the Scope of Reason

By the way, the abovementioned dynamic Way is beyond the reach of human reason, for the Way is the dynamic and concrete truth of being (*dharmata*) integrating and transcending all dualities, whereas reason abstracts everything and makes one see all things dualistically. For instance, reason opposes subject to object, separates activeness from passiveness, distinguishes between Enlightenment and Delusion, and detaches Practice from Enlightenment, with the result that some regard the Way as lying in the objective world and Buddhist Patriarchs as living in the subjective world. Therefore, they inadvertently isolate the Way from Buddhist Patriarchs and imagine it to exist in the realm of idea apart from the historical world to which Buddhist Patriarchs belong. Further, those who depend on reason think that the Way is the active motive power and Buddhist Patriarchs are the passive receivers of the Way. Thus they regard the Way as something heterogeneous to the Buddhist Patriarchs. But, as it is, we are one in doing with the all-leveling "Act (Actualization)" of the Way[9] when we act and see things on the scenes of Practice, so that we are capable of seeing reality as it truly is, where subject and object, or activeness and passiveness, are one sharing the fundamental source of the Way. It is Wisdom or *Bodhi* that enables us to see things in their integration with the Way. It is called "*Prajñā*" in the scriptures of Primitive Buddhism.

It is very difficult for us to understand this Wisdom (*Prajñā*) by means of reason. So I would like to explain this Wisdom indirectly in comparison with Eckhart's mysticism which offers the closest analogy to it. Eckhart is a medieval German theologian who has attained the innermost depth of man's soul, that is, being in itself, by going beyond the scope of reason and will.[10] There he becomes one with God. In his oneness with God, according to Eckhart, "the ground of the God is my ground".[11] It is there that God's "eye" becomes his "eye" and he comes to see all things with "God's eye". As God's "Act (Actualization)" is beyond distinction between subject and object, and activeness and passiveness, his doings integrated with God's "Act (Actualization)" go beyond the distinction between subject and object and opposition between activeness and passiveness. It is clear that Eckhart sees into his oneness with God through his very practice on the standpoint of doing, far from discussing the essential oneness of God with man objectively with the use of reason. He is dynamically perceptive in his oneness with the all-embracing act of the Holy Spirit. It is like an artist's perception of his art work from within in integration with it while working at it instead of just externally appreciating a work of art. This artistic perception of Eckhart's is very similar to our "notion" of the Way.

Such is the activity of reason that we must try to reduce it and stop its dualistic operation in Zazen in order to be one with the "Act (Actualization)" of the Way. It is for this reason that Dogen remarks as follows:

> "We must learn in Zazen that the Buddhist Way does not lie in thought, discretion, divination, contemplation, sense perception, and intellectual understanding. . . . "

"In learning the Way we should not resort to thought, discretion, and so on." (*Gakudōyōjinshū*)

It goes without saying that such words as "thought" and "discretion" in the above passage pertain to the activities of reason.

Now, if the vast, infinite, and absolutely free "Act (Actualization)" of the Way is beyond the reach of human reason, there is no other way left for us but to "know the Way with our whole body and mind", thoroughly permeated with the "Act (Actualization)" of the Way. Therefore, Dogen teaches us that the "Way of Buddhism is to be entered without fail through Practice."

According to Dogen, it is because of the Way that man is urged to apply himself diligently to the Practice of the Way through his awareness of the inconstancy of this world. So his sincere Practice is already activated by the "Act (Actualization)" of the Way, which varies in dimensions, ranging from the motivating dimensions that urge him to practice the Way to the intrinsic dimensions that vitalize him, including the abovementioned dimensions of Wisdom (Enlightenment). These dimensions may be regarded as various facets of the constant "Act (Actualization)" of the Way. Therefore, the Practice of the disciplinant activated by the Way is penetrated by Buddha's life, so that Wisdom (Enlightenment), the intrinsic Motive Power, comes to reveal itself by dint of his Practice. Hence Dogen's teaching, "Enlightenment lies in Practice on the part of the disciplinant" (*loc. cit.*). This sentence, "Enlightenment lies in Practice", means that Enlightenment, which is already intrinsic to the Practice vitalized by the Way, externally reveals itself through it.

What we must note here is that by "Proof (Enlightenment) intrinsic to Practice" Dogen does not mean the participation of Proof (Enlightenment) in Practice as part of the Way, but the inherence of the entire Proof (Enlightenment) of the "Act (Actualization)" of the Way in Practice. The Way is simple and indivisible. It is the dynamic totality capable of integrating all into itself. Therefore, its "Act (Actualization)" is nothing other than its entity. As for the "Act (Actualization)" of the Way, in addition, it is always intrinsic as a whole to the disciplinant in its working. Wisdom (Proof) as a facet of its "Act (Actualization)" also works, always intrinsic as a whole to the disciplinant. Dogen calls this "Act (Actualization)" of the Way "*honsho* (True Proof)". Like the Way "*honsho* (True Proof)" is "*musho fushu* (omnipresent)", permeating our historical world as it activates and vitalizes everything in this whole universe.

This True Proof, however, never reveals itself except through right Practice in conformity to the Way even though it permeates this whole world. Dogen calls this revelation of the Way "genjō". Dogen made much of "genjō (the Actualization of the Way)" to such an extent that in his late years, when he was compiling his essays into *Shobogenzo*, he started with the volume of *Genjōkoan* (The Actualization of Enlightenment) of all his writings. Dogen's fundamental standpoint lies in Practice; he stands on the very scene where the True Proof reveals itself. This True Proof is "what everybody is fully endowed with but what remains concealed until he comes to practice the Way". (*Bendowa*). Dogen owes the above words of his to his own viewpoint, based on such Practice.

Practice in the true sense of the word should be vitalized by True Proof. Dogen calls such a kind of Practice "*shojo no shu* (Practice in Proof)". As I have mentioned above, moreover, True Proof as a whole is inherent in the Practice of the beginner, so that he should not search for Enlightenment outside of Practice; True Proof as a whole is already definitely inherent and at work in Practice. The disciplinant is already" in the Way" (*Gakudōyōjinshū*).

Then, it is no longer necessary for the disciplinant to keep on practicing in search of Enlightenment, for the Practice of the beginner is already the "whole of True Proof". He need not pursue Proof outside of Practice. He should only take care to practice in conformity to the Way, well aware of his own standpoint in Practice.

Ordinary people of the world think that Zazen is for the sake of Enlightenment and that it is futile to do Zazen without being enlightened. As long as one is obsessed with such a thought, Enlightenment for him is the aim of Practice and Practice is the means of Enlightenment. Consequently the disciplinant practices Zazen in pursuit of such a gain as Enlightenment. Before he knew where he was he finds himself in the darkness of delusion at the mercy of the "calculation impulse of gain and loss" which is the source of Delusion.

Thoroughly smashing the idea of doing Zazen for the sake of Enlightenment, Dogen turned to Zazen vitalized by True Proof, that is, "Practice in Proof". He calls this view of Zazen "*shusho fuji* (oneness of Practice and Proof)". From the fundamental viewpoint of the Way, the oneness of Practice and Proof means that Practice and Proof or Practice and Enlightenment are one as well as two, and two as well as one. This is because the Way is the fundamental force capable of integrating Practice and Proof by vitalizing them beyond their opposition. The Way develops itself as True Proof, and by dint of its "Act (Actualization)" it gives birth to Practice in the disciplinant, and by so doing it dynamically integrates Proof and Practice. It goes without saying that from the viewpoint of this integrating "Act (Actualization)" of the Way, "there is no distinction between the Zazen of the Buddhist Patriarchs, in whom the Buddhist Law is concretely embodied, and that of us beginners, for both represent the whole of True Proof", as Master Sokuo Eto writes.[13] Indeed Master Eto is right, but in the Zazen of the Buddhist Patriarchs they are completely permeated with True Proof both in body and mind, with their firm belief that "they are originally in the middle of the Way", whereas in the case of the beginner, he is not yet satisfied by True Proof in body and mind, his "eyes are not yet opened to the Way", and the belief that "he is in the middle of the Way" is not yet fully established in him.

As is clear from what I have just mentioned above, Dogen's Proof is very different from what ordinary people deem as enlightenment, which is a kind of intuition by means of consciousness. Dogen's Proof (Enlightenment), on the contrary, pertains to the "Act (Actualization)" proved by the Way beyond human consciousness. The Chinese character for *chêng* (*sho*) meaning proof or the assurance of absolute rightness, as exemplified by such phrases as *shoko* (evidence), *shomei* (certificate), and *shonin* (witness).[14] Dogen's True Proof, then, means that the Way assures of the rightness of the Practice of the disciplinant who is disciplined to practice in conformity to the Way. Speaking from the angle of the disciplinant, Dogen's True Proof means that if the disciplinant calms down the activities of reason and goes beyond all kinds of duality to "enter the Way of Buddha, with body and mind in harmony", he is assured of his capability of practicing the Way in integration with the "Act (Actualization)" of the Way beyond the dualities of activeness and passiveness and of the absolute rightness of his Practice by the "Act (Actualization)" proved by the Way. As he is one with the Proof of the Way, there is indeed no room for any doubt within him. As a result, he attains the absolute belief that "he is in the middle of the Way as he has ever been originally".

NOTES

1. About Dogen's life and thought, cf. H. Dumoulin S.J., *Geschichte des Zen-Buddhismus*, Band II: Japan (Bern 1986).
2. Dogen Zenji, Shobogenzo, transl. by Kosen Nishiyama (Tokyo: Nakayama Shobo), Vol. I (1975); Vol. II (1977); Vol. III (1983); Vol. IV (1983).
3. Tanabe Hajime, Shobogenzo no tetsugaku shikan [A Personal Reflexion on Shobogenzo] (Tokyo: Iwanami Shoten, 1940).
4. Watsuji Tetsuro, Nihon Seishinshi Kenkyu [A Study of the History of Japanese Philosophy] in the *Collected Works* IV (Tokyo: Iwanami Shoten, 1962), pp.156–246.
5. Tetsuji Morohashi, *Dai kanwa jiten [Great Chinese—Japanese Dictionary]* and *Ko Kanwa Jiten [Large Big Chinese—Japanese Dictionary]*. (Taishu Kan), Vol. 11, p. 124.
6. As I think that Taoism did not influence directly on Dogen's thought, here I do not discuss the Tao of Taoism in detail.
7. Martin Heidegger, *Gelassenheit* (Pfüllingen: Verlag Günther Neske, 1959).
8. Cf. *ibid.* Heidegger says: "Das besinnende Nachdenken, das den Sinn nachdenkt, der in allem waltet, was ist."
9. It is similar to Meister Eckhart's thought: "Das Wirken und das Werden aber ist eins. Wenn der Zimmermann nicht wirkt, wird auch das Haus Nicht.. . . Gott und ich, wir sind in solchen, er wirkt, und ich werde." (J. Quint, Meister Eckhart (München: Carl Hanser Verlag, 1969), S. 186, DWI, 113. Cf. S 402. DW III, 194).
10. R.S. § II 4 art, 13; DW I.S. 152, 3; Sermon. 247.
11. "Hier ist Gottes Grund mein Grund und mein Grund Gottes Grund." Quint, *ibid.* S. 180, DW I, S. 90.
12. LW III, n.6, p. 8.
13. See Shobogenzo Josetsu — Bendowa gikai [An introduction to the treasure house of the Eye of the True Dharma — A comment on a Treatise on the practice of the way] (Tokyo: Iwanami Shoten), p. 216.
14. See Master Eto, Shuso To Shite No Dogen [Dogen as Buddhist patriarch], p. 226.

Die Idee der Freiheit in Japan

KANJI TANEMURA
Universität Kagoshima

1. Die Vorstellung der Freiheit in der Zeit vor der Meiji-Ära

Heutzutage entspricht in Japan das Wort '*Jiyu*' der 'Freiheit' auf Deutsch oder 'liberty' auf Englisch. '*Jiyu*' war eigentlich ein chinesisches Wort. Seit es in Japan eingeführt wurde, herrschte überhaupt die Bedeutung "wie es beliebt" oder "eigensinnig", die es enthält — das heiβt die Nuance der Freiheit innerhalb des privaten Gebietes; obwohl '*Jiyu*' im Buddhismus "für sich selbst" bedeutet und nicht nur die private Freiheit, sondern auch die Selbständigkeit des Individuums ausdrückte.

Im *Japanisch-Portugiesisches Wörterbuch*, das Anfang des 17. Jahrhunderts veröffentlicht wurde, um den christlichen Missionaren das Japanische zugänglich zu machen, gab es den Begriff '*Jiyu*', der als 'Liberdade' auf Portugiesisch mit sich "frei und ungezwungen verhalten" erklärt wurde. Während der feudalistischen Edo-Ära erkannte man im allgemeinen an, daβ das Wort '*Jiyu*' "nach Belieben" bedeutete. Von der Mitte bis zum Ende der Edo-Ära entwickelten sich die Warenproduktion und -zirkulation rasch, und dadurch blühte die Kaufmannskultur, so daβ sich die Wirtschaftskraft der Kaufleute sprunghaft steigerte. Das Wort '*Jiyu*' trat also sehr oft in den literaristischen Werken der Kaufmannsschriftsteller auf, z.B. bei Saikaku Ihara, Sanba Shikitei und Ikku Jippensha. Von ihnen wurde die Freiheit positiv bewertet, weil sich in der Kaufmannswelt bereits die Lebensanschauung ausbreitete, daβ die Lebensweise "nach Belieben und frei handeln" wünschenswert und menschlich sei. [63] Die Ritter wurden damals als die Ausführer der feudalen Ethik gezwungen, ihre natürlichen Bedürfnisse zu unterdrücken; dagegen konnten die Kaufleute leicht ihre mit dem menschlichen Gefühl übereinstimmende, natürliche Moral entfalten. Die freie Geschlechtsliebe, der freie Handel und der freie Gelderwerb konnten erst in der Kaufmannswelt anerkannt werden. [59]

In den Werken der Kaufmannsschriftsteller kann man sowohl Gedanken über "die Freiheit des Körpers" als auch über das freie Verhalten im sozialen Leben finden. Im strengen japanischen Klassensystem "Ritter, Bauern, Handwerker und Kaufleute" fanden sich oft ungerechte Unterdrückung und die Forderung des bedingungslosen Gehorsam von seiten der Ritter, der obersten Klasse, gegenüber den Kaufleuten. Darum war ihre Forderung nach Befreiung aus dem Zwang und nach der Freiheit des eigenen Körpers immer stark.

Contemporary philosophy. A new survey. Vol. 7, 315–339.

Die positive Auffassung von *'Jiyu'* war aber überhaupt nur in der Kaufmannsgesellschaft gültig. In der Feudalzeit verstand man unter *'Jiyu'* hauptsächlich Selbstsucht, Eigenwilligkeit und Luxus. Der Wunsch nach Freiheit wurde sehr oft als Sünde angesehen. Dennoch brach am Ende der Edo-Ära der Wunsch nach Freiheit auch bei den Bauern hervor. Nach den letzten Studien über die japanische Geschichte, gab es bei der Bauernbewegung im Kinki-Distrikt den Gebrauch von *'Jiyu'* '*Fu-Jiyu* (Unfreiheit)', die dem modernen Begriff ähnlich ist. In der Bittschrift, die die Bauern bei der Vogtei einreichten, kann man den Anspruch auf die Freiheit der wirtschaftlichen Tätigkeit finden. Ebenfalls, daβ die Behörde die Bauern schützen soll, damit sie "ohne *Fu-Jiyu*" des Handels "ausgedehnt" und "nach Belieben" Produktion und Handel vollziehen können. [2] Diese Freiheit entsprach in ihrer Bedeutung schon der europäischen bürgerlichen Freiheit und stand, obwohl nicht adäquat, in Beziehung mit dem Begriff 'Recht'; insofern war der Begriff 'Freiheit' nicht völlig ein Fremdbegriff aus Europa.

Es gibt nun noch einen Inhalt, der in die japanische 'Freiheit' einflieβt. Es ist der Sinn der Selbständigkeit oder Unabhängigkeit. Die Freiheit bedeutet eigentlich auch in der japanischen Gesellschaft die Freiheit von der Macht oder dem Zwang, aber es gab noch eine andere Art von Freiheit im Denken Japans dieser Zeit, nämlich die Freiheit, etwas Positives zu schaffen. Das, was am bedeutendsten zur Verbreitung der "Freiheit als Selbständigkeit oder Autonomie" beitrug, war der Geist des Rittertums (*Bushido*).

Die Ritterklasse hatte eigentlich ein starkes Herrschaftsbewuβtsein, insofern, daβ sie im Rang über drei Volksschichten stand. Durch ihr stolzes Klassenbewuβtsein formten sie durch fortdauernde Abhärtung im zivilen Leben und beim Militär den Geist der Unabhängigkeit oder Selbständigkeit und das Gefühl der Selbstachtung oder Verantwortlichkeit. [62] Bei der japanischen Feudalgesellschaft banden sich die Ritter an einen bestimmten Gebieter in einem Verhältnis von Gnade und Dienst. Sie erhielten vom Gebieter irgendein Territorium (oder erbliche Besoldung) und zum Ersatz dafür hatten sie sowohl in Friedens- als auch Kriegszeiten, dem Herrscher treu zu sein und eventuell sogar ihr Leben für ihn zu opfern. Innerhalb dieser festen Senkrechtbeziehung versammelten sich die Ritter alle rings um den Herrn. Sie verpflichteten sich keineswegs zueinander, und deswegen konnten sie manchmal ihre eigene Selbständigkeit oder ihr Alleinsein untereinander behaupten; obwohl dies auf das Fehlen einer demokratischen öffentlichen Gemeinsamkeit hinweist. [9]

In einer Krisenzeit des feudalen Klans und Shogunats muβten sie ins Feld ziehen und stets den Tod vor Auge haben. Keine Furcht und Gefaβtheit vor dem Tode zu zeigen, die Bereitschaft zur Selbstaufopferung für den Herrn, die bei einer groβen Tragödie keineswegs zu erschütternde Gelassenheit und die Beherrschung des eigenen Gefühls im Glück oder Unglück usf., galten als die unerläβlichen Tugenden des Ritters. [43] Zum Unterschied von den Rittern in der Kamakura- und Muromachi-Ära, die unaufhörlich Feldzüge mitmachten, konnten die Ritter in der Edo-Ära friedlich leben und muβten deshalb nur selten schreckliches Unglück erleiden. Trotzdem wurden die Tugenden 'Selbstbeherrschung' und 'Selbstüberwindung' als der Kernbegriff des japanischen Rittertums auf immer der Nachwelt überliefert.

Überhaupt verabscheute der Ritter den Handelsgeist des Kaufmannes und achtete vor allem die Landwirtschaft, weil er als feudaler Herrscher die Agrarwirtschaft als das Fundament des Staates sah. Da die Produktionskraft der japanischen Landwirtschaft niedrig und unbeständig war, wurde der Ritter, dessen ökonomisches Leben vom Bauern abhing, zu Armut und Sparsamkeit getrieben. [9] Der Stolz auf die Armut und das Lob der Geduld, die ihren Ursprung in der niedrigen Produktionskraft und dem armen Leben der Feudalenzeit hatten, machten die Grundlage des ritterlichen Geistes, gemeinsam mit der Selbstbeherrschung und der Selbstüberwindung, aus. Im japanischen Rittertum setzte sich, wie oben erwähnt, der stoische Charakter und der damit verbundene hohe Wert der Tugenden, z.B. Unabhängigkeit, Selbstüberwindung, Enthaltsamkeit, Verantwortlichkeit usw., durch. [43] Sie sind zwar feudale Tugenden und abgesetzt von der Form der Freiheit, die die neueren Bürger zu erlangen suchten. Teilweise aber entstand darin jedoch zugleich die Idee, die der bürgerlichen Enthaltsamkeit bei B. Franklin und der persönlichen Autonomie bei I. Kant gemeinsam ist.

Das freie eigenständige Leben, das der Wunsch der Kaufleute war, der Anspruch auf die Freiheit der wirtschaftlichen Tätigkeit, der sich bei den Bauern fand, der Geist der Selbständigkeit und Selbstüberwindung, der im Rittertum wirkte, usw. spielten eine große Rolle bei der Übernahme der europäischen Idee der Freiheit. Mit anderen Worten, in Japan gab es bereits eine ziemlich reife Basis, auf der viele Japaner die europäische Freiheit auf den japanischen Begriff "*Jiyu*" übertragen konnten. Dennoch war das Bewußtsein, daß die Freiheit ein ursprüngliches Recht sein soll, in der japanischen Feudalzeit sehr schwach ausgeprägt. [63] Das Verständnis und die Überzeugung, daß die menschlichen Freiheiten der menschlichen Natur entspringen und fundamentale Menschenrechte sind, entstanden erst bei den Aufklärungsdenkern in der Meiji-Ära, besonders bei Yukichi Fukuzawa und Chomin Nakae.

2. Die Idee der Freiheit bei Yukichi Fukuzawa

2.1 *Die Aufgabe der Aufklärung von Fukuzawa und seine Freiheitsidee (1)*

Als das Tokugawa-Shogunat zusammenstürzte und die Meiji-Regierung unter Aufrechterhaltung des Tenno-Systems die neue japanische Gesellschaft aufzubauen begann, war es Yukichi Fukuzawa, der sich für die richtige Modernisierung Japans am energischsten aussprach und am meisten die damalige öffentliche Meinung leitete. Fukuzawa war der größte Denker der Aufklärung, den Japan im 19. Jahrhundert hervorbrachte. Dadurch, daß seine Idee über die Freiheit natürlich durch den Kontakt mit der europäischen Kultur geprägt wurde, hatte sie den besten demokratischen Charakter in der Meiji-Ära. Er entwickelte jedoch keine neue Ansicht über die Freiheit, unabhängig von den anderen Aufgaben. Seine Freiheitstheorie beschäftigte sich nur mit den Aufgaben, die Japan damals dringend zu lösen hatte, besonders die Frage der Verbindung der japanischen Zivilisation mit der europäischen. Um die Bedeutung seiner Freiheitsidee richtig begreifen zu können, benötigt man zunächst das Verständnis dessen, was sein Aufklärungsgedanke bezweckte und aufzulösen versuchte.

Die Abwendung von der Abschließungspolitik zur Öffnungspolitik in der letzten Periode der Edo-Ära war keine Folge des freiwilligen Anspruchs von Japan selbst. Der militärische wirtschaftliche Druck der europäischen Mächte rüttelte Japan aus dem friedlichen Schlaf und nötigte es, sich nach dem Modell der europäischen Modernisierung zu verändern. Die überwältigende Macht der westlichen Zivilisation, die seit dem Opiumkrieg noch deutlicher und unanzweifelbarer wurde, bewirkte ein Krisengefühl in den japanischen Regierenden und Intellektuellen. Es stellte sich die Frage, ob ihr Vaterland, wie bisher, seine Unabhängigkeit erhalten könne. Auch für Fukuzawa, der durch den Kontakt mit der anderen Kultur für die internationale Situation sehr feinfühlig war, gab es selbstverständlich keine eigene Gedankentätigkeit ohne die Aufgabe dieser Unabhängigkeit.

Er nahm keineswegs Partei für die bornierten Nationalisten, die wegen der Erhaltung der japanischen Unabhängigkeit den Verkehr mit den Ausländern abbrachen und die Einführung der europäischen Zivilisation haßten. Um die Mängel der japanischen Zivilisation zu ersetzen und die nationale Unabhängigkeit zu erreichen, dachte er, müssen Japaner vielmehr möglichst schnell die Vorzüge der europäischen Zivilisation absorbieren. Was hielt er also für die Mängel der japanischen Zivilisation und die Stärke der europäischen Zivilisation? Es waren "der selbständige freie Geist" und "die festgelegte Mathematik und Naturwissenschaft".

In Japan fehlte nach Fukuzawa vornehmlich der selbständige freie Geist oder die unabhängige Geisteskraft. Dieses negative Phänomen wurde einige Jahrhunderte lang in der Geschichte der "Vorherrschaft des Starken" gebildet und nach dem Ende des Feudalismus noch als "die Hochschätzung der Regierung und die Geringschätzung des Volkes" oder als die Abhängigkeit des Volkes von der Regierung erhalten. Fukuzawa, der sich mit dieser Wirklichkeit auseinandersetzte, behauptete, daß der wahre Zweck der Zivilisation überhaupt "der Fortschritt der menschlichen Bequemlichkeit und Würde" oder "die Entwicklung der menschlichen Weisheit und Tugend" sei, [6] und anstatt der traditionellen Ethik, die auf dem Konfuzianismus beruht, riet er dem Volk, sich europäische Wissenschaften und Techniken anzueignen. Das Volk soll dadurch die kritischen Einsichten in die von der Macht und Konvention beherrschten irrationalen Gesellschaftsbeziehungen haben, so daß sie den wissenschaftlichen Geist ausbilden können, der objektiv und ohne Vorurteile die Natur und die Gesellschaft aufzufassen versucht, und dazu sollen sie die unabhängigen Geisteskräfte erlangen, die bewußt die alten schlechten Gewohnheiten ausrotten könnten. — Dies war seine größte Hoffnung.

Seine Ansichten über die Freiheit, beeinflußt von den anglo-amerikanischen Philosophien, die am Anfang der Meij-Ära in Japan eingeführt wurden, waren im weiteren Sinne von utilitaristischem Charakter. Er meint, daß die Freiheit im individuellen Gebiet vor allem im freien Gebrauch des menschlichen Körpers besteht, nämlich, "ein Mensch kann insofern frei und ungezwungen seinen Körper in Bewegung setzen, als er kein Recht des anderen Menschen stört". [5] Diese Freiheit des Körpers ist natürlich zugleich die Freiheit, nach der jeder an seinen Lieblingsort geht, am Lieblingsplatz bleibt, frei arbeitet und ungezwungen spielt. Indem Fukuzawa als Aufklärungsdenker auf den tätigen Menschen, der mittels seines eigenen Körpers seine Bedürfnisse realisiert, anfmerksam machte, erkannte er nicht "die Willensfreiheit" als die Elementarform der Freiheit, sondern "die Handlungsfreiheit". Von diesem Standpunkt aus, klagte er leidenschaftlich an, wie stark das feudale Klassensystem "Ritter, Bauern, Handwerker und Kaufleute" die Freiheit des Körpers und die Handlungen des untergeordneten Volkes unterdrückte, und wie oft Frauen im Familienleben von den übermächtigen Männern gedemütigt wurden.

Außerdem war es die ausgezeichnete Auffassung vom Zusammenhang der Freiheit mit dem Wissen, die das hohe Niveau seines Begriffes der Freiheit zeigte. Die Schlußfolgerung vorweggenommen, war Fukuzawa der erste Japaner, der verstand und behauptete, daß der freie Geist aus dem energischen Verstand entsteht; mit anderen Worten, daß die Selbständigkeit und die Freiheit des Individuums erst durch die Erwerbung wissenschaftlicher Kenntnisse von der objektiven Welt ausgebildet und erhalten werden. Seine Kritik des Konfuzianismus ist in dieser Hinsicht das beste Beispiel.

Die Lehre von Konfuzius und Menzius spielte schon lange eine wichtige Rolle, die ideologisch die feudale Fürst-Untertan-Beziehung stützte. Indem der Konfuzianismus diese Klassenbeziehung für die ursprüngliche Beziehung hielt, die von Natur aus bestimmt ist, beharrte er immer bei der Lehre, daß die Ethik der Fürst-Untertan-Beziehung auf der natürlichen Anlage der Menschen beruht. Der Unterschied zwischen Fürst und Untertan ergab sich selbstverständlich erst nach der Geburt der Menschen und der Bildung der Gesellschaft und war also gar nicht die ursprüngliche unwandelbare Menschennatur.

Diese falsche Lehre, dachte er, ebenso wie die ptolemäische Theorie in der Astronomie, besteht in der Verwechslung des Wesens der Sache mit deren Erscheinung. Ihr Grundmangel war nämlich, daß sie von Anfang an die wissenschaftlichen Kenntnisse von der Natur und der Gesellschaft ablehnte und somit nicht die Wahrheit der Dinge begriff. Nach dieser Überlegung sagte er: "Die Dinge sind vorhanden, und erst danach kann die Ethik sein. Kein Ding entsteht, nachdem die Ethik vorhanden war. Es ist nicht erlaubt, mittels der Vermutung zuerst den ethischen Charakter der Sachen zu erklären und durch diese Ethik der Wahrheit der Sache zu schaden". [6]

Nach Fukuzawa kann die passende Beziehung oder Kommunikation unter den Menschen nur durch die positive Untersuchung der Natur und der Gesellschaft bestimmt werden. Die der bürgerlichen Gesellschaft entsprechenden demokratischen Beziehungen der Menschen werden nämlich nur innerhalb der Natur- und der Gesellschaftsanschauung ausgebildet, die ganz unabhängig vom konfuzianistischen Moralismus durch die neuen wissenschaftlichen Einsicht rekonstruiert wurden.

Er verwarf natürlich nicht einseitig die Ethik überhaupt. Es war vielmehr sein wichtigstes Ziel, eine neue bürgerliche Ethik zu schaffen. Das, was er hier stark verneinte, war die Lebenshaltung, die vor der vernünftigen Erkenntnis von der Welt kritiklos die bestehende traditionelle Ethik oder fixierte Wertmaßstäbe bestimmt. Er nannte diese Haltung "die blinde Ergebenheit". Diese ist der Gipfel der

Unselbständigkeit oder der Unfreiheit. Zum Beispiel war es die blinde Ergebenheit in die alte Gewohnheit, daβ der Ritter trotz Friedenszeiten noch nach der feudalen Zeit am Schwerttragen festhielt und sein Geld verschwendete. Es war auch nichts anderes als die blinde Ergebenheit der falschen Würde, daβ die Regierung zwischen Herrscher und Volk künstliche Unterschiede machte und die Norm von den Hohen und Niedrigen hinsichtlich des Ranges in der Kleidung und der Sprache festsetzte. [6]

Dagegen ist das, was die blinde Ergebenheit in diese ungereimten überlieferten Sitten überwindet, nichts anderes als die Tätigkeit des Verstandes, der den fixierten Weltmaβstab in Zweifel stellt und beharrlich seinen Weg der Ergründung der Wahrheit (der Natur- und Sozialwissenschaft) zu gehen versucht. Der freie Geist macht nur insofern Fortschritte, als er sich fest mit diesem energischen Verstand verbindet. Nach Fukuzawa war also die subjektive praktische Freiheit der Menschen und die wissenschaftliche Erkenntnishaltung, die das Prinzip oder die Wahrheit der Dinge zu erforschen sucht, absolut unzertrennlich.

2.2 *Fukuzawas Freiheitsidee (2)*

Wir wollen Fukuzawas Ansichten über die Freiheit auf dem sozialen Gebiet weiter betrachten. Seine Lehre von der sozialen Freiheit übertrifft zwar nicht den europäischen Aufklärungsgedanken an Originalität, aber seine entschlossene Zeitkritik und seine positiven Vorschläge zur Modernisierung glänzten am herrlichsten auf diesem Gebiet. Auf Guizots *Geschichte der Zivilisation* beruhend, führte er nun die Entstehung der Freiheit in Europa auf die geschichtliche Tatsache zurück, daβ die verschiedenen Lehren von der Menschenbeziehung gleichzeitig existierten und sich gegenseitig diskutierten, so daβ die eine die andere nicht beherrschen konnte. Nach seiner Meinung ist Europas Zivilisation eine Geschichte, in der der Verstand und der freie selbständige Geist des Volkes im Verlauf der Veränderungen oder des Wettbewerbs der vielfältigen politischen Systeme befruchtet wurde. Die mannigfaltigen Gedanken und die pluralistischen Weltanschauungen, die in dieser lebendigen Geschichte entstanden, sind besonders wertvoll. In Wahrheit sind sie es, die in Japan am meisten fehlten. Wie aber entstanden solche Umstände?

Im Vergleich mit Europa zeigte Fukuzawa auf, daβ die japanische Zivilisation lange Jahre als charakteristisches Merkmal ein "Miβverhältnis der Macht" aufwies. Unter dem über Jahrhunderte dauernden feudalen System war die Macht immer auf der Seite der höheren Stände innerhalb aller Bereiche der Gesellschaft, z.B. Männer und Frauen, Eltern und Kinder, Lehrer und Schüler, Gebieter und Untertanen, groβe Klane und kleine Klane, Haupttempel und Nebentempel usw. Auch in der Meiji-Ära blieb diese Tendenz in der Rangordnung der Staatsbeamten und zwischen den Beamten und den Bürgern hartnäckig erhalten. Fukuzawa sagte, "In Japan gibt es eine Regierung, aber keine Nation". [6] Der Vorzug seiner Kritik besteht in der Einsicht, daβ das Miβverhältnis der Macht unbedingt "die Ungleichheit der Menschenrechte" erzeugt.

Dieses Mißverhältnis der Macht entstand durch die Unwissenheit und das Fehlen eines selbständigen Geistes im Volk, die die Gemütsanlage der ganzen Nation ist, die dem Herrscher die Autokratie überläßt. Fukuzawa kritisierte in dieser Hinsicht nicht nur scharf die autokratische Regierung, sondern auch das abhängige Volk, das nicht freiwillig an den Staatsangelegenheiten teilzunehmen und sich stets auf die Autorität zu stützen versucht. Wie soll man nun "das Mißverhältnis der Macht" als einen undemokratischen Zustand überwinden? Was ist denn die Triebkraft dafür?

Kurz gesagt, war es die Erschaffung der mannigfaltigen Urteile und der pluralistischen Wertvorstellungen, die sich auf "die Geschäftigkeit und den Wortwechsel" gründeten. [6] Fukuzawa verstand unter Geschäftigkeit und Wortwechsel konkret z.B. den selbständigen privaten Betrieb, den sich ein Vermögen erwerbenden Intellekt, die den Willen zur Wahrheit und den freien Geist bildenden Wissenschaften, und weiter die lebhaften öffentlichen Diskussionen über Politik und Wirtschaft. Sein Begriff von der sozialen Freiheit bedeutete in diesem Sinne vor allem die Befreiung vom Zwang der Regierung oder Macht, mit anderen Worten, die Sicherstellung der freien Rede und der sozialen Handlungen der Zivilbevölkerung.

Wenn Fukuzawa nun über den freien Geist des 'Volkes sprach, muß man beachten, daß er das Menschenbewußtsein als soziales Bewußtsein, sozusagen als den Zeitgeist verstand. Er hatte nämlich die Einsicht, daß die Entwicklung der menschlichen Geschichte auf das in dieser Zeit herrschenden Bewußtsein des Volkes als der Zeitströmung begründet ist. Verschiedene wichtige Ereignisse in der japanischen Geschichte, z.B. die Entstehung oder der Untergang der Regierung von Ashikaga, Oda, Toyotomi und Tokugawa, hatten ihre Ursache nicht im Verdienst oder Versagens eines bestimmten Helden. Sie waren vielmehr die Folge der Zeitströmung und sind der vorherrschenden sozialen Gemütsart des Volkes zuzuschreiben. Auf diese Weise drückte Fukuzawa, wie Hegel mit seinen eigentümlichen Worten, die Einsicht in die Notwendigkeit der Geschichte aus. Indem er die geschichtliche Bedeutung von der Macht des Publikums ausreichend erkannte, mußten für ihn die Freiheit und die Unabhängigkeit bis zur Gemütsart des Volkes als dem objektiven Zeitgeist erhöht werden. Also muß man aufs neue erkennen, daß Fukuzawas Aufklärungsarbeit mit dieser ausgezeichneten Geschichtsauffassung diese großartige soziale Perspektive hatte.

3. Die Idee der Freiheit bei Chomin Nakae

3.1 *Nakaes Freiheitsidee (1)*

Chomin Nakae, einer der führenden Denker der frei-demokratischen Bewegung in den achtziger Jahren des 19. Jahrhunderts, legte in seinem kurzen Leben von 55 Jahren den Grundstein zur späteren Entwicklung der modernen Philosophie und des fortschrittlichen Denkens in Japan. Es ist nicht übertrieben festzustellen, daß seine

Abhandlungen über Freiheit, Recht, Gleichheit und Frieden, obwohl nicht systematisch, doch in Hinsicht auf den progressiven tiefsinnigen Inhalt die Arbeit Fukuzawas übertrafen. Es war Nakaes Ziel, in der Gesellschaft der Meiji-Ära den Rousseauismus zu realisieren und fixieren, der ein theoretischer Ausdruck der radikalen Volksdemokratie in Europa war. Zu diesem Zweck, als Journalist, als Politiker oder als Philosoph, hoffte er, im Volk Selbstbewußtsein und ein Gefühl von Freiheit und Recht zu wecken und drang fortwährend und eindringlicher als Fukuzawa auf deren Verwirklichung in der Praxis.

Seine Aufklärungsarbeit, ebenso wie Fukuzawa, immer im Bewußtsein der Wichtigkeit der nationalen Unabhängigkeit Japans von den europäischen Mächten, zielte jedoch hauptsächlich auf "den Aufschwung des Rechtes der bürgerlichen Freiheiten" im Lande. Sein konsequenter Anspruch auf die Verfassung des Sozialkontrakts, die dem Volk innewohnende Staatshoheit, die allgemeine Wahl und die Freiheit der Rede, der Veröffentlichung, der Versammlung und des Vereins usw. waren die bedeutenden Beweise.

Seine Demokratie wurde jedoch oft von der Meiji-Regierung eingeschränkt, die zwecks der Vergrößerung der Staatsgewalt, Maßregeln zur Bereicherung des Landes zu treffen begann. Mitten in dieser freien-demokratischen Bewegung schlug er das Verfahren vor, daß man zunächst nicht die von oben gegebene "Demokratie als das kaiserliche Geschenk" ablehnen und sie bis zum gleichen Niveau wie die von der Basis gewonnene "Demokratie als der Wiederherstellung" bringen sollte. [31] Dies war gerade das Resultat seiner qualvollen, aber flexiblen Wahl. Nakaes Ideal scheiterte zwar schließlich aufgrund der Unreife der Demokratie in der Meiji-Ära, aber man darf nicht übersehen, daß seine Ideen von Freiheit, Gleichheit und Menschenrechten, einen wichtigen Beitrag zur Bildung der späteren Taisho-Demokratie [64] [14] und des sozialistischen Gedankengutes [18] [19] [11] leisteten.

Was nun Nakaes Freiheitslehre betrifft, so basiert sein Versuch einer Begriffsbestimmung der Freiheit auf der westlichen Philosophie und zwar besonders auf der französischen Philosophie. Dies war der Ausdruck des ihm eigentümlichen philosophischen Bewußtseins, das damals in keinem anderen Demokraten entstand.

In einem Leitartikel der *Toyo-Jiyu-Zeitung* beschrieb er z.B. die zwei Arten der Freiheit: er unterschied zwischen 'liberté morale' und 'liberté politique'. Jene Freiheit bedeutet, daß "der persönliche Geist sich selbst erschöpfend ohne die Fesselung der anderen entwickelt". Diese ist nichts anderes als die Freiheit, mit der man für sich handeln kann und mit den anderen Menschen etwas vollbringen kann, sagte er andererseits und weiter führte er konkret an: die Freiheit des Körpers, die Gedankenfreiheit, die Redefreiheit, die Versammlungsfreiheit, die Vereinsfreiheit, die Freiheit der Beschäftigung und die Freiheit der politischen Teilnahme usw. [29]

Die Freiheit des Geistes (liberté morale) ist hier nicht die reine Willensfreiheit, die der Idealismus bisher behauptet hat, sondern die geistige Freiheit, die auch die Willensfreiheit im engeren Sinne enthält. In bezug auf die Beziehung zwischen den beiden Arten von Freiheit erwähnte Nakae die Natur der geistigen Freiheit als die ursprünglichere aller Freiheiten. Das bedeutet: "Indem die Freiheit des Geistes die angeborene Grundlage ist, entspringen daraus die zweite Freiheit der Handlung und auch alle anderen Arten von Freiheit". [29] Hierin besteht ein bestimmter Unterschied zu Fukuzawa, der zu allererst auf die Freiheit des eigenen Körpers und die Handlungsfreiheit achtet.

Nakae behauptete ebenfalls, daβ die Freiheit des Geistes tief in der menschlichen Natur wurzelt und noch ursprünglicher ist als die anderen, und daβ sie in alle Gebiete, d.h. Religion, Wissenschaft, Kunst, Industrie und Politik usw. reicht. Von dieser Behauptung aus kann man verstehen, daβ er die Notwendigkeit zu begründen versuchte, daβ alle bürgerlichen Freiheiten selbstverständlich allen Menschen gehören und sich die Freiheiten als Rechte mit dem Fortschritt der menschlichen Vernunft unbegrenzt erweitern sollen.

Wenn Nakae nun die Freiheit erörterte, dachte er sie meistens als die Freiheit, die nicht nur individuell ist, sondern eigentlich alle Menschen besitzen sollen. Dem lag nämlich der Gedanke vom für jedermann gleichen Recht zur Freiheit oder Recht zur Gleichheit zugrunde. Er verehrte ebenso die Gleichheit wie die Freiheit, ja, die Gleichheit mehr als die Freiheit. Seine Lehre von der Freiheit vollendet sich erst dadurch, daβ sie in sich die Lehre von der Gleichheit einfügt.

In seinem Hauptwerk *Sansuijin-Keirin-Mondo* [Gespräch über die Staatsverwaltung unter drei Betrunkenen], z.B., sagt Herr Gentleman, der Nakaes Ideal vertritt, daβ man im Konstitutionalismus viele moderne Rechte haben kann, aber daβ dieser Konstitutionalismus wegen der Unwirklichkeit der Gleichheit zur Hälfte fehlerhaft ist. Anschlieβend behauptet er, daβ die wahre Demokratie sich erst vollendet, indem man mit der Freiheit auch die Gleichheit erlangen kann, d.h. "Gleich und frei zu sein ist also das oberste Gesetz der politischen Verfassung". [31] Dies ist ein Beispiel, in welchem sich Nakaes Hochschätzung der Gleichheit deutlich zeigt. Hier ist der Grund, warum die von ihm aufgestellte Demokratie noch stärker die Tendenz zur Volksdemokratie als die zur bürgerlichen Demokratie hatte.

In der Lehre von der Pöbelemanzipation (*Burakumin*-Emanzipation), die in der Abhandlung "die Welt neues Volkes" vorgeschlagen wurde, war seine Gleichheitsidee am glänzendsten ausgedrückt. [32] Am Beginn der Meiji-Ära wurde den *Burakumin* [japanischen Pöbel] der Name "neues gemeines Volk" gegeben, einer Gruppe, die während der feudalen Zeit als die niedrigste Klasse unaufhörlich unterdrückt worden war. Infolge der neuen Bezeichnung löste sich der Unterschied scheinbar auf. In der Tat aber war ein versteckter strengerer Unterschied als früher hinsichtlich der beruflichen Möglichkeiten und der Heirat usf. geblieben. In diesem Zusammenhang wandte sich Nakae an die Öffentlichkeit und verlangte, daβ man *Burakumin* keineswegs für eine andere Art von Mensch oder niedrigeren Mensch halten solle, weil es keinen Unterschied in der Struktur oder Funktion des Körpers zwischen allen Menschen gibt und das gesamte Volk Japaner sind. Diese Aufforderung brachte die Tatsache an den Tag, daβ der damalige Meiji-Staat des Tenno-Systems wie der feudalistische Staat mit dem Opfer der niedrigsten Schicht "neuem gemeinem Volk" auf dem Unterschied und der Unterdrückung aufbaute, und daβ die sogenannte "Gleichheit der vier Klassen" eine Illusion war.

Nakae zufolge, kann "die Welt des neuen Volkes", die ohne Unterschied zwischen gemeinem Volk und neuem gemeinem Volk alle Menschen subsumiert, wie ein großer Kreislauf werden, worin es weder Klassengegensätze noch Unterordnung gibt, und nur in dieser paritätischen Welt kann man den sozialen Wahn auflösen und sich seelisch und moralisch retten. Seine zukunftsweisende Kritik am Pöbelunterschied und den Vorschlag der Pöbelemanzipation kann man nicht genug hochschätzen, besonders wenn man Nakae mit Fukuzawa [7] vergleicht, der teilweise den Pöbelunterschied anerkannte.

3.2 *Nakaes Freiheitsidee (2)*

Als Philosoph nun, nicht als Aufklärungsdenker, zeigte Nakae dann sehr klar seine Fertigkeiten als er das Thema von "Willensfreiheit" oder "Moral und Freiheit" untersuchte, das das traditionelle Grundproblem in der Philosophie war. Das bedeutet, daß die philosophische Freiheitsfrage sozusagen erst ernsthaft von Nakae in Japan erforscht wurde.

Nach allgemeiner Ansicht soll Chomin Nakae die Willensfreiheit vom Standpunkt des Materialisten verneint haben. Wenn man aber seine Darstellung sorgfältig liest, wird es klar, daß er in Wahrheit "die unbedingte Willensfreiheit" kritisierte und nicht die Willensfreiheit selbst völlig verneinte. Der Grund, warum er nicht der naiven Verneinung der Willensfreiheit zustimmte, ist, daß er den Inhalt der Willensfreiheitslehre vom französischen Spiritualismus völlig verstand und ihren philosophischen Sinn klar erkannte.

In einem philosophischen Buch *Rigaku-Kogen* [Die Einleitung in die Philosophie] stellte Nakae vorurteilslos die folgende Ansichten dar, z.B. die Aktivität des Willens, seine Funktion der selbständigen Wahl, seine ursprüngliche Spontaneität, die auch das Böse aufnehmen kann, die Selbstverständlichkeit seiner Realität und die Grundlegung der Moralität durch die Willensfreiheit usw., die in der idealistischen Freiheitslehre ausgedrückt wurden. Auch beim Kritisieren des freien Willens lehnte er nicht alle hier erwähnten Ansichten als falsch oder unfruchtbar ab. Er versuchte hingegen teilweise den rationalen Erfolg von den idealistischen Theorien aus anzunehmen. [30]

Obgleich Nakae die Aktivität des Geistes rechtfertigte, lenkte er stets zugleich die Aufmerksamkeit auf die Wichtigkeit des Motivs selbst, das diese Aktivität hervorruft — ein Aspekt, den die idealistische Freiheitslehre zu vernachlässigen neigt. Sein Materialismus berücksichtigte also am höchsten die Realität von Bedürfnis, Wissen und Motiv, die die Vorbedingung der Willensbestimmung sind. Was die Willensfreiheit betrifft, verneinte er sie nicht ganz sondern bezog sie auf die menschliche Fähigkeit, was bedeutet, daß man ohne äußere Störung einem bestimmten Motiv nachfolgen kann. [30] In der Tat kann der geistige Bereich für sich unabhängig vom Körper existieren. Indem der Geist sich fest mit der körperlichen Fähigkeit verbindet, ist es auch für ihn unmöglich, das zu leisten, was der Körper nicht kann. Außerdem kann die Wirkung jedes Geist nicht willkürlich die intellektuelle moralische Fähigkeit überschreiten, die in der Erziehung und dem sozialen Leben gebildet wurde. Der menschliche Geist wird vom eigenen Körper, den materiellen Bedingungen und der kulturellen Umgebung bedingt, und die Willensfreiheit ist in diesem Sinne nur relativ. — Dies war es, was Nakae erklärte.

Am Ende seines Lebens stellte Nakae seine originelle Willensfreiheitslehre in seinem Buch *Fortsetzung, Anderthalb-Jahre* von demselben Standpunkt aus dar. [33] Dort behauptete er ebenfalls, daß die absolute Willensfreiheit auf schwachen Füßen stehe und schwer zu unterstützen sei, und daß der Wille für sich ohne Beziehung zu Bedürfnis und Gewohnheit keine Handlung bestimmen könne. Aber existiert nicht deshalb kein freier Willie, weil der Mensch, zum Unterschied vom zum Magnet angezogenen Eisen, ein spontanes Bewußtsein hat. Der Mensch kann in seinem täglichen Leben die guten Gewohnheiten annehmen, indem er bewußt "sich gut benimmt und seine Handlung entwickelt". Folglich gab Nakae der Willensfreiheit eine folgende neue relativierte Bestimmtheit. "Wenn man beim Wählen des Zwecks die Freiheit des Willens haben sollte, bedeutet das nicht, daß man hinsichtlich allen Tuns und Lassens frei sei, aber daß man nichts anderes als die Freiheit der Bestimmung im Hinblick darauf habe, was man durch die gewöhnliche Erfahrung bekam". [33]

Beachtet "die Freiheit der Bestimmung im Hinblick darauf, was man durch die gewöhnliche Erfahrung bekam", warnte Nakae, daß man entfernt vom realen Lebensprozeß nur auf den Augenblick achtet, worin sich der Wille entschließt. Er forderte vielmehr, daß man fortdauernd erklären sollte, unter welchen Bedingungen, was der subjektive Wille bis zur Zeitpunkt des Entschlusses erfuhr und bekam. Wie man von dieser Hinsicht aus verstehen kann, kritisierte Nakae darum "die reine Willensfreiheit", weil er den, dem Materialisten eigentümlichen, praktischen Geist hatte. Wenn man nur den unbedingt freien Willen für die Bestimmungsursache der Handlung hielte, würden "das sich täglich Bilden", "die sich einschließenden Lebensumstände" und "die Gewohnheiten und der Geist in der Zeit" ganz unnütz und sinnlos, [33] — so dachte er.

Nakaes philosophische Freiheitslehre erreichte jedoch leider nicht die Stufe, die logisch das Verhältnis zwischen Freiheit und Notwendigkeit versteht. [28] Für diesen Zweck würde sicher "das dialektische Denken" notwendig sein. Nakae war zwar ein standhafter Materialist, blieb jedoch lebenslang von der logischen Dialektik entfernt. Die Untersuchung dieser Aufgabe wurde in Japan ernsthaft erst nach der späteren Einführung der deutschen Philosophie, besonders Kant und Hegel, begonnen. Der geschichtliche Sinn von Nakaes Theorie besteht in der Vorwegnahme späterer Untersuchungen.

4. Die Idee der Freiheit bei Kitaro Nishida

4.1 *Die Eigenart der Nischida-Philosophie und seine Freiheitslehre (1)*

Von Ende des 19. Jahrhunderts bis zum Anfang des 20. Jahrhunderts waren die Wiederherstellung der Sinologie und die Untersuchung der deutschen Philosophie wichtige Richtungen in der japanischen Geisteswelt. Auf dem philosophischen Gebiet untersuchte man besonders die Kantsche Philosophie beeinflußt vom

Neukantianismus und danach galt der deutsche Idealismus, Fichte und Hegel usf., als der wichtigsten Studiengegenstand an den japanischen Kaiserlichen Universitäten.

Diese Richtung wurde gerade dadurch bestimmt, daβ die freie-demokratische Bewegung in der Mitte der Meiji-Ära scheiterte und das absolutistische Meiji-Staatssystem durch das Gelingen der Politik der Bereicherung und Verstärkung des Landes errichtet wurde. [11] Das Interesse der Intellektuellen und Denker begann sich nun hauptsächlich von der Politik und dem sozialen Problem zu dem innerlichen Ich und der Bildung der privaten Persönlichkeit zu wenden. [16], [17] Da die Lehre von der nationalen Moral [42] und der japanischen Nationalismus damals mächtiger wurden, beschleunigte sich die Tendenz des Spiritualismus und Personalismus in der Gedankenwelt noch stärker. Kitaro Nishida, der als der gröβte Philosoph in der japanischen Neuzeit anzusehen ist, begann seine Arbeit in dieser wissenschaftlichen Situation.

Er kam in Berührung mit der Erhabenheit der europäischen Erkenntnistheorie und Ontologie durch W. James in England, Kant und Hegel in Deutschland, und zugleich absorbierte er die bürgerliche Freiheitsidee und die progressive Geschichtsauffassung, die diese Philosophien durchdrangen. Das Verständnis und die Aufnahme der westlichen Philosophie ist aber nicht alles, was er tat. Das *Zen*-Erlebnis und die Sympathie für das japanische Schönheitsgefühl in seiner Jugend veranlaβten ihn zur Vereinigung der westlichen Logik mit dem östlichen Pantheismus oder japanischen Naturalismus. Die Erforschung des Ursprünglichen, das dem Unterschied zwischen Subjekt und Objekt vorangeht, wurde lebenslang Nishidas unverändertes Motiv. Die Begriffe "die reine Erfahrung"[35], der absolut freie Willie" [36], "der Ort" [37], das absolute Nichts" [38] [39], "das dialektische Allgemeine" [40], die sich ausdrückende Welt"[41], und "die geschichtliche reale Welt" [41] usf., die im Prozeβ der Entwicklung seiner Gedanken nacheinander veröffentlicht wurden, waren alle die Erfolge seines philosophischen Denkens, das "die Gestalt des Gestaltlosen", "die Stimme des Stimmlosen" in den östlichen Gedanken logisch zu begreifen suchte.

Darüber hinaus erhob er Zeit seines Lebens den stärksten ethischen Anspruch, daβ der Philosoph die Frage, wie man leben soll, ausdrücklich zu beantworten hat. Die Originalität seiner Philosophie besteht also darin, daβ sie das ursprüngliche Wesen der Welt erforschte und zugleich das Verhältnis zwischen Welt und Individuum begriff, und zwar, daβ sie innerhalb dieser Welt den Sinn des menschlichen Lebens und die Freiheit der Persönlichkeit erklärte. In diesem Sinne kann man bestimmen, daβ Nishidas Philosophie eine Weltphilosophie ist, die sich auf den japanischen Idealismus und westlichen Liberalismus gründet.

Wegen seines Naturells, seiner wissenschaftlichen Erfahrung und der damaligen ihn umgebenden Gedankensituation behandelte er keineswegs unmittelbar die politische soziale Freiheit, die Fukuzawa und Nakae untersucht hatten. Die moralische persönliche Freiheitsfrage und die philosophische Frage nach dem Verhältnis zwischen Notwendigkeit und Freiheit waren vielmehr die wichtigsten Themen seiner Freiheitslehre. Das zeigt einerseits die Schwäche in seiner Philosophie, aber andererseits insofern ihre Stärke, weil die Engheit des Themas sich im Gegenteil positiv auf die theoretische Vertiefung in den Begriff Freiheit fördernd, auswirkte.

In Nishidas Erstlingswerk *Untersuchung über das Gute*, das gleichzeitig das erste originale philosophische Buch in der japanischen Neuzeit war, wurde die Freiheit als eines der wichtigsten Themen behandelt. Er machte sich hier die hergebrachten Lehren von der Willensfreiheit und der mechanischen Notwendigkeit zum Gegenstand und kritisierte beide. Einerseits übte er Kritik am Verfechter der Willensfreiheit aus dem Grund, weil er eine mysteriöse Fähigkeit annimmt, die ohne allen Grund willkürlich ein Motiv bestimmen kann. Andererseits griff er den Verfechter der mechanischen Notwendigkeit in Hinsicht darauf scharf an, daß er das Bewußtseinsphänomen, das von idealem Charakter sein soll, dem Naturphänomen gleichsetzt, das der Notwendigkeit unterliegt. [35]

Auf solche Kritik hin bestimmte Nishida noch einmal die Freiheit des Willens. Der Mensch fühlt sich am freiesten, wenn die Ursache des Motivs aus dem tiefsten Inneren seines Selbst entsteht. "Das Bewußtsein ist also frei, nicht, weil es das Naturgesetz übertritt und zufällig wirkt, aber weil es vielmehr der Natur des Selbst folgt. Es ist weiter frei, nicht, weil es ohne allen Grund wirkt, aber weil es den Grund gut erkennt". [35] Es ist klar, daß Nishida hier wie Hegel "die Freiheit als die Einsicht in die Notwendigkeit" behauptet; obwohl diese Notwendigkeit nicht das Gesetz in der materiellen Welt, sondern die innerliche Natur des Geistes bedeutet. Er erkannte im menschlichen Geist die eigentümliche Natur oder Gesetzmäßigkeit an, die stets die endlose Bildung des Selbst förderte, und die Freiheit nannte er nur die Wirkung des Bewußtseins, die mit dieser Natur übereinstimmt und sie richtig entwickeln kann. Insofern stand er der Lehre von der Willensfreiheit näher als der Lehre von der Notwendigkeit. Man kann daran erkennen, daß Nishida sich genau der Schwächen der Lehre von Willensfreiheit bewußt war und sie durch die theoretische Ergänzung idealistisch verfeinerte.

Es war in den Jahren 1913 bis 1916, daß Nishidas zweites philosophische Buch *Anschauung und Reflexion im Selbstbewußtsein* veröffentlicht wurde. Damals, wenn auch nur ungenügend, setzte sich die Parteienregierung in Japan durch, und die sozialistischen Gedanken verbreiteten sich rasch. In der Gedankenwelt beeinflußten am stärksten die Demokratie von Sakuzo Yoshino, [64] die Bildungslehre der Shirakaba-Schule [27] und die Persönlichkeitslehre von Jiro Abe [1] usf. Intellektuelle und junge Männer. Es war die Blütezeit der sogenannten "Taisho-Demokratie".

Durch dieses Buch versuchte Nishida der Form des Systems des Selbstbewußtseins gemäß alles Wesen zu begreifen und die Vereinigung des Wertes mit dem Sein oder der Bedeutung mit der Tatsache zu erörtern, die seit *Untersuchung über das Gute* sein Hauptthema gewesen war. Indem er den Standpunkt des Selbstbewußtseins des transzendentalen Ichs, das mit Fichtes 'Tathandlung' Ähnlichkeit hat, bis zur realen Welt ausdehnte, erkannt er endlich den transzendenten Willen als Grundlage des Weltalls an. Er faßte nämlich den Vereinigungspunkt des Ideals mit der Wirklichkeit als 'den absolut freien Willen' auf. [36] Dies zeigt, daß er die Einheitskraft des Wesens in *Untersuchung über das Gute* zur Richtung des Voluntarismus noch stärker vorwärts trieb.

Nach seiner Meinung jedoch ist der freie Wille des Individuums nicht unabhängig von diesem absolut freien Willen. Die beiden stehen nicht im Gegensatz, sondern im Verhältnis zwischen dem Einzelnen und dem Allgemeinen. Das Einzelne ist frei nur innerhalb des absolut freien Willens, und auch dieser Wille ist erst frei für sich, indem er die Selbständigkeit des Einzelnen zuläßt. Trotzdem verneinte Nishida nicht, daß der einzelne Wille in der natürlichen Welt existiert. Er erkannte also einerseits an, daß der Wille unter dem Naturgesetz steht, aber er sicherte andererseits dadurch das Gebiet der Freiheit, das er auf Kants Behauptung von "der intelligiblen Welt" gründete. Er sagte, "Der menschliche Wille gehört schlechthin in seinem Grunde zur Welt des noch tieferen Erlebnisses oder zur intelligiblen Welt, wie Kant sagt. In dieser Welt soll das Ganze eins sein, und zugleich jenes Einzelne frei sein". [36]

Seine Freiheitsidee, in der sich die individuelle Freiheit auf die absolute Freiheit bezieht, besaß denselben Geist wie der damalige Humanismus und die Persönlichkeitslehre, weil sie verstanden, daß es gut sei, die persönliche Freiheit und die Individualität vollständig zu entwickeln, und zwar daß diese Freiheit von der Wirkung des dynamischem Allgemeinen, genannt "große Natur" oder "universales Leben", unterstützt wird. Nishidas liberalistische Philosophie, die sich in solchen Gedankenumständen bildete, war nichts anderes als der logische metaphysische Ausdruck des Taisho-Liberalismus.

4.2 *Nishidas Freiheitsidee (2)*

Die originale Nishida-Philosophie soll mit dem Werk *Vom Wirkenden zum Anschauenden* zum ersten Mal zum Ausdruck gekommen sein. Darin übte er Selbstkritik an der bisherigen mystischen Tendenz, so daß er den absoluten Willen zu Grund des Wesens legte, und er kam zur Ersinnung des Begriffs "das Ort des Nichts" als das Allgemeine, dadurch verschiedenes spezifisches Sein stehen kann.

Das Selbstbewußtsein, das den Gegensatz zwischen "erkennendem Ich" und "erkanntem Ich" in sich enthält, ist "das Ich als Ort". Dieses Ich als Ort ist nicht das Seiende, und deswegen läßt es sich wohl "Nichts" nennen. [37] Ebenfalls, gemäß Aristoteles' Definition, daß das Individuum etwas bedeutet, was zum Subjekt wird, aber nicht zum Prädikat, hielt Nishida die Natur des Individuums für das wahrhaft Freie, was nicht vom Allgemeinen bestimmt wird, sondern sich für sich selbst bestimmt. Aber er brachte das Allgemeine, das auch dieses selbständige Individuum in sich subsumiert und unterstützt, zur Richtung des Prädikats im Urteil vorwärts, so daß dieses Allgemeine etwas heißt, was zum Prädikat wird, aber nicht zum Subjekt, und schließlich faßte er es als "den Ort des absoluten Nichts" auf. [38] [39] Nach Nishida also, muß man das Individuum als die Selbstbestimmung des Ortes des absoluten Nichts auffassen, und "man kann den wahrhaft freien Willen nur im Ort des absoluten Nichts anschauen".

Darin, daß er das Unsichtbare, das allem Wesen zugrunde liegt, das schöpferische "absolute Nichts" nannte, zeigt sich zwar der Einfluß orientalischen

Gedankengutes, aber er gab nicht dem orientalischem Gedanken selbst die logische Form. Er versuchte vielmehr den europäischen Liberalismus oder Individualismus, der im selbständigen freien Individuum seinen Ursprung hat, durch eine orientalische Weltanschauung ontologisch zu erklären.

Nishida, der mit der Logik "des Ortes des Nichts" der neuen japanischen Metaphysik einen Weg zu bahnen begann, schritt nun zur Deutung der Welt und des Menschen, die auf "der sich ausdrückenden Welt" oder "der geschichtliche realen Welt" beruht. Es waren nicht nur die Anregung des Marxismus, der damals einen überwältigenden Einfluß auf die japanische Gedankenwelt ausübte, die ihn zu dieser Richtung drängte, sondern auch teilweise die Einflüsse von Heideggers hermeneutischer Phänomenologie und Bergsons Lebensphilosophie. Die berühmten jungen Philosophen, die Nishidas Schüler waren, z.B. Kiyoshi Miki [25] und Jun Tosaka, [58] zeigten starkes Interesse an der sozialen Wirklichkeit, und sie widmeten sich daher leidenschaftlich der Erforschung des Marxismus. Nishida wurde oft von ihnen kritisiert und dadurch stark direkt und indirekt stimuliert. Er gab dadurch keineswegs seinen bisherigen Idealismus auf. Indem er sich aber ernstlich mit der Geschichtsauffassung des Marxismus auseinandersetzte, konnte er seine eigene Handlungstheorie und die Lehre vom geschichtlichen Wesen errichten. [40] [41]

In Hinsicht auf den praktischen Charakter des Menschen, erklärt Nishida ihn mit der Logik "Vom Gebildeten zum Bildenden". Die materiellen Dinge und die tierischen Wesen werden von der Welt gebildet, aber bilden sich nicht für sich selbst. Nur Menschen werden in der Geschichte gebildet und zugleich bildet die Geschichte bewußt. Menschliche Gesellschaft und Kultur sind die Welt der sozialen Herstellung und die Welt des geschichtlichen Körpers. Der Mensch bildet sich in der Handlung, die die Geschichte bildet. Er nimmt gerade an der Bildung der Geschichte durch die Herstellung oder die Schöpfung der Dinge teil.

Nishida nannte das herstellende handelnde Selbst "die Spitze der Schöpfung der Welt" oder "das schöpferische Moment der Welt". Unser geschichtlich gebildeter Körper wird in der Welt entwickelt und auch diese Welt bildet sich schöpferisch nur durch die eigene Praxis und Herstellung des körperlichen Selbst. Jedes handelnde Selbst nimmt an der Schöpfung der Welt teil und wirkt für sich als ein Bestandteil der Welt. Nishidas Menschenauffassung "Vom Gebildeten zum Bildenden" war in diesem Sinne zugleich die Logik der Entwicklung der geschichtlichen Welt.

Das Selbst und die Welt sind, bei Nishida, zwar von derselben Qualität und beziehen sich unmittelbar aufeinander, aber es ist klar, daß hier "die Logik des dynamischen Verhältnisses zwischen dem Einzelnen und dem Allgemeinen in der geschichtlichen Welt" vorhanden ist. Nishidas Idealismus, der zeigte, daß die unersetzbare Subjektivität des Individuums, die sich in der Herstellung ausdrückte, mittels der Gesellschaft zugleich sich mit der Bildung der ewigen Geschichte vereinigt, legte das Fundament für dieses Denken und wies den Weg in die Zukunft für die liberalistischen Intellektuellen, die während der Taisho-Demokratie heranwuchsen. [57]

Die Nishida-Philosophie, wie schon erwähnt, war eine Philosophie von sehr ethischem Charakter. Es ist der Beweis dafür, daβ er auch in seinen späteren Jahren fortdauernd die Einsicht von Tatsache und Wert, Logik und Ethik, suchte.

Der oben erklärte Widerspruch des menschlichen Wesens ist nicht nur ein logischer oder struktureller Widerspruch; er wird im Grunde auch als der ethische Widerspruch erneut verstanden. Nach Nishida, wenn der Mensch als die Spitze der Schöpfung der Welt unter der gegebenen geschichtlichen Umgebung etwas Neues zu schaffen versucht, verursachen sicher die expressiven Dingen, die in der Welt vorhanden sind, die Handlung des Menschen. Die in der Welt seienden Dinge bedeuten keineswegs bloβen Widerstand oder Gegenstand sondern haben vielmehr die Expressivität, die das menschliche Leben anruft und den Menschen zur Entscheidung zwingt. "Das objektive Sollen", das uns befiehlt, daβ wir dem alten System und der bestehenden Ethik nicht folgen aber eine neue Idee verwirklichen sollen, ist nicht der Anspruch von innen im Menschen selbst, sondern "der Ruf von auβen" der expressiven Dinge. Nur innerhalb dieser sich ausdrückenden Welt entsteht der Charakter des Menschen "das Gebildete und zugleich das Bildende", und zwar das, was gebildet wird und bildet, sind nicht nur Dinge oder Körper, sondern auch Persönlichkeit, Wert und Idee. [41]

Indem das handelnde Selbst bei Nishida auβerdem das Subjekt der Schöpfung des Wertes war, wurde seiner Handlung stets ein persönlicher ethischer Sinn gegeben. In der Handlung des Menschen, der gebildet wird und zugleich bildet, spiegelten sich also die Leiden des menschlichen Lebens und der ethische Widerspruch der menschlichen Natur sehr deutlich wider. Er sagt, "In uns sollen darum unendliches Bemühen und Mühe sein, weil der Widerspruch die Wahrheit des realen Lebens ist. Daraus entsteht also unendliches Sollen. Der Widerspruch ist die Tatsache des menschlichen Lebens. Dies bedeutet die Frage über 'Leben oder Sterben'. Hierin besteht gerade die geschichtliche Wirklichkeit". [41]

Deswegen kann man aufs neue bestätigen, daβ die philosophische Logik, die die innerliche Natur des Wesens, sei es der Mensch oder die Welt, erklärt, bei Nishida nicht die wertfreie Logik der Tatsache war, sondern auch die Ethik, die das Leben und das Sterben des Menschen ernsthaft behandelt. Nishida-Philosophie war die Metaphysik über die Welt und Geschichte, die mit der ursprünglichen Freiheit des Menschen begann, und zugleich die Ethik, die in dieser Metapysik gerade das Problem des Lebens "wie man leben soll" zu beantworten suchte.

Aber während seines ganzen Lebens interessierte er sich nicht für die wissenschaftlichen Erkenntnisse über die Staatsstruktur und das politische System in Japan. Dies kommt daher, weil sein Denken hauptsächlich auf die Deutung des Sinnes der Welt überhaupt beschränkt war, und ihm die sozialwissenschaftliche Methode fehlte, mit der er die japanische Wirklichkeit realistisch analysieren und kritisieren hätte können. [57] Seine kulturelle-liberalistische Philosophie konnte also gegen den Ultra-Nationalismus und den Militarismus, die nachher die japanische Gesellschaft beherrschten, keinen Widerstand leisten. Dennoch, indem seine Weltanschauung die westliche Philosophie (besonders den deutschen Idealismus) genug verarbeitete und auch das traditionelle Gemüt der Japaner geschickt ausdrückte, konnte sie eine repräsentative Philosophie des 20. Jahrhunderts in Japan werden.

5. Die Idee der Freiheit bei Hajime Tanabe

5.1 *Die Aufgabe der Philosophie Tanabes und "die Logik der Art"*

Hajime Tanabe ist ein bedeutender Philosoph, der in der japanischen Neuzeit Kitaro Nishida an die Seite gestellt wird. Mit Nishida, seinem Vorgänger, war Tanabe ein Führer der Kyoto-Schule, die damals in der philosophischen Welt am einflußreichsten war. Er begann zuerst ein Studium der Wissenschaftslehre [48] und der mathematischen Philosophie [49], darauf beschäftigte er sich mit der eigentümlichen Interpretation der Hegelschen Dialektik. [50] Durch die kritische Nachfolge der Nichts-Philosophie von Nishida strebte er weiterhin allmählich danach, die originale Tanabe-Philosophie zu errichten.

Die dreißiger Jahre des zwanzigsten Jahrhunderts, in der er seine neue Philosophie, genannt "die Logik der Art", entwickelte, waren die Zeit, als Japan durch den Eroberungskrieg gegen China in den Pazifischen Krieg einzutreten begann. Im Lande herrschte der Ultra-Nationalismus, der den selbstaufopfernden Dienst am Kaiserreich erzwang und die private Freiheit des Individuums völlig negierte, [21] [22] während sich zugleich im philosophischen Bereich der fanatische Japanismus ausbreitete. Der Philosoph Tanabe, der für die Geistesströmungen und Politik der Zeit feinfühlig war, nahm diese Neigung teils kritisch, teils positiv, in seine philosophische Logik auf. Die Tanabe-Philosophie war zu allererst eine Sozialphilosophie oder soziale Ontologie, ehe sie, wie bei Nishida, die allegemeine Ontologie der Welt wurde.

In seinen Gedanken zeigte sich besonders deutlich der Einfluß des völkischen Nationalismus, der damals auf der ganzen Welt mächtiger wurde. "Die Einheit des Volkes und die Regulierungskraft des Staates kann man nicht vom Standpunkt aus verstehen, das die Gesellschaft schlechthin als das Wechselverhältnis der Individuen aufzufassen sucht", behauptete er, [52] und deswegen versuchte er die Logik und, den *raison d'être* der Art oder Spezies zu erörtern, die im Volk, dem Staat und der Klasse ausgedrückt wird.

Innerhalb der Sozialphilosophie Tanabes, muß man beachten, spiegelte sich die Gegenüberstellung mit der Staatstheorie des Marxismus. Ungeachtet der wiederholten Unterdrückungen endete die Sympathie der Intellektuellen für die materialistische Geschichtsauffassung in der damaligen Wirklichkeit nicht. [15] [23] [20] Tanabe konnte jedoch nicht dem Marxismus zustimmen, der den Klassencharakter des Staats behauptete, und er strebte danach, "den vernünftigen Grund des Staates als der praktischen Einheit der Wirklichkeit mit dem Ideal" zu entdecken. Von diesem Gesichtspunkt aus zielte er im Bereich der Philosophie auf die Einheit von Kants Moralität und Hegels Sittlichkeit. [51] Im Bereich der Politik nahm er weder für den individualistischen Liberalismus noch für den nationalistischen Totalitarismus Partei, doch er verfolgte bis zum Ende die Logik der Gemeinschaft als die Einheit der beiden, die auf dem japanischen Idealismus beruht. [53]

Die Freiheitslehre von Hajime Tanabe wurde in der Logik der Art als der sozialen Ontologie, und zwar damit vereinigt, dargestellt. Was war dann also seine "Logik der Art"?

Die menschliche Gesellschaft ist keine Beziehung, die nach dem Individuum entsteht, oder gleichzeitig mit diesem entsteht, dachte er. Indem sie vor dem Individuum steht und die Individuen vereinigen kann, hat sie das Substrat, das nicht abhängig von Werden und Verschwinden des Individuums ist, sozusagen "das Artsubstrat". Die gesellschaftlichen Wesen, Staat, Volk, Klasse usf., liegen gerade auf der Stellung der 'Art' gegen das Ganze als Gattung und das Einzelne als Individuum, und sie enthält in sich dieses Ganze und das Einzelne als ihre Momente der Vermittlung.

Das Individuum muβ die Art zur Voraussetzung haben, anders als Aristoteles, der das Individuum mit der Substanz identifizierte. Tanabe sagt, "Die Art, die dem Individuum innewohnt, ist die Ursache des einzelnen Wesens. Deswegen muβ man behaupten, daβ das wahre Sein ursprünglich nicht das Individuum ist, sondern die Art". Auf diese Weise verfocht er den Begriff 'Art' als das Allgemeine, das die Unsterblichkeit der menschlichen Gesellschaft erhält und darin die einzelnen Wesen organisch vereinigt. [51]

In der ersten Periode, in der Tanabe die Logik der Art vorzuschlagen begann, bedeutete 'die Art' nur die unmittelbare Einheitskraft, durch die die Individuen vereinigt werden. Er führte hiermit den Totemismus in der unkultivierten Gesellschaft als ein Beispiel an und hielt das dem Totemismus innewohnende "Gesetz der Teilnahme [loi de participation]", das Gemeinschaftlichkeit in der Gemeinschaft ausdrückt, für einen Typus der Art. Das Individuum vertritt die Logik der Absonderung, die das Gemeinschaftliche auflöst, und die Art die Logik der Teilnahme, die das Gemeinschaftliche erhält.

Der Kritik anderer Erforscher und der Änderung der Weltlage entsprechend, wurden danach die Begriffe 'Art' und 'Individuum' teilweise abgeändert und die Gattung-Art-Individuum-Beziehung wurde aufs neue definiert. Die Art, die früher als das Substrat gegen das einzelne Subjekt nur eine unmittelbare Einheitskraft besitzen sollte, soll nun das dynamische Allgemeine als die Selbstnegativität bedeuten, die in sich die Spaltung und den Gegensatz enthält. Das Individuum, das früher nur als das selbständige eigensinnige Ich galt, soll nun das Vermittelnde der Umwandlung bedeuten, wodurch sich die Art als der Gegensatz zur Gattung als der absoluten Einheit wenden kann. [52] Warum fand jedoch eine solche Abänderung statt?

Weil Tanabe wohl bemerkt hatte, daβ die reale Art, Staat, Volk, Klasse usf., weder das mit sich Identische noch das Ruhige oder das Stabile war, sondern daβ sie das Widerspruchsvolle, das den strengen Gegensatz des politischen wirtschaftlichen Interesse in sich enthält, und auch das sich wegen des Widerspruchs Verändernde war. Noch dazu ist die Art in der Wirklichkeit nicht das bloβ Ideelle, sondern sie hat eine natürliche materielle Grundlage. Indem Tanabe nämlich teilweise die Gesellschaftsauffassung des historischen Materialismus aufnahm, die auf den Widerspruch zwischen der Produktionskraft und der Produktionsverhältnisse achtet, führte er ins Substrat der Art das Moment "die Materie als Selbstentfremdung des Absoluten" ein. Das heiβt z.B. die Produktionskraft, die ein wichtiger Faktor der Klassenspaltung ist, und die völkische Einheit, die sich in "Blut und Boden" zeigt, usw. [52] Die Art wurde also nicht der Logik der Identität gemäß, sondern der Logik der Selbstentfremdung gemäß noch einmal begriffen.

5.2 *Tanabes Freiheitslehre*

Die Tanabe-Philosophie war zwar ebenfalls von stark nationalistischem Charakter, gab aber nicht einfach den Standpunkt auf, der die individuelle Freiheit verteidigt, da er schon die Taufe der deutschen Willensfreiheitslehre der Neuzeit empfangen hatte. Wie man die wahre individuelle Freiheit innerhalb der Logik der Art oder des Staats vollbringen soll, war seine wichtigste Fragestellung.

Am Beginn der Entwicklung seiner Individuumslehre sollte das Individuum "das freie Sein, das über die Möglichkeit des Nichtseins existieren kann", sein. Sein Wesen sollte darin bestehen, daß es willkürlich sogar das Nichtsein wählen oder Selbstmord begehen kann. [51] Außerdem kann das Individuum in die Zukunft blicken und das gegenwärtige Selbst von der Zukunft aus freiwillig bestimmen. Hinsichtlich der Beziehung auf die Art, wird es nicht nur von der Art bestimmt. Indem es die Bestimmung von der Art für sich vermittelt, negiert es die Art und wendet sie in die Gegenrichtung. "Das pluraldimensionale Sein, das die Verwandlung der Richtung und die Einheit des Gegensatzes ermöglicht"—dies soll das echte Individuum sein. [51] Auf diese Weise wurde der freie Wille des Individuums darin erkannt, daß er die Bestimmung der Art verwandelt und diese Bestimmung für sich verwaltet. Insofern kann man verstehen, daß das Individuum von Tanabe als ein revolutionäres Wesen aufgefaßt wurde, das die alte Einheit der Art zerstören und eine neue Einheit schaffen kann.

Als er später die Ursprünglichkeit der Art zu behaupten begann, wurde erklärt, daß das einzelne Ich, wenn es selbständig ist, nicht das echte Individuum sei. Das Einzelne muß vom Allgemeinen vermittelt werden und im Allgemeinen verneint und zugleich bejaht werden. Das echte Individuum soll erst dadurch entstehen, daß es gerade in der Art dialektisch verneint wird. Hier wurde die neue Interpretation über die Bedeutung der subjektiven Handlung des Individuums vorgeschlagen.

Die Handlung des Individuums bildet nämlich nicht die Art. Durch die Selbstverneinung dieser Art entsteht jene Handlung und deshalb können sich auch das Individuum und die Gattungseinheit erhalten. Trotzdem wird die spontane Freiheit des Individuums somit nicht ausgelöscht. Die freie individuelle Handlung, unterstützt von der Einheitskraft der Art, kann nach Tanabe zugleich die negative Art zur Allgemeinheit der Gattung reinigen. [52] Dies zeigt konkret Tanabes Hoffnung, daß das soziale Substrat 'die Art', die die Klassengegensätze und die wirtschaftlichen Widersprüche in sich enthält, durch die uneigennützigen politisch-sozialen Taten der Individuen bis zum vollständig Allgemeinen, das Gegensätze und Widersprüche aufhebt, erhöht werden soll.

Die Handlung des Individuums, die Tanabe behandelte, hatte also an sich den der Art dienenden politisch-ethischen Inhalt und den sich vom privaten Interesse befreienden religiösen Charakter — zum Unterschied von der kulturellen künstlerischen Handlung Nishidas. Hierin liegt ein Grund, warum er behauptete, daβ die individuelle Handlung innerhalb der Art Selbstaufopferung und zugleich Selbstverwirklichung sei. Indem er besonders das wahre Selbstbewuβtsein, das in der Handlung entsteht, die Entscheidung, die auf Leben und Tod fällt, und die religiöse Selbstlosigkeit, die den Egoismus der Individuen überwindet, usw. beachtete, konnte er sagen, daβ es nur in der menschlichen Handlung das 'Nichts' gibt, das das Grundprinzip des Subjektes ist. Es war für Tanabe unmöglich, die "Nichts-Philosophie" anzuerkennen, in der Nishida der Wirkung des schaffenden Nichts gemäβ, die geschichtlich-reale Welt erklärte, weil das 'Nichts' nicht in der Welt, sondern in der Handlung steht. [52]

Am Ende der dreiβiger Jahre, vom faschistischen Absolutismus stark beeinfluβt, identifizierte Tanabe schlieβlich 'die Art' mit dem Staat des Tenno-Systems and erklärte öffentlich das Staatswesen als das "Substrat gleich Subjekt" der wirklichen Welt. Nach seiner Ansicht ist der Staat der Entstehungsgrund der geschichtlichen Welt, und die wahre Freiheit des Individuums kann sich also nur durch den Dienst am Staat verwirklichen. Folglich sagt er, "Die Autonomie und Freiheit in der Ethik verschwinden nicht, sondern ermöglichen sich vielmehr innerhalb des Dienstes am Staat und des Gehorsams gegen den staatlichen Befehl.... Die Freiheit kann sich nur durch die Notwendigkeit, die ihre Negation ist, verwirklichen. In diesem Sinne muβ die wahre Freiheit die Freiheit der Selbstverneinung sein". [53] Indem "die Logik der Art" von Tanabe die ursprüngliche Absolutheit des Staates verfocht, löste sie am Ende, gegen seinen eigenen Willen, die individuelle Freiheit im Nationalismus auf. Mit anderen Worten, unterwarf sich der Liberalismus als zerbrechliche Idee dem absolutistischen Tenno-System als der hartnäckigen Wirklichkeit.

Nach dem zweiten Weltkrieg, als der japanische Faschismus unterging, veröffentlichte Tanabe sogleich seine Gedankenhaltung, so daβ er seine bisherigen Ansichten über die Absolutheit des Staates und die Assimilation des Individuums an den Staat selbst streng kritisieren muβte.[54] Nämlich, daβ es "das radikale Böse" nicht nur im Individuum gibt, sondern auch im Staat, und daβ man sich auf den religiösen Standpunkt stellen muβ, der dem transzendenten Geschichtsgericht Gottes folgt und Buβe tut, um dieses Böse zu überwinden. "Die Philosophie als Buβe (*Zangedo*)", die das Christentum und Shinrans Buddhismus vereinigte, war seine letzte These. [55]

Tanabes Freiheitstheorie, die sich auf "der Logik der Art" gründet, war das Resultat seines philosophischen Denkens, das das eigentliche Wesen, des die Zeit überwältigenden Staates oder Volkes, und die Idee der individuellen Freiheit möglichst logisch zu begreifen versuchte. Aber man muβ erkennen, daβ die Herrschaft und der Bruch dieser Theorie zugleich der Ausdruck des Glorias und des Elendes oder der Stärken und der Schwächen der liberalistischen Intellektuellen in der ersten Hälfte des 20. Jahrhunderts war. Die Niederlage des japanischen Faschismus und der neue Aufbau der japanischen Kultur nach dem Krieg enthüllten die Mängel der spekulativen Freiheitslehre der Philosophen und verlangten nach einer neuen Philosophie über die Freiheit.

6. Der Begriff der Freiheit nach dem zweiten Weltkrieg

Es war "die Verfassung des japanischen Staates", die 1946 verkündet wurde, die nach dem Krieg beim japanischen Volk zu einer Umwandlung seiner Wertvorstellung führte und den wahren Sinn der Freiheit lehrte. Diese Verfassung, die drei Grundsätze, d.i. die dem Volke innewohnende Souveränität, den Verzicht auf den Krieg und die Achtung der fundamentalen Menschenrechte, in sich enthält, manifestierte zum erstenmal, sowohl nominell als auch wirklich, "die Freiheit als das Recht", die die freie-demokratische Bewegung der Meiji-Ära gewünscht und bezweckt hatte.

Auch in der Verfassung des Kaiserreiches, die bis zur Niederlage des Krieges über das japanische Volk geherrscht hatte, wurden zwar die Begriffe '*Kenri* (das Recht)' '*Jiyu*' dargelegt, aber ihr Inhalt war weit entfernt von der modernen Bestimmung. Nämlich, in der Klausel namens "Recht und Pflicht des Untertanen" wurden zuerst die Militär- und Steuerpflicht angeführt und danach wurden verschiedene Rechte und Freiheiten unter der Bedingung der Erfüllung der Pflicht vom Tenno als dem Souverän gewährt. Außerdem gab es darin auch die Bestimmung, daß sogar diese beschränkten Rechte und Freiheiten im Notfall von der höchsten Gewalt des Tennos entzogen werden könnten. Der Begriff '*Kenri*' auf Japanisch war bei dieser Verfassung noch nicht als 'das Richtige', das das Recht eigentlich in sich enthält, aufgefaßt.

Die neue Verfassung Japans kehrte nun die Recht-Pflicht-Beziehung der alten Verfassung um und gab dem Begriff '*Kenri*' die eigentliche Bedeutung 'das Richtige'. Das Recht wurde nicht darum anerkannt, weil es als das Entgelt für den Dienst am Staat gilt, sondern weil es der unentbehrliche richtige Anspruch der Nation ist. Die neue Verfassung, die auf der Naturrechtslehre in der Neuzeit begründet ist, [8] erklärte das fundamentale Menschenrecht für "das unverletzlich ewige Recht" und zugleich ermahnte sie zu keinem Mißbrauch des Rechts und erlegte der Nation die Verantwortlichkeit auf, daß sie für das öffentliche Wohl das Recht anwenden solle.

Obwohl die Idee der Verfassung jedoch hoch und hehr war, war es nicht der Fall, daß sie nicht zeitweise umgangen wurde. Dies gilt von der Zeit ihrer Einführung bis in die Gegenwart. Hinsichtlich der Anwendung dieser Verfassung wurden in der Tat die Menschenrechte der Minderheiten, besonders Daseinsrecht und Umgebungsrecht, unter dem Vorwand des öffentlichen Wohles sehr oft verletzt. Indem das japanische Gericht im allgemeinen die für die Regierung günstigen Urteile fällte, widersetzte es sich oft dem Anspruch der Nation auf die Freiheit. [8] In Japan besteht noch kein völliges Einverständnis über das Wesen der sozialen bürgerlichen Freiheit, weil der Klassengegensatz von Arbeit und Kapital und der Widerspruch der Interessen von Regierung und Nation fest verankert im Lande bleiben. Heute noch ergibt sich eine strenge Auseinandersetzung in Hinsicht auf die Frage, was man für das wahre Recht der Nation halten soll, oder was überhaupt das öffentliche Wohl sei.

Im Bereich der Philosophie und Gedankenwelt jedoch, war es wie in der Vorkriegszeit für die japanischen Gelehrten unmöglich, keine Beziehung mit der

Tendenz europäischen Gedankengutes zu haben. Was die Frage der Freiheit betrifft, wurden die Intellektuellen und die Studenten hauptsächlich von Existentialismus und Marxismus stark beeinflußt und so nahmen sie die Ideen von Jaspers, Sartre, Merleau-Ponty usw. einerseits, von Stalin, Lukács, Lefebre usw. andererseits positiv auf. Durch den Gegensatz, die Verwicklung und die partielle Verschmelzung – der existentialistischen Freiheitslehre, die die subjektive Freiheit und den 'Entwurf' des Individuums verficht, [10] und der marxistischen Freiheitslehre, die für die soziale Freiheit und die Klassenpraxis des Proletariates eintritt — wurde der Kern der japanischen Freiheitslehre nach dem Krieg geformt.

Es war "die Disputation über die Subjektivität", die auf diesem Wege des Entwicklungsprozesses entstand. Diese Disputation, die auf die Überwindung der Unreife der Demokratie und der Unvollständigkeit des modernen Ichs in Japan zielte, begann zuerst in der literarischen Welt behandelt zu werden, und sprang danach in den philosophischen Bereich über. Die damaligen bedeutenden Philosophen und Sozialwissenschaftler nahmen an der philosophischen Disputation teil, und es waren besonders Katsumi Umemoto und Kazuto Matsumura, die eine Hauptrolle spielten. Umemoto, dessen Philosophie sich auf dem subjektiven Materialismus gründete, hob hervor daß die Aufklärung über die Vereinigung "der wissenschaftlichen Einsicht in die materiellen Bedingungen der Menschenemanzipation" mit "der subjektiven Freiheit als der existentialen Stütze des Menschen" die theoretische 'Lücke' des bisherigen Marxismus war, und, um diese Lücke zu schließen, schlug er vor, daß man die Frage der ethischen Subjektivität des Individuums, das sich zwecks der Menschenemanzipation der Klassenpraxis hingeben kann, erörtern solle. [60] [61] Dagegen behauptete Matsumura vom Standpunkt des orthodoxen Marxismus, daß Marxismus vor allem auf dem Klasseninteresse beruhen muß und "seine subjektive Seite das Interesse der Arbeiterklasse, das Bedürfnis nach ihrer Emanzipation und die Klassen- oder Vereinigungsbewußtsein ist, die sich während des gemeinsamen Kampfes bilden", und er kritisierte, daß die Subjektivierung des Marxismus von Umemoto nichts anderes als Revisionismus sei, der den heterogenen moralistischen Faktor einbrächte. [24]

Danach wurde die Gegenbehauptung von Umemoto veröffentlicht und viele Philosophen nahmen dazu Stellung, aber diese Disputation ließ allmählich nach, ohne viel Erfolg zu zeigen. Dennoch, der Vorschlag Umemotos wies zwar die Schwäche auf, daß er die ethische Subjektivität vor der wissenschaftlichen Analyse der Geschichte und Gesellschaft verfocht, aber er zeigte zugleich darin den geschichtlichen Sinn, daß er die Notwendigkeit lehrte, die Bildung des umwälzenden Subjektes in der Gesellschaft im Hinblick auf das Bewußtsein und die Moral des Individuums theoretisch zu erklären. Die Frage der Subjektivität und der existentialen Freiheit des Individuums, das auf die Menschenemanzipation zielt, ist heutzutage unter dem Thema "die Bildung der Persönlichkeit als des Souveräns" von verschiedenen Gesichtspunkten aus behandelt worden. [45][13][26][46] In diesem Sinne ist sie heute noch eine der wichtigsten Fragen für die japanischen Denker.

Die Vorstellung der Freiheit in Japan veränderte sich selbstverständlich nach dem

Krieg (ebenso wie andere Gedanken) durch die politische wirtschaftliche Lage im In- und Ausland. Die Kritik an Stalin in der Sowjetunion, die spätere Auseinandersetzung zwischen der Sowjetunion und China, die negativen Phänomene der Unterdrückung der Menschenrechte in einigen sozialistischen Ländern usf. lösten bei den japanischen Intellektuellen und Arbeitern großes Entsetzen und Enttäuschung aus und zwangen sie aufs neue, das Verhältnis zwischen der bürgerlichen Freiheit und der proletarischen Freiheit, und zwischen der sozialen Freiheit und der privaten Freiheit zu untersuchen. [47] [44] [56] Die Verstärkung der sozialen Kontrolle und der Glauben an die wissenschaftliche Technik unter dem nationalen Monopolkapitalismus, die sich nach dem hochgradigen Wirtschaftswachstum in Japan rasch ausbreiteten, erzeugten das Phänomen der tiefen Entfremdung der Arbeit, dem Leben und der Menschenbeziehung usf. gegenüber und drängten die Philosophen zur Entwicklung einer neuen Freiheitslehre, die den Weg zeigen soll, diese Entfremdung zu überwinden. [4] [12] [34]

Es ist nun bemerkenswert, daß viele neue Vorschläge über die Freiheitslehre in den siebziger Jahren bis Anfang der achtziger Jahre hauptsächlich von den Materialisten gemacht wurden. Ein Forscher behauptete, daß die Freiheit bisher in der Beziehung auf die Notwendigkeit und das Gute definiert worden sei, aber daß sie in Wahrheit weder Handlung noch Fähigkeit bedeutet. Nach ihm besteht das Wesen der Freiheit darin, daß es am Hindernis und am Zwang fehlt, die irgendeine Tat stören, und die soziale Freiheit und die ethische Freiheit haben eine Gemeinsamkeit in Hinsicht auf die Befreiung vom Zwang, sei dieser die politische Macht oder das sinnliche Bedürfnis. [4] Ein anderer stellte fest, daß die philosophische Freiheit, die durch die Ansicht von F. Engels "Freiheit besteht in der auf Erkenntnis der Naturnotwendigkeiten gegründeten Herrschaft über uns selbst und über die äußere Natur" repräsentiert wird, und die soziale bürgerliche Freiheit, die das Lebens-, Vermögensrecht und die Gedanken-, Glaubens-, Rede-, Verlags-, Versammlungsfreiheit usw. umfaßt, in ihre innere Einheit gebracht werden soll, und zwar, daß jene Freiheit auch die notwendige Bedingung der Entwicklung dieser Freiheit ist. [3] Eine andere Meinung, die diesem Vorschlag zustimmt, kritisiert, daß die bisherigen Freiheitslehren nur unter den Kategorien von Notwendigkeit-Zufälligkeit behandelt worden seien. Ansch ließend wird erklärt, daß sich die philosophische Freiheit und die soziale Freiheit unter dem Verhältnis von Wesen und Erscheinung oder Inhalt und Form stellen soll, weil die Phase der beiden different ist, auch daß man in diesen Begriff 'Notwendigkeit' nicht nur Kausalität, sondern auch 'Entwicklung' einbringen muß. [2]

Durch diese Ansichten ist noch keine bestimmte Linie gefunden worden, mit der die verschiedenen Forscher übereinstimmen können. Diese Vorschläge sind alle jedoch unzweifelhaft vom modernen Theorieninteresse durchdrungen, so daß man auch den Sinn der sozialen politischen Freiheit, die sich heute in Japan differenziert und umgestaltet, philosophisch erörtern soll, und daß man hiermit jetzt Vorzüge und Mängel der herkömmlichen Freiheitslehren der Philosophie oder Ethik richtig auffassen muß. Die Freiheit enthält selbstverständlich mannigfaltige Phasen und Bereiche, z.B. die Freiheit als die Einsicht in die Notwendigkeit, die Wahlfreiheit im Willen, die ethische autonome Freiheit, die persönliche Freiheit, die soziale bürgerliche Freiheit usw. Um diese Freiheiten total und organisch zu begreifen, ist noch eine mühevolle Untersuchung und eine konstruktive gegenseitige Kritik notwendig. Auch in Japan muβ es wohl die theoretische Zusammenarbeit, besonders zwischen Materialismus und Idealismus oder Marxismus, Existentialismus und Pragmatismus, sein, die zum Fortschritt der Freiheitstheorie beitragen kann.

BIBLIOGRAPHIE

Abe, Jiro [1] *Der Personalismus*. In *Sämtliche Werke* IV. Kadokawa-Verlag, 1922.

Ajisaka, Makoto [2] *Über die Freiheit*. Otsuki-Verlag, 1981.

Akima, Minoru [3] "Die Aufgabe der philosophischen Freiheitslehre im Marxismus". *Wissenschaft und Gedanke*, Nr. 15, 1975.

Awata, Kenzo [4] *Die Freiheit und der Wert im Marxismus*. Aoki-Verlag, 1975.

Fukuzawa, Yukichi [5] *Die Empfehlung der Wissenschaft*. In *Sämtliche Werke* III Iwanami-Verlag, 1872.

—— [6] *Der Umriβ der Zivilisationslehre*. Ebd. IV, 1875.

—— [7] *Die Autobiographie Fukuzawas*. Ebd. VII, 1899.

Hasegawa, Masayasu [8] "Die Verfassung des japanischen Staates und die Freiheit". *Wissenschaft und Gedanke*, Nr. 15, 1975.

Ienaga, Saburo [9] *Die Geschichte der moralischen Gedanken in Japan*. Iwanami-Verlag, 1954.

Iijima, M. und Yoshizawa, D. (Hrsg.) [10] *Die Freiheit in Existentialismus*. Riso-Verlag, 1972.

Ishikawa, Takuboku [11] *Der gegenwärtige Zustand der eingesperrten Zeit*. In *Sämtliche Werke* X. Iwanami-Verlag, 1910.

Iwasaki, Chikatsugu [12] *Die menschliche Freiheit und der wissenschaftliche Geist*. Chobun-Verlag, 1976.

Kawai, Akira [13] *Die Bildung der demokratischen Persönlichkeit*. Aoki-Verlag, 1972.

Kawakami, Hajime [14] *Die Erzählung der Armut*. In *Sämtliche Werke* IX. Iwanami-Verlag, 1917.

—— [15] *Die zweite Erzählung der Armut*. Ebd. X VII, 1930.

Kitamura, Tokoku [16] *Was bedeutet der Beitrag zum Menschenleben?* In *Sämtliche Werke* II. Iwanami-Verlag, 1893.

—— [17] *Die Lehre vom innerlichen Leben*. Ebd. II, 1893.

Kotoku, Shusui [18] *Der Imperialismus*. In *Sämtliche Werke* II. Meiji-Bunken, 1901.

—— [19] *Das Wesen des Sozialismus*. Ebd. IV, 1903.

Kozai, Yoshishige [20] *Die gegenwärtige Philosophie* (Sammlung I). Keiso-Verlag, 1937.

Kultusministerium Japans [21] *Der wahre Sinn der Staatsform*. 1937.

—— [22] *Die Pflicht des Untertans*. 1941.

Kushida, Tamizo [23] *Die materialistischen Geschichtsauffassung*. In *Sämtliche Werke* I. Kaizo-Verlag, 1935.

Matsumura, Kazuto [24] *Die Dialektik und die Problematik des Wendepunkts*, Hoseidai-Verlag, 1950.

Miki, Kiyoshi [25] *Die materialistische Geschichtsauffassung und das gegenwärtige Bewuβtsein*. In *Sämtliche Werke* III. Iwanami-Verlag, 1928.

Motojima, Kunio [26] *Die Theorie über die Bildung des umwälzenden Subjektes*. Aoki-Verlag, 1977.

Mushanokoji, Saneatsu [27] *Meine Lebensanschauung*. In *Sämtliche Werke* XXIII. Shincho-Verlag, 1920.

Nagata, Hiroshi [28] *Die Geschichte des Materialismus in Japan*. Hoseidai-Verlag, 1936.

Nakae, Chomin [29] *Aufsätze der "Toyo-Jiyu-Zeitung"*. In *Sämtliche Werke* XIV. Iwanami-Verlag, 1881.

—— [30] *Rigaku-Kogen* [Die Einführung in die Philosophie]. Ebd. VII, 1886.

—— [31] *Sansuijin-Keirin-Mondo* [Gespräch über die Staatsverwaltung unter drei Betrunkenen] Ebd. VIII, 1887.

—— [32] *Aufsätze der "Toun — Zeitung"*. Ebd. X I, 1888.

—— [33] *Fortsetzung, Anderthalb Jahre*. Ebd. X, 1901.

Nakano, Tetsuzo [34] *Marxismus und die menschliche Freiheit*. Aoki-Verlag, 1977.

Nishida, Kitaro [35] *Untersuchung über das Gute*. In *Sämtliche Werke* I. Iwanami-Verlag, 1911.

—— [36] *Anschauung und Reflexion im Selbstbewußtsein*. Ebd. II, 1917.

—— [37] *Vom Wirkenden zum Anschauenden*. Ebd. IV, 1927.

—— [38] *Das selbstbewußte System des Allgemeinen*. Ebd. V, 1930.

—— [38] *Die selbstbewußte Bestimmung vom Nichts*. Ebd. VI, 1932.

—— [40] *Die Grundfrage in der Philosophie*. Ebd. VII, 1933.

—— [41] *Die philosophische Abhandlungssammlung* I-IV. Ebd. VIII-XI,1935–1946.

Nishimura, Shigeki [42] *Die Lehre von der japanischen Moral*. Iwanami-Verlag, 1887.

Nitobe, Inazo [43] *Bushido* [Das Rittertum]. Iwanami-Verlag, 1899.

Okamoto, Seiichi [44] *Die Problematik der Freiheit*. Iwanami-Verlag, 1959.

Shibata, Shingo [45] *Die Theorie über die menschlichen Natur und die Persönlichkeit*. Aoki-Verlag, 1961.

Shoji, Kokichi [46] *Die soziale Veränderung und das umwälzende Subjekt*. Todai-Verlag, 1980.

Takakuwa, Sumio [47] *Über die menschliche Freiheit*. Iwanami-Verlag, 1956.

Tanabe, Hajime [48] *Eine Einführung in die Wissenschaft*. In *Sämtliche Werke* II. Chikuma-Verlag, 1918.

—— [49] *Untersuchung über die mathematischen Philosophie*. Ebd. II, 1925.

—— [50] *Hegelsche Philosophie und Dialektik*. Ebd. III, 1931.

—— [51] *Die Logik des sozialen Wesens*. Ebd. VI, 1934–1935.

—— [52] *Die Erklärung des Sinnes der Logik der Art*. Ebd. VI, 1937.

—— [53] *Die Logik des Staatswesens*. Ebd. VII, 1939.

—— [54] *Die Philosophie als Zangedo* [Buße]. Ebd. IX, 1946.

—— [55] *Die Dialektik der Logik der Art*. Ebd. VII, 1947.

Tanaka, Shoji [56] *Die Freiheit der Gegenwart*. Ochanomizu-Verlag, 1983.

Tanemura, Kanji [57] "Die praktische Anschauung in der Nishida-Philosophie und die Praxis des Marxismus". *Wissenschaft und Gedanke*, Nr. 43/44, 1982.

Tosaka, Jun [58] *Die Lehre von den japanischen Ideologien*. In *Sämtliche Werke* II. Keiso-Verlag, 1935.

Tsuda, Sokichi [59] *Untersuchung über den nationalen Gedanken in der japanischen Literatur* I–IV. Iwanami-Verlag, 1916 –1921.

Umemoto, Katsumi [60] *Die materialistische Geschichtsauffassung und die Moral* (Sammlung I). Sanichi-Verlag, 1949.

—— [61] *Die Lehre vom Menschen*. Ebd. II / III, 1961.

Watsuji, Tetsuro [62] *Die Geschichte der ethischen Gedanken in Japan*. In *Sämtliche Werke* X II / X III. Iwanami-Verlag, 1952.

Yamada, Kou [63] *Die ethischen Gedanken des japanischen Sozialismus*. Aoki-Verlag, 1978.

Yoshino, Sakuzo [64] *Der wahre Sinn des Konstitutionalismus* (Sammlung I). Shinkigen-Verlag, 1916.

L'hybridité de culture et l'imagination créatrice: Le cas de Tetsurô Watsuji comme penseur du Japon moderne

MEGUMI SAKABE

1.

Depuis que les pays asiatiques ont subi l'influence écrasante de la culture occidentale, on s'y trouve toujours face au problème qui est de savoir comment assimiler ou s'accommoder à cette culture nouvelle venue sur la base de la tradition culturelle de son propre pays. Parfois on se sent même plus ou moins menacé par la crise de sa propre identité, soit culturelle soit personnelle. Car, une fois pris comme but ce qu'on appelle la "modernisation" se modelant sur la civilisation occidentale, on s'est engagé presque inévitablement, d'un côté, dans l'ethnocentrisme européen et, d'un autre côté, comme en réaction à celui-ci, à l'autre extrémité de l'ethnocentrisme, qui s'enorgueillie de la culture traditionnelle de son propre pays ou de sa propre race etc.. Pour la plupart des intellectuels des pays asiatiques dans leur époque de "modernisation", il a donc été toujours très difficile d'échapper à cette sorte d'oscillation ambivalente entre les deux extrémités de l'ethnocentrisme. Commençant au niveau politique et idéologique pour se manifester einsuite au niveau linguistique, mental etc., il y a eu presque partout le danger d'une perte d'identité culturelle ou personnelle allant de pair avec des conflits culturels, qui, eux aussi, ont été présents non seulement dans la vie extérieure mais dans la vie intérieure. En certaines circonstances, comme cela s'est manifesté aussi dans d'autres régions du monde, on a été forcé d'employer la foi religieuse empruntée de l'Occident justement pour lutter contre la menace d'une perte d'identité émanante de la culture occidentale.

Dans cette situation toujours menaçante au fait d'une oscillation ambivalente et inquiète ou inquiétante, on s'est trouvé, comme une conséquence inévitable, devant un problème crucial: comment doit-on traiter le rapport identité-différence ou plutôt identité-altérité surtout au niveau des problèmes culturels? Si l'on voulait se libérer de cette oscillation ambivalente entre deux extrémités de l'ethnocentrisme, on ne pourrait pas, et ne devrait pas, éviter le problème du pluralisme culturel, ou plus exactement, celui de l'hybridité culturelle. Puisque, au moins une fois entré dans l'époque des hautes cultures, toutes les cultures ne pourraient pas échapper à un certain degré d'hybridité et, par conséquent, on ne pourrait parler de "purisme" culturel, sous quelle forme que ce soit, sauf comme une idéologie, consciemment employée ou inconsciemment adoptée.

Le problème à résoudre serait ainsi le suivant: sans se livrer à ce mouvement

Contemporary philosophy. A new survey. Vol. 7, 341–356.

oscillant entre deux extrémités de l'ethnocentrisme, où l'on ne pourrait pas échapper à certains traits du "purisme" culturel, comment trouver le point d'équilibre réciproque et dynamique juste au centre du conflit identité-altérité entre les cultures différentes?

Par la nature même de ce problème, il se manifestera un des aspects les plus fondamentaux ou les plus radicaux concernant le champ dit "intersubjectif" de l'être humain.

Or, prenant un exemple parmi les penseurs de la modernité japonaise, on pourrait considérer Tenshin OKAKURA (1862–1913) comme un type de penseur d'une sorte d'hybridité culturelle. Prenant avantage des tensions comparativement faibles existant entre les cultures de l'Occident et de l'Orient à l'aube de la formation de la nation moderne du Japon au début de l'époque de Meiji, Okakura, comme fondateur de l'École nationale des beaux arts, estimant la valeur des arts orientaux aussi bien que celle des arts occidentaux, indiqua pour les beaux-arts japonais la voie à prendre pour créer un style convenant à l'époque de la modernisation du Japon: le style par lequel l'imagination créatrice, n'hésitant jamais à accepter son hybridité et adoptant activement des éléments des arts occidentaux qu'elle regarde comme nécessaires, essaie de fomer un nouveau art principalement sur le fondement de la tradition artistique du Japon. Et plus tard comme directeur de la section des arts orientaux de musée de Boston, il soutint une sorte d'Orientalisme culturel et politique, qui, contre l'invasion des pouvoirs d'Europe en tant qu' elle se caractérise par la civilisation moderne de la technologie et d l'impérialisme, en appelait à l'éveil de l'Est sans aucune trace de l'obsession de purisme ethnocentrique. Lors de ces considérations sur la culture de Japon ou celle de l'Orient, Okakura fut toujours conscient de l'hybridité de la culture en général dans tous ses ordres: par exemple, il a souligné l'importance du rôle joué par les éléments du Sud (Taoisme ou Bouddhisme du sudd l'Inde etc.) respectivement à l'intérieur des cultures de Chine et d'Inde.

On pourrait trouver au moins quelque affinité entre la situation dans laquelle Okakura se trouvait et celle existant en Allemagne entre les années 1750 et 1780 (celles de Lessing à Herder): dans les deux cas, on trouve la floraison d'une sorte de cosmopolitisme ouvert (exceptionnellement ouvert, e.g. dans le cas de l'Allemagne, à l'égard du monde juif), qui, prenant avantage des tensions comparativement faibles existant entre la culture dominante étrangère et celle indigène à l'aube de la formation de la nation moderne, va parallèlement avec un respect élevé de la tradition de sa propre culture.

Or, retournant au sujet de la pensée japonaise, la situation est devenu plus en plus sévère après l'époque où Okakura exercait ses activités. Surtout à l'époque de la Deuxième Guerre mondiale où l'impérialisme du Japon allait vers son extrémité, les tensions de l'ethnocentrime et du pluralisme culturel atteindront un point culminant. Je voudrais examiner donc dans les chapitres suivants le cas de Tetsurô WATSUJI (1889–1960) dont l'itinéraire de la pensée montrera quelques aspects caractéristiques du problème du rapport entre l'ethnocentrisme et le pluralisme culturel qui se manifestait et parfois menaçait le maintien de l'identité, soit culturelle, soit personnelle, dans la modernité développée du Japon.

2.

Watsuji Tetsurô, né en 1889 comme fils de médecin pratiquant dans le village Nibuno situé dans une plaine fertile pas loin de Kôbé et d'Osaka, après être sorti du lycée de Tôkyô, fit ses études de philosophie à l'Université de Tôkyô, où il deviendrait professeur environ vingt ans après. S'intéressant beaucoup à la littérature et au théâtre comme étudiant de lycée et d'Université, alors qu'il avait choisi la philosophie comme spécialité, il s'engagea dans un mouvement théâtral et fut membre d'un groupe d'amateurs de la littérature. Il écrivit quelques essais, par exemple, sur Blake, Strindberg, Ibsen etc. Dans le domaine des recherches académiques, il fut surtout influencé par le Dr. R. Koebel, professeur de l'Université de Tôkyô, qui, d'origine Russe et formé philosophiquement en Allemagne, introduisait au Japon non seulement la philosophie occidentale en général, mais aussi quelques traits principaux de la tradition humaniste incluant la philologie classique et la philosophie de Grèce ancienne. Son mémoire de licence fut "Sur le pessimisme et la thèorie du salut chez Schopenhauer".

Attiré par quelques penseurs remarquables de tendances post-Hégéliennes, il devint un pionnier dans le domaine des recherches détaillées sur Nietzsche et Kierkegaard au Japon. Dans ses livres: "Études sur Nietzsche" (1913) et "Søren Kierkegaard" (1915), Watsuji présenta, quoique d'un point de vue différent de celui de la philosophie existentielle qui serait à la mode une décennie après, les portraits philosophiques de ces deux penseurs d'un point de vue qu'on pourrait caractériser comme celui de la philosophie de la vie.

Focalisant son attention sur la culture traditionnelle au Japon dans les années suivantes, Watsuji publia en 1919 le livre intitulé "Koji Junrei" (Pèlerinage des vieux temples). En présentant brillamment dans le style d'un essai libre ses impressions sur quelques statues fameuses, éléments d'architécture, jardins de vieux temples bouddhiques à Nara et Kyôto, ce livre s'est qualifié comme un des plus populaires parmi ses publications. Révisé par l'auteur apprès la Seconde Guerre mondiale, il attire même aujourd'hui un vaste public, surtout de jeunes lecteurs.

Ce qui caractérise spécifiquement cet essai sur les traditions culturelles, c'est une sorte d'ouverture de cœur sur les cultures du monde, orientales aussi bien qu'occidentales. Ici dans ce livre, Watsuji porte son attention sur les arrière-plans historiques situés derrière les statues et éléments d'architecture bouddhique qu'il rencontre en suivant l'itinéraire de son pèlerinage: par exemple, derrière quelques statues et éléments d'architecture bouddhique du 7ème et 8ème siècle, il évoque une série d'influences stylistiques qui ont pour origine l'art Gandara d'Inde ancienne ou, dans certains cas, même les arts occidentaux de l'époque hellénique. Ce trait remarquable d'ouverture de cœur ou d'universalisme concernant la culture a probablement quelque chose à voir avec l'atmosphère pacifiste qui dominait dans la société japonaise de cette époque juste après la Première Guerre mondiale. Prenant son origine dans cette courte période imprégnée de l'atmosphère de la démocratie et des lumières, "Koji Junrei" a joué un rôle correspondant, en un sens, à celui joué par le "Bildungsroman" à l'époque du romantisme à son premier stade.

Aprè la publication de "Koji Junrei", entrant définitivement dans la carrière académique, (professeur à l'Uiversité de Tôyô en 1921, professeur adjoint d'éthique à l'Université de Kyôto en 1925, professeur à l'Université de Kyôto en 1932, professeur à l'Université de Tôkyô en 1934), Watsuji se consacra à des recherches sur l'histoire culturelle, japonaise ainsi qu'orientale, et publia quelques œuvres importantes dans ce domaine. Ensuite, en 1935, Watsuji publia "Fùdo — Ningen-gaku teki Kôsatsu" (Le climat — Une étude anthropologique), un de ses livres les plus connus à côté de "Koji Junrei". Ce livre-là a pour origine les propres expériences de Watsuji lors de son voyage de 1927 pour l'Europe à travers des climats variés, qui, eux, ont suscité chez lui beaucoup d'intérêt pour le problème de l'espace terrestre comme demeure des êtres humains qui en même temps et à son tour, exerce une certaine influence sur leurs cultures.

Visant à compléter l'analyse imparfaite de l'existence humaine (Dasein) de Heidegger, qui, à son avis, met trop l'accent sur le temps et par conséquent a tendance à négliger l'importance du problème de l'espace, Watsuji essaie dans ce livre d'élucider quelques aspects du climat qui, ayant une qualification ni exclusivement subjective ni exclusivement objective, donnent forme à des types variés de culture ou de mentalité selon les traits caractéristiques du climat de chaque région de la terre. Tout conscient de sa dette en particulier envers l'opinon de Herder concernant l'influence du climat sur la vie et la culture humaine, il souligne l'importance méthodologique de la "compréhension" (Verstehen) dans l'analyse de la relation réciproque entre le climat et la vie culturelle humaine; le climat comme un des moments déterminants de la vie culturelle humaine ou de ses types ne peut pas être traité comme simple objet de l'observation telle qu'elle est employée dans le domaine des sciences naturelles, mais, pour ainsi dire, comme une sorte de "physionomie" qui ne pourrait "être comprise" qu'en étant vécue de l'intérieur ou au moins en étant vécue par l'intermédiaire de simulations imaginaires.

Ainsi, fondamentalement au point de vue du pluralisme culturel à la manière de Herder, Watsuji clarifie non seulement les conditions auxquelles les types variés de la culture humaine se construisent mais aussi celles auxquelles, en dépit de la variété de ces types, la "compréhension" mutuelle ou l' "intelligibilité" entre eux se réalise. On pourrait facilement remarquer ici une sorte d'élargissement de la portée de la considération sur l'espace ou le champ "intersubjectif" vers le domaine dans lequel un moment au delà de l'être humain, i.e. climat, ni seulement subjectif ni objectif, se manifeste comme quelque chose à la portée de la "compréhension" humaine. (On pourrait rappeler ici la notion de "chiasme" chez Merleau-Ponty.)

En tout cas, ici, considérant l'"espace intersubjectif" élargi au delà du cercle de l'humanité, on atteint un des points caractéristiques de la pensée de Watsuji.

3.

En 1934, l'année précédant la publication de "Fûdo", Watsuji publia un livre intitulé "Ningen no Gaku to shite no Rinri-gaku" (Ethique comme science de Ningen) présentant les idées fondamentales de sa pensée éthique qui s'élaboreraient quelques années après dans trois volumes de "Ethique" (1937–1949). Dans ce livre-là, en appliquant la méthode de la phénoménologie herméneutique de Heidegger à l'analyse de quelques termes de base concernant la structure du champ intersubjectif dans la langue japonaise: Rinri, Ningen, Hito, Aida-gara, Sonzai, Yû etc., Watsuji présente ses notions fondamentales sur l'interprétation du monde dans lequel l'être humain se trouve. Je présenterai aussi brièvement que possible les points principaux de son analyse.

Rinri倫理, qui correspond à l'"éthique", veut dire les lignes cohérentes déterminant la structure de base de la communauté humaine. Selon Watsuji, on peut trouver un fil conducteur pour l'élucidation de cette structure de base dans l'analyse du mot Ningen人間, parce que ce mot veut dire l'homme individuel aussi bien que "entre" [aida, ma] 間 ou ce qui s'appellerait un champ intersubjectif entre les hommes individuels. Dans l'usage originel de ce mot en Chinois, le sens était borné à ce dernier, i.e.le champ intersubjectif, ou simplement le "monde" (Seken世間). L'expansion ou la mutation de sens vers le premier i.e. l'homme individuel a eu lieu au cours du développement historique de la langue japonaise, d'abord par un malentendu fortuit concernant la traduction de soûtras bouddhiques chinois: Pour l'économie des lettres, déjà dans les textes chinois, les termes désignant les stades successifs de la métempsychose de l'âme: Jigoku-chû (地獄中 dans les enfers), Gaki-chû (餓鬼中 entre les mauvais démons), Chikusyô-chû (畜生中 entre les animaux), Nin-gen (人間 entre les hommes) et Ten-jyô (天上au ciel) se réduisaient tous à des mots de deux lettres: Jigoku (地獄), Gaki (餓鬼), Chikusyo (畜生), Ningen (人間), Tenjyô (天上). Comme conséquence de cette réduction, le mot Nin-gen en venait à être mal compris comme désignant l'homme individuel par analogie avec Chikusyô qui veut dire exclusivement et simplement l'animal individuel.

Ce malentendu, cependant, loin d'être une simple méprise, s'enraciné profondément dans la compréhension implicite latente dans la langue Japonaise concernant la structure du champ intersubjectif ou, plus proprement, de sa relation avec le dit ego-sujet: Le mot "Hito" veut dire à la fois l'homme en général (人) et autrui ou les autres (他人). Watsuji interprète ce fait justement comme indiquant le fait que dans la structure même de l'Ego comme personne il y a la relation avec une autre personne ou d'autres. En d'autres termes, on pourrait dire qu'il n'y a aucune division nette entre ce qu'on appelle la personne individuelle et le champ intersubjectif dans lequel on évolue au milieu des relations je-tu, je-vous, je-ils, mais dans la structure de je comme personne il existe toujours quelques traits fondamentaux de "persona-masque".

Commençant par cette analyse de la structure fondamentale de "Nin-gen" qu'il caractérise aussi comme un champ de relations mutuelles pour l'action humaine au sens le plus large de ce mot, Watsuji mène ses considérations ensuite vers le rapport entre "Nin-gen" et Être (Yû 有, Sonzai存在). Yû veut dire "avoir", tandis que Son 存 Zai 在 veulent dire respectivement être ou vivre dans le monde caractérisé pour celui-là par le temps et pour celui-ci par l'espace. Ainsi le sens du mot Sonzai存在peut s'interpéter comme indiquant le fait même que "Ningen" a lui-même au milieu du monde caractérisé par la structure de "entre" ou "distance-conjonction" (Ma 間) de l'espace et du temps.

Ce rapport etroit entre "Ningen" et Être jette quelque lumière sur le rapport également etroit qui existe entre "Ningen" et le langage, parce que le langage ou la pratique linguistique ne peut être regardée comme rien d'autre qu'une manifestation de la structure fondamentale ou de l'articulation du champ dans lequel "Ningen" se trouve.

Le verbe "aru" ("ari"), qui correspond à peu pràs à "Etre" avec sa fonction comme copule ainsi que comme verbum substantivum, s'interprète chez Watsuji comme l'index d'un rapport profondément enraciné entre chaque pratique verbale et le fait fondamental que le "Nin-gen" a soi-même. Ainsi, à chaque phrase, l'existence d'une sorte d'omniprésence (au moins au niveau de sa structure de base) de ce mot "ari" est supposée.

"Ainsi, généralement, "dé aru" comme copule n'est rien d'autre qu'une détermination de "ari" au sens de "ga aru" (être là [i.e. verbum substantivum]). Que veut dire cette "détermination"? Une idée ou une chose *est* là. Mais, elles *sont* simplement là, sans agir (ou se comporter) comme quelque chose qui se détermine. Il n'y a rien d'autre que "Nin-gen" (homme) qui agit (ou se comporte). Si l'on dit quelque chose comme "le rouge à sa qualité de rouge", on ne fait que projeter à ce rouge l'action de "Nin-gen" (homme) par laquelle on détermine le rouge comme le rouge et le distingue d'avec d'autres idées. C'est anisi, par l'action de "Nin-gen" (homme), qu'une idée ou une chose se déterminent comme *étant* telle ou telle. J'ai dit que "quelque chose est là" veut dire "quelque chose *est* eu par "Nin-gen" (homme)(="Nin-gen" a quelque chose)" et cet être se fonde dans le "Sonzai" de "Ningen" [le fait que "Ningen" (homme) a lui-même".] Par conséquent, la détermination de "ari" est la détermination de la manière d'avoir par laquelle "Nin-gen" (homme) a une idée ou une chose."Nin-gen" (homme) a le vent, c'est-à-dire, le vent *est* là (le vent souffle, il fait du vent). "Nin-gen" (homme) a ce vent comme quelque chose dont on sent la douceur, c'est-à-dire, le vent est doux. Ou, "Nin-gen" a le vent comme quelque chose qui est discriminé comme vent, c'est-à-dire, le vent *est* le vent même. Ces manières variées d'avoir ne sont rien d'autre que les manières du "Sonzai" (être) du "Nin-gen" (homme). La distinction entre "de aru" (être tel ou tel) et "ga aru" (être là) est, donc, une distinction à l'intérieur du "Sozai" (être) du "Nin-gen" (homme). Conséquemment, s'il y a quelque chose comme le "Yû" (être) fondamental au fond de ces deux "ari", ce ne peut être que le "Sonzai" (être) du "Nin-gen" (homme). Strictement parlant, seulement à ce niveau fondamental peut-on dire que "ari" veut dire le "Sonzai". Ainsi, on pourrait regarder le mot "ari", qui se sépare dans son emploi en "ga aru" et "de aru", comme l'"apophansis" du "Sonzai" (être) du "Nin-gen"(homme).

Manifestant les articulations fondamentales du champ de "Nin-gen" et en même temps servant de moyen indispensable pour que le "Nin-gen" *ait* lui-même, le language fonctionne comme une ex-plication (Aus-legung) (opérée spontanément ou consciemment) de "Nin-gen" en tant qu'il est un champ de relations réciproques d'actions profondément enracinées dans son "Sonzai".[1]

Il est certain, je pense, que, dans l'interprétation présentée ci-dessus par Watsuji sur "ari", "Sonzai", "Yû" etc., il n'y a presque aucune trace de l'idéalisme qui réduit les formes aussi bien que les matériaux de notre savoir au sujet épistémologique, soit singulier, soit collectif. Mais, néanmoins, on ne peut pas nier qu'il reste quelque ambiguïté concernant le rapport entre "être" et "avoir" et le statut de "Nin-gen" par rapport à l'un et l'autre. Quel est le rapport entre le singulier et le pluriel (ou collectif) dans "être" et "avoir" de "Nin-gen" et dans ceux des choses intra-mondaines? Ne faut-il pas distinguer une couche profonde d'une autre moins profonde du "Sonzai" de "Nin-gen": (1) l'une, couche "poïétique" d'imagination créatrice dans laquelle l'identité du singulier n'étant pas encore fixée dans le mouvement dynamique du "chiasme" entre les personnes, entre les choses et entre les personnes et les choses, une personne se transpose (par un transfert de sentiments) dans une autre (même dans une autre chose) et un glissement de sens s'effectue par transposition métaphorique et (2) l'autre, couche de (la vie quotidienne) dans laquelle on *a* sa propre identité de personne et le mot garde son sens propre? Et dans cette couche-là d'imagination créatrice, n'est-il pas inévitable que ce qu'on pourrait appeler "entre-champ" de "Nin-gen" ait un certain degré de caractère "hybride" dans tous les aspects de la culture?

4.

Dans un article écrit à la même époque que "Ningen no Gaku to shite no Rinrigaku" et publié dans le volume "Jinkaku to Jinrui-sei" (Personne et humanité) (1938), interprétant un paragraphe de la "Critique de la raison pure" dans lequel Kant traite du rôle de l'espace et du temps dans le fonctionnement de l'aperception transcendantale, Watsuji dit comme suit:

> "Il va sans dire que l'union des deux formes, de l'espace et du temps, ne se limite pas au cas de l'union de l'âme et du corps. Tandis que le moi-objet est seulement la représentation dans le temps, le phénomène extérieur en général se définit comme la représentation dans l'espace aussi bien que dans le temps. Ainsi, le temps est 'la condition formelle du phénomène en général' ("Critique de la raison pure" B50). S'il s'agit d'un objet soit intérieur soit extérieur, l'acte de le représenter se trouve déjà soumis à la condition nécessaire du temps. Tous les corps matériels sont donc représentation dans le temps et dans l'espace. Il n'y a aucune représentation exclusivement spatiale. Ainsi, tous les corps matérials ont le double caractère d'être intérieurs/extérieurs. *On peut obtenir 'l'éveil cosmique du moi' (Ji-kaku) à l'occasion de n'importe quelle expérience concernant les corps matériels. Par conséquent, tous les corps matéiels peuvent provoquer l'éveil de la personne. Pourtant, nous venons de voir que nous ne trouvons jamais cette sorte de notion de la personnalité dans la philosophie de Kant.* Parmi les représentation dans le temps et l'espace, Kant estime entre autres celle se situant exclusivement dans le temps où se lient étroitement le corps et l'âme. C'est parce qu'il remarque l'unité intrinséque de cette sorte de représentation qui a en même temps un côté intérieur et

> un côté extérieur. Prenant ainsi la conscience de soi dans cette représentation spatiale, la personnalité transcendantale englobe le corps en elle-même. Parmi les corps matériels, chez Kant, seul le corps humain est capable de s'unir avec la personne.[2] (souligné par Sakabe)

Watsuji parle ici de la possibilité grâce à laquelle on peut obtenir l'éveil cosmique du moi (Ji-kaku) à n'importe quelle occasion de son expérience concernant les corps, soit vivants soit non-vivants (e.g. obtenir l'éveil cosmique en voyant des fleurs ou des nuages etc.). Ainsi, sans aucune référence spéciale ou aucun rapport à son corps qui serve de médium indispensable entre son propre moi et l'univers, il peut y avoir une voie directe par laquelle on devient conscient de son vrai moi dans l'expérience de n'importe quelle chose qui se trouve dans l'univers. On pourrait facilement discerner ici la même manière d'élargir la portée de notre "compréhension" hors du champ de ce qu'on appelle intersubjectivité que nous avons vue dans le cas de "Fûdo": Selon l'opinion de Watsuji, qui s'enracine profondément dans la tradition de la philosophie orientale (ou, plus précisément, bouddhique), il peut y avoir une sorte de relation réciproque (qui aurait au moins une certaine similarité avec les notions Merleau-Pontiennes de "chiamse" ou "omniprésence de la chair") entre le moi de l'homme et les choses qui se trouvent dans l'univers.

Quelques pages plus avant, Watsuji cite (séparément) ces deux textes de Kant:

> "Un objet empirique s'appelle l'objet extérieur, s'il se représente dans l'espace, et il s'appelle l'objet intérieur, s'il se représente seulement dans le rapport temporel; l'espace et le temps, cependant, doivent tous les deux se trouver seulement *en nous*." ("Critique de la raison pure", A173)
>
> "L'objet transcendantal, qui se situe au fond des phénomènes extérieurs ainsi que de l'intuition intérieure, n'est ni matière ni un être pensant en soi-même, mais un fondement inconcevable du phénomène, qui fournit le concept empirique de la première aussi bien que de la deuxième sorte." (ibid., A379–80)

Remarques de Watsuji sur ce dernier texte:

> "Inconcevable (unbekannt) voulant dire ici 'inobjectivable', le fondement inconcevable n'est donc que le sujet qui se situe au fond de toutes les objectivations. Autrement dit, il n'est que la personnalité transcendantale. Le phénomène extérieur est 'la manière d'être représenté' (A387) de cet objet, qui est non-objet, par le sens extérieur dans l'espace. Il est la manière d'être objectivé de la personnalité transcendantale. Ainsi, on pourrait regarder l'espace en tant que forme d'intuition comme la manière dont la personnalité transcendantale s'extériorise. Bien entendu, cette extériorisation ne se trouve pas hors de la personnalité transcendantale elle-même. L'extériorisation ne peut avoir lieu que dans l'intérieur d'elle-même.... L'espace étant la manière de s'extérioriser que nous venons d'expliquer, généralement, c'est par l'espace que la personnalité transcendantale s'extériorise ou devient le corps matériel. Il n'y a

aucune différence entre ce fait et le processus par lequel cette même personnalité devient le moi-objet sous l'action du temps. *Mais, comme dans le cas du moi-objet, il s'ensuit que le corps matériel qui objective la personnalité transcendantale doit être ce qui fait la personne comme telle en 'remplissant' cette personnalité. Une telle notion de la personnalité comme constituée par des corps matériels n'est pas nécessairement bizarre chez nous [orientaux]. Mais, si on la cherchait dans la philosophie de Kant, ce serait extrêmement bizarre.* Car, si n'importe quel corps matériel pouvait devenir le moment de réalisation de la personnalité, la distinction kantienne entre le corps et la personne perdrait toute sa signification."[3] (Souligné par Sakabe)

Ce qui menait Watsuji ainsi au delà du domaine de la pensée philosophique kantienne ou, plus généralement, au delà d'une sorte d'anthropo-centrisme de la tradition Judéo-chrétienne, a été d'un côté l'influence de la philosophie de Kitarô NISHIDA (1870–1945) qui, suivant la tradition du BouddhismeZen, soulignait la primauté du champ cosmique englobant par rapport au soi atomique qui en découle, et d'un autre côté l'influence de Kant-interpétation de Heidegger, qui tentait de trouver au fond de l'unité de l'aperception transcendantale une structure extase/intégration du temps. On peut trouver une trace évidente de l'influence de ce dernier dans les lignes suivantes:

"Alors qu'on devient conscient du temps comme le champ où l'on est conscient de soi et aussi comme appartenant à l'unité de son soi, le soi vacant de l'aperception qui a le temps *en lui-même* se remplit du soi-objet qui est *dans le temps*. Autant que le soi s'objective, il est déjà *dans le temps*; mais, précisément parce que le temps est dans tel soi qui n'est jamais objectivable, ce n'est rien d'autre que le temps dans lequel le soi, étant vacant au sens objectif, s'objective. Ainsi, dans 'la conscience dans le temps' trouve-t-on la clef pour comprendre comment la personnalité transcendantale devient une personne. Nous ne pouvons pas trouver chez Kant l'analyse détaillée de cette sorte de temps. Ce que Kant analysait a été le temps en tant qu'il rend possible l'objet de la nature. Mais, ce qu'il faut traiter maintenant, c'est le temps qui rend possible la personne comme telle. Nous ne devons pas considérer le temps comme une forme de l'intuition mais comme une structure de la personnalité transcendantale. Traiter ce problème voudrait dire aller au delà de Kant. Je me borne, cependant, à faire remarquer que Kant est déjà venu très proche de ce problème."[4]

Bien que, comme montré par ces lignes, Watsuji ait été influencé par Kant-interpétation de Heidegger surtout par celle concernant le problème du temps, assez étrangement, il ne mentionne jamais le problème de l'imagination (créatrice) qui joue un rôle crucial chez Heidegger. De même, le problème de "poïêsis", qui dans la philosophie de Nishida est regardé comme une médiation indispensable entre le champ cosmique englobant et le soi individuel, n'occupe guère de place dans le cadre de la considération de Watsuji sur le champ de "Nin-gen".

Ainsi, la pensée de Watsuji a, pour ainsi dire, au moins une certaine tendance à établir un court-circuit entre le champ cosmique englobant et le soi individuel (ou la totalité des individus: "Nin-gen") sans médiation de l'imagination créatrice, poïêsis etc., qui, elle, pourrait ou devrait enraciner sa pensée dans la couche profonde de l'entrecroisement et de la transposition métaphorique entre des personnes, entre des choses et entre les personnes et les choses. En d'autres termes, quoiqu'il s'intéresse beaucoup au champ de l'"entre" ("Aida" 間), Watsuji a quelque tendance à confondre la logique propre de la couche profonde de ce champ (qui serait celle de "la structure faible" au sens mathématique de ce terme) avec la logique de l'autre couche (de la vie quotidienne) qui est moins profonde et dans laquelle l'identité propre de chaque personne individuelle et de chaque chose est définitivement établie.

D'où vient (1) la tendance de Watsuji à ce qu'on pourrait appeler "Nin-gen-centrisme", qui, sans tenir compte du glissement indéfini de l'imagination vers d'autres personnes et d'autres choses, objective le champ entier de "Nin-gen" comme une totalité établie et (2) la limitation de la portée de sa pensée principalement au domaine de l'éthique, où, sans se soucier de l'entrecroisement des personnes et des choses dans la couche profonde de la pensée imaginaire, l'identité de "Nin-gen" (comme les personnes aussi bien que comme leur totalité) peut se traiter comme déjà établie en sorte qu'il peut "avoir" lui-même.

Et c'est précisément aussi sur le même fondement de cette logique propre (ou plutôt la confusion entre les logiques différentes) de Watsuji que, abandonnant le pluralisme culturel, il cédera considérablement à un purisme ethnocentrique à l'époque de la Deuxième Guerre mondiale.

5.

Sur le fondement posé par "Nin-gen no Gaku to shite no Rinri-gaku" (Ethique comme science de Nin-gen"), Watsuji commenca à construire son propre système de l'éthique dont les premiers deux volumes furent publies en 1934 et 1942. Commençant par l'explication de concepts-clés et de la méthode de l'éthique, il développe dans ces volumes ses vues sur (1) la structure fondamentale de "Nin-gen Sonzai" (aproximativement: être [et avoir] du champ personnel/interpersonnel), (2) les structures d'espace et de temps du champ "Nin-gen" et (3) les systèmes moraux de la communauté.

Le troisième chapitre sur les sytèmes moraux de la communauté se divise en paragraphes traitant successivement la famille, la communauté (incluant la communauté régionale, l'oganisation civile et la race) et l'état. Dans cette construction hiérarchique des communautés humaines, qui, influencée considérablement par la philosophie sociale de Hegel, tend vers une notion même plus "organique" de la société ou de la communauté que celle de Hegel, Watsuji cède, au moins à un certain degré à l'idéologie dominante de cette époque: l'ultra-nationalisme. Voici quelques lignes présentant le plus manifestement cette sorte de tendance de Watsuji à cette époque-là:.

> "Parmi toutes les totalités, la plus haute et la plus ultime est celle de l'État, qui, incluant en elle-même toutes les totalités plus bases qu'elle, n'est inclue elle-même en aucune totalité finie. Cette sorte de totalité a été, c'est vrai, déjà comprise dans la totalité d'une race comme quelque chose 'sacré' i.e. ayant le pouvoir suprême. Mais, c'est dans l'État qu'on est évidemment conscient de cette sainteté et de ce pouvoir suprême exprimés légalement sous la forme la souveraineté. Les pouvoirs de l'État trouvent leur expression concentrée dans la souveraineté, de sorte que la souveraineté, étant non seulement le pouvoir de législation mais aussi le pouvoir suprême d'exécution, a sa racine dans la sainteté. La souveraineté est, en outre, le pouvoir suprême non seulement exécutant sa souveraineté à l'intérieur de l'État mais aussi déniant toute interférence qui vient de l'extérieur. Elle est donc le pouvoir le plus puissant qui n'admet jamais l'existence d'aucun pouvoir plus élevé que lui. C'est justement ce que veut dire le droit de la souveraineté. La souveraineté n'est rien d'autre que l'expression du fait que la totalité de l'État est la totalité ultime de l'être humain fini."[5]

Parallèlement à cette manifestation résolue de la théorie totalitaire de l'État, Watsuji donna dans son livre "Sonnô Shisô to sono Dentô" (L'idée royaliste du Japon et sa tradition) (1943) un panorama historique qui accentue considérablement quelques traits du purisme ethnocentrique (sans parler du racisme) ou de l'impérialisme culturel dont on peut trouver des traces dans l'histoire de la pensée japonaise. Dans ce livre, par exemple, Watsuji cite les lignes suivantes de Norinaga MOTOORI (1730–1801), un des fondateurs de l'école de Kokugaku:

> "Parce qu'elle appartient à notre propre pays, on doit appeler la science de notre Empire simplement la science et discriminer la science chinoise comme chinoise.... On ne pourrait dire quelque chose comme 'la science japonaise' que si l'on se situait en dehors de notre propre Empire. Il n'y aura aucune raison d'adopter une telle sorte de qualification, si l'on n'est pas dans un pays etranger, e.g. Chine, Corée, Pays-bas etc.."

Remarque de Watsuji sur ces lignes:

> "C'est un argument qui est tout à fait naturel et parfaitement bon. Mais, même aujourd'hui, 150 ans après qu'il ait été formé, cet argument tout évident n'est pas parfaitement compris; il y a toujours ceux qui ne sont pas contents s'ils n'ajoutent pas l'adjectif 'japonais' aux noms de toutes sortes de sciences. Ils ne sont pas encore libre d'une idée fixe qui veut que la science soit entendue toujours comme la science européenne ou l'éthique, l'éthique européenne. Selon Norinaga, l'éthique n'est que la science de la vraie 'voie', c'est-à-dire, la voie par laquelle l'Empereur règne sur le monde sublunaire, et conséquemment on ne doit pas l'appeler éthique 'japonaise'. Par contre, c'est justement l'éthique des Européens qui a, strictement parlant, besoin d'être déterminé comme occidentale ou européenne."[6]

Watsuji présente davantage de traits caractéristiques de la "Voie" de l'Empire japonais, qui, selon lui, doit se situer à un niveau plus élevé que toutes autres religions du monde:

> "La volonté manifestée à travers l'ancètre-Dieu de la famille Royale ou l'Empereur lui-même comme incarnation de cette divinité ne se détermine pas comme celle de tel ou tel Dieu ou de quelque Être absolu. L'Un fondamental n'est rien d'autre que l'origine de la sainteté infiniment fluide qui pourrait s'appeler le néant absolu. La race comme une totalité vivante que je viens de mentionner n'est pas non plus l'Un absolu. Elle obtient sa qualification comme une totalité du mystère indéterminé de derrière. En d'autres termes, la race comme un totalité vivante se fonde sur une totalité plus élevée qui est absolue. Par conséquent, l'expression de la totalité de la race par l'ancètre Dieu ou Empereur-Dieu est en même temps la totalité absolue exprimée par l'intermédiaire de la race. Ici nous trouvons l'expression de la totalité ultime. Il est tout à fait clair, donc, que celui qui incarne cette totalité ultime a une souveraineté sacrée et on comprendra donc pourquoi cette souveraineté sacrée peut comprendre en elle-même toutes les autres autorités religieuses. Tout ce qui se constate ici vient du même fait qui veut que la totalité absolue ne se détermine pas comme tel ou tel Dieu défini."[7]

Quel amalgame étrange d'une idéologie sans réserve d'impérialisme ou totalitarisme et de la philosophie traditionnelle orientale du Néant! Ou, quelle mixture étrange de la manifestation de la logique même de la volonté du pouvoir et de celle de la logique de "participation" au sens que Lévy-Bruhl donne à ce terme!

On pourrait bien facilement discerner ici la même logique du court-circuit entre la totalité (indéfinie) et l'individu qui est mise en œuvres dans le cas de Kant-interprétation de Watsuji. Une fois pris dans cette logique du court-circuit ou, plus proprement, le court-circuit de différentes sortes de logique, qui, sans réflexion méthodologique suffisante sur la médiation possible, perversement confond la logique de la raison avec celle de l'imagination (cette dernière étant opèrante, à mon avis, dans le domaine de la pensée mythique, art, littérature et religion, par quelque nom (logique de la "participation", pensée sauvage, transposition métaphorique etc.) que l'on veuille l'appeler), on serait, tôt ou tard, consciemment ou inconsciemment, victime de l'exploitation idéologique du totalitarisme, impérialisme, ethnocentrisme (purisme ethnique), racisme, fascisme etc..

En fait, le même Watsuji, qui, comme nous l'avons vu, avait une fois présenté une série d'activités de l'imagination créatrice qui pourrait s'entrecroiser avec d'autre sous diverses formes — autres personnes, autres cultures, autres environnements écologiques etc. —, et, tout en admettant l'hybridité de chaque culture, développer certaines vues brillantes sur le pluralisme des cultures du monde, ici à l'époque de la Deuxième Guerre mondiale, fut victime d'un ethnocentrisme fanatique prenant son origine dans le Kokugaku et du totalitarisme du groupe ethnique des Japonais.

Cette sorte de court-circuit ou perversion que j'ai mentionnée tout à l'heure, de la logique de l'imagination par celle de la raison ou de la vie quotidienne se manifestait, certainement souvent, dans quelques pays à l'époque du fascisme ou du totalitarisme. Comme Hanna Arendt l'a clairement montré dans "L'Impérialisme", l'exploitation idéologique du totalitarisme a une tendance générale à pénétrer dans le domaine du sacré et jusqu'au véritable bout de la concience individuelle. Mais, néanmoins, en ce qui concerne ce court-circuit ou plutôt cette perversion de la pensée, je pense qu'on peut remarquer quelques traits caractéristiques de la culture Japonaise: les Japonais, qui pourraient se caractériser, d'un côté, par leur sensibilité raffinée dans le domaine de l'art et de la religion ou, en un mot, dans le domaine de l'imagination créatrice, pourraient probablement avoir, d'un autre côté, à cause du manque d'une logique appropriée de la médiation, un degré spécifiquement faible de résistance à l'exploitation idéologique fondée sur une telle sorte le court-circuit ou de perversion comme je viens de le préciser. On pourrait rappeler ici le cas de Norinaga MOTOORI (qui fut un excellent philologue de la littérature classique japonaise et en même temps un penseur de l'ethnocentrisme fanatique) ou (quoique un peu anachronique et même comique) le cas de Yukio MISHIMA, dans lesquels (tous les deux) on trouve la combinaison curieuse d'une sensibilité extrêmement raffinée avec une manifestation barbare d'ethnocentrisme ou même une certaine sorte de glorification de la violence.

6.

Après la défaite du Japon dans la Deuxième Guerre mondiale, qui lui avait donné un choc sévère, Watsuji commença à modifier et reconstruire sa pensée pendant le temps de la guerre. Parallèlement à la révision et la publication de quelques livres écrits dans sa jeunesse (y compris "Koji Junrei"), il révisa "Rinri-gaku" (Ethique) vol. 2, et publia le vol. 3 du même ouvrage (1949), dans lesquels, modifiant considérablement son ethnocentrisme ou ultra-nationalisme du temps de guerre, il présenta un système d'éthique suivant la ligne du pluralisme culturel de Herder, dont l'influence, comme nous l'avons vu, remonte jusqu'à l'époque de "Fûdo". Puis à peu près selon la même ligne, reformulant et augmentant les matériaux une fois traités dans "Sonnô Shisô to sono Dentô" (L'idée royaliste du Japon et sa tradition), il publia "Nihon Rinri Shisô-shi" (Histoire de la pensée éthique du Japon) (deux vols., 1952)

Cependant, à mon avis, même par cette série de révisions durant la première décennie suivant la deuxième guerre mondiale des ouvrages originellement écrits pendant le temps de la guerre, il ne pouvait pas accomplir la déconstruction complète du cadre de l'ethnocentrisme ni la reconstruction nouvelle de sa pensée au point de vue du véritable plurarisme culturel. On pourrait donner comme une preuve de cette sorte d'inconsistance de pensée une des œuvres majeures de cette même décennie: "Sakoku" (Isolement national), qui a connu un vaste public à l'époque de sa publication et désormais est regardé comme une des plus populaires parmi ses publications. Dans ce livre, suivant les vicissitudes des relations entre Japon et pays européens depuis le 16ème siècle, Watsuji essaie de trouver la cause de la défaite du Japon dans la Deuxième Guerre mondiale principalement dans la fermeture du pays (Sakoku) au début du 17ème siècle par le gouvernement (shogoun) de Tokugawa, qui a entrainé dans les siècles suivants un manque de tradition de pensée rationnelle et scientifique dans le Japon moderne. Il est certainement indéniable que Watsuji a converti ici dans cette œuvre sa pensée ethnocentrique du temps de guerre en un "rationalisme" de la modernité occidentale. Il pourrait être, cependant, aussi indéniable que, par cette conversion, pour ainsi dire, il ait simplement déplacé le centre et se soit converti à l'envers de la pensée ethnocentrique: l'ethnocentrisme de l'Europe moderne en tant qu'il se caractérise par la science moderne, la technologie, le capitalisme, le colonialisme etc.. Je pense, conséquemment, qu'il y a dans la pensée de Watsuji de cette décennie, au moins un certain degré d'inconsistance au regard du rapport entre ethnocentrisme et pluralisme culturel subsistant malgré sa conversion au pluralisme culturel.

Après la première décennie après la Deuxième Guerre mondiale, durant laquelle, en dépit de la productivité dans le domaine des recherches systématiques et historiques, il restait quelque chose d'inconsistant dans la pensée de Watsuji comme nous venons de le voir tout à l'heure, Watsuji est entré dans une phase nouvelle et dernière de sa pensée, qui a été plutôt négligée même parmi les spécialistes de la pensée philosophique japonaise, mais me semble extrêmement importante pour la compréhension de la profondeur de la pensée de Watsuji aussi bien que des problèmes en face desquels la philosophie japonaise se trouve aujourd'hui.

Cette nouvelle phase de la pensée de Watsuji que je viens de mentionner est représentée par son dernier ouvrage volumineux: "Kabuki to Ayatsuri-jôruri" (Kabuki et le théâre de poupées de Jôruri), dans lequel, se rappelant l'impression profonde causée dans son enfance par une scène de Kabuki présentant la séparation d'une mère d'avec son enfant, et cherchant l'origine de cette impression, Watsuji finalement la trouva dans quelques vieilles pièces de Jôruri à l'époque de Mouromachi (14ème, 15ème siècle) qui se jouaient à l'aide de poupées et représentaient quelque fusion hybride entre les traditions spirituelles Bouddhique et Chrétienne.

Ainsi, dans cette dernière phase de l'itinéraire de sa pensée, pour ainsi dire, en se réconciliant avec la réminiscence de son enfance ainsi qu'avec celle du moyen âge du Japon où, avant la formation de la pensée ethnocentrique de l'époque d'Édo, l'imagination créatrice avait toujours gardé son pouvoir d'entrecroisement (hybride) avec d'autres cultures, d'autres espèces d'animaux, d'autres genres de "personnes" comme des poupées ou des masques etc., et en se libérant de perversions de la logique de l'imagination et de celle de la raison, Watsuji a finalement réussi à intégrer dans sa pensée la couche profonde de la pensée imaginaire où, sans l'identité fixée de la personne et de la chose, il peut y avoir un libre entrecroisement entre le moi et l'autrui, l'être humain et l'autre être vivant, la personne et la poupée etc.

7.

Dans le cas de Watsuji comme penseur du Japon moderne, comme on l'a vu, il se manifeste quelques traits, disons, "pathologiques" et pervers dont, peut-être dans la plupart des cas sans s'y engager activement et consciemment, néanmoins, Watsuji et beaucoup d'autres intellectuels du Japon ont été victimes surtout pendant la Deuxième Guerre mondiale. Selon mon analyse faite ci-dessus, ces traits-là, dont l'origine remonte à la deuxième moitié du 18ème siècle (à l'époque où on donnait le fondement définitif de Kokugaku), pourrait étre regardés comme une manifestation de la logique de perversion ou plutôt de la perversion de différentes sortes de logique, qui devraient appartenir respectivement à la région de l'imganiation créatrice et celle de la raison ou de la pensée de la vie quotidienne. Une fois pris par cette logique de perversion ou perversion de la logique, il serait presque inévitable que l'on s'engage, soit consciemment, soit inconsciemment, dans une tendance à rendre absolue l'instance de l'éthique ainsi qu'à diviniser sa propre culture tout en pervertissant sa pensée culturelle selon la logique perverse de l'ethnocentrisme.

Mais, cependant, même juste au milieu de l'époque de la Deuxième Guerre mondiale, il y a eu quelques penseurs qui, n'étant pas complètement pris par ce mouvement de perversion de la pensée et de purisme ethnocentrique, développaient quelques pensées qui ont gardé à leur manière le véritable équilibre entre la pensée imaginaire et celle dite "raisonnable". Comme exemple de cette sorte de penseurs, j'aimerais citer d'abord Kitarô NISHIDA, philosophe célèbre, qui, comme on l'a vu, influençait considérablement la pensée éthique de Watsuji, mais cependant à la différence de ce dernier, tout en gardant sa logique de prédicat et celle de "champ" qui ne s'engage jamais par avance vers l'identité fixée du sujet au sens grammatical ainsi que philosophique de ce terme, pouvait s'enraciner toujours dans la couche profonde de la pensée poétique ou plutôt "poïétique" avec un plus haut degré de résistance à la perversion ethnocentrique de la pensée.

J'aimerais citer, deuxièment comme un autre exemple de ces penseurs, Sôétsu YANAGI (1889–1961), théoricien et historien des arts, qui, d'un côté s'intéressant beaucoup à la tradition spirituelle et mystique de l'Occident aussi bien que du Bouddhisme, et d'un autre côté surtout, s'engageant comme un des fondateurs du mouvement pour estimer et encourager les arts populaires traditionells à la manière de William Morris, a montré un respect élevé tout à fait libre d'ethnocentrisme, pour les arts de Corée (une des colonies du Japon de cette époque) etc..

Comme le troisième exemple, je voudrais citer les noms de Kunio YANAGIDA (1875–1962) et Shinobu ORIGUCHI (1887–1953), ethnologues et hommes de lettres, qui, tous les deux comme fondateurs de l'école du "Nouveau Kokugaku", prenant beaucoup d'intérêt à collectionner des récits folkloriques toujours vivants dans la tradition orale et à éclairer la structure de la culture de base du Japon, ont montré une mentalité presque complètement libre de l'idéologie dominante ethno centrique de cette époque et essayé d'évoquer l'intérêt pour imagination vive et ouverte manifesté dans la tradition orale et le dialecte employé dans la vie quotidienne.

NOTES

(Je cite œuvres de Watsuji sans mentionner le nom d'auteur selon la pagination de ses Œuvres Complètes publiées par Iwanami Shoten, dont les chiffres romains indiquent le volume.)

1. "Ningen no Gaku to shite no Rinri-gaku", IX, pp. 148–149.
2. "Jinkaku to Jinruisei", IX, p. 350.
3. *Ibid.*, pp. 345–346.
4. *Ibid.*, pp. 333–334.
5. "Rinri-gaku" (le texte de la première édition de 1942), XI, pp. 419–420.
6. "Sonnô Shisô to sono Dentô", XIV, pp. 232–233.
7. *Ibid.*, pp. 43–44.

Deutsche Philosophie in Japan

CHUKEI KUMAMOTO
(Hiroshima)

1. Die Entwicklung der deutschen Philosophie in Japan

1. Die europäische Philosophie wurde in den sechziger Jahren des 19. Jahrhunderts von Amane NISHI(1829–97) und Yukichi FUKUZAWA (1835–1901) in Japan eingeführt. Da die japanische Intelligenz stärkeres Interesse an der englischen und französischen Philosophie zeigte, herrschten damals hauptsächlich die Gedanken von MILL, COMTE, ROUSSEAU u.a. Vor 1882 wurde eine politische Delegation nach Deutschland gesandt, um die Landesverhältnisse kennen zu lernen, und 1889 wurde von ihnen die japanische Verfassung, welche die deutsche nachahmte, entworfen. Das hat naturgemäss stark auf die japanische Gedankenwelt eingewirkt.

Als die Universität Tokio 1887 gegründet wurde, wurde Ernest Francisco FENOLLOSA (1853–1908) als Professor eingeladen, wo er eine Vorlesung über europäische Philosophie hielt. Schon im Jahre 1884 wurde die "Philosophische Gesellschaft" in Japan gegründet und 1887 erstmals die *Philosophische Zeitschrift* herausgegeben. 1887 wurde Ludwig BUSSE (1845–1923) eingeladen. 1890 kehrte Tetsujiro INOUE (1855–1944) aus Deutschland zurück, wo er sechs Jahre Philosophie studiert hatte. Sie alle lehrten hauptsächlich deutsche Philosophie an der Universität Tokio. So erhielt die deutsche Philosophie eine führende Stellung in der damaligen japanischen Gedankenwelt. Damals wurden einerseits die Werke von Kant und Hegel untersucht und andererseits wurde Nietzsche von der Intelligenz sehr geschätzt.

2. Ab 1910 wirkte der Neukantianismus so stark auf die japanische Gedankenwelt ein, dass die Kantforschung sehr gefördert wurde und einige bemerkenswerte Arbeiten erschienen, z.B. die von Sanjuro TOMONAGA (1871–1951), Genyoku KUWAKI (1874–1946), Yoshishige ABE (1883–1966), Hajime TANABE (1885–1962) u.a. Es ist sehr wichtig, dass damals Kitaro NISHIDA (1870–1945), durch die Synthese der europäischen Philosophie mit dem Buddhismus, ein eigenes philosophisches System, das sich bis 1945 entwickelte, begründete.

In den zwanziger Jahren des 20. Jahrhunderts setzte sich durch die Nachwirkung der Russischen Revolution der Marxismus in Japan durch. Einige Kantforscher versuchten einen Zusammenhang zwischen Kant und Marx zu finden. Und einige Marxisten beabsichtigten, über den Marxismus zur Hegelforschung zu kommen.

Contemporary philosophy. A new survey. Vol. 7, 357–375.

In den dreissiger Jahren des 20. Jahrhunderts war die japanische Gedankenwelt sehr vielfältig. Nicht nur Kant und Hegel sondern auch Fichte, Heidegger, Jaspers u.a. wurden immer gründlicher untersucht. Als bedeutende Kantforscher sind Teiyu AMANO (1984–1980), Wakichi MIYAMOTO (1883–1792) und Masaaki KOSAKA (1900–69), als Hegelforscher Hajime TANABE, Iwao KOYAMA und Takezo KANEKO, als Fichteforscher Kenji KAWASE (1889–1973), Motomori KIMURA (1895–1946) zu nennen. Vor allem TANABEs Arbeit ist sehr wichtig. Er hat aufgrund der Philosophie NISHIDAs die europäischen Gedanken und den Buddhismus dialektisch vereinigt. Die Philosophien von NISHIDA und TANABE sind jetzt der Stolz der japanischen Gedankenwelt.

3. Kurz nach Ende des zweiten Weltkriegs blühten der Marxismus und der Existentialismus auf, was zweifellos die damalige soziale und gedankliche Unordnung in Japan beweist. Aber ab 1950 wurden Kant und Hegel wieder wissenschaftlich untersucht, sowie viele andere deutsche Denker: z.B. Fichte, Schelling, Husserl, Heidegger und Jaspers. Jetzt hat, wie mir scheint, die japanische Gedankenwelt das deutsche Niveau erreicht.

Kantforscher:	Tasuku HARA (1916–76), Takeo IWASAKI (1913–76), Ichigu TAKAMINE, Yukio MIWATARI u.a.
Hegelforscher:	Kinshiro KASHIYAMA (1907–77), Takezo KANEKO, Hajimu NAKANO, Osamu SAKAI u.a.
Fichteforscher:	Chukei KUMAMOTO, Akira OMINE u.a.
Husserlforscher:	Yoshihiro NITTA, Gen KIDA u.a.
Heideggerforscher:	Goichi MIYAKE, Jiro WATANABE, Koichi TSUJIMURA, Yoshio KAYANO u.a.
Jaspersforscher:	Saburo SUZUKI, Masao KUSANAGI u.a.

2. Immanuel Kant

1. Es ist keine Übertreibung, wenn man sagt, dass die Entwicklung der deutschen Philosophie in Japan nichts anderes als die der Kantforschung ist. Kants Philosophie hat nicht nur in der Meiji-Periode den Ausgangspunkt der japanischen Philosophie gebildet, sondern sie bestimmt auch nach wie vor die japanischen Gedanken.

Ab 1878 hielt FENOLLOSA an der Universität Tokio Vorlesungen über europäische Philosophie. BUSSE und KOEBER lehrten hauptsächlich deutsche Philosophie. Tetsujiro INOUE, Rikizo NAKAJIMA (1858–1918) und Yujiro MOTORA (1858–1912), von den obigen angeleitet, untersuchten auch die deutsche Philosophie, und besonders Kant und Hegel. 1889 erschien NAKAJIMAs Aufsatz über Kant:

Rikizo NAKAJIMA: Kants kritische Philosophie. 1892.

1896 erschien eine wunderbar strenge Kantstudie von Tsutomu KIYONO (1853–1904), die eine direkte Untersuchung der Kantschen Werke darstellt:

Tsutomu KIYONO: Die Erklärung von Kants Kritik der reinen Vernunft. 1896.

KIYONO gibt darin einerseits eine genaue Erklärung und einen Kommentar und andererseits eine eigene Interpretation. Seine Arbeit war nicht nur damals sehr bedeutend, sondern ist auch für die Gegenwart noch wissenschaftlich wertvoll.

Die *Kant-Studien*, 1896 von Vaihinger gegründet, wirkte bald auf die japanische Gedankenwelt ein. Ab 1900 erschienen die zwar wenigen, aber sehr strengen und bemerkenswerten Kantstudien:

Seiichi HATANO: Humes Einwirkung auf Kant. 1900.
Wakichi MIYAMOTO: Die Entwicklung des Kantischen Kritizismus. 1909.

2. Der Neukantianismus, der Ende des 19. Jahrhunderts in Deutschland auftrat, begann 1910 auf japanische Kantforscher einzuwirken. Damals erschienen einige wertvolle Kant-Schriften, die sich hauptsächlich nach ihm richteten. Die Schriftsteller waren alle führende Denker in Japan, die bis heute die japanische Philosophie geleitet haben:

Sanjuro TOMONAGA: Geschichte der Sich-selbst-bewusst-werdung des Ich in der neueren Zeit. 1916.
Genyoku KUWAKI: Kant und die gegenwärtige Philosophie. 1917.
Sanjuro TOMONAGA: Kants Lehre vom Frieden. 1922
Yoshishige ABE: Kants praktische Philosophie. 1923.
Masujiro OSEKI: Studien zu Kant. 1923.
Hajime TANABE: Kants Teleologie. 1923.

TOMONAGA (1871–1951) hat unter dem Einfluss der Süd-west-deutschen Schule die Entwicklung des Begriffs des Ich in der neueren Zeit, und dadurch die Stellung von Kant erklärt. Vor allem seine Einsicht in Kants Lehre vom Frieden ist als sehr wertvoll anzusehen. KUWAKIs (1874–1946) Schrift, die nach der Süd-west-deutschen Schule einerseits aufklärerisch, andererseits wissenschaftlich dargestellt wurde, enthält eine eigene Einsicht in Kants kritische Methode sowie in das Ding-an-sich. Sie wurde bis zum zweiten Weltkrieg als eine Arbeit angesehen, die die japanischen Kantianer unbedingt gelesen haben mussten. ABEs (1883–1966) Schrift ist hauptsächlich aufklärerisch, ist aber von den japanischen Sittenlehrern hoch geschätzt. TANABE (1885–1962), der vom Standpunkt der Marburger Schule aus viele Kant-Aufsätze, und zwar "Über das Bewusstssein überhaupt" (1919), "Das Problem des Erkenntnissubjekts" (1919–22) u.a. veröffentlichte, systematisierte in seiner Schrift vermittelst der Hegelschen Dialektik Kants Teleologie durch den Begriff der sich selbst bewussten Zweckmässigkeit. OSEKIS Schrift, die mehr als 1,000 Seiten enthält, versuchte Kants Kritizismus als ein System zu interpretieren, und ist in der Geschichte der Kant-Interpretation auch in Japan sehr bedeutend. Nach OSEKI besteht das System der Kantschen Philosophie in der Idee der moralischen Anschauung.

Der Begriff der Freiheit, die den Kernpunkt der Moralität bildet, werde erstens in der Kritik der reinen Vernunft als transzendentale Freiheit aufgestellt, und in der

Kritik der praktischen Vernunft werde er einerseits als *ratio essendi* des Moralgesetzes postuliert, andererseits im Zusammenhang von empirischem und intelligiblem Charakter realisiert. Eben dadurch werde das Moralgesetz vollends festgestellt. Die objektive Zweckmässikeit, die in der Urteilskraft als die Einheit von Natur und Freiheit aufgestellt werde, bedeute eine Ordnung, wodurch sich die moralische Handlung in der Sinnenwelt realisieren solle.

In dieser Zeit wurden einige von Kants Werken ins Japanische übersetzt; *Prolegomena, Kritik der praktischen Vernunft und Grundlegung zur Metaphyssik der Sitten.* Ab 1921 wurden Kants Werke in 18 Bänden im Iwanami-Shoten Verlag herausgegeben.

Zugleich wurden auch einige Arbeiten über Kant ins Japanische übertragen: K. Fischer, Buchenau, G. Simmel, F. Paulsen u.a., ebenso bedeutende Aufsätze aus den *Kant-Studien.* 1921 hat Ernst Cassierer für die *Philosophische Zeitschrift* einen Aufsatz: "Kant und die neuere Physik" beigetragen.

1924, d.h. zum zweihundertjährigen Geburtstag Kants, wurde im Iwanami-Shoten Verlag eine Sondernummer der Zeitschrift *Der Gedanke* herausgegeben. Darin wurde Kants Philosphie von allen Seiten, d.h. von Seiten der Erkenntnistheorie, der Moralphilosphie, der Geschichtsphilosophie, der Religionsphilosophie u.a. untersucht und kritisiert.

3. In den dreissiger Jahren des 20. Jahrhunderts gab es einen Versuch, die Kantsche Philosophie ganzheitlich zu erfassen. Drei Arbeiten sind bemerkenswert:

Yusuke YAMAGUCHI: Der Begriff der Philosophie bei Kant. 1931.
Yoshinori ONISHI: Die Erforschung der Kritik der Urteilskraft Kants. 1931.
Masaaki KOSAKA: Kant. 1940.

YAMAGUCHI versucht Kant nicht vom Standpunkt der Notwendigkeit der Kantschen Gedankenentwicklung, sondern vom Standpunkt der Forderung der philosophischen Idee her einheitlich zu erfassen. Die Forderung der Idee macht, nach YAMAGUCHI, die Bewegungskraft der Entwicklung des philosophischen Gedankens aus. Sie habe zwei Momente, d.h. erstens die systematische Forderung, alle Probleme zu umfassen, zweitens die wissenschaftliche Forderung, die Gewissheit der philosophischen Erkenntnis zu sichern. Kant habe diese zwei Momente bewusst harmonisch vereinigt.

ONISHI (1888–1959) erfasste die Kritik der Urteilskraft nicht äusserlich als Verbindung der ersten und zweiten Kritik, sondern vielmehr innerlich als Erklärung ihres eigenen Problems.

KOSAKAs (1900–69) Buch über macht einen Gipfelpunkt der japanischen Kantforschung vor dem zweiten Weltkrieg aus. KOSAKA versucht Kants Philosphie als grosses anthropologisches System zu erfassen. Ihm zufolge hat Kant einerseits die Aufklärung vollendet, andererseits aber ist er der Schöpfer der Geistesphilosophie; denn, dadurch, dass er durch die Untersuchung des menschlichen Verstandes die denselben übersteigende Transzendenz findet, hat er den Verstand als den Gesetzgeber für die Natur aufgestellt. Darin hat Kant den Übergang von der Philosophie des Verstandes zu der des Geistes vollzogen.

KOSAKA erklärt erstens durch die Untersuchung der Kritik der reinen Vernunft die Gesetzgebung des Verstandes für die Natur und zweitens durch die Untersuchung der Kritik der praktischen Vernunft die Gesetzgebung für die Moral. Drittens erfasst er die Kritik der Urteilskraft nicht nur als blosses Verbindungsmittel der beiden genannten Kritiken, sondern auch positiv als die teleologische Welt, die den Menschen auf den Endzweck vorbereiten soll. Nach KOSAKA ist die Zweckmässigkeit der Kultur und Geschichte als der letzte Zweck der Natur eben die allgemeine Zweckmässigkeit der Sinnenwelt und der intellgiblen Welt, die von KOSAKA die geschichtliche Zweckmässigkeit genannt wird. Auf diese Weise zieht sich die ganze Kantsche Philosophie in der Urteilskraft zusammen.

Neben den obigen Arbeiten sind noch folgende wichtige Studien zu nennen:

Teiyu AMANO: Kants Kritik der reinen Vernunft. 1935.
Tetsuro WATSUJI : Kants Kritik der praktischen Vernunft. 1935.
Wakichi MIYAMOTO: Kantforschungen—Zur transzendentalen Apperzeption. 1941.

4. Nach dem zweiten Weltkrieg nahm die Kantforschung rasch zu, und Kant wurde von verschiedenen Seiten aus untersucht. Ab 1965 wurden Kants Werke in 17 Bänden im Riso-Sha Verlag herausgegeben, und 1975 wurde die japanische Kant-Gesellschaft gegründet, die seitdem jährlich eine Tagung abhält. Bis heute erschienen etwa 30 Kantschriften und rund 400 Aufsätze. Die wichtigsten unter ihnen sind:

Tasuku HARA: Systematische Interpretation der Kantischen Philosophie. 1947.
Sueo TAKEDA: Kantstudien zum Problem der Analogie. 1957.
Yukio MIWATARI: Die Struktur des Kantischen Kritizismus. 1957.
Derselbe: Die Struktur des Kantisch en Kritizismus (Fortsetzung). 1960.
Takeo IWASAKI: Kant. 1958.
Derselbe: Studien zu Kants Kritik der reinen Vernunft. 1965.

Nach HARA (1916–76) hat Kant der alten Metaphysik eine neue entgegengesetzt. Kant hat nämlich eine Metaphysik der Natur sowie der Moral aufgestellt und beide in der teleologischen Weltanschauung als Metaphysik der Metaphysik vereinigt. Das bedeutet die Systematisierung der beiden Metaphysiken in der Kritik der Urteilskraft. Kants Weltanschauung ist nämlich nichts anderes als eine moralische Teleologie, in der einerseits der Mensch und die Natur harmonisch vereinigt wird, und andererseits der Mensch Gott begegnen kann. Nach HARA ist nämlich die Philosophie die Arbeit der menschlichen Vernunft, die sich das für uns Absolute als das an sich Absolute aufzustellen anmasst.

TAKEDA geht der Frage nach, was für ein Logos der Kantschen Philosophie zugrunde liegt, und er stellt die neue Interpretation auf, Kant als Ganzes zu erfassen. Nach TAKEDA wird in der Kritik der reinen Vernunft die Grenze des Logos der tranzendentalen Logik gezeigt und zur Logik der Analogie übergegangen. Kants transzendentale Logik kann nur auf die Natur angewandt werden. Der Mensch als metaphysiches Wesen aber sucht eine ganz andere Logik der Analogie, die in der Teleologie der Kritik der Urteilskraft gegeben wird.

MIWATARI nennt in seinen zwei Kantschriften die Struktur der kritischen Philosophie einen "Mittelweg". Der Mittelweg bedeute nicht bloss den Weg zwischen dem Rationalismus und dem Empirismus, oder zwischen dem Dogmatismus und dem Skeptizismus, sondern in Wahrheit einen die Gegensätze umfassenden höheren Standpunkt, d.h. die Dreieinigkeit, die als analytisch-synthetisch-dialektisch, oder kritisch-transzendental-experimental zu charakterisieren sei. Die Dreieinigkeit gründet sich schliesslich auf der Vereinigung der Apperzeption und der Einbildungskraft. Von diesem Standpunkt her hat MIWATARI auch alle anderen kantischen Probleme erklärt.

IWASAKIs (1913–76) Arbeit bildet zweifellos einen Gipfelpunkt der gegenwärtigen Kantforschung in Japan. Er hat eine eigene, epochemachende Einsicht entwickelt. IWASAKI zufolge bedeutet Kants kopernikanische Wendung nicht, wie bisher behauptet, die Vertretung eines erkenntnistheoretischen Subjektivismus, sondern vielmehr die Feststellung des endlichen, menschlichen Standpunkts. Kant habe erstens in der theoretischen Philosophie die Mathematik und die Naturwissenschaft begründet, um nicht Dinge an sich, sondern Erscheinungen zu erkennen. Zweitens habe er in der Moralphilosphie gegen den bis dahin geltenden Standpunkt, der das Moralgesetz auf Gott begründet, den menschlichen Standpunkt, der sich auf die menschliche Freiheit gründet, aufgestellt. Kant habe nämlich die Wendung vom göttlichen zum menschlichen Standpunkt vollzogen.

3. Johann Gottlieb Fichte

1. Obgleich das Interesse an Fichte nicht so gross ist wie an Kant und Hegel, ist Fichte nach wie vor gründlich untersucht. Fichtes Philosphie wurde kurz vor 1900 in Japan eingeführt. Die ersten Jahre des 20. Jahrhunderts bildeten also die Periode der Aufklärung über Fichte. Damals erschienen einige einleitende Aufsätze über Fichte:

Tsukiko MINE: Fichte. 1900.
Danjo EBINA: Fichtes Überzeugung. 1901.
Hiroto TOMIZU: Fichtes Wissenschaftslehre. 1906.

Ab 1900 wurde Fichte von einigen Denkern genauer untersucht. Bis zum zweiten Weltkrieg erschienen rund 100 Aufsätze und mindestens 6 Bücher über Fichte. Vor allem die Arbeiten von Sanjuro TOMONAGA (1871–1951) sind sehr bemerkenswert:

Sanjuro TOMONAGA: Die Entstehung des praktischen Idealismus bei Fichte. 1914.
Derselbe: Die Entwicklung der Religionslehre bei Fichte. 1916.

Derselbe: Fichtes Atheismusstreit. 1927.

TOMONAGA untersuchte die europäische Philosophie vom Standpunkt des Neukantianismus und veröffentlichte Bücher über Descartes und Kant. Er interessierte sich vor allem für Fichtes Religionslehre, die ihm zufolge verschiedene Religionslehren in sich enthält. Er behandelte nicht nur umfassend die frühe sondern auch die späte Religionslehre Fichtes.

Masao KUBO behandelte nicht nur die Wissenschaftslehre, sondern auch die praktische Philosophie erschöpfend. Er veröffentlichte ein Buch und wenigstens 7 Aufsätze über Fichte. Die folgenden Arbeiten sind hoch zu schätzen:

Masao KUBO: Leben und Spekulation bei Fichte. 1916.
Derselb: Fichtes Geschichtsphilosophie, 1920.
Derselbe: Religion und Metaphysik. 1921.

Kenji KAWASE (1889–1973) betrachtete hauptsächlich Fichtes Wissensschaftslehre und Religionslehre. Er stellte die bedeutende Einsicht auf, das absolute Ich, das den Grund des theoretischen sowie des praktischen Ich ausmacht, sei als Grenzbegriff zu erfassen. Der Grenzbegriff, der betreffs der Reihenglieder einerseits als ein Glied der Reihe anzusehen sei, hat eine Kontinuität mit denselben, doch als Idee überschreitet er andererseits die Reihe. Das absolute Ich sei entweder real, d.h. als realisiertes Ideal vom Nicht-Ich ganz getrennt oder ideal d.h. es habe als Prinzip des Strebens des Ich einen Zusammenhang mit dem Nicht-Ich. Es gelte die Forderung, alles Besondere praktisch oder theoretisch zu verallgemeinern. Von diesem Standpunkt aus erklärte KAWASE auch die Fichtesche Religionslehre. Die Entwicklung derselben sei nichts anderes als die Manifestation der im absoluten Ich schon verborgenen Momente. KAWASEs Aufsätze über Fichte sind:

Kenji KAWASE: Betrachtungen zu Fichtes Wissenschaftslehre. 1925.
Derselbe: Betrachtungen zu Fichtes Religionslehre. 1925.

Motomori KIMURA (1895–1946) untersuchte besonders die frühe Wissenschaftslehre sehr genau und erklärte das Wesen Fichtes im Gegensatz zu Hegel. Er unterscheidet Fichtes Dialektik von Hegels dadurch, dass die Synthesis bei Hegel in der Versöhnung von Gegensätzen, bei Fichte dagegen im steten Streit bestehe, und dass sich die dialektische Entwicklung bei Hegel in der letzten Synthesis vollende, bei Fichte aber der endlose Fortschritt zur Idee sei. Hegels Philosophie sei pantheistisch, Fichtes aber praktisch. KIMURAs Aufsäze sind:

Motomori KIMURA: Überlegungen zum Grundprinzip von Fichtes Philosophie. 1934.
Derselbe: Das Wesen der Wissenschaftslehre Fichtes. 1934.

Überdies wurde Fichtes Sittenlehre von Kenjuro YANAGITA als Überwindung des Formalismus der Kantschen Moralphilosophie erfasst und von Kazuto

TAKAYAMA, dem Historiker, Fichtes Anschauung über Deutschland genauer erklärt.

Damals gab es einige wichtige Schriften über Fichte:

Masao KUBO: Die Philosophie Fichtes. 1926.
Genyoku KUWAKI: Fichtes Wissenschaftslehre. 1935.
Motomori KIMURA: Fichte. 1937.

Im allgemeinen beschränkte sich damals die Fichteforschung auf die frühe Philosophie, und fast die Hälfte seiner Werke in der frühen Periode wurde ins Japanische übertragen: z.B. "Über den Begriff der Wissenschaftslehre, oder der sogenannten Philosophie"; "Grundlage der gesamten Wissenschaftslehre"; "Erste Einleitung in die WL."; "Zweite Einleitung in die WL."; "Versuch einer neuen Darstellung der WL."; "Einige Vorlesungen über die Bestimmung des Gelehrten"; "Über das Wesen des Gelehrten und seine Erscheinungen im Gebiete der Freiheit"; "Die Bestimmung des Menschen"; "Die Wissenschaftslehre 1804"; "Die Anweisung zum seligen Leben, oder auch Religionslehre"; "Reden an die deutsche Nation" u.a.

2. Nach dem zweiten Weltkrieg wurde Fichtes Philosophie noch eingehender als zuvor untersucht. Vor 1960 wurde die Philosophie des frühen Fichte erforscht. Es erschienen rund 40 Aufsätze. Damals behandelte man, wie mir merkwürdig scheint, meistens die praktische Philosophie. Die wichtigsten Studien sind:

Gisen MISAKI: Die Bedeutung des Fichteschen Atheismusstreits. 1950.
Hakuai AKIYAMA: Fichtes Staatslehre. 1952.
Shokan SUGAYA: Religion und Wissenschaftslehre. 1953.
Kazuto TAKAYAMA: Kulturstaat und Vernunftstaat. 1951.
Toyojiro AKABANE: Fichtes Lehre vom geschlossenen Handelsstaat. 1958.

1959 erschien eine bedeutende Studie zu Fichtes Philosophie der Politik von Shigeru NANBARA (1889–1978), Professor an der Universität Tokio und später Richter am internationalen Gerichtshof.

Shigeru NANBARA: Fichtes Philosophie der Politik. 1959.

Der zweihundertjährige Geburtstag Fichtes war auch in Japan ein Wendepunkt in der Fichteforschung, denn einige Forscher begannen nun den späten Fichte zu beachten, was der Fichterenaissance in Deutschland entspricht.

In den sechziger Jahren erschienen wertvolle Aufsätze über den späten Fichte:

Chukei KUMAMOTO: Die Struktur des absoluten Wissen beim späten Fichte. 1961.
Derselbe: Das absolute Wissen und das Absolute. 1965.
Akira OMINE: Sein und Freiheit. Eine Untersuchung zur "Darstellung der Wissenschaftslehre 1801". 1970.
Derselbe: Metaphysik des Lichts und Horizont des Bewusstseins. 1971.

Giichi SAITO: Studien zur Fichteschen Philosophie. 1968.

Es finden sich heute zwei Schriften, die die ganze Entwicklung der Fichteschen Philosophie von 1794 bis 1814 genau erfassen:

Chukei KUMAMOTO: Die Erforschung der Wissenschaftslehre Fichtes. 1970.
Akira OMINE: Fichteforschungen. 1976.

1985 wurde die japanische Fichte-Gesellschaft gegründet und seitdem wurde jährlich eine Tagung abgehalten. Dies war sicher epochemachend, denn die Deutsche Fichte-Gesellschaft war damals noch nicht gegründet.

4. Georg Wilhelm Friedrich Hegel

1. An der Universität Tokio erklärte FENOLLOSA ab 1887 Hegels Philosophie an dem Kommentar von WALLACE im Zusammenhang mit der Entwicklungstheorie DARWINs, was auf die Hegelforscher in Japan stark eingewirkt hatte. Yujiro MIYAKE (1860–1945), der erste Hegelforscher, der von FENOLLOSA geleitet wurde, schrieb eine Geschichte der Philosophie:

Yujiro MIYAKE: Tropfen der Philosophie. 1889.

Er sah dort das Verhältnis von Sein und Nichts in Hegels Logik als den Gedanken an und behauptete zugleich, dass die Entwicklung des Gedankens nichts anderes sei als die Entwicklung der Welt. Hegels Logik sei zugleich als Ontologie zu fassen.

Manshi KIYOZAWA (1863–1903) nahm durch FENOLLOSA grosses Interesse an Hegel. Er stellte in seiner Schrift Hegels Philosophie dar und sah die dialektische Einheit von Gegensätzen für wertvoll an, die er im Zusammenhang mit dem Buddhismus erfasste.

Manshi KIYOZAWA: Vorlesungen über die Geschichte der europäischen Philosophie. 1891–93.

Ab 1887 lehrten BUSSE und ab 1892 KOEBER und ab 1890 INOUE die deutsche Philosophie. Die wissenschaftliche Tätigkeit von Hajime ONISHI (1964–1900) war damals prächtig. Bei der japanischen Intelligenz war Kant am populärsten und Hegel stand neben ihm. In den neunziger Jahren des 19. Jahrhunderts begann man Hegels Philosophie nicht mittelbar durch den Kommentar, sondern direkt durch Hegels Werke zu studieren. Auf diese Weise vollendete ONISHI sein Buch:

Hajime ONISHI: Die Geschichte der europäischen Philosphie. 1903.

Rikizo NAKAJIMA (1858–1918) ist der erste Philosoph, der Kant und Hegel genau kritisiert hat. Seine Schrift ist:

Rikizo NAKAJIMA: Hauptzweck der Hegelschen Dialektik. 1890.

Er nannte Hegels Philosophie den logischen Idealismus, und charakterisierte ihn folgendermassen: die Identität vom Denken und Sein bilde ihre Grundlage, und die Identität sei nicht blosse Einfachheit, sondern die Einheit, die durch die Negation und die Negation der Negation sich entwickelt.

2. Anfang des 20. Jahrhunderts erschienen bemerkenswerte Hegel-Schriften.

Yujiro MOTORA: Über Hegels Ontologie. 1909.
Sakuzo YOSHINO: Grundlage der Hegelschen Rechtsphilosophie. 1905.
Shinkichi UESUGI: Hegels Stellung in der Geschichte der Staatslehre. 1905.

MOTORA (1858–1912) untersuchte Hegels Lehre des Seins in der "Wissenschaft der Logik" und kritisierte sie. YOSHINO (1878–1933) und UESUGI (1878–1929) schätzten Hegels Lehre des objektiven Geistes sehr hoch. 1905–06 übersetzte Tadayoshi KIHIRA (1874–1949) mit Ryotaro ODAGIRI den ersten Teil von der "Enzyklopädie" ins Japanische und veröffentlichte einige Aufsätze. Vor allem erfasste KIHIRA Hegels Denken sehr genau. Das Denken wolle die allgemeine Notwendigkeit begreifen, die in Wahrheit Produkt des Denkens selbst ist. Das Denken also müsse das Andere setzen, um dieses in sich einzunehmen und in sich zurückzukommen. Das heisse dialektisch. Die Philosophie sei nämlich die Sich-selbst-bewusst-werdung des Denkens. Eben darin bestehe eine Ähnlichkeit von Hegels Denken und dem Buddhismus.

Zur gleichen Zeit war es Kitaro NISHIDA (1870–1945), der durch die Vereinigung der europäischen Philosophie, besonders Hegels, mit dem Buddhismus ein eigenes grosses System der Philosophie aufbaute. Obgleich er Hegel in seiner frühen Periode kaum nennt, liegt doch seiner Ansicht Hegels Denken zugrunde. Und in seiner späten Periode, wo er sich hauptsächlich nach der Logik von Aristoteles zu richten scheint, bildet Hegels Dialektik nach wie vor sein Leitmotiv. Hegel und der Buddhismus machen also sozusagen die zwei Zentren der Ellipse seiner Philosophie aus.

3. Ab 1910 wirkte der Neukantianismus auf die japanische Gedankenwelt ein. Fast zwanzig Jahre lang verfiel darum die Hegelforschung in Japan ganz, während die Kantforschung eine schnelle Entwicklung nahm. Erst 1930 trat wieder eine echte Hegelforschung auf, welche der Hegelrenaissance in Deutschland und dem damals weltweit verbreiteten Marxismus entsprach.

Im Zuammenhang mit der Hegelrenaissance wurde 1931, d.h. hundert Jahren nach Hegels Tod, in Tokio der internationale Hegel-Bund gegründet. Im gleichen Jahr begann der Iwanami-Shoten Verlag Hegels grosse Werke herauszugeben, wodurch sich die Hegelforschung sehr schnell entwickelte. Unter rund zehn Bücher, die damals erschienen, sind die folgende sehr bemerkenswert:

Hajime TANABE: Hegels Philosophie und die Dialektik. 1932.
Risaku MUDAI: Hegelstudien. 1935.
Iwao KOYAMA: Hegel. 1936.
Takezo KANEKO: Hegels Staatslehre. 1944.

Am wichtigsten ist die Hegelinterpretation von TANABE. TANABE, der zuerst Kants Philosophie untersuchte und viele bedeutende Aufsätze veröffentlicht hatte, studierte Mitte der zwanziger Jahre des 20. Jahrhunderts intensiv Hegels Philosophie. Er erreichte einerseits durch die Bestreitung des Marxismus, andererseits durch die Auseinandersetzung mit NISHIDA den Standpunkt der absoluten Dialektik, der Hegels Dialektik völlig überwindet. Nach TANABE ist die wahre Dialektik der Standpunkt, der sich nicht auf dem Geist wie bei Hegel, sondern auf dem buddhistischen Nichts gründet. TANABEs Philosophie ist neben NISHIDA die bedeutendste Arbeit, die von einem Japaner verfasst wurde.

Damals wurde viel Literatur über Hegel ins Japanische übersetzt, z.B. J.K.F. Rosenkranz, W. Dilthey, N. Hartmann, K. Fischer, B Croce, J. Royce u.a.

4. Nach dem zweiten Weltkrieg wurden zuerst der Marxismus, dann der Existentialismus am populärsten. Der Marxismus entsprach der damaligen politischen und ökonomischen Verwirrung, und der Existentialismus dagegen der psychologischen Verwirrung. Marxistische Hegelforscher wie Hiroomi SAEGUSA, Taketo TAKECHI, Shinichi FUNAYAMA und andere junge Studenten begannen Ansehen zu gewinnen. Ihre wichtigen Arbeiten sind:

Taketo TAKECHI: Die Welt der Hegelschen Logik, 3 Bde. 1947.
Shinichi FUNAYAMA: Entwicklung und Struktur des Hegelschen philosophischen Systems. 1963.
Shuro HONDA: Hegel und die Naturdialektik. 1976.

Die unparteiischen Hegelforscher arbeiten dagegen auf einem höheren Niveau. Vor allem KANEKOs Übersetzung und Kommentar der "Phänomenologie des Geistes" ist sehr bemerkenswert. Er hat von 1925 bis 1979 den ganzen Text zweimal ins Japanische übertragen und einen sehr ausführlichen Kommentar von 1950 Seiten verfasst. Es finden sich noch folgende andere bedeutende Hegelstudien:

Kinshiro KASHIYAMA: Studien zu Hegels Phänomenologie des Geistes. 1961.
Derselbe: Studien zu Hegels Logik. 1970.
Hajimu NAKANO: Hegel. 1970.

KASHIYAMA (1907–77) hat die "Phänomenologie des Geistes" ins Japanische übersetzt. NAKANO und Osamu SAKAI sind gegenwärtig Spezialisten für die Hegelsche Philosophie.

Bis jetzt sind noch einige andere Bücher und viele Aufsätze über Hegel erschienen. Meiner Ansicht nach hat die japanische Hegelforschung jetzt dasselbe Niveau wie das in Deutschland erreicht.

5. Friedrich Wilhelm Nietzsche

1. Was die Einführung der Nietzscheschen Philosophie in Japan betrifft, so ist es ungewiss, ob ungefähr 1895 entweder KOEBER an der Universität Tokio erstmals eine Vorlesung über Nietzsche gehalten hat oder ob INOUE 1900 nach der Heimkehr aus Deutschland über die Beliebtheit von Nietzsche in jenem Land gesprochen und seinen Studenten die Nietzscheforschung als Dissertationsthema gegeben hat. Aber es steht fest, dass ein Aufsatz über Nietzsche erst 1898 in einer Zeitschrift erschien:

Seichi YOSHIDA: Nietzsches Philosophie. 1898.

Dann erschienen Aufsätze von Chikufu TOBARI (1873–1955) und Chogyu TAKAYAMA (1871–1902), die erstmals eine bestimmte Ansicht zu Nietzsche vertraten:

Chikufu TOBARI: Über die gegenwärtige Literatur in Deutschland. 1900.
Derselbe: Über Friedrich Nietzsche. 1900.
Chogyu TAKAYAMA: Literat als Kulturkritiker. 1900.

TOBARI und TAKAYAMA stimmen darin überein, dass sie Nietzsches Gedanken vom Übermensch als wertvoll ansehen, denselben Gedanken als Darstellung der Persönlichkeit Nietzsches betrachten, und dass sie Nietzsches Theorie vom Genie als den Charakter des radikalen Individualismus ansehen. Das war die verbreitete Nietzsche-Interpretation der damaligen Intelligenz. Nietzschestudien von Philosophen erschienen erstmals 1902:

Genyoku KUWAKI: Allgemeine Übersicht zu Nietzsches Ethik 1902.

KUWAKI bewundert Nietzsche nicht leidenschaftlich wie TOBARI und TAKAYAMA. Indem er den Grundriss von "Zarathustra" scharf analysiert, sieht er Nietzsches Gedanken in bezug auf die systematische Philosophie als sehr unvollständig, seine Einsicht als einseitig und sein Erlebnis als beschränkt an. Nietzsche wurde von ihm wissenschaftlich nicht hoch geschätzt.

Das war die frühe Nietzsche-Interpretation ungefähr im Jahre 1900, als Nietzsches Philosophie erst in Japan eingeführt wurde. Dagegen erschien 1913 eine streng wissenschaftliche Studie von Tetsuro WATSUJI, die eine epochemachende Arbeit genannt werden kann.

Tetsuro WATSUJI: Nietzsche-Studien. 1913.

Es gab damals in Deutschland keine bestimmte Nietzscheforschung. WATSUJI, der erst 24 Jahre alt war, wollte eben durch die direkte Berührung mit Nietzsches Werken das Wesen derselben erfassen. Nach seiner Interpretation besteht Nietzsches Gedanke in dem Willen zur Macht, der durch die Negation des

Menschlichen und durch die Verzweiflung des Lebens gewonnen werden sollte. Der, der die Grösse des Lebens zeigen will, solle durch die Überwindung des individuellen Willens zum Standpunkt des übermenschlichen Willens zur Macht kommen. Das wurde zur damaligen allgemeinen Nietzsche-Interpretation.

Jiro ABE (1883–1959), der eine ähnliche Ansicht wie WATSUJI vertrat, war ein führender Denker der damaligen Intelligenz, die sich selbst bilden wollte.

Bis zum zweiten Weltkrieg erschienen Nietzsches sämtliche Werke zweimal in japanischer Übersetzung: 1916–29 wurden sie in 10 Bänden im Shincho-sha Verlag herausgegeben und 1942–56 in 12 Bänden im Nihonhyoron-sha Verlag. Vor dem zweiten Weltkrieg erschienen allmählich mehr als 20 Bücher und 50 Aufsätze über Nietzsche.

Nach dem zweiten Weltkrieg hat die Nietzscheforschung in Japan wie in Deutschland sehr grosse Fortschritt gemacht. Erstens ist zu bemerken, dass 1950, d.h. 50 Jahre nach seinem Tod, Sondernummern von *Das Ideal* und der *Philosophischen Zeitschrift* herausgegeben wurden. Zweitens gab es in der zweiten Hälfte des 20. Jahrhunderts die Tendenz, durch den Einfluss der Heidegger- und Jaspersforschung den Nihilismus bei Nietzsche zu sehen. NISHITANIs Einsicht scheint mir am wichtigsten:

Keiji NISHITANI: Der Nihilismus bei Nietzsche. 1951.

NISHITANI zeigt zwei Seiten des Nihilismus auf. Erstens wird der Nihilismus an sich in der allgemeinen Bedeutung als Langweiligkeit, Ermüdung des Lebens oder Schwäche des Willens zum Leben angesehen. Zweitens ist er für sich der Wille zum Nichts, wo die Schwäche des Willens zum Leben auf sich reflektiert und sich selbst verfolgt. Wenn man in den zweiten Nihilismus aufgehe und sich seines Standpunkts bewusst werde, erscheint der Nihilismus in der strengen Bedeutung des Wortes, d.h. der sich selbst bewusste Nihilismus. Eben das sei Nietzsches Nihilismus.

Von 1950 bis heute wurden Nietzsches Werke viermal herausgegeben. 1. Nietzsches Werke in 6 Bänden. Mikasa Verlag. 1950–51. 2. Nietzsches Werke in 12 Bänden. Shincho-sha Verlag. 1950–53. 3. Nietzsches Werke in 23 Bänden. Kadokawa Verlag. 1950–. 4. Nietzsches Werke in 17 Bänden. Riso-sha Verlag. 1962–.

Es finden sich heute etwa 60 Bücher und 130 bemerkenswerte Aufsätze über Nietzsche. Die folgenden Titel sind sehr wichtig:

Bücher

Tasuku HARA: Nietzsches Weltanschauung. 1950.

Shozo SHINTA: Nietzschestudien. 1956.

Denzaburo YOSHIZAWA: Pascal und Nietzsche. 1964.

Aufsätze

Masaaki KOSAKA: Die Probleme von Leben, Kultur und Existenz bei Nietzsche. 1948.

Eiho KAWAHARA: Übermensch, Existenz. 1953.

Shozo SHINTA: Der Nihilismus und die ewige Wiederkunft. 1961.
Derselbe: Nietzsche als Existenzphilosoph. 1961.

6. EDMUND HUSSERL

1. 1911 machte der japanische Philosoph Kitaro NISHIDA in einem Aufsatz Husser erstmals bekannt:

Kitaro Nishida: Über die Lehre der Schule des reinen Logismus in der Erkenntnistheorie. 1911.

In diesem Aufsatz stellte NISHIDA die Gedanken des Neukantianismus und Husserls vor und kritisierte sie von seinem Standpunkt aus. Husserl wurde von den damaligen japanischen Denkern, die schon Kant und Hegel untersucht hatten, schnell aufgenommen und interessierte wegen seiner Denkweise viele Studenten. Junge Philosophen hofften, bei Husserl zu studieren. Z. B. waren Hajime TANABE 1925 bei Husserl, Tokuryu YAMANOUCHI 1923, Satomi TAKAHASHI 1925 und Risaku MUDAI 1926 u.a. Sie alle wurden später führende Denker in Japan. Husserls Philosophie wurde von ihnen genau erforscht und in Japan eingeführt. Ihre Bücher über Husserl erschienen um 1930 und spielten eine führende Rolle:

Tokuryu YAMANOUCHI: Einführung in die Phänomenologie. 1929.
Satomi TAKAHASHI: Die Phänomenologie Husserls. 1931f.
Yukichi SHITAHODO: Husserl. 1936.
Risaku MUDAI: Studien zur Phänomenologie. 1940.

Zur selben Zeit wurden einige wichtige Schriften von Husserl ins Japanische übersetzt: "Philosophie als strenge Wissenschaft", "Ideen zu einer reinen Phänomenologie und phänomenologischen Wissenschaft" u. a.

Im Zusammenhang mit japanischen Kollegen hat Husserl selbst 1919–20 in der Zeitschrift *Neugestaltung* folgende Aufsätze veröffentlicht:

E. Husserl: Erneuerung, ihr Problem und ihre Methode. 1923.
Derselbe: Erneuerung als individualistisches Problem. 1924.
Derselbe: Die Methode der Wesensforschung. 1924.

Die Herausgabe der *Husserliana*, die 1950 begann, leitete in Japan wie auch in Deutschland eine Epoche der Husserlforschung ein. Husserl wurde damals besonders geachtet. Yoshihiro NITTA ist heute als führender Denker tätig. Er war zeitweise im Husserl-Archiv der Universität Köln und ist Berater des "World Institute for Advanced Phenomenological Research and Learning". Sein Buch über Husserl ist klar und überzeugend:

Yoshihiro NITTA: Was ist Phänomenologie?. 1968.
Derselbe: Die Phänomenologie. 1978.

Im ersten Buch erklärt er die gegenwärtige Lage der Phänomenologie Husserls und betrachtet die Probleme vom Bewusstsein und Sein, von Ding und Welt, von Wissenschaft und Erfahrung u.a.

Neben ihm sind Hirotake TATEMATSU und Gen KIDA als führende Husserlforscher zu nennen:

Gen KIDA: Die Philosophie der Gegenwart. 1969.
Derselbe: Die Phänomenologie. 1970.

In der Zwischenzeit wurden viele Schriften Husserls ins Japanische übersetzt. "Logische Untersuchungen", "Kartesianische Meditationen", "Die Idee der Phänomenologie", "Vorlesungen zur Phänomenologie des inneren Zeitbewusstseins", "Die Krisis der europäischen Wissenschaften und die transzendentale Phänomenologie" u.a.

Es ist anzumerken, dass 1979 die Japanische Phänomenologische Gesellschaft von NITTA und TATEMATSU gegründet wurde, wo seitdem jährlich eine Tagung abhalten wird.

7. Martin Heidegger

1. Was Untersuchungen über Heidegger betrifft, gibt es heute etwa 40 Bücher und fast tausend Aufsätze. Das sind die meisten Arbeiten nach der Kantforschung, und übertrifft die Hegelforschung vollends, was das sehr grosse Interesse der Japaner an Heidegger beweist. Heideggers Denken hat grosse Ähnlichkeit mit dem japanischen Denken, und besonders mit dem Buddhismus. Darum interessierten sich für Heidegger, wie mir scheint, nicht nur Philosophen sondern auch Literaten und die Intelligenz. Viele Japaner neigen dazu, Heidegger im Zusammenhang mit dem Buddhismus zu verstehen.

Heideggers Philosophie wurde hauptsächlich von japanischen Denkern, die bei ihm studiert oder ihn persönlich kennengelernt hatten, in Japan eingeführt. Hajime TANABE, der 1923 in Freiburg Husserl und Heidegger kennenlernte, machte erstmals 1923 der japanischen Gedankenwelt Philosophie Heideggers bekannt:

Hajime TANABE: Eine neue Wendung der Phänomenologie.
Heideggers Lebensphänomenologie. 1923.
Derselbe: Erkenntnistheorie und Phänomenologie. 1925.

Kiyoshi MIKI, der 1923 bei Heidegger in Marburg studierte und seine Freundschaft gewann, bekannte, dass Heidegger das ganze Leben hindurch stark auf sein Denken eingewirkt hatte. Seine Aufsätze über Heidegger sind sehr wichtig:

Kiyoshi MIKI: Grundbegriff der hermeneutischen Phänomenologie. 1927.
Derselbe: Die Ontologie Heideggers. 1930.

Shuzo KUKI und Tetsuro WATSUJI, die während ihres Aufenthalts im Deutschland Heideggers Philosophie studierten, haben unter dem Einfluss derselben ihr Denken entwickelt:

Shuzo KUKI : Struktur des "IKI". 1930.
Tetsuro WATSUJI: Ethik als Wissenschaft vom Menschen. 1930.

Da Heidegger auf diese Weise die japanische Intelligenz stark interessierte, erschienen bis zum zweiten Weltkrieg schon etwa 6 Bücher und 90 Aufsätze über ihn. Aber die meisten Untersuchungen, die damals noch unreif waren, sahen Heideggers Gedanken als Anthropologie an und legten das Gewicht ganz und gar auf die Fundamentalontologie. Davon sind nur zwei Bücher wertvoll:

Eiichi KITO: Die Ontologie Heideggers. 1935.
Hideyoshi NAKAGAWA: Heidegger. 1943.

2. Nach dem zweiten Weltkrieg, wurden Heideggers Werke in 24 Bänden von 1952 bis 1968 im Riso-sha Verlag herausgegeben. Zugleich wurden viele Schriften von Heidegger einzeln ins Japanische übertragen. Übersetzer sind Seishiro OE, Sadao HOSOYA, Tomio TEZUKA, Koichi TSUJUMURA, Tsutomu KUWAKI, Yutaka KAMEI, Atsuya KAKIHARA, Yoshio KAYAMA u.a. Dadurch wurde die Heideggerforchung in Japan so gefördert, dass sich das Wesen von Heideggers Philosophie leichter erfassen liess und ein genaues Heidegger-Bild entstand. Wertvolle Bücher über Heidegger sind:

Goichi MIYAKE: Die Philosophie Heideggers. 1950.
Jiro WATANABE: Heideggers Denken zur Existenz. 1962.
Derselbe: Heideggers Denken zum Sein. 1062.
Koichi TSUJIMURA: Heideggerstudien. 1971.
Yoshio KAYANO: Die Entwicklung der Philosophie beim frühen Heidegger. 1972.

Die Heideggerforchung in Japan hat, meiner Ansicht nach, dasselbe Niveau wie in Deutschland erreicht. Ja, japanische Denker übertreffen in der Heidegger-Interpretation sogar die deutschen Denker.

8. Karl Jaspers

Man sagt, dass das Interesse an Jaspers' Denken und der Einfluss desselben in Japan am grössten in der Welt sind. Das kommt daher, dass Jaspers' Philosophie mit dem Buddhismus sehr verwandt ist. Jaspers hat in seiner Jugendzeit grosses Interesse an Buddhismus gehabt und denselben vor allem während des zweiten Weltkriegs studiert, und daher später Buddha und Nagarjuna in seinem Werke: "Die grossen Philosophen" klar dargestellt. Die Grundstruktur seines Denkens ist eigentlich buddhistisch. Vor allem seine Lehre vom "Umfassenden" hat ganz denselben Charakter wie im Buddhismus. Dieser Zusammenhang wurde erstmals durch Hajime TANABE hervorgehoben. TANABE stellte 1938 fest, dass sich Jaspers aus der Tradition der europäischen Philosophie herausgehend zur buddhistischen Anschauung der Nichtigkeit näherte. Darin stimmen heute alle Jaspersforscher überein.

So wird Jaspers nach wie vor intensiv erforscht. Die Jaspersforschung in Japan scheint jetzt sowohl ihrer Qualität als ihrer Quantität nach die Forschung in allen Ländern zu übertreffen. Unter etwa 20 Bücher über Jaspers, die bis heute erschienen, sind folgende hervorzuheben:

Saburo SUZUKI: Die Existenzphilosophie von Jaspers. 1947.
Derselbe: Jaspersstudien. 1953.
Takezo KANEKO: Die Philosophie der Existentialvernunft bei Jaspers. 1953.
Takeo SAITO: Die Struktur und Entwickung des absoluten Bewusstseins bei Jaspers. 1961.
Derselbe: Jaspersstudien. 1962.
Masao KUSANAGI: Die Grundprobleme der Existenzphilosophie. 1962.
Shinji HAYASHIDA: Die Existenzphilosophie von Jaspers. 1971.

Es gibt ungefähr 200 Aufsätze, die Jaspers' Philosophie unter verschiedenen Aspekten untersuchten. Die wichtigsten sind:

Setsuro KOMATSU: Die Philosophie der Existenz bei Jaspers. 1935.
Eiichi KITO: Die Existenzphilosophie von Jaspers. 1949.
Masaaki KOSAKA: Das Umfassende bei Jaspers. 1949.
Takezo KANEKO: Der Charakter der Existenzphilosphie bei Jaspers. 1950.
Yoshio KAYANO: Über den philosophischen Glauben bei Jaspers. 1951.
Takeo SAITO: Über die Kommunikation bei Jaspers. 1953.
Shinji HAYASHIDA: Jaspers' Lehre der Politik — Die auf Ethos gegründete Politik. 1961.

1955 wurden Jaspers' Werke in 26 Bänden im Riso-sha Verlag herausgegeben. Viele Schriften von Jaspers wurden einzeln ins Japanische übersetzt. Die Zahl der ins Japanische übersetzten Schriften von Jaspers ist grösser als in allen anderen Ländern. 1951 wurde die Japanische Jaspers-Gesellschaft gegründet, die seitdem jährlich eine Tagung abhält.

9. Leibniz, Schelling, Schopenhauer, Hartmann

Gottfried Wilhelm Leibniz

Die Hauptschriften von Leibniz werden in Japan zwar im grossen und ganzen studiert, aber es gibt kaum sreng wissenschaftliche Studien. Folgende Arbeiten sind bemerkenswert:

Torataro SHIMOMURA: Leibniz. 1938.
Makoto YAMAMOTO: Studien zur Philosophie von Leibniz. 1915.
Hiroshi NAGAI: Leibniz. 1954.

Heute sind wenigstens zwei Schriften von Leibniz ins Japanische übertragen: "Discours de métaphysique" und "La monadologie".

Friedrich Wilhelm Joseph Schelling

Im Zusammenhang mit dem deutschen Idealismus wird Schellings Philosophie zwar studiert, doch gibt heute kaum Spezialisten. Folgende Arbeiten sind wichtig:

Moriichi KATSUTA: Schelling. 1936.
Motomichi AKAMATSU: Schellingstudien. 1943.
Tomio NISHIKAWA: Studien zur Philosohie von Schelling. 1960.
Kenji FUJITA: Schelling. 1962.
Kiyoyuki YAMAMOTO: Die Philosophie der Schellingschen Freiheitslehre. 1970.
Makoto KONO: Studien zur praktischen Philosophie von Schelling. 1973.

Einige Schriften von Schelling sind ins Japanische übersetzt: "System des transzendentalen Idealismus", "Bruno", "Philosophische Untersuchungen über das Wesen der menschlichen Freiheit", "Vorlesungen über die Methode des akademischen Studiums", und "Darstellung des philosophischen Empirismus".

Arthur Schopenhauer

Schon in den zwanziger Jahren des 20. Jahrhunderts war Schopenhauer bei der japanischen Intelligenz bekannt und sehr populär. Interessante Schriften wurden daher oft ins Japanische übersetzt, z.B. "Die beiden Grundprobleme der Ethik", "Aphorismen zur Lebensweisheit". 1972–74 wurden Schopenhauers Werke in 15 Bänden in japanischer Übersetzung im Hakusuisha Verlag herausgegeben. Es finden sich nach wie vor aber kaum streng wisssenschaftliche Studien.

Nicolai Hartmann

Hartmann wurde hauptsächlich im Zusammenhang mit Kant und dem deutschen Idealismus studiert. Seine Aufsätze über Kant wurden von japanischen Kantforschern sehr hoch geschätzt und schon in den zwanziger Jahren des 20. Jahrhunderts ins Japanische übertragen. Vor allem seine "Ethik" wurde im Zusammenhang mit der Kantischen Ethik von Kant- und Hegelforschern sehr genau untersucht. Aber die Grundstruktur seines Denkens ist noch nicht genau erforscht worden. Es ist anzumerken, dass seine Hauptschriften schon in den vierziger Jahren des 20. Jahrhunderts ins Japanische übertragen wurden, und zwar "Zur Grundlage der Ontologie", "Möglichkeit und Wirklichkeit", "Der Aufbau der realen Welt, Grundriss der allgemeinen Kategorienlehre" und "Neue Wege de Ontologie". In den letzten zehn Jahren erschienen einige wichtige Aufsätze über das Wesen der Philosophie Hartmanns:

Masanori KUMAGAI: Die Ontologie von Nikolai Hartmann. 1970 u. 1973.

Political philosophy in modern Japan

MASATOSHI YOSHIDA
Hosei University, Tokyo

1. Japan's modernization and the role of intellectuals

This essay examines Japanese political philosophy in the modern age, that is, in the period since the Meiji Restoration of 1868. The first observation we may make is that works of political philosophy in modern Japan are neither systematic nor metaphysical, unlike Thomas Hobbes' *Leviathan* (1651), Hegel's *Philosophy of Right of Law* (1821), and other Western philosophical treatises. On the whole, the Japanese works are situational and practical, falling rather in the broad category of political discourse. This characteristic may be attributed first to the turbulent history of Japan's modernization and, second, to the fact that political theses were advanced not only by philosophers but by journalists, political scientists, and activist intellectuals. I shall discuss the ideas and activities of certain of these opinion leaders with a view to illustrating this peculiar brand of political philosophy in modern Japan.

First, let us consider what the main concerns of the intellectual elite in modern Japan were and how they dealt with these concerns. The century of Japan's modernization produced many outstanding thinkers and important groups of activist intellectuals. During the Meiji era (1868–1912), there were "enlightenment" thinkers, champions of freedom and popular rights, and socialists. In the 1910s and '20s there were the humanitarian literary group, Shirakaba-ha, and the liberalist proponents of the "Taisho democracy". In the prewar Shōwa period (1926–45), there were the ultra-nationalists and the materialists, and in the postwar era radical liberals, Marxists, and more recently a wide array of modernists. I do not intend here to try to deal with the question of how all these individuals or groups presented their ideals or sought to put them into practice in the context of historical reality. Nevertheless, I believe it is safe to say that the common theme that preoccupied members of the intellectual elite was modern Japan itself. And the problems concerned persons struggled with were: how to modernize Japan, how to evaluate Japanese modernity, and what program to adopt to achieve the ideal society.

What, to begin with, do we mean by "the modern age" in Japan? Essentially, "modernity" is what distinguishes a capitalist society from a feudal/traditional society; modernization may be described as the emergence of a capitalist economy and concomitant cultural and social change. The modernization process differed in every country where it took place, but nowhere else was it so complex a process and so drastic in the changes it wrought as in Japan. One of the features of Japan's

Contemporary philosophy. A new survey. Vol. 7, 377–395.

modernization was that it was sparked by external pressure from the Western powers, when the steam whistle of Commodore Matthew Perry's "black ships" shattered the country's centuries-long seclusion. In the latter half of the nineteenth century, the age of imperialism was well underway; remote but not unnoticed in the far corner of the East, Japan was a backward nation for which modernization could make the difference between independence and subjugation.

In the West, the modern age began when the bourgeoisie overthrew the feudal system, and the dynamic that lay behind those forces was the modern spirit and modern civilization. Inasmuch as modern civil society was built on the premise of anti-feudal ideals, it had to be based on principles of modern thought like freedom and democracy; there ideals, in turn, led to progress in science and technology and the flowering of material civilization. Because Japan was a late starter, however, modernization began as Westernization, but the process was not directed toward "democratization" or "the autonomy of the individual" through the internalization of progressive modern thought so much as toward "industrialization." Its goal was to catch up with the West by attaining the level of its material civilization. This is testified to by the drastic policies adopted to "enrich the nation and strengthen its arms" and "increase production and promote industrial development" thrust forward by the newly formed absolutist regime. The "civilization and enlightenment" movement along this line became the main axis of Japan's modernization.

The agony and struggle of the intellectual elite of modern Japan begins here. They found themselves caught up on the horns of a bewildering ideological dilemma: between popular rights and the authority of the state; between tradition and modernization; Japan and the West, Japan and the rest of Asian, *ad infinitum*. Modernization was being implemented from the top down at the initiative of absolutist authority under the emperor system; the problem that confronted most concerned intellectuals was: what course should Japan take to best serve the interests of the people?

Under the aegis of the absolutist regime, Japan opted to concentrate on the acquisition of Western material culture, becoming in the end an imperialist power that perpetrated wars of aggression all over Asia. The advent of the modern age in Japan meant civilization without democratization, and ended in authoritarian rule at home and imperialism abroad. This phase of modern history only came to an end with the demise of the fascist militarist regime at the end of World War II.

During the early phase of the Pacifics War, there were some groups of intellectuals — the Japanese Romantic school, the Bungakkai writers, and the Kyoto school philosophers — among whom "conguest of modernity" was the main theme of discussion. Their ultimate aim was to rationalize the war which has resulted from the modernization process begun in the mid-nineteenth century by overcoming Western modernity and formulating a Japanese alternative. Although they tried to avoid criticizing Japan's modernization program, their arguments laid bare its inherent contradictions. In a paper prepared for a round-table discussion on the theme "Conquest of Modernity" held in 1942, one participant wrote as follows: "Overcoming modernity means... superseding democracy in the political sphere, capitalism in the economic sphere, and liberalism in the realm of thought" (Suzuki

[10], p. 42). The overall aims of thinkers of this persuasion, while they might differ somewhat in nuance from person to person, revolved around the following: 1. To separate in modern Western civilization the materialistic aspect (technology, industry) and the philosophical aspect (liberalism, democracy, etc.). 2. To attribute contradictions in Japan's modernization effort to Western modernity. 3. To resolve the dilemmas and contradictions by overcoming modernity through the Japanese—East Asian—spirit. In the final analysis, the arguments of the advocates of "overcoming modernity" were a declaration of the bankruptcy of the entire process of Japan's modernization drive that had been aimed solely at modernization after the Western model. It was little wonder that Japan's modernization policy collapsed completely with the defeat, despite their effort to absolve and affirm it. The defeat, moreover, was not simply that of modern Japan; it was also the defeat of the modern Japanese intelligentsia. Of course, this does not apply to those who had maintained a firmly critical stance *vis-à-vis* modern Japan, but to the far greater number who had tried to compromise with reality and give it *ipso facto* approval.

So, after the end of the war, how did Japanese intellectuals evaluate the course of Japan's modernization and its catastrophic results? In the postwar period, the ideology and institutions of the prewar period were discarded and dismantled through the efforts of the people themselves, and the country began to build a new system dedicated to peace and democracy. Opinion leaders too—centering mainly around those of liberal or Marxist persuasion—devoted themselves to the creation of the new system. Their efforts contributed greatly in promoting popular understanding of the democratic rights that had been denied the people before the war. So, after the war, the ideals of peace and democracy were firmly set in place in the institutional, private and intellectual life of the nation. But it was not long, once the rehabilitation process had run its course, until in the 1960s, when the revived monopolistic capitalist state had rebuilt a powerful politico-economic monolith. Claiming that modernization had been attained (despite its fictitious nature), the monolith proceeded to weave a new and equally unyielding framework of state control over the populace. Clasped within this revised power structure, we are being pushed irrevocably away from the "modern" to the "post-modern" perspective. In this transition, great discord had broken out among our intellectuals over the issues of how to interpret contemporary Japan and what course of reform it should take in the future. The solidarity of democratic-minded opinion leaders has long been lost, and an increasing number are beginning to abandon their once-professed principles and even to withdraw completely from political activism.

Given these circumstances, we must seriously ask what really qualifies people to be intellectuals in our time, what issues they should confront and in what way. In order to fulfill this task, we need to know what main themes of discussion were addressed by the leading thinkers from the modern to the contemporary ages in this country. We need to know how they tackled these matters as issues of personal commitment.

I shall focus on three themes of discussion among thinkers and opinion leaders in modern Japan, citing the ideas of individuals who illustrate each of the three. The first theme was what direction Japan's modernization should take and how to achieve its goals. Fukuzawa Yukichi (1835–1901)* was a leading figure who confronted this problem as educator and journalist. Second, once the course for Japan's modernization had been laid out, how did our intellectuals respond to it given the severe historical circumstances in which the country found itself. We can see this prototype in the ideas of the nationalist philosopher, Watsuji Tetsurō (1889–1960). The third theme is how to achieve genuine democratization after the war. We may approach this theme through Maruyama Masao (1914–), leading liberal thinker of the postwar period. It is my hope to show the fundamental character and nature of modern Japanese political philosophy through an examination of the thought of these three important figures in intellectual history.

2. On the road to modernization

What were the guiding principles according to which Japan set out to modernize itself? Prominent among the thinkers who enunciated and put into practice those ideas was Fukuzawa Yukichi. Fukuzawa was active from the very beginning of the Meiji period as an exponent of the "civilization and enlightenment" movement and his writings and pronouncements testify to the urgent need at that time to build a basic framework for modern Japan. The development of Fukuzawa's thought, indeed, echoes the actual process through which modern Japan emerged.

> "It is said that Heaven created no human being higher or lower than another. This means that Heaven made people in such a way that all are of equal standing; they are born without distinction of high or low, noble or mean.... and they are completely free, causing no interference to others and each living in peace and tranquility" (Fukuzawa [4], p. 29).

Gakumon no susume (Encouragement of Learning; 1872–76), which begins with this passage, was probably the first work to introduce in Japan the Western concept of democracy and liberalism. People welcomed his discussion of the inherent rights of man and the equality of all people as the enlightened thought that would usher in a new era. Two decades later, when the Sino-Japanese War (1894–95) broke out, Fukuzawa observed:

> "War has broken out between Japan and China, but if we examine the root cause of the conflict, we can see that it is a clash, not so much between the two countries as between the forces striving for civilization and enlightenment and the forces determined to repel them ... That some distasteful violence should develop in the process of eliminating obstacles to the progress of civilization in the world is inevitable ..." (Fukuzawa [5], pp. 491–92)

* Surnames are given first for all Japanese names referred to in this paper, in accordance with Japanese practice.

These lines clearly show Fukazawa as a defender of imperialist Japan. The pursuit of civilization and enlightenment as he saw it led him ultimately to affirm war as an unavoidable and inevitable result.

So we are led to wonder: what did Fukuzawa mean by "civilization" and what was "enlightenment" to him? Or, what did they have to be?

The answer to these questions will help reveal the nature of modern Japan as the logical consequence of Fukuzawa-style enlightenment. It is generally accepted that Fukuzawa underwent a conversion in the early 1880s, shifting from champion of popular right (*minken*) to leader of the authority of the state (*kokken*). I would question, however, whether it was really a conversion. Fukuzawa's thought was focused unswervingly on the modernization of the country and the advance of civilization. His view of civilization is crystalized in his early manifestoes, *Encouragement of Learning* and *Bunmei-ron no gairyaku* (Outline of Civilization, 1875). An examination of the ideas on civilization set forth in these works will provide a better picture of the nature of his thought.

In general, Fukuzawa associated civilization with modernization modelled after the West, and he presented specific policies and programs for achieving that goal. Some see his way of thinking as basically pragmatic, but in the overall orientation of his thought, he was far from flexible. His goal was firmly fixed. He introduced the concept of dichotomy in order to clarify whether the goal should be "civilization" or "barbarism" by saying "Light or heavy, long or short, good or evil, right or wrong—all these words stem from dichotomous thinking ... I propose to make what has proven important and good the basis of argument" (Fukuzawa [1], p. 9).

Fukuzawa distinguished between "barbarian," "semi-civilized," and "civilized" in the historical development of civilization. The civilization process, however, was unending, and he believed that by the standards of the future, the stage of advancement of civilization in Western countries of his time would be considered quite backward. Nevertheless, he accepted the West as the model for civilization for lack of a better alternative at the time. Those who seek to promote the progress of civilization of their country must "make Western civilization their goal and center their discussion on it", he said (Fukuzawa [1], p. 19). Flexibility in his thinking appears only as he sought concrete ways to achieve the ultimate goal of modernization on the Western model.

How should the goal of creating a society as advanced and civilized as the West be realized? *Encouragement of Learning* begins with the notion that all people are equal. If there is a distinction among people, he says, it is in the matter of "learning". So, at first, he posited learning as the very core of civilization. Now, let us examine what Fukuzawa meant by learning. He did not mean learning in the sense taught by the Confucianists or the National Learning scholars, but learning that was more or less universal: the practical learning developed in the West. His abhorrence for feudal Japan — "the feudalistic status hierarchy and factionalism based on family connections are my sworn enemies" and "It is as foolish for a loyal retainer to commit suicide for the sake of his lord as for a servant to kill himself for having lost household funds while on an errand" — led him to deny tradition in general, and to reach out for the new and the Western.

What was at the core of Western practical learning that Fukuzawa considered necessary for his country? It consisted of two things, as he points out in his autobiography: "A comparison of the Confucian tradition of the East and the civilization of the West shows us that the East is lacking in two things: in the tangible realm, science, and in the intangible realm, the spirit of independence" (Fukuzawa [3], p. 167). On this point, Fukuzawa's thought remained basically unchanged through the last years of his life. At the outset, moreover, he emphasized that the spirit of civilization had to be well established before a suitable form could be created (Fukuzawa [1], p. 21). Indeed, for Fukuzawa the core of civilization lay in this "spirit of independence". However, it is apparent that he saw independence, or the spirit of civilization, as having a layered structure. For the people, for example, he describes it as no more than an abstract goal; in explaining independence in the context of the relationship between the government and the people, or between one nation and another, he is somewhat more specific. He talks of national independence in terms of the cooperative effort of government and the people, or of the duties of the people to the state and the law of the land.

In *Encouragement of Learning*, he defines human equality as "equal, rights and justice". He begins with a discussion of the relationship of the farmers and townsmen to the samurai of feudal Japan and explains the proper relationship between government and people as follows: "The government and the people... differ only in the degree of power; they are not different in respect to rights" (Fukuzawa [4], p. 39). The real difference is in "duties", and the people have no need to feel indebted to the government. However, "the people are bound by a contract with their government to abide by its laws". (Fukuzawa [4], p. 40) Indeed, he said, that is their duty as citizens. Oppressive government in a country is not only caused by tyrants and tyrannical officials but by people's ignorance of their duties. Here is where learning comes in; it is necessary for people to learn their duty.

The spirit of independence, therefore, is the goal of learning. This viewpoint is even more clearly explained in *Outline of Civilization*, which argues for "civilization by and for the state". We may note that he expressed these views around the time the freedom and popular rights movement (*jiyū minken undō*), calling for the creation of a parliament, was beginning. Here again, Fukuzawa saw no contradiction in shedding the "attachment to old customs and practices" and adopting the Western "spirit of civilization" on the one hand and maintaining and strengthening the national polity with the emperor at its apex on the other. On the contrary, he believed it to be a basic factor in the process of acquiring Western civilization that the national polity based on the "imperial line" "continued unto this day unbroken from time immemorial" be preserved. He writes, "Since nothing can strengthen the national polity and increase the glory of our imperial tradition more than the civilization of the West, how can we hesitate to adopt it? We must by all means adopt Western civilization" (Fukuzawa [1], p. 33). The approach to enlightenment Fukuzawa espoused did not conflict with the establishment of an absolutistic system. His encouragement of learning was a means of improving the intellectual caliber of the people, the agents of change entrusted to carry on the task of building such a civilization. In dealing with the question of national independence,

Fukuzawa's views on civilization were even more graphic. The whole of *Outline of Civilization* is, in a sense, focused on the topic of the autonomy of the state.

> "The concern the people of this country show today about their nation's independence testifies to the facts that Japanese civilization is just at a level where such concern is warranted, and that the spirit of civilization has spread to the point where the people's attention is focused solely on this problem, to the exclusion of other considerations" (Fukuzawa [1], p. 183).

Fukuzawa's perspective on civilization converged closely with the matter of independence; in fact it was relevant only insofar as it served the interests of national independence. It is true that he viewed the concept of the state in relative terms. For example, citing the Christian "brotherly love for all people on the earth", he pointed out the absurdity of drawing national boundaries. However, he allowed realism to take precedence for the time being, saying, "When we look at what is going on in the world today, we can see that nations are being built everywhere and governments are necessary in building nations... Though this may be contrary to the religious spirit, it is unavoidable given the forces of power in the world" (Fukuzawa [1], p. 190). And thus he puts aside the idealism of brotherly love, and goes on to make a sharp distinction between religion as a "personal ethic" and the "spirit of nation building and independence". It is clear that he was far more concerned with how to assure Japan's independence than with how to create an ideal society.

What underlay Fukuzawa's realism — or perhaps we should say his sense of crisis — *vis-à-vis* national autonomy was his astute perception of world affairs. He saw how the Western powers had hot stopped at simply exporting goods overseas; they had already entered the imperialist stage of development, establishing colonies abroad and investing large sums of capital there. He noted that their schemes were directed at "the backward, undeveloped countries"; Japan, too, had been forced to sign unequal treaties with the Western powers. He gave examples of their infringements on the rights and national independence of the indigenous peoples of North America, and of Persians, Indians, Chinese, and so on. His theory of civilization is encapsulated in the following passage.

> "Our paramount task is to set a goal and proceed toward civilization. What is our goal? It is to draw clear destinctions between our country and the rest of the world, and maintain our national independence... Therefore our goal is national autonomy, and the spread of civilization among the people is the means of achieving this goal" (Fukuzawa [1], p. 207).

Here he does an about-face from his position advocating "independence of the individual and autonomy of the nation" which we find in *Encouragement of Learning*, relegating the former to a mere means of achieving the latter. Fukuzawa's preoccupation with nationalism and state authority is made all the clearer by his criticism of "those today who advocate people's rights and equality", that is, of the abstract arguments and unrealistic view of the world among freedom and popular rights activists and advocates of a national diet (parliament). For Fukuzawa, civilization meant Westernization, which in turn was but a means of preserving Japan's independence. He encouraged, nature, especially such intangible learning as would foster the spirit of independence. Once it was established that intangible learning was intended for the purpose of acquiring civilization and enlightenment, the next concern was how to put "tangible learning", i.e., the sciences, to practice to build a strong nation. From the outset, Fukuzawa was little interested in the establishment of people's rights and this made his philosophy of enlightenment well tailored as the mainstay of the Meiji ideology underlying the policies for "enriching the nation, strengthening its arms" and for "increasing production and promoting industrial development". But Fukuzawa was wise enough to qualifcy his statements on civilization. "The civilization I have described is not civilization in its true sense. The first step is to secure national independence; then we can move on at some later stage to true civilization" (Fukuzawa [1], p. 209).

Outline of Civilization was published in 1875, and the subsequent years brought several new historical developments — the upsurge of the freedom and popular rights movement, the development of the dispute over the Korean peninsula, and other foreign affairs issues. The tasks Fukuzawa had reserved for the future were soon rendered totally unattainable. Indeed, Fukuzawa himself was drawn further and further away from the "civilization in its true sense" as his arguments for unified effort on the part of the government and the people in enhancing national authority gave way to outright imperialism.

And thus did Fukuzawa ultimately end up an apologist for Japan's imperialist aggression in Asia. In his famous essay "Datsu-A ron" (Dissociating from Asia), published in 1885, he took the stand that Japan "has already pulled itself out of the shell of Asian conservatism and shifted toward Western civilization". And, he goes on, "Japan cannot afford to wait for its neighbors to awaken to civilization and work together with them to revive Asia. It should dissociate from other Asian countries and move along with the civilized nations of the West... and deal with its neighbors as the Western nations do" (Fukuzawa [2], pp. 239–40). The ultimate conclusion of his theory of modernization on the Western model had transformed itself into vindication of Western-style imperialism.

Fukuzawa admittedly struggled with a most perplexing theme — how his diminutive, vulnerable country, located in a remote corner of Asia, should build a modern state in the latter half of the nineteenth century while being pressed upon on all sides by Western nations that had already entered the imperialistic stage of development. It is not surprising that his conclusion was that Japan had to establish firm national authority so that it could attain a strength on a par with the Western powers. To the intangible aspects of Western learning, like the idea of popular rights which he had first thought would play a central role in the growth of civilization, he later added the parallel concept of national authority, and by this means he found it easy to relegate popular rights to a subordinate role.

Fukuzawa had witnessed the strength of Western countries first hand during two trips abroad in his youth. Watching events unfold in Japan, he felt that the very survival of the nation was at stake. For him, therefore, maintaining national integrity and autonomy was a matter of paramount concern, and he was convinced that there was no other way to achieve this than through Westernization. The very gravity of the task meant that government and people, public and private sector — i.e., everyone — had to work together in a concerted effort.

Fukuzawa's thought is imbued throughout with a spirit of progressiveness and independence. Nevertheless, while recognizing that adoption of advanced civilization was vital for the state as it faced the imminent threat of subjugation, and though he had firm ideas about the need to rid the country of old customs and backward practices, he did not ask himself that basic question: what kind of modernization would be best for the Japanese people? As a result, he never saw the expansion and consolidation of freedom and popular rights for the people as an end in itself. He called for the education of the people, but only in the hope that they would thereby qualify as members of an ideal, Western-style society. Ultimately, the Westernization of Japan as an ideal society was carried out only in line with the "enrich the nation and strengthen its arms" and "increase production and promote industrial development" slogans and later the policy of imperialism.

3. Under the absolutist emperor system

The leaders of modern Japan under the nationalist policy espoused by Fukuzawa firmly established an absolutist regime centered on imperial authority and pushed steadily in the direction of militarism and imperialism. These tendencies became extreme in the 1930s and '40s as Japan, having joined the ranks of the fascist nations, indulged in oppression at home and plunged deeper into wars of aggression in Asia and armed hostilities with other imperialist powers. The end, of course, was the catastrophic defeat at the end of World War II. During this age of crisis, how did thinkers and opinion leaders in Japan react?

In addressing this question, let us consider Watsuji Tetsurō, leading Japanese philosopher of the establishment school, who took a generally transcendental stance during this period. Watsuji was exceptional in that he attempted to describe the culture and ethics peculiar to Japan using Western philosophical concepts and methodologies. Through these efforts, he contributed to the maintenance of the absolutist emperor system.

Watsuji graduated from the philosophy department of Tokyo Imperial University in the last year of the Meiji period (1912). From this time on into the Taishō period (1912–1926), Japanese philosophers were eagerly absorbing and building upon the neo-Kantian philosophy of the Baden, or southwest German, school. By the Taishō period, Japan had already emerged as an imperialist power, albeit a late starter, and had entered a relatively stable period of capitalist development. Intellectual currents at the time reflected this phase, and the center of discussion began to shift from the nationalism of the Meiji period to individualism and liberalism. These changes paved the way for the acceptance of neo-Kantian philosophy as a philosophy of values and culture. There were movements to "safeguard constitutional government" and to gain universal suffrage among the people, and against this backdrop the "Taishō democracy" movement among the intellectual elite unfolded.

World affairs underwent an abrupt change with the stock market crash of 1929, and Japan in the early Shōwa period (1926–1989) began to move down the path to a fascist state leading ultimately to war. Watsuji accepted a teaching position at Kyoto Imperial University in 1925, just as the reign of the new Shōwa emperor was beginning. As a professor of ethics at Kyoto and later at Tokyo Imperial University, he devoted himself to the development of a nationalistic system of ethics as befitted his position as leader of academic philosophy in Japan.

Thus, it was following his appointment to the faculty of Kyoto Imperial University that Watsuji seriously set about the task of constructing a Japanese system of ethics. His most important publications of this period were *Fūdo* (Climate), *Ningen no gaku to shite no rinrigaku* (The Significance of Ethics As the Study of Man), and his masterpiece, *Rinrigaku* (Ethics). The first of these consisted of a discussion of specific Japanese traits, while the latter two were theories of moral philosophy that generalized and applied the ideas of the first. All three were attempts to establish a new system of ethics by providing a theoretical foundation for a peculiarly Japanese morality and pattern of human relations.

In 1927, Watsuji travelled to Germany, where he studied for about one year and a half. This sojourn provided the inspiration for writing *Fūdo*, published in 1935. The book took shape chiefly around ideas that were sparked by the experience of unfamiliar climates encountered during his voyage by sea to Germany and his reading of Martin Heidegger's *Being and Time* upon arrival in Berlin. In its pages, climate is defined as the structural moment of human existence. As such, it is not a "natural environment", but a representation of subjective human existence. (Watsuji [77], p. 1) Whereas Heidegger saw time as the framework for human existence, Watsuji attempted to shed light on time and historicity by focusing on space instead. But the methodological underpinnings of his approach were none other than phenomenology in the tradition of Edmund Husserl and hermeneutics of the Heideggerian school.

The first question posed in *Fūdo* is whether climate can simply be viewed as a natural phenomenon. Watsuji decides that in the relationship between humankind and climate, coldness and heat do not have independent being, but are known to exist only through sensation and as such are "intentional experiences". He then goes on to state that it is not "I" but "we"—the "we" of interpersonal relationship—who experience these aspects of climate (Watsuji [11], p. 10).

Based on this theory of climate, Watsuji divides the world's climatic types into three broad categories labeled "monsoon", "desert", and "pastoral". The first refers to the climate typical of the monsoon region including Japan, the Asian continent (China, India, etc.) and the Indian Ocean area; the second, to that of the desert region encompassing the Arab countries, Africa, and Mongolia; and the third, to that of the grassy meadows of Europe. The monsoon-type climate is characterized by an extreme humidity resulting from a combination of heat and moisture carried into the region by the periodic winds of summer. Watsuji describes the personality structure of the people who inhabit the monsoon region as "receptive/submissive" (Watsuji [11], p. 26). The desert is harsh and lifeless, and here Watsuji sees the chief characteristic of the inhabitants' mode of existence as "aridity", (Watsuji [11], p. 48) implying a confrontational and combative relationship between human beings and the world, as well as between fellow humans. The pastoral-type climate is a combination of humidity and aridity, which Watsuji sees as characterized by "submissiveness to nature"; climate here, he claims, is both "natural and rational" (Watsuji [11], p. 64).

Having thus derived three patterns of culture, Watsuji proceeds to analyze two subtypes of the monsoon-type climate, China and Japan. By so doing, he attempts to elucidate the "distinctively Japanese mode of existence." His conclusion was that the "passive-receptive" character of the monsoon region inhabitant was in Japan compounded by "tropical/arctic" and "periodic/spasmodic" dimensions. The national character of the Japanese could be thus described as "quiet passion and combative indifference" (Watsuji [11], p. 138).

Working from this definition of character type, he deduces the importance of the *ie*, family or household, in human relations in both Japan and China. He contraposed the *ie* to the husband — wife unit in the European *polis* and the tribe in desert cultures. Europeans see human beings as individuals, but individual distinctions are eliminated within the *ie*. The preponderance of the *ie* is reflected even in the structure of dwellings. That the *ie* is strongly symbolic of the subordinate relationship of the individual to the whole is extrapolated to the peculiarities of the Japanese people's mode of existence as a nation. According to Watsuji, the totality of the Japanese nation had traditionally been understood in religious, or mythological, terms, but this communal bond, somewhat akin to a religious order, gave rise to "a national community fused together by emotional identity in which awareness of the individual is unnecessary" (Watsuji [11], p. 149). In this union of the people into a quasi-religious fraternity, he saw the basis of Japanese morality. As one of the most prominent features of this thought, Watsuji posits "reverence for the emperor" that historically symbolizes "devotion to the cult of totality".

Watsuji's approach in *Fūdo* was to try to interpret human existence in climatic terms, drawing heavily upon the methodologies of modern European philosophy. Concretely speaking, the treatise was the author's particularistic definition of the cultural and spiritual realities of Japan and East Asia. But even though he may have succeeded in understanding man *in the context of* climate through a hermeneutical methodology that makes climate the framework for human existence, he did not see man *in opposition to* climate. As a consequence, perception (interpretation) of individual climatic phenomena is equated with reality, and the particularistic definition of climatic types becomes a determinant of the lives and thought of a given people. Therefore, the philosophy expounded in *Fūdo* becomes a climatic fatalism, in which political and economic structures and other structural factors of human existence are completely disregarded, as though "we should grasp the meaning of the fate of having been born in a certain climate and love that climate" (Watsuji [11], p. 204).

The first work in which Watsuji presented his unique system of Japanese ethics was the abovementioned *Significance of Ethics As the Study of Man*, published in 1934. In it, he strove to establish the basic concepts and methodology of his moral philosophy, and, indeed, it provides an outline of his system of ethics. As in *Fūdo*, Watsuji employed a thoroughgoing hermeneutical methodology, beginning his discussion with a consideration of the meaning of such words as *rinri* (ethics) and *ningen* (human being). When attempting to define *rinri*, he begins by examining the significance of the *substance* that is represented by the word (Watsuji [14], p. 7).

The word *rin* (J. reading) was derived from Chinese and has the same meaning as the indigenous Japanese word *nakama*. *Nakama* can indicate a relationship—being *between people*—on the one hand, as well as the *people* party to a certain relationship on the other. Thus, Watsuji defines ethics not as individual consciousness aloof from communal ties, but as "a field of study that attempts to clarify the fundamental order and rationale of human relations and, hence, of the human community" (Watsuji [14], p. 13). He next points out that the Japanese word *ningen*, though usually translated as "human being(s)" or "people", implies "society" or "the world" as well. To Watsuji, it is highly significant that the Japanese word for man has this additional connotation. The dual meaning of the word *ningen* is a product of the "entirety of Japanese history". The fact that *ningen*, which means "humankind as a whole" or "the world", can also mean an individual "person" is also attributable to the "dialectic relationship between the whole and its parts" (Watsuji [14], p. 19). Watsuji notes that the German and French words for "people" do not have this dual character.

How are we to evaluate these basic concepts of Watsuji's ethics? His method is based throughout on an analysis and semantic interpretation of elements of the Japanese language. But analysis of words that have to do with ethics is not the same as the analysis of ethical realities, and the ethical import of the words that can be used within the confines of the Japanese language cannot be universal. Watsuji's moral philosophy is no more than an expression of his desire to present and rationalize a native Japanese system of ethics.

In any case, Watsuji, through these methods, was able to unite, internally, "human existence" as a rudiment with all social organizations, as "manifestations" of it. Let us examine the formative development from the individual to the nation and the nature of the nation that is the terminal point of this evolutionary process in the third chapter of *Rinrigaku* (1937–49). To Watsuji, the "human organization" develops concentrically around the axis of the "family" as a "communal unit of people in close, day-to-day cooperation" (Watsuji [15], p. 432). The developmental process begins with the family, after which kinship groups, local communities, economic organizations, and cultural associations follow to form a hierarchical order. It is from these that the nation finally takes shape as a "human collectivity of all human organizations" that "is conscious of and assures their developmental sequence". But as long as all human organizations are expressions of human existence, this process is only an inevitable development with its basis in the inherent nature of these entities. This is because "human existence takes place as a human group and must rise to as high an order as possible insofar as this can be accomplished in accordance with a humane order. We must recognize the significance of this human existence as *something sacred*" (Watsuji [15], p. 617).

We may ask what is the nature of the nation of which Watsuji speaks. In his words, it is a "thoroughly 'public' community in which the 'self' is completely overcome" (Watsuji [15], p. 595). The source of the nation's power is the "authority of totality," and the real strength of the nation is "moral vigor". Watsuji asserts that from this one can deduce that the proper conduct of the nation's members is that described in the imperial rescript: "revere the Constitution, obey the nation's laws, and render loyal and courageous service to the public in a national crisis" (Watsuji [15], p. 620). Not only in emergencies, but in ordinary times as well, the rescript on education — especially its key injunction to "guard and maintain the prosperity of the imperial throne" — is the compass upon which loyal subjects are expected to depend for guidance.

It was from this standpoint that Watsuji's "theory of national morality" developed. In his 1932 work *Gendai Nihon to chōnin konjō* (Modern Japan and the Townsman's Mentality) (Watsuji [12], he states that beginning in the Sengoku (Warring States) period [ca. 1480– ca. 1570], the ethos of the Japanese townsman had something in common with the "bourgeois spirit" of Western Europe, but it was transmuted by Fukuzawa Yukichi into a spirit of rationalization and individualism. This line of thought held that samarui should be liberated from the totality of their masters' houses, and that the individual took precedence over the totality of society or the state. This "utilitarian individualism" was instrumental in building a modern Japan, but now that the townsman's mentality has become the dominant spirit of our time, we are facing a "modern-day crisis". Watsuji regarded the proletariat as dangerous in that it was a "class united through economic interests". This was precisely identical, he believed, with the ethos of the townsman in the bourgeois Gesellschaft. "Class struggle" was actually an internal quarrel among those imbued with this mentality. Given such a deplorable situation, Watsuji urged the elevation of Japanese society to a self-conscious moral community through the restoration of the Japanese "attitude of self-sacrifice". He cautioned against class struggle, which was an inevitable reality of the times, and advocated harmony within the community of the nation.

In *Nippon no shindō* (The Way of the Loyal Subject in Japan) (Watsuji [13]), written in 1944, just before the end of the Second World War, Watsuji directly confronts the problem of reverence for the emperor. He maintains that *bushidō*, the samurai's feudal-military code of behavior, formed a "doctrine of reverence for the sovereign" within the traditions of Buddhism and Confucianism, and asserts that this should be cultivated afresh today as a "doctrine of reverence for the emperor". According to him, Japanese, from of old, have defined that ultimate authority not as an absolute god, but as an infinite entity. There is, in other words, no "devotion to the Absolute" save "devotion to the Throne", and this is the highest level of commitment in the world. The duty of the impersonal subject is thus defined as "selfless service to the sovereign", i.e., the sacrifice of personal interest to the public good.

As an advocate of a "national morality" enjoining the populace to overcome the mercenary spirit of the townsman ("class consciousness") and pursue the way of the loyal subject, Watsuji ultimately won a place for himself as a philosopher of the state, and he discharged the duties of this role admirably. It is true that he made use of the methodologies of Western philosophy, but he did this in order to offer hermeneutical "explanation" of the negative traditions and realities that prevailed in prewar Japan. Concretely, he counterposed the Japanese *Gemeinschaft* to the Western Gesellschaft, establishing and rationalizing its "superiority". It is by virtue of such word that Watsuji became the dean of establishment philosophy in wartime Japan.

4. The postwar period

After opening its doors to the outside world in the mid-nineteenth century, Japan vigorously pursued modernization policies under the absolutist emperor system. These policies led to wars of aggression under a fascist state, and after horrendous sacrifices on the part of the people, it collapsed in prostrate defeat. After the war, the people of Japan attained, for the first time in their history, the political conditions whereby they could give concrete expression to liberal, democratic, and pacifist ideas in their lifestyles and institutions. Now let us examine how intellectuals of the postwar period thought this new task on the national agenda should be addressed and in what way they contributed to this endeavor.

Among the most prominent intellectuals of the postwar era is political scientist Maruyuma Masao (1914–), formerly professor at the University of Tokyo. No other scholar has been more active in the effort to implant the ideals of peace and democracy in Japan after 1945. An examination of his intellectual endeavor will reveal a great deal that is of direct practical value to us today.

In Maruyama's *Nihon no shisō* (Japanese Thought) [Iwanami Shoten, 1961] there is a chapter entitled *"De aru" koto to "suru" koto* ("Being" and "Doing"). Based on a lecture, its contest is concrete and lucidly expressed. It offers a wealth of insights, and is invaluable as a concise expression of Maruyama's thought. By contrasting the logic and value of Being with those of Doing, and by tracing the transition from one to the other, the author explores such problems as the transition from premodern to modern society, the institutional and functional aspects of democracy, and the perversion of the two sets of logic and value in the present day. Maruyama first sees the shift from Being to Doing as a movement away from the authority of social status and the dogma prevailing in a status-oriented society toward a "dynamic of the modern spirit" which values actual functions and effect. Secondly he stresses the necessity, in light of this change from "ascriptive" value to "functional" value, of having the people monitor and criticize the workings of democracy, instead of becoming fixated on or making fetishes out of "institutions" and "doctrines". Thirdly, he criticizes the perversion of the two types of logic in the midst of the massification of postwar Japanese society—particularly the growing emphasis on "vulgar efficacy and immediate utility" in the realm of culture. As mentioned before, these three points are in themselves an accurate summation of Maruyama's thought.

Nihon seiji shisōshi kenyū (Studies in the Intellectual History of Tokugawa Japan; Tokyo: The University of Tokyo Press, 1952), a collection of scholarly studies that Maruyama wrote during the war, is a study, through examination of the ideas of Edo-period Confucian scholar Ogyū Sorai, of the "development" of Confucianism as the governing philosophy of the feudal system—i.e., the emergence of "modern thinking" within the process of its internal decay. The classic postwar work *Gendai seiji no shisō to kōdō* (Thought and Behavior in Modern Japanese Politics; Tokyo: Miraisha, 1956–57) is an incisive explication and critique of the premodern nature of politics and thought in Japan before the Second World War. It also contains theoretical studies aimed to promote the realization of genuine modernity in postwar Japan through the firm establishment of democracy, and offers a prescription for the future based on an analysis of the complex political situations and ideological trends in Japan and the rest of the world after the war. The perspective from which Maruyama explored such intellectual issues has consistently been the logic/value of Doing, as opposed to Being. This assumed an axial position in Maruyama's thought process as a "modern", as opposed to premodern, spirit and a "functional", as contrasted with "ascriptive", logic and value. Let us now look at how Maruyama approached the solution of these problems with a "modern" method geared toward the attainment of "modern" goals.

The central theme of *Nihon seiji shisōshi kenkyū* was the tracing of the "process of the internal disintegration of the structure of Confucian thought", which was the ruling ideology of the rigid feudal regime. Maruyama posited this process in the conflicting development of logic in the teachings of Chu Hsi and Sorai, with the concepts of "nature" and "invention" as indices, and he attempted to make clear that it involved "the world-historical problem of the conflict between the medieval view of social and state institutions and the modern bourgeois view" (Maruyama [8], p. 197).

According to Maruyama, the Sorai school's central theme was the transformation of the Confucian theory of "natural order" to "autonomous invention". That meant, first of all, breaking the continuum between cosmic nature—"the Way of Heaven" or "the Principle of Nature"—and human nature—man's "Original Nature"—and excluding this cosmic nature from the Way of the Sages, regarding the former as analogous to the latter. Secondly, Sorai assigned the role of absolute inventors of the Way to the Sages (the Early Kings) and consistently maintained the "absolute primacy" of the Early Kings' act of invention, through which nature is separated from history. Sorai was able, by drawing an analogy with the universal validity of the Sages' invention and the monarchs (rulers) of each age, to formulate "forceful political measures" for surmounting the prevailing social disorder and "laying a new basis for the ultimate standards of feudal society". Although the position assumed by Sorai was a "reactionary" stance that took shape in the critical stage of the erosion of the various feudal relationships, Maruyama sees that "the Gesellschaft logic was embedded in his [Sorai's] theory of invention" (Maruyama [8], p. 222). Sorai's position is contrasted with the individual's awareness of his autonomy in the transition from "status to contract" and "Gemeinschaft to Gesellschaft" that took place in Europe from the end of the medieval period through the early modern period. In other words, the roles played by God and the absolute monarchs in the shift in Europe from the "idea of natural order and organic theory" to the complete form of the idea of invented order, i.e., "the social contract theory or the mechanistic outlook", were played by the Sages in the Sorai school in Japan (Maruyama [8], p. 238). Thus, Sorai's theory of autonomous invention was seen as fulfilling two separate political functions in feudal society: first, it could serve as "a logical weapon to transform the feudal order and establish a new one", and second, it could "deprive feudal social relations and their ideological cement of any *raison d'être*" (Maruyama [8], pp. 247–8). The former is the positive implication, and the latter the negative. It was actually the latter of these that Maruyama regarded as having affected more significantly the history of thought in the second half of the Tokugawa period.

With regard to intellectual perspectives of this work, which marks the starting point in Maruyama's research, we should note, first, a harsh rejection of "all that was premodern" and an intense desire to see "that which is modern" created. Second, the concept of "autonomous invention" that functioned as the mediator and promoter of the transition from premodern to modern is regarded as extremely important. This point is where Maruyama's own autonomous stance is especially pervasive. I may venture to point out, however, that the "things modern" that he brings up here are modern only in comparison with "things premodern", and that, therefore, the autonomy capable of effecting the transition is actually confined to that of "superhuman" individuals like God or monarchs. In spite of this, Maruyama's perspective was one which, in wartime and under a fascist state, rejected "fascist historiography" by "turning attention to the 'modern' aspects of the Meiji Restoration—even in the midst of loud calls for the 'conquest' and disavowal of modernity—and by extension to the maturation of modern elements in Tokugawa society" (Maruyama [8], p. 372). This was indeed an outstanding scholarly perspective and profound insight as well. We must now ask how the modernist position and methodology of Maruyama functioned in Japan after World War II.

The principal fruits of Maruyama's theoretical activity in the postwar period are contained in *Gendai seiji no shisō to kōdō* (1964), a work that can be broadly divided up into critical analyses of the prewar fascist system and thought, and treatises designed to help build a democratic system in the postwar period. In either category of writing, Maruyama makes a lucid presentation of his ideas as a "free, autonomous individual".

Among his critical explications of the prewar system is an insightful study of the structure of ultra-nationalism. (Maruyama [6], pp. 11–28) He argues here that the prewar system in which national sovereignty monopolized the "spiritual authority and political power" was such that "the whole national order was composed of links, with the emperor—absolute value—in the center and that the authority of rulers gradually changed in degree of value in proportion to their distance from the emperor". Even though the emperor was the "center of all authority" and "the fountainhead of all virtue", it would not be correct to conclude that "he alone enjoyed subjective freedom" in this "system of progressive reliance" on values. He, too, was secured only by the "infinity of the perpendicular axis ("the prosperity of the Imperial Throne coeval with heaven and earth")". The "thought and behavior patterns of Japan's wartime leaders", as discussed in "*Gunkoku shugisha no seishin keitai*" (Maruyama [6], pp. 88–130), were also vividly portrayed as a "colossal system of irresponsibility" where there was total lack of autonomy on the part of Japanese officials. By examining testimonies they gave as defendants in the war crimes trial, Maruyama illustrates how weak their awareness of subjective responsibility was. He defines this psychology as "surrender to the *fait accompli*" and "escape into competence", placing it in the "the system of irresponsibility" in Japanese fascism. It is these thought and behavior patterns that are compared with the Nazis, who controlled the state authority the way outlaws do "as outlaws". Even in such an analysis of the structure and ideology of Japan's prewar state system, Maruyama remains severely critical of their premodernity from the perspective of "absence of autonomy".

It is natural, then, that the establishment of autonomy became a central concern for Maruyama in fostering a democratic system and liberal thought in post-World War II Japan. From that viewpoint he vigorously came to grips with issues that emerged soon after the series of postwar democratic reforms, such issues as the Red Purge (1950), the May Day Incident (1952), the Korean War (1950–53), the San Francisco Peace Treaty (signed in 1951), and more recently the anti-U.S.—Japan security treaty movement (1960).

His theoretical stance in the early postwar years is most plainly stated in "*Aru jiyūshugisha e no tegami*" (Letter to a Certain Liberalist). (Maruyama [6], pp. 131–51) Here, he presents "strategies and tactics" as a "nonpartisan intellectual" to deal with the "political and ideological panic" in the early 1950s that was accompanied by fanatic anti-communism. Calling himself a "pragmatist" in the "world of political judgment", Maruyama affirms his opposition to dogmatization of any ideology, and shows his basic attitude: "The thing to do immediately, which would be beneficial not only to the freedom of intellectuals but also to expansion of political and economic rights of the masses, is not to defend the *existing* democracy but to promote and expand *from now* the democracy that has finally begun to take root. This democracy is not a 'Soviet-type democracy' but a Western bourgeois democracy. The democracy 'in this sense' is still a 'task' not a 'reality' for us" (Maruyama [6], p. 146). From the standpoint of this bourgeois democracy, he defended the Japan Communist Party for its "role in democratization in the Western European sense" and opposed the totalitarian tendency to suppress and weaken the party.

However, Maruyama's own thought and behavior in the effort toward creation and establishment of postwar democracy began to come under attack from both the right and left in the wake of the protests against the Japan—U.S. security treaty. Ultra-leftists claimed that postwar democracy was a fiction, and rightists pronounced the "end of ideology". Naturally, the appearance of these ideologies was attributable to new materialistic developments, that is, the growing economic and political independence of Japanese monopolistic capital and its governing apparatus, the state, and the resultant evolution of "modern society" and "mass society". Maruyama warned the danger of totalitarian tendencies rising from the bottom with the emergence of a kind of mass society in Japan in the 1960s.

However, because of the complexity of modern Japan as an advanced capitalist nation, it was exceedingly difficult to create the bourgeois democracy Maruyama proposed. It was around that time, in the early 1970s, that he returned to scholarly research on Japanese intellectual history.

One of Maruyama's most important writings of the 1970s is "Rekishi ishiki no 'kosō'" (The Deep Layer of Historical Consciousness; 1972). It shows the present focus of his intellectual concern. What he means by the "deep layer" of historical consciousness is the "conception and style of description" abstracted from Japanese myths of cosmogony and the *Kojiki* (Records of Ancient Matters) and the *Nihon shoki* (Chronicles of Japan), both partly dealing with the mythical age of the gods. In the "deep layer" he discerns "the pattern of thought that has been a persistent undercurrent in the various aspects of the development of historical consciousness up to modern times" (Maruyama [9], p. 3). He defines the thought pattern as a "spontaneous flow of impersonal dynamic energy" (*tsugitstugi ni nariyuku ikioi*). This thought pattern was a *Basso ostinato vis-à-vis* the ideas introduced to ancient Japan from the Asian continent—Confucianism, Buddhism, and Taoism—and to modern Japan after the Meiji Restoration from the West. Each influential school of foreign thought played the melody at one time or another in Japanese history after it was introduced here. The persistent *Basso ostinato*, meanwhile, altered each melody to suit the Japanese scale, so to speak. Maruyama says, for example, that the "process of transformation or modification of the Confucianist law of nature" in the early-modern intellectual history of Japan was "not so much a shift from orthodoxy to heterodoxy as the resurfacing of the persistent 'deep layer'.

In other words, he understands that "the historical dynamism of the Edo period consists not so much of one-way 'modernization' as of the conflict and synergism of two moments, modernization and the rise of the 'deep layer'" (Maruyama [9], p. 31). This viewpoint clearly diverges from that presented in *Nihon seiji shisōshi kenkyū*, which discerns modern elements in the "development" of Confucianism itself. Obviously, Maruyama has made some concession to indigenous, traditional thought. In his treatment of the "deep layer" at least, it is hard to find the reformist attitude toward the status quo that Maruyama's theory and practice once had. His 1972 work may be an honest statement of Maruyama's current thought *vis-à-vis* the fictitious modernity that Japan has reached today. It also reflects the complexity and difficulty of our time.

5. Conclusion

As the above discussion shows, the problem of Japan's modernization has been the fundamental concern of Japanese political philosophy since the beginning of the Meiji Restoration. It still is. The debate was led by anti-democratic nationalism during the prewar and wartime periods, and even today genuine modernization is difficult as the contradictions stemming from capitalism grow in number and intensity. A distinctive feature of the study of Japanese modernization is that comparison has always been made with Western modernity or with Japan's premodernity.

Japan today is on a par with the West in economic and social development, and it is beset with many contradictions that are as serious as, sometimes more serious than, those plaguing European nations or the United States. There is now a strong need for a political philosophy relevant to both Japan and the West, that sublimates modernity itself.

Bibliography

Fukuzawa Yukichi. [1] *Bumei ron no gairyaku* [Outline of Civilization]. In Keio Gijiku (Ed.), *Fukuzawa Yukichi zenshū* [Collected Works of Fukuzawa Yukichi], Vol. 4. Tokyo: Iwanami Shoten, 1957.

[2] "Datsu-A ron" [Dissociating from Asia]. In Keio Gijuku (Ed.), *Fukuzawa Yukichi zenshū* [Collected Works of Fukuzawa Yukichi], Vol. 10. Tokyo: Iwanami Shoten, 1957.

[3] *Fukuō jiden* [The Autobiography of Fukuzawa Yukichi]. In Keio Gijuku (Ed.), *Fukuzawa Yukichi zenshū* [Collected Works of Fukuzawa Yukichi], Vol. 7. Tokyo: Iwanami Shoten, 1957.

[4] *Gakumon no susume* [Encouragement of Learning]. In Keio Gijuku (Ed.), *Fukuzawa Yukichi zenshū* [Collected Works of Fukuzawa Yukichi], Vol. 3. Tokyo: Iwanami Shoten, 1957.

[5] "Nisshin no sensō was bun'ya no sensō nari" [The Sino-Japanese War: A Battle Between Civilization and Barbarism]. In Keio Gijuku (Ed.), *Fukuzawa Yukichi zenshū* [Collected Works of Fukuzawa Yukichi], Vol. 14. Tokyo: Iwanami Shoten, 1957.

Maruyama Masao. [6] *Zōho-ban Gendai seiji no shisō to kōdō* [Thought and Behavior in Modern Japanese Politics], enlarged edition. Tokyo: Miraisha, 1964. (English translation edited by Ivan Morris under the title *Thought and Behavior in Modern Japanese Politics*. London: Oxford University Press, 1963.)

[7] *Nihon no shisō* [Japanese Thought]. Tokyo: Iwanami Shoten, 1961.

[8] *Nihon seiji shisōshi kenkyū* [Studies in the Intellectual History of Tokugawa Japan]. Tokyo: The University of Tokyo Press, 1952. (*Studies in the Intellectual History of Tokugawa Japan*, translated by Mikiso Hane. Tokyo: The University of Tokyo Press, 1974.)

[9] "Rekishi ishiki no 'kosō'" [The Deep Layer of Historical Consciousness]. In *Rekishi shisōshū* [Japanese Views of History], *Nihon no shisō* [Japanese Thought], Vol. 6. Tokyo: Chikuma Shobō, 1972.

Suzuki Shigetaka. [10] "'Kindai no chōkoku' oboegaki" [Notes on "Overcoming Modernity"]. In *Bungakkai* [Literary World]. Tokyo: Bungei Shunjūsha, 1942.

Watsuji Tetsurō. [11] *Fūdo* [Climate]. In *Watsuji Tetsurō zenzhū* [The Complete Works of Watsuji Tetsurō], Vol. 8. Tokyo: Iwanami Shoten, 1962.

[12] *Gendai Nihon to chōnin konjō* [Modern Japan and the Townsman's Mentality]. In *Watsuji Tetsurō zenshū* [The Complete Works of Watsuji Tetsurō], Vol. 4. Tokyo: Iwanami Shoten, 1962.

[13] *Nihon no shindō* [The Way of the Loyal Subject in Japan]. In *Watsuji Tetsurō zenshū* [The Complete Works of Watsuji Tetsurō]. Vol. 14. Tokyo: Iwanami Shoten, 1962.

[14] *Ningen no gaku to shite no rinrigaku* [The Significance of Ethics as the Study of Man]. In *Watsuji Tetsurō zenshū* [The Complete Works of Watsuji Tetsurō], Vol. 9. Tokyo: Iwanami Shoten, 1962.

[15] *Rinrigaku* [Ethics]. In *Watsuji Tetsurō zenshū* [The Complete Works of Watsuji Tetsurō], Vol. 10. Tokyo: Iwanami Shoten, 1962.

Index of names

(Note: The names of historical Chinese persons are given in the common Westernised forms, where available [Confucius, Mencius], or in Wade–Giles romanisation [Lao Tzu, Chuang Tzu], even where the individual authors have used other forms [Lao-tze] or Pinyin romanisation [Zhuangzhi]. All other names are given as they appear in the text. Names between parentheses commonly denote noms-de-plume. All names are indexed by the family name. Oriental names thus appear in the normal written order, without a comma, and Western names in 'inverted' order, with a comma, wherever possible. Furthermore, for the sake of rapid access to the index, all accents and diacritics have been omitted.)

Contemporary philosophy. A new survey. Vol. 7, 397–404.

Index of subjects

(Note: Due to the necessity of indexing technical terms in a variety of Oriental languages, and to provide rapid access to this index, accents, diacritics and italic type have been avoided. Capitalisation broadly follows the authors' usage.)

Contemporary philosophy. A new survey. Vol. 7, 405–411.

GPSR Compliance

The European Union's (EU) General Product Safety Regulation (GPSR) is a set of rules that requires consumer products to be safe and our obligations to ensure this.

If you have any concerns about our products, you can contact us on ProductSafety@springernature.com

In case Publisher is established outside the EU, the EU authorized representative is:

Springer Nature Customer Service Center GmbH
Europaplatz 3
69115 Heidelberg, Germany

Batch number: 10816188

Printed by Printforce, the Netherlands